AF595518

Inflammation and Natural Products

Inflammation and Natural Products

Editors

Azahara Rodríguez-Luna
Salvador González

Basel • Beijing • Wuhan • Barcelona • Belgrade • Novi Sad • Cluj • Manchester

Editors
Azahara Rodríguez-Luna
Universidad Loyola
Andalucía
Sevilla, Spain

Salvador González
Alcalá de Henares University
Madrid, Spain

Editorial Office
MDPI
St. Alban-Anlage 66
4052 Basel, Switzerland

This is a reprint of articles from the Special Issue published online in the open access journal *Life* (ISSN 2075-1729) (available at: www.mdpi.com/journal/life/special_issues/20TEX8M6C0).

For citation purposes, cite each article independently as indicated on the article page online and as indicated below:

Lastname, A.A.; Lastname, B.B. Article Title. *Journal Name* **Year**, *Volume Number*, Page Range.

ISBN 978-3-0365-9477-4 (Hbk)
ISBN 978-3-0365-9476-7 (PDF)
doi.org/10.3390/books978-3-0365-9476-7

Cover image courtesy of CANTABRIA LABS

© 2023 by the authors. Articles in this book are Open Access and distributed under the Creative Commons Attribution (CC BY) license. The book as a whole is distributed by MDPI under the terms and conditions of the Creative Commons Attribution-NonCommercial-NoDerivs (CC BY-NC-ND) license.

Contents

About the Editors

Azahara Rodríguez-Luna

Azahara Rodríguez-Luna is a professor and researcher in the Faculty of Health Sciences at Universidad Loyola Andalucía in Seville, Spain. Her primary research areas include inflammation and natural products, with a specific focus on skin pathologies and photoprotection. Throughout her research career, she has extensively explored various natural products derived from microalgae and terrestrial plants, with a particular emphasis on polyphenols and carotenoids.

Salvador González

Salvador González works within the Department of Medicine and Medical Specialties, Alcalá de Henares University, Madrid, Spain. He is an accredited Full Professor and Associate Professor of Dermatology at the University of Alcalá de Henares, Madrid. He is an internationally renowned expert in in vivo confocal microscopy and clinical research areas, including cutaneous oncology, photoprotection, and photodermatology, among others. Throughout his research career, González has conducted extensive investigations in both basic and clinical research related to the treatment and prevention of skin cancer and inflammation using natural products.

Preface

This reprint embarks on a journey to gather fresh perspectives on compounds and extracts derived from natural products and their remarkable biological activities, with a primary focus on anti-inflammatory pathways. Its contents comprise a compilation of full papers and review articles published in a 2023 Special Issue of *Life*, titled 'Inflammation and Natural Products.' These publications delve into the world of secondary metabolites from plants, showcasing their remarkable antioxidant and anti-inflammatory properties and their clinical significance across various pathological conditions.

Azahara Rodríguez-Luna and Salvador González
Editors

life

Article

In Vitro and In Vivo Anti-Psoriasis Activity of *Ficus carica* Fruit Extracts via JAK-STAT Modulation

Jeong Hwa Lee [1] and Mi-Young Lee [1,2,3,*]

1 Department of Medical Science, Soonchunhyang University, 22 Soonchunhyang-ro, Asan 31538, Republic of Korea; 20237438@sch.ac.kr
2 Department of Medical Biotechnology, Soonchunhyang University, 22 Soonchunhyang-ro, Asan 31538, Republic of Korea
3 Eshel Biopharm Co., Ltd., Soonchunhyang-ro, Asan 31538, Republic of Korea
* Correspondence: miyoung@sch.ac.kr; Tel.: +82-41-530-1355

Abstract: Psoriasis, a chronic and autoimmune inflammatory disorder of the skin, has been often underdiagnosed and underestimated despite its prevalence and considerable negative effects on the quality of life. In this study, the anti-inflammatory activity of *Ficus carica* fruit extract (FFE) was investigated against LPS-stimulated RAW 264.7 cells. The in vitro results showed that FFE reduced the production of nitric oxide (NO) and iNOS expression. Moreover, FFE reduced the level of β-hexosaminidase released with histamine in allergic reactions. However, the MAPK and NFκB signaling molecules associated with the inflammatory response were not significantly regulated by FFE. In contrast, the phosphorylation of JAK1 and STAT3 in the JAK–STAT signaling pathway was dramatically reduced by FFE treatment. Psoriasis-like skin lesions were induced in BALB/c mice using imiquimod (IMQ) to test the feasibility of FFE as a treatment for psoriasis. The efficacy of FFE was evaluated based on phenotypic and histological features. FFE was effective in relieving the symptoms of psoriasis-like skin lesions, such as erythema, dryness, scales, and thick epidermis. Notably, STAT3 modulation was also contributable to the in vivo ameliorative activity of FFE. Taken together, FFE with anti-psoriasis activity in vitro and in vivo through the JAK–STAT modulation could be developed as a therapeutic agent against psoriasis.

Keywords: psoriasis; *Ficus carica* fruit extract; JAK–STAT modulation

Citation: Lee, J.H.; Lee, M.-Y. In Vitro and In Vivo Anti-Psoriasis Activity of *Ficus carica* Fruit Extracts via JAK-STAT Modulation. *Life* **2023**, *13*, 1671. https://doi.org/10.3390/life13081671

Academic Editors: Azahara Rodríguez-Luna and Salvador González

Received: 13 June 2023
Revised: 26 July 2023
Accepted: 26 July 2023
Published: 31 July 2023

Copyright: © 2023 by the authors. Licensee MDPI, Basel, Switzerland. This article is an open access article distributed under the terms and conditions of the Creative Commons Attribution (CC BY) license (https://creativecommons.org/licenses/by/4.0/).

1. Introduction

Psoriasis is a chronic, non-communicable, painful, disfiguring, and disabling disease that negatively affects the quality of life of patients [1]. It can occur at any age but is most common in those aged 20–30 years and 50–60 years [2]. The reported worldwide prevalence of psoriasis ranges from 0.09 to 11.4%, making it a serious global problem [3,4]. The psoriasis lesions are associated with swollen papules and silvery-white scales probably due to overactive immunity in keratinocytes provoked by IL-23 and IL-17 [5]. It has been demonstrated that Th1 overactivation can induce psoriasis, and Th17 cells play a key role in its pathogenesis and severity [6,7]. Based on the pathogenesis of this disease, psoriasis can be broadly characterized by abnormal keratinocyte division and differentiation. Symptoms, such as itching, scaling, erythema, burning, and edema, have been observed in the skin lesions of psoriasis.

In general, signal transduction pathways regulate various immune and inflammatory reactions and are involved in cell proliferation, differentiation, and apoptosis. It has been reported that an alteration in the NFκB, JAK–STAT, Akt, or Wnt signaling pathways induces psoriasis [8]. NFκB is the fundamental modulator of physiological systems, such as inflammation, development, cell cycle, proliferation, and cell death [9]. NFκB consists of the Rel family, and the major Rel/NFκB complex is the p50/p65 heterodimer [10]. Inactive +NFκB dimers are located in the cytoplasm and bind with the IκB family of inhibitory proteins, including IκBα [11]. The phosphorylation of IκBα by the IKK complex triggers liberation of the NFκB dimer, leading to the entering of NFκB into the nucleus and triggering the inflammation cascade. Psoriasis, as an inflammatory skin disorder, is marked by elevated and activated NFκB via phosphorylation [12].

The mammalian family of mitogen-activated protein kinases (MAPKs) include extracellular signal-regulated kinase (ERK), p38, and c-Jun NH2-terminal kinase (JNK) [12]. The MAPK pathways are activated by diverse extracellular and intracellular stimuli, including cytokines, hormones, and various cellular stressors that include oxidative stress and endoplasmic reticulum stress [13]. The potential role of the p38 MAPK pathway has been suggested to be involved in the inflammatory pathogenesis of psoriasis [14].

Members of the mammalian Janus kinase (JAK) and signal transducer and activator of transcription (STAT) families have been extensively analyzed in mouse and human systems [15]. Four JAKs—JAK1, JAK2, JAK3, and Tyk2—and seven STATs—STAT1, STAT2, STAT3, STAT4, STAT5A, STAT5B, and STAT6—were identified [15]. Upon phosphorylation of JAK, STAT and other cytokines are subsequently phosphorylated. Drugs that inhibit the JAK–STAT signaling pathway are known as JAK inhibitors, including tofacitinib, ruxotinib, and baricitinib [8]. Psoriasis is associated with the overexpression of JAK1 and STAT3 [8]. Thus, the JAK and STAT inhibitors can be used as therapeutics to treat psoriasis [16].

Ficus carica is used in traditional medicine for a wide range of ailments related to digestive, endocrine, reproductive, respiratory, gastrointestinal, and urinary tract infections [17,18]. *Ficus carica* contains a variety of coumarins and flavonoids in its fruits and leaves, and these compounds are known to have antioxidant, anti-cancer, and anti-inflammatory effects [19,20]. Psoralen is one of the components of *F. carica* that has been used as a drug for skin disease [21]. Psoralen is currently an FDA-approved treatment under the name PUVA therapy [22] and is an effective treatment for ailments, such as eczema, psoriasis, and vitiligo, when co-treated with UV-A. As a photosensitizer, psoralen must be exposed to UV-A radiation for therapeutic efficacy.

The leaves and fruits of *F. carica* differ not only in chemical composition but also in efficacy and toxicity [17]. Studies on the efficacy of *F. carica* fruit extract on inflammation in psoriasis have not been available until now. In this study, the in vitro ameliorative effects of *F. carica* fruit extract were analyzed in LPS-activated RAW 264.7 cells. The in vivo suppressing effect of the extract on the inflammatory response was also investigated using a psoriasis-like mouse model.

2. Materials and Methods

2.1. In Vitro Study

2.1.1. Plant Extracts

The methanolic extracts of *F. carica* fruit used in this study were obtained from the Korea Plant Extract Bank at the Korea Research Institute of Bioscience and Biotechnology (Daejeon, Republic of Korea). The fruit (31 g), which was dried in the shade and powdered, was added to 1 L of methyl alcohol 99.9% (HPLC grade) and extracted through 30 cycles (40 KHz, 1500 W, 15 min ultrasonication and 120 min standing per cycle) at room temperature using an ultrasonic extractor (SDN-900H, SD-ULTRASONIC CO., LT., Seoul, Republic of Korea). After filtration (Qualitative Filter No. 100, HYUNDAI MICRO CO., LT., Seoul, Republic of Korea) and drying under reduced pressure, *F. carica* extract (8.4 g) was obtained and diluted to a 5% stock solution using dimethyl sulfoxide (DMSO).

2.1.2. DPPH Free Radical Scavenging Assay

Diluted *Ficus carica* fruit extract (FFE) at various concentrations was added to the DPPH ethanolic solution, and the volume of the reaction mixture was adjusted to 200 μL. Each mixture was suspended vigorously and incubated in the dark at 37 °C for 30 min. The absorbance was measured at 517 nm. The DPPH free radical scavenging activity was calculated according to this formula:

$$\text{Scavenging activity}(\%) = \frac{1 - (\text{Abs}_{\text{FFE}} - \text{Abs}_{\text{blank}})}{\text{Abs}_{\text{control}}} \times 100$$

where Abs_{FFE} is the absorbance of the FFE, $\text{Abs}_{\text{blank}}$ is the absorbance of the blank, and $\text{Abs}_{\text{control}}$ is the absorbance of the control. Ascorbic acid was used as the positive control.

2.1.3. Determination of Total Polyphenol Content

Total polyphenol content was determined using the Folin–Ciocalteu assay with some modifications [23]. Briefly, 750 µL of the samples and gallic acid were mixed with 150 µL Folin–Ciocalteu's phenol reagent and incubated for 5 min at room temperature. A total of 150 µL of 20% Na_2CO_3 was then added and incubated for 30 min at 40 °C. After incubation, the absorbance of the reaction mixture was measured at 750 nm. A gallic acid curve was used to quantify the total polyphenol content. The results were expressed as milligrams (mg) of gallic acid equivalents (GAE). The total phenolic contents in the FFE were calculated using this formula:

$$C = cV/m$$

where C = total phenolic content mg GAE/g of FFE, c = concentration of gallic acid obtained from the calibration curve in mg/mL, V = volume of the extract in mL, and m = mass of the extract in grams.

2.1.4. Determination of Total Flavonoid Content

Total flavonoid content was determined using the aluminum chloride ($AlCl_3$) colorimetric method with some modifications [24]. The entire process was performed at room temperature. Briefly, 400 µL of the samples and quercetin were mixed with 30 µL of 5% $NaNO_2$ for 5 min. Thirty microliters of 10% $AlCl_3$ was added, and the reaction mixture was incubated for 5 min. A total of 400 µL of 4% NaOH was added and incubated at room temperature for 15 min. Finally, the mixture was adjusted with distilled water to a final volume of 1 mL and gently mixed. The absorbance was measured at 510 nm. A quercetin curve was used to quantify the total flavonoid content. The results were expressed as milligrams (mg) of quercetin equivalents (QE). The total flavonoid contents in the FFE were calculated using this formula:

$$C = cV/m$$

where C = total flavonoid content mg QE/g of FFE, c = concentration of quercetin obtained from the calibration curve in mg/mL, V = volume of the extract in mL, and m = mass of the extract in grams.

2.1.5. LC–MS/MS

Chemical components were analyzed using a liquid chromatography–tandem mass spectrometry (LC–MS/MS) system, specifically employing the Thermo TSQ Altis system coupled to a Thermo 3000 RSLC with a quaternary gradient pump and autosampler (Thermo Fisher Scientific, Waltham, MA, USA) integrated with the Vanquish Flex UHPLC system. Chromatographic separation was carried out on an Agilent C18 column. The mobile phase was comprised of distilled water (DW) and acetonitrile (ACN), both with added acetic acid. The column temperature and sample organizer were maintained at 40 °C and 15 °C, respectively. Elution from the column was accomplished using a gradient method at a constant flow rate of 0.3 mL/min with the following conditions: from 5% to 95% of solvent B over 2 min, a 2 min wash with 100% B, and a 2 min re-equilibration with 5% B. The injected sample volume was 1.0 µL. Analyses were performed in the optimized data-dependent acquisition (DDA) mode via negative ion mode electrospray ionization (ESI-). The analyses constituted a full MS survey scan within the m/z range of 100–2000 Da (scan time: 150 ms) and MS/MS scans for the three most intense ions. Collision energy was applied in a stepped manner from 30 V to 100 V. ESI parameters were set as follows: capillary voltage at 2.9 kV, cone voltage at 40 V, source temperature at 120 °C, desolvation temperature at 350 °C, cone gas flow at 50 L/h, and desolvation gas flow at 800 L/h.

High-purity nitrogen was used as the nebulizer and auxiliary gas, while argon served as the collision gas.

2.1.6. Nitric Oxide Assay

RAW 264.7 cells, a murine macrophage cell line, were seeded 1×10^6 cells/well on a 6-well plate and incubated for 24 h. FFE was added into the wells with the LPS (1 μg/mL) and incubated for 18 h at 37 °C in a CO_2 incubator. The amounts of nitrite formed were measured using Griess reagent (0.2% naphthylethylenediamine dihydrochloride and 2% sulfanilamide in 10% H_3PO_4). Briefly, 100 μL of cell supernatant was mixed with 150 μL of Griess reagent. The absorbance was measured at 540 nm using a microplate reader (Biotek, Synergy HTX, Winooski, VT, USA). The concentration of nitrite was determined from a sodium nitrite standard curve.

2.1.7. Western Blot

The cells were lysed by adding RIPA buffer (50 mM Tris-HCl, pH 7.4, 150 mM NaCl, 1% Triton X-100, 0.5% sodium deoxycholate, 0.1% SDS, and 2 mM EDTA) with 1% (*v*/*v*) protease inhibitor cocktail (PIC) and 1% (*v*/*v*) phenylmethylsulfonyl fluoride (PMSF) and then sonicated twice for 20 s at 10 s intervals. Then, the cells were centrifuged at 15,000 rpm at 4 °C for 50 min. The supernatant was collected for the Bradford protein quantification assay [25]. Equal amounts of protein were loaded onto a 10% sodium dodecyl sulfate-polyacrylamide 10% SDS-PAGE gel and electrophoretically separated for 1 h. The proteins were then transferred to gels and nitrocellulose (NC) membranes for 1 h at 400 mA. The transferred NC membranes were blocked in 5% skim milk in TBST buffer (Tris-buffered saline, 0.1%) at 4 °C overnight. The membranes were then incubated for 2 h at room temperature with individual primary antibodies and washed three times with TBST buffer. Then, the membranes were incubated for 1 h at room temperature with secondary antibodies and developed using ECL reagents. The relative intensities of the Western blot bands were quantified using β-actin.

2.1.8. β-Hexosaminidase Assay

RBL-2H3 cells, a murine basophil cell line, were seeded 2×10^5 cells/well on a 6-well plate and incubated for 24 h. The cells were then treated with 2,4-dinitrophenyl (DNP)-IgE (100 ng/mL) overnight. To remove the excess DNP-IgE, the cells were washed twice with the Siraganian buffer (119 mM NaCl, 5 mM KCl, 5.6 mM glucose, 25 mM PIPES, 0.4 mM $MgCl_2$, 1 mM $CaCl_2$, 0.1% BSA, and pH 7.2). FFE diluted in the Siraganian buffer was added to the cells and incubated for 1 h. The cells were stimulated with DNP-BSA (100 ng/mL) for 2 h. After incubation, the supernatant was collected and incubated with 2 mM p-nitrophenyl-N-acetyl-β-d-glucosaminide (p-NAG) in 0.1 M citrate buffer (pH 4.5). The enzyme reaction was terminated by adding 0.1 M sodium carbonate buffer (pH 10), and the absorbance was measured at 405 nm.

2.2. *In Vivo Study*

2.2.1. Animals

Six-week-old male BALB/c mice were purchased from ORIENT Inc. (Sungnam, Republic of Korea). The mice were kept in cages and maintained at 25 °C under a 12 h light/12 h dark cycle during the entire experiment. The animal experiments were approved by the Institutional Animal Care and Use Committee (SCH22-0053).

2.2.2. Psoriasis-like Mouse Model

The dorsal hair of each group of five mice was removed with hair removal cream. The mice were divided into six groups ($n = 5$): normal (non-treated group), IMQ (negative control group), dexamethasone (DEX, positive control group, 5 mg/mL dexamethasone in autoclaved DDW), FH (high-dose *Ficus carica* fruit extract, 10 mg/mL), FM (moderate-dose *Ficus carica* fruit extract, 5 mg/mL), and FL (low-dose *Ficus carica* fruit extract, 1 mg/mL).

Inflammation was induced by local administration of 62.5 mg of IMQ to the dorsal skin and ear once every 7 d. Eight days after the first administration, all groups were sacrificed, and the dorsal skin, ear, and spleen were collected.

2.2.3. Western Blot

Mouse dorsal skin was lysed by adding RIPA buffer (50 mM Tris-HCl, pH 7.4, 150 mM NaCl, 1% Triton X-100, 0.5% sodium deoxycholate, 0.1% SDS, and 2 mM EDTA) with 1% (v/v) protease inhibitor cocktail (PIC) and 1% (v/v) phenylmethylsulfonyl fluoride (PMSF) and then sonicated four times for 20 s at 5 s intervals. Then, the lysate was centrifuged at 15,000 rpm at 4 °C for 50 min. The subsequent processes were the same as those used in Section 2.1.7.

2.2.4. Histological Analysis

The dorsal skin was collected and fixed with 10% formalin. After fixation, the skin was dehydrated and embedded in paraffin. The tissues were cut into 3 μm sections. Hematoxylin and eosin (H&E) staining was performed to measure epidermal thickness.

2.3. Statistical Analysis

All results are expressed as the mean ± SD and evaluated by one way ANOVA to confirm the significance between groups. Comparisons of three or more groups were performed using Scheffe's post-hoc test. Data were considered significantly different at $p < 0.05$.

3. Results

3.1. Antiradical Activity and Chemical Constituents

Table 1 shows the antiradical activity of *Ficus carica* fruit extracts determined using the DPPH assay. The total polyphenol and flavonoid contents in FFE are also indicated. The value at which the concentration scavenges 50% of free radicals (IC_{50}) was 626.52 ± 24.75 μg/mL, showing significant radical scavenging activity. The total polyphenol and flavonoid contents were 56.94 ± 0.01 mg GAE/g and 14.71 ± 0.25 mg QE/g, respectively. The results showed that the antiradical activity may arise from the presence of polyphenols and flavonoids. The phytochemicals in the extract were directly analyzed by LC–MS/MS. The sixteen identified compounds are listed in Table 2.

Table 1. Antiradical activities of *F. carica* fruit extracts.

Plant Parts	Antiradical Activity	Total Phenolics (mg GAE/g)	Total Flavonoids (mg QE/g)
	DPPH IC_{50} (μg/mL)		
Fruit	626.52 ± 24.75	56.94 ± 0.01	14.71 ± 0.25

Values are given as mean ± SD of the triplicate experiments.

Table 2. Chemical constituents of *F. carica* fruit extracts by LC–MS/MS.

Compound	Concentration (mg/kg)	Efficacy [Ref.]
Rutin	2254.58	Anti-radical [26]
Chlorogenic acid	1974.34	Anti-radical [27]
Protocatechuic acid	1244.37	Anti-radical [28]
Psoralen	1037.41	Anti-psoriasis [29]
Schaftoside	860.55	Anti-inflammation [30]
Orientin	400.83	Anti-radical [31]
Bergapten	355.51	Anti-inflammation [32]
Caffeoylmalic acid	326.92	N.D
Vitexin	278.8	Anti-inflammation [33]
Fraxin	190.38	Anti-IR injury [34]
Cichoriin	137.55	Anti-radical [35]
Sinapic acid	119.06	Anti-inflammation [36]
Loganic acid	93.73	Anti-radical [37]
Sweroside	56.24	Anti-inflammation [38]
Salidroside	51.51	Anti-inflammation [39]
Nodakenetin	24.85	Anti-inflammation [40]

N.D (Not determined).

3.2. In Vitro Anti-Inflammatory Effect of Ficus carica Fruit Extract in LPS-Stimulated RAW 264.7 Cells

To investigate the in vitro anti-inflammatory efficacy of FFE, FFE dissolved in DMSO at the non-cytotoxic dose was applied to the LPS-stimulated RAW 264.7 cells. As a result of the NO assay using Griess reagent through the cell supernatant, FFE inhibited NO production in the RAW 264.7 cells in a concentration-dependent manner (Figure 1). During the inflammatory reaction, NFκB enters the nucleus, and various inflammatory molecules, including iNOS and COX-2, are produced. iNOS catalyzes the formation of NO from L-arginine, leading to NO production in response to the inflammatory reaction [41]. As shown in Figure 2, Western blot was performed to investigate the expression of the inflammatory marker proteins in the RAW 264.7 cells. LPS-induced inflammatory marker proteins, such as iNOS and COX-2, were reduced by treatment with FFE (Figure 2).

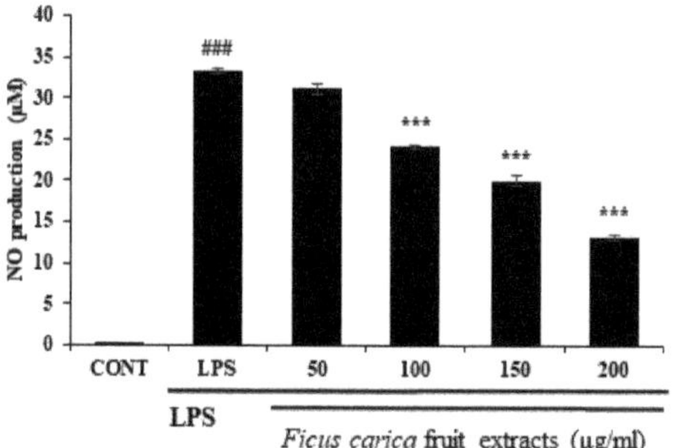

Figure 1. Effect of *F. carica* fruit extracts on nitric oxide production of RAW 264.7 cells. *Ficus carica* fruit extracts significantly inhibit NO production of LPS-stimulated RAW 264.7 cells in a dose-dependent manner. All data are expressed as mean ± SD *** $p < 0.001$ compared with control group, ### $p < 0.001$ compared with LPS group. All experiments were performed in triplicate.

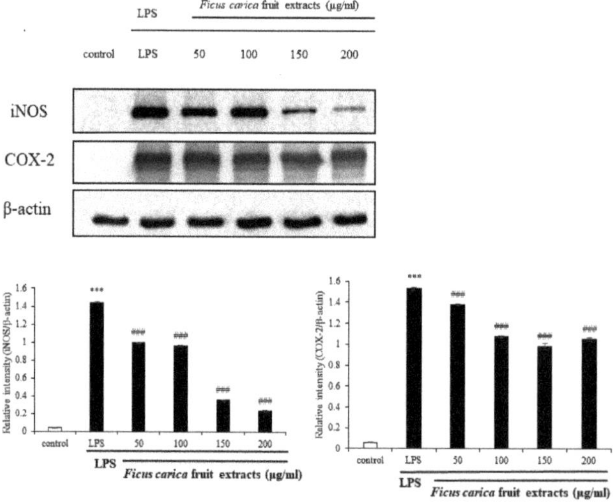

Figure 2. Effect of *F. carica* fruit extracts on iNOS and COX-2 expressions of RAW 264.7 cells. Analysis of the expression levels of iNOS and COX-2 were performed with ImageJ and normalized against β-actin. All data are expressed as mean ± SD *** $p < 0.001$ compared with control group, ### $p < 0.001$ compared with LPS group. All experiments were performed in triplicate.

3.3. *Effect of F. carica Fruit Extract on the MAPK and NFκB Signaling Pathways in RAW 264.7 Cells*

Western blotting of proteins belonging to the inflammatory pathways was performed to examine the underlying mechanism that relieves inflammation. The MAPK signaling pathways include ERK, JNK, and p38 [42]. ERK and p38 phosphorylation was slightly reduced by FFE treatment (Figure 3B). This result suggests that the anti-inflammatory efficacy of FFE may not be closely related to the MAPK signaling pathway.

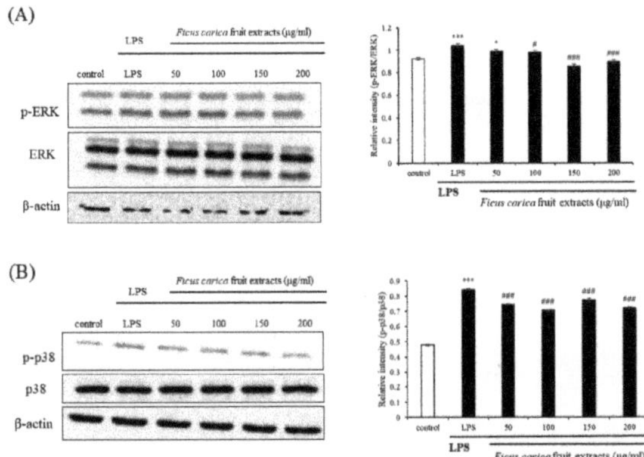

Figure 3. Effect of *F. carica* fruit extracts on the phosphorylation of mitogen-activated protein kinase (MAPK) cascade (p-ERK 1/2 and p-p38) in LPS-stimulated RAW 264.7 cells. The expressions of p-ERK 1/2 (**A**) and p-p38 (**B**) were analyzed with ImageJ and normalized against β-actin. All data are expressed as mean ± SD * $p < 0.05$, *** $p < 0.001$ compared with control group, # $p < 0.05$, ### $p < 0.001$ compared with LPS group. All experiments were performed in triplicate.

In the physiologically normal state (absence of inflammation), cytoplasmic IκBα prevents NFκB from entering the nucleus. However, once IκBα is phosphorylated due to inflammation, it cannot properly function as an NFκB inhibitor, allowing NFκB to freely enter the nucleus. Then, the NFκB that enters the nucleus acts as a transcriptional regulator, releasing various inflammatory cytokines and chemokines [43]. Figure 4 shows that FFE inhibits the phosphorylation of NFκB and IκBα to some extent.

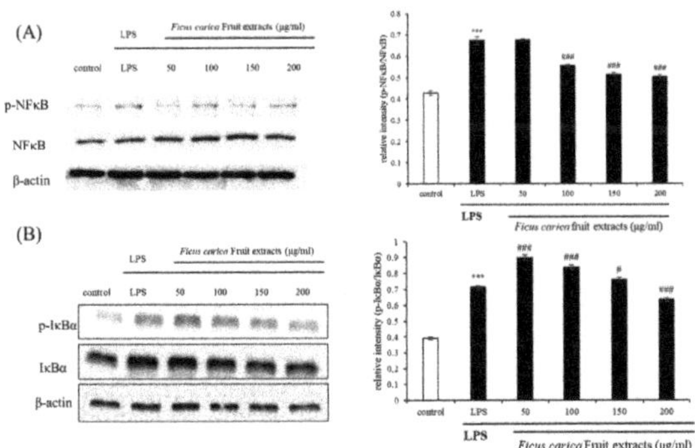

Figure 4. Effect of *F. carica* fruit extract on the LPS-stimulated activation of phosphorylation of NFκB (**A**) and IκBα (**B**) in RAW 264.7 cells. The expression of p-NFκB and p-IκBα was analyzed with ImageJ and normalized against β-actin. All data are expressed as mean ± SD *** $p < 0.001$ compared with control group, # $p < 0.05$, ### $p < 0.001$ compared with LPS group. All experiments were performed in triplicate.

3.4. Regulatory Effect of Ficus carica Fruit Extracts on the JAK–STAT Signaling Pathway

Currently, the JAK–STAT signaling pathway is emerging as a new inflammatory pathway following the classical and traditional MAPK and NFκB signaling. When cytokines, such as IL-6, IL-19, and IL-22, bind to their receptors, JAK is phosphorylated and binds to the cytokine receptor to make a receptor complex, resulting in the phosphorylation of downstream STAT3 at Tyr705 and Ser727, and the phospho-STAT3 meet to form dimers. These dimers bind to the nuclear DNA and regulate the expression of inflammatory proteins [44,45]. Western blotting showed that the expression levels of LPS-induced p-JAK1 and p-STAT3 were drastically reduced following FFE application (Figure 5). These results suggest that FFE sufficiently inhibits inflammation in vitro by acting as an inhibitor of the JAK–STAT pathway rather than as the classically accepted inhibitor of the MAPK and NFκB pathways.

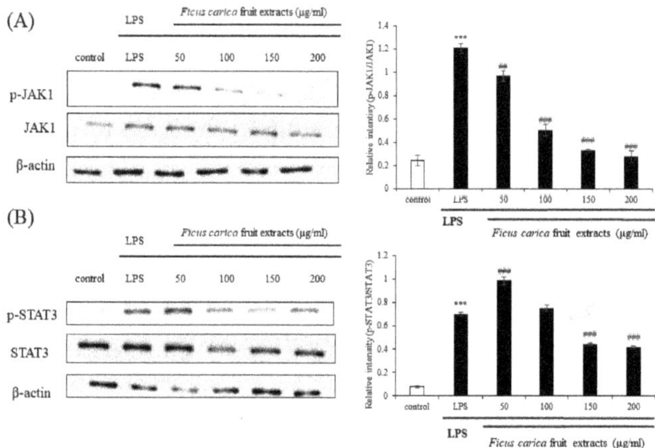

Figure 5. Effect on *F. carica* fruit extracts on p-JAK1 and p-STAT3 expressions of RAW 264.7 cells. The expressions of p-JAK1 (**A**) and p-STAT3 (**B**) were analyzed with ImageJ and normalized against β-actin. All data are expressed as mean ± SD *** $p < 0.001$ compared with control group, ## $p < 0.01$ compared with LPS group, ### $p < 0.001$ compared with LPS group. All experiments were performed in triplicate.

3.5. Effect of Ficus carica Fruit Extracts on the β-Hexosaminidase Release

The β-hexosaminidase assay was performed to investigate the anti-allergic activity of FFE. β-hexosaminidase is an enzyme that is released together with histamine and is involved in allergic inflammation. β-hexosaminidase could be used as a biomarker for an allergic reaction since it is released with histamine [46]. To investigate the release of β-hexosaminidase at the cellular level, we used the rat basophilic leukemia cell line, RBL-2H3. The cells were sensitized using DNP-IgE, and DNP-BSA was used as an antigen to induce allergic reactions. The amount of secreted β-hexosaminidase bound to the substrate was determined. Figure 6 shows that the negative control group, treated with only DNP-IgE, elevated β-hexosaminidase release by approximately 2.4 times compared to the control group. However, upon treatment with FFE, the release of β-hexosaminidase was downregulated compared to that in the negative control group. These data suggest that β-hexosaminidase release is inhibited by FFE, resulting in a reduction in inflammation.

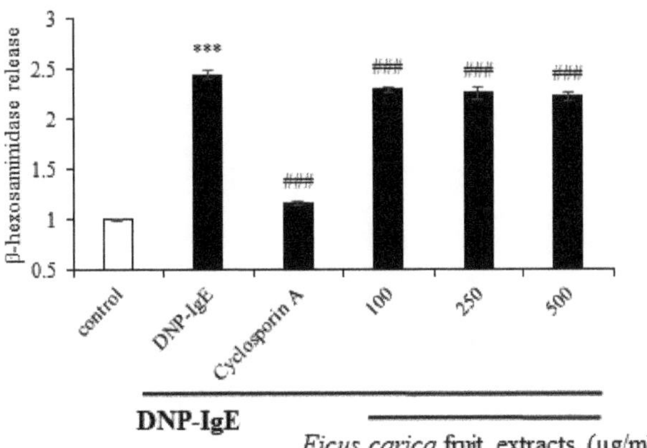

Figure 6. Effect of *F. carica* fruit extracts on β-hexosaminidase release in DNP-IgE-stimulated RBL-2H3 cells. The data are the mean ± SD. *** $p < 0.001$ compared with control group, ### $p < 0.001$ compared with DNP-IgE. All experiments were performed in triplicate.

3.6. IMQ-Induced Psoriasis in BALB/c Mice

Figure 7 shows the scheme of the animal experiments conducted to investigate the psoriasis-ameliorating effect of FFE. Psoriatic skin lesions were excessively induced by daily treatment with 62.5 mg of IMQ on the dorsal skin of mice for 7 days. At 5, 6, and 7 d, IMQ cream was applied in the morning to develop and maintain psoriasis, and then FFE was applied in the afternoon. FFE was divided into three doses: high, moderate, and low. Figure 8 shows the phenotypical results of the animal experiments. Simple phenotypic features, such as dryness, scales, and erythema, were excessively induced in the psoriasis-induced group using only IMQ. However, symptoms including scales and erythema were alleviated in a concentration-dependent manner in the FFE group, and an almost superior improvement was observed in the high dose (FH) group similar to that in the dexamethasone-treated group.

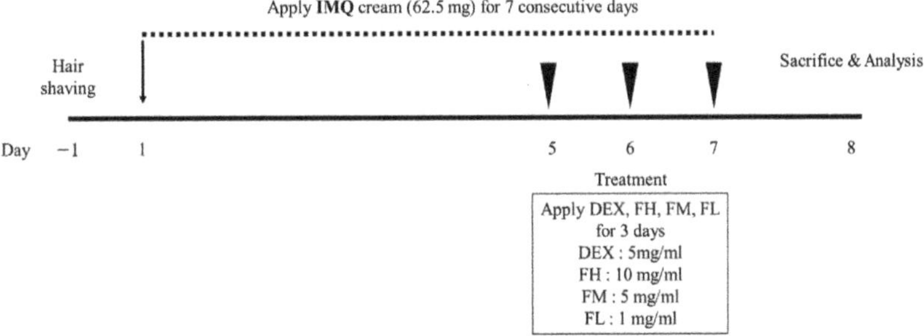

Figure 7. The scheme of animal experiments conducted to investigate the psoriasis-improvement effect of *F. carica* fruit extracts. IMQ, Imiquimod; DEX, dexamethasone; FH, high-dose *Ficus carica* extract; FM, moderate-dose *Ficus carica* extract; FL, low-dose *Ficus carica* extract.

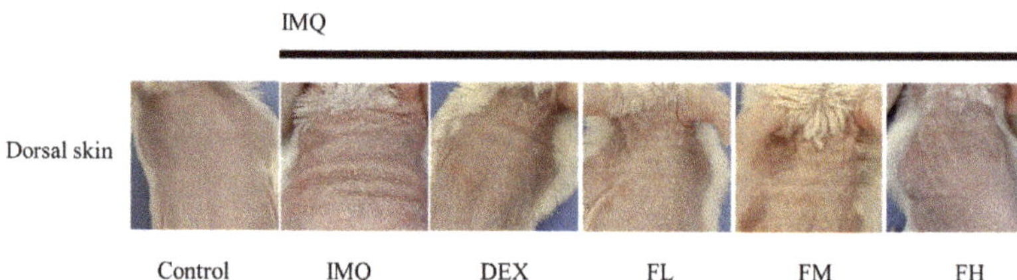

Figure 8. Phenotypical observations of dorsal skin in mice with IMQ-induced psoriasis-like skin lesions. Control mice with daily topical application of Vaseline on the shaved dorsal skin. Test mice with daily IMQ-treated (62.5 mg) dorsal skin on day 3 after IMQ treatment showing psoriasis-like inflammation and erythema lesions and silvery-white scales. IMQ, imiquimod; DEX, dexamethasone; FH, high-dose *Ficus carica* extract, FM, moderate-dose *Ficus carica* extract; FL, low-dose *Ficus carica* extract. All experiments were performed in triplicate.

3.7. Measurement of PASI Score in BALB/c Mice

The PASI score evaluates the severity of psoriasis by measuring the degree of redness (erythema), thickness, scales of the psoriasis lesion, and the extent of psoriasis spread. Figure 9 shows that all indicators, such as erythema, scales, and dorsal thickness, steadily increased in the IMQ-treated group. However, the scores for erythema, scales, and dorsal skin thickness decreased after FFE treatment. In particular, the high-concentration FFE treatment group showed an improvement similar to that of the positive control group (dexamethasone).

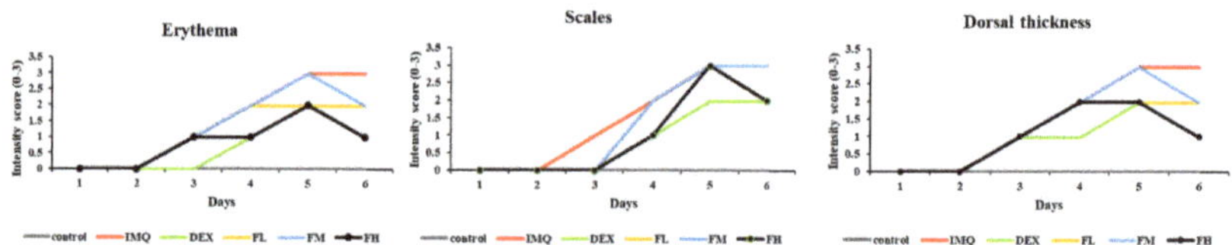

Figure 9. PASI scores showing intensity of erythema and scales of the control and treated mice dorsal skin on a 0–3 point scale. All experiments were performed in triplicate.

3.8. Histological Analysis of Effect of *F. carica* Fruit Extracts on Psoriasis-like Skin Lesions in BALB/c Mice

Figure 10 shows that the thickness of the stratum corneum increased in the IMQ-treated group, which caused psoriasis. However, FFE treatment reduced the epidermal thickness, and the high-dose FFE treatment reduced the level of scales comparable to that of dexamethasone.

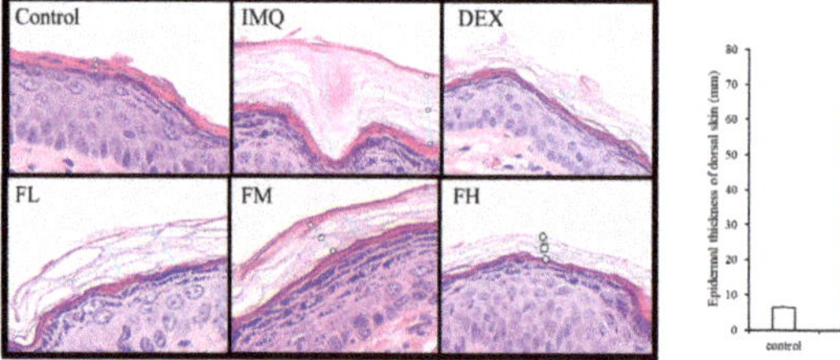

Figure 10. Histological examinations stained with hematoxylin and eosin (H&E). H&E-stained dorsal skin of control and IMQ-treated mice. Images magnified 40×. *** $p < 0.001$ compared with control group, ### $p < 0.001$ compared with IMQ. All experiments were performed in triplicate.

3.9. *Effect of Ficus carica Fruit Extracts on the Spleen Weight*

The spleen is a representative immune organ, and an enlarged spleen indicates excessive inflammation in the body [47]. The size of the spleen significantly increased when only IMQ was applied. However, the weight of the spleen was significantly reduced in the high-dose FFE group compared to that in the IMQ group (Figure 11).

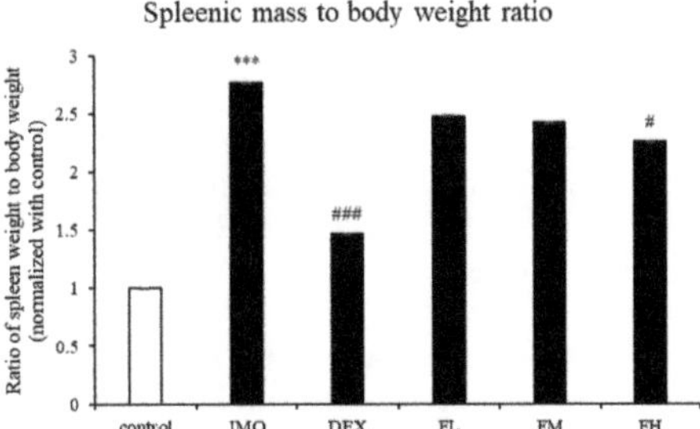

Figure 11. Effect of *F. carica* fruit extracts on the ratio of spleen weight to body weight. Mice were sacrificed, and the ratio of spleen weight to body weight was determined at 24 h after the final administration. Data presented are mean ± SD (n = 5) *** $p < 0.001$ indicates a statistically significant difference from the control group. # $p < 0.05$, ### $p < 0.001$ indicates a statistically significant difference from the IMQ group. All experiments were performed in triplicate.

3.10. *Effect of F. carica Fruit Extracts on the Phosphorylation of STAT3 in Mouse Dorsal Skin*

To investigate the mechanism underlying the in vivo ameliorative effect of FFE on psoriasis-like skin lesions, the p-STAT3 expression pattern was investigated. Figure 12 shows that p-STAT3 was overexpressed in the group treated with IMQ alone. However, the IMQ-induced p-STAT3 expression was reduced by FFE treatment. These results indicated that FFE improved the symptoms of psoriatic skin lesions by regulating STAT3 expression.

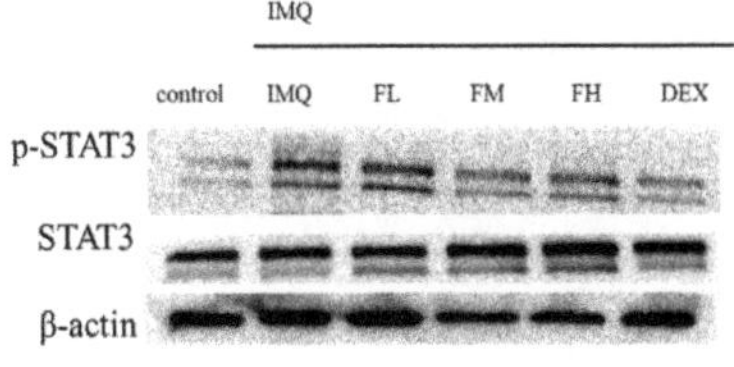

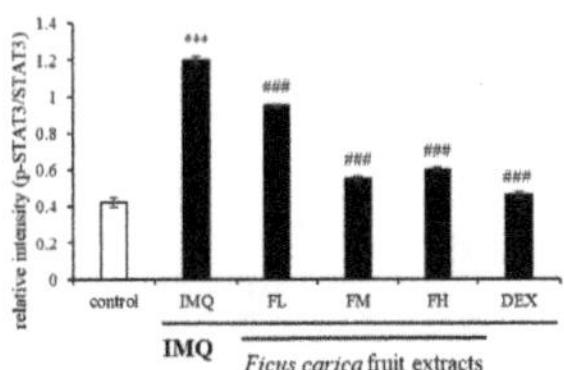

Figure 12. Effect of *F. carica* fruit extracts on the phosphorylation of STAT3. The expression of p-STAT3 was analyzed with ImageJ and normalized against β-actin. All data are expressed as mean ± SD *** $p < 0.001$ compared with control group, ### $p < 0.001$ compared with IMQ group. All experiments were performed in triplicate.

4. Discussion

Psoriasis is an immune-mediated, long-lasting inflammatory skin disease [48] characterized by great risk of morbidity, chronicity, disability, and associated comorbidities [49]. The common occurrence of psoriasis on the skin primarily involves an inflammatory response involving IFN-α, IFN-γ, IL-1, IL-6, IL-17, IL-22, and IL-23 from dendritic cells, macrophages, and helper T cells (Th cells) [50,51]. Activation of NF-κB-mediated inflammatory events involving iNOS and COX2 is known to be associated with psoriasis initiation and progression. The pathogenesis of psoriasis is complex, and the contribution of both innate and adaptive immunity might help manage this complex disease [52,53].

An extensive literature survey revealed that *F. carica* is a sacred and important medicinal plant used for the ethnomedicinal treatment of anemia, bronchitis, constipation, diabetes, fever, hemorrhoids, inflammation, liver disorders, infectious diseases, and many other ailments worldwide [7,17]. Pharmacological studies on fresh plant materials, crude extracts, and isolated components of *F. carica* have provided evaluations of their anti-bacterial, anti-cancer, anti-fungal, anthelmintic, anti-inflammatory, anti-mutagenic, anti-pyretic, anti-spasmodic, anti-platelet, antiviral, cytotoxic, hepatoprotective, hypoglycemic, hypolipidemic, and immunostimulant activities [15].

In this study, the in vitro and in vivo anti-psoriasis effect of *F. carica* fruit extract (FFE) was investigated. The in vitro results showed that FFE reduced the production of nitric oxide (NO), iNOS, and COX-2 from LPS-stimulated RAW 264.7 cells. This anti-inflammatory activity is thought to be due to the basic antioxidant activity of the extract, probably owing to the presence of polyphenols and flavonoids.

The MAPK and NFκB signals, which are typical inflammatory pathways, were not significantly regulated by FFE. Currently, the JAK–STAT signaling pathway is emerging as the main mechanism underlying psoriasis pathogenesis. Interestingly, phosphorylation of JAK1 and STAT3 in the x JAK–STAT signaling pathway, which has been highlighted as a new inflammatory signal, was dramatically modulated by FFE. In addition, FFE reduced the release of β- hexosaminidase in the anti-allergic experiments.

When psoriasis was induced in BALB/c mice with IMQ, the dorsal thickness of the mice increased. However, this increased dorsal thickness was significantly reduced by FFE treatment. Moreover, the PASI score was reduced with FFE. In addition, histological observations revealed that the dermal thickness of the epidermis was decreased. The size and weight of the spleen were also reduced. Notably, FFE reduced the phosphorylation of STAT3 in the psoriasis-like mouse model.

Taken together, *F. carica* fruit extract might show a superior ameliorative effect on psoriatic skin lesions via anti-inflammatory effects associated with JAK–STAT modulation. Thus, the extract could be developed as a candidate material for the treatment of psoriasis. In further experiments, *F. carica* fruit extract will be fractionated to find the ingredient with the highest anti-psoriasis efficacy and developed as a drug with high specificity.

Author Contributions: Investigation, J.H.L.; Supervision, M.-Y.L.; Writing—original draft, J.H.L.; Writing—review and editing, M.-Y.L. All authors have read and agreed to the published version of the manuscript.

Funding: This research received no external funding.

Institutional Review Board Statement: The animal study protocol was approved by the Animal Research Ethics Committee of Soonchunhyang University (SCH22-0053) for studies involving animals.

Informed Consent Statement: Not applicable.

Data Availability Statement: Data is unavailable.

Acknowledgments: This study was supported by Soonchunhyang University.

Conflicts of Interest: Author Mi-Young Lee is the CEO of Eshel Biopharm Co., Ltd. and a professor at Soonchunhyang University and declares no conflict of interest.

References

1. World Health Organization. Global Report on Psoriasis. World Health Organization, 2016. Available online: https://apps.who.int/iris/handle/10665/204417 (accessed on 1 April 2022).
2. NHS . Available online: https://www.nhs.uk/conditions/psoriasis/ (accessed on 8 April 2022).
3. Gibbs, S. Skin disease and socioeconomic conditions in rural Africa: Tanzania. *Int. J. Dermatol.* **1996**, *35*, 633–639. [CrossRef]
4. Danielsen, K.; Olsen, A.; Wilsgaard, T.; Furberg, A.-S. Is the prevalence of psoriasis increasing? A 30-year follow-up of a population-based cohort. *Br. J. Dermatol.* **2013**, *168*, 1303–1310. [CrossRef]
5. Mahmoudi, M. (Ed.) *Challenging Cases in Allergic and Immunologic Diseases of the Skin*; Springer Science & Business Media: Berlin, Germany, 2010. [CrossRef]
6. Cai, Y.; Fleming, C.; Yan, J. New insights of T cells in the pathogenesis of psoriasis. *Cell. Mol. Immunol.* **2012**, *9*, 302–309. [CrossRef]
7. Mansoor, H.U.H.; Ahmed, A.; Rasool, F. The inhibitory potential of chemical constituents of *Ficus carica* targeting interleukin-6 (IL-6) mediated inflammation. *Cell Biochem. Funct.* **2023**, *41*, 573–589. [CrossRef]

8. Woo, Y.R.; Cho, D.H.; Park, H.J. Molecular mechanisms and management of a cutaneous inflammatory disorder: Psoriasis. *Int. J. Mol. Sci.* **2017**, *18*, 2684. [CrossRef]
9. Christian, F.; Smith, E.L.; Carmody, R.J. The regulation of NF-κB subunits by phosphorylation. *Cells* **2016**, *5*, 12. [CrossRef]
10. Wan, Y.; Huang, C. The biological functions of NF-κB1 (p) and its potential as an anti-cancer target. *Curr. Cancer Drug Targets* **2009**, *9*, 566–571. [CrossRef]
11. Oeckinghaus, A.; Ghosh, S. The NF- B family of transcription factors and its regulation. *Cold Spring Harb. Perspect. Biol.* **2009**, *1*, a000034. [CrossRef]
12. Xu, F.; Xu, J.; Xiong, X.; Deng, Y. Salidroside inhibits MAPK, NF-κB, and STAT3 pathways in psoriasis-associated oxidative stress via SIRT1 activation. *Redox Rep.* **2019**, *24*, 70–74. [CrossRef]
13. Kim, E.K.; Choi, E.-J. Pathological roles of MAPK signaling pathways in human diseases. *Biochim. Biophys. Acta (BBA) Mol. Basis Dis.* **2010**, *1802*, 396–405. [CrossRef]
14. Mavropoulos, A.; Rigopoulou, E.I.; Liaskos, C.; Bogdanos, D.P.; Sakkas, L.I. The Role of p38 MAPK in the aetiopathogenesis of psoriasis and psoriatic arthritis. *J. Immunol. Res.* **2013**, *2013*, 569751. [CrossRef]
15. Kiu, H.; Nicholson, S.E. Biology and significance of the JAK/STAT signalling pathways. *Growth Factors* **2012**, *30*, 88–106. [CrossRef]
16. Rezagholizadeh, L.; Aghamohammadian, M.; Oloumi, M.; Banaei, S.; Mazani, M.; Ojarudi, M. Inhibitory effects of Ficus carica and Olea europaea on pro-inflammatory cytokines: A review. *Iran. J. Basic Med. Sci.* **2022**, *25*, 268–275. [CrossRef]
17. Badgujar, S.B.; Patel, V.V.; Bandivdekar, A.H.; Mahajan, R.T. Traditional uses, phytochemistry and pharmacology of *Ficus carica*: A review. *Pharm. Biol.* **2014**, *52*, 1487–1503. [CrossRef]
18. Hajam, T.A.; Saleem, H. Phytochemistry, biological activities, industrial and traditional uses of fig (*Ficus carica*): A review. *Chem. Interact.* **2022**, *368*, 110237. [CrossRef]
19. Abe, T.; Koyama, Y.; Nishimura, K.; Okiura, A.; Takahashi, T. Efficacy and Safety of Fig (*Ficus carica* L.) Leaf Tea in Adults with Mild Atopic Dermatitis: A Double-Blind, Randomized, Placebo-Controlled Preliminary Trial. *Nutrients* **2022**, *14*, 4470. [CrossRef]
20. Elghareeb, M.M.; Elshopakey, G.E.; Hendam, B.M.; Rezk, S.; Lashen, S. Synergistic effects of Ficus Carica extract and extra virgin olive oil against oxidative injury, cytokine liberation, and inflammation mediated by 5-Fluorouracil in cardiac and renal tissues of male albino rats. *Environ. Sci. Pollut. Res.* **2021**, *28*, 4558–4572. [CrossRef]
21. Ren, Y.; Song, X.; Tan, L.; Guo, C.; Wang, M.; Liu, H.; Cao, Z.; Li, Y.; Peng, C. A Review of the Pharmacological Properties of Psoralen. *Front. Pharmacol.* **2020**, *11*, 571535. [CrossRef]
22. Doppalapudi, S.; Jain, A.; Chopra, D.K.; Khan, W. Psoralen loaded liposomal nanocarriers for improved skin penetration and efficacy of topical PUVA in psoriasis. *Eur. J. Pharm. Sci.* **2017**, *96*, 515–529. [CrossRef]
23. Vl, S. Analysis of total phenols and other oxidation substrates and antioxidants by means of folin-ciocalteu reagent. *Methods Enzymol.* **1999**, *299*, 152–178.
24. Chang, C.-C.; Yang, M.-H.; Wen, H.-M.; Chern, J.-C. Estimation of total flavonoid content in propolis by two complementary colometric methods. *J. Food Drug Anal.* **2002**, *10*, 3. [CrossRef]
25. Bradford, M.M. A rapid and sensitive method for the quantitation of microgram quantities of protein utilizing the principle of protein-dye binding. *Anal. Biochem.* **1976**, *72*, 248–254. [CrossRef]
26. Enogieru, A.B.; Haylett, W.; Hiss, D.C.; Bardien, S.; Ekpo, O.E. Rutin as a potent antioxidant: Implications for neurodegenerative disorders. *Oxidative Med. Cell. Longev.* **2018**, *2018*, 6241017. [CrossRef]
27. Xu, J.-G.; Hu, Q.-P.; Liu, Y. Antioxidant and DNA-protective Activities of chlorogenic acid Isomers. *J. Agric. Food Chem.* **2012**, *60*, 11625–11630. [CrossRef]
28. Zhang, S.; Gai, Z.; Gui, T.; Chen, J.; Chen, Q.; Li, Y. Antioxidant effects of protocatechuic acid and protocatechuic aldehyde: Old Wine in a new bottle. *Evid. Based Complement Altern. Med.* **2021**, *2021*, 6139308. [CrossRef]
29. Lam, J.; Polifka, J.E.; Dohil, M.A. Safety of dermatologic drugs used in pregnant patients with psoriasis and other inflammatory skin diseases. *J. Am. Acad. Dermatol.* **2008**, *59*, 295–315. [CrossRef]
30. Zhou, K.; Wu, J.; Chen, J.; Zhou, Y.; Chen, X.; Wu, Q.; Xu, Y.; Tu, W.; Lou, X.; Yang, G.; et al. Schaftoside ameliorates oxygen glucose deprivation-induced inflammation associated with the TLR4/Myd88/Drp1-related mitochondrial fission in BV2 microglia cells. *J. Pharmacol. Sci.* **2019**, *139*, 15–22. [CrossRef]
31. Praveena, R.; Sadasivam, K.; Deepha, V.; Sivakumar, R. Antioxidant potential of orientin: A combined experimental and DFT approach. *J. Mol. Struct.* **2014**, *1061*, 114–123. [CrossRef]
32. Liang, Y.; Xie, L.; Liu, K.; Cao, Y.; Dai, X.; Wang, X.; Lu, J.; Zhang, X.; Li, X. Bergapten: A review of its pharmacology, pharmacokinetics, and toxicity. *Phytother. Res.* **2021**, *35*, 6131–6147. [CrossRef]
33. Venturini, C.L.; Macho, A.; Arunachalam, K.; de Almeida, D.A.T.; Rosa, S.I.G.; Pavan, E.; Balogun, S.O.; Damazo, A.; Martins, D.T.D.O. Vitexin inhibits inflammation in murine ovalbumin-induced allergic asthma. *Biomed. Pharmacother.* **2018**, *97*, 143–151. [CrossRef]
34. Topdağı, Ö.; Tanyeli, A.; Akdemir, F.N.E.; Eraslan, E.; Güler, M.C.; Çomaklı, S. Preventive effects of fraxin on ischemia/reperfusion-induced acute kidney injury in rats. *Life Sci.* **2019**, *242*, 117217. [CrossRef]
35. Khalil, H.E.; Abdelwahab, M.F.; Ibrahim, H.-I.M.; AlYahya, K.A.; Altaweel, A.A.; Alasoom, A.J.; Burshed, H.A.; Alshawush, M.M.; Waz, S. Cichoriin, a Biocoumarin, Mitigates Oxidative Stress and Associated Adverse Dysfunctions on High-Fat Diet-Induced Obesity in Rats. *Life* **2022**, *12*, 1731. [CrossRef]

36. Lee, E.H.; Shin, J.H.; Kim, S.S.; Seo, S.R. Sinapic acid controls inflammation by suppressing NLRP3 inflammasome activation. *Cells* **2021**, *10*, 2327. [CrossRef]
37. Abirami, A.; Sinsinwar, S.; Rajalakshmi, P.; Brindha, P.; Rajesh, Y.B.R.D.; Vadivel, V. Antioxidant and cytoprotective properties of loganic acid isolated from seeds of *Strychnos potatorum* L. against heavy metal induced toxicity in PBMC model. *Drug Chem. Toxicol.* **2019**, *45*, 239–249. [CrossRef]
38. Wang, J.; Cai, X.; Ma, R.; Lei, D.; Pan, X.; Wang, F. Anti-inflammatory effects of sweroside on LPS-induced ALI in mice via activating SIRT1. *Inflammation* **2021**, *44*, 1961–1968. [CrossRef]
39. Hu, R.; Wang, M.-Q.; Ni, S.-H.; Liu, L.-Y.; You, H.-Y.; Wu, X.-H.; Wang, Y.-J.; Lu, L.; Wei, L.-B. Salidroside ameliorates endothelial inflammation and oxidative stress by regulating the AMPK/NF-κB/NLRP3 signaling pathway in AGEs-induced HUVECs. *Eur. J. Pharmacol.* **2020**, *867*, 172797. [CrossRef]
40. Lin, Y.; Chen, Y.; Zeng, J.; Li, S. Nodakenetin Alleviates Inflammatory Pain Hypersensitivity by Suppressing NF-κB Signal Pathway. *Neuroimmunomodulation* **2022**, *29*, 486–492. [CrossRef]
41. Aktan, F. iNOS-mediated nitric oxide production and its regulation. *Life Sci.* **2004**, *75*, 639–653. [CrossRef]
42. Johnson, G.L.; Lapadat, R. Mitogen-activated protein kinase pathways mediated by ERK, JNK, and p38 protein kinases. *Science* **2002**, *298*, 1911–1912. [CrossRef]
43. Baker, R.G.; Hayden, M.S.; Ghosh, S. NF-κB, inflammation, and metabolic disease. *Cell Metab.* **2011**, *13*, 11–22. [CrossRef]
44. Banerjee, S.; Biehl, A.; Gadina, M.; Hasni, S.; Schwartz, D.M. JAK–STAT signaling as a target for inflammatory and autoimmune diseases: Current and future prospects. *Drugs* **2017**, *77*, 521–546. [CrossRef]
45. Xu, H.; Yu, A.-L.; Zhao, D.-P.; Meng, G.-Y.; Wang, L.; Shan, M.; Hu, N.-X.; Liu, Y.-L. Ursolic acid inhibits Th17 cell differentiation via STAT3/RORγt pathway and suppresses Schwann cell-mediated Th17 cell migration by reducing CXCL9/10 expression. *J. Endotoxin Res.* **2022**, *28*, 155–163. [CrossRef]
46. Fukuishi, N.; Murakami, S.; Ohno, A.; Yamanaka, N.; Matsui, N.; Fukutsuji, K.; Yamada, S.; Itoh, K.; Akagi, M. Does β-hexosaminidase function only as a degranulation indicator in mast cells? The Primary Role of β-Hexosaminidase in Mast Cell Granules. *J. Immunol.* **2014**, *193*, 1886–1894. [CrossRef]
47. Balato, N.; Napolitano, M.; Ayala, F.; Patruno, C.; Megna, M.; Tarantino, G. Nonalcoholic fatty liver disease, spleen and psoriasis: New aspects of low-grade chronic inflammation. *World J. Gastroenterol.* **2015**, *21*, 6892–6897. [CrossRef]
48. Girolomoni, G.; Griffiths, C.E.M.; Krueger, J.; Nestle, F.O.; Nicolas, J.-F.; Prinz, J.C.; Puig, L.; Ståhle, M.; van de Kerkhof, P.C.M.; Allez, M.; et al. Early intervention in psoriasis and immune-mediated inflammatory diseases: A hypothesis paper. *J. Dermatol. Treat.* **2015**, *26*, 103–112. [CrossRef]
49. Griffiths, C.E.M.; Armstrong, A.W.; E Gudjonsson, J.; Barker, J.N.W.N. Psoriasis. *Lancet* **2021**, *397*, 1301–1315. [CrossRef]
50. Xu, Y.; Shi, Y.; Huang, J.; Gu, H.; Li, C.; Zhang, L.; Liu, G.; Zhou, W.; Du, Z. The Essential Oil Derived from *Perilla frutescens* (L.) Britt. Attenuates Imiquimod–Induced Psoriasis-like Skin Lesions in BALB/c Mice. *Molecules* **2022**, *27*, 2996. [CrossRef]
51. Xu, T.; Wang, X.; Zhong, B.; Nurieva, R.I.; Ding, S.; Dong, C. Ursolic acid suppresses interleukin-17 (IL-17) production by selectively antagonizing the function of RORγt protein. *J. Biol. Chem.* **2011**, *286*, 22707–22710. [CrossRef]
52. Eteraf-Oskouei, T.; Allahyari, S.; Akbarzadeh-Atashkhosrow, A.; Delazar, A.; Pashaii, M.; Gan, S.H.; Najafi, M. Methanolic extract of *Ficus carica* Linn. leaves exerts antiangiogenesis effects based on the rat air pouch model of inflammation. *Evidence-Based Complement. Altern. Med.* **2015**, *2015*, 760405. [CrossRef]
53. Pastwińska, J.; Karaś, K.; Sałkowska, A.; Karwaciak, I.; Chałaśkiewicz, K.; Wojtczak, B.A.; Bachorz, R.A.; Ratajewski, M. identification of corosolic and oleanolic acids as molecules antagonizing the human RORγT nuclear receptor using the calculated fingerprints of the molecular similarity. *Int. J. Mol. Sci.* **2022**, *23*, 1906. [CrossRef]

Disclaimer/Publisher's Note: The statements, opinions and data contained in all publications are solely those of the individual author(s) and contributor(s) and not of MDPI and/or the editor(s). MDPI and/or the editor(s) disclaim responsibility for any injury to people or property resulting from any ideas, methods, instructions or products referred to in the content.

 *life*

Review

A Review of the Potential Benefits of Herbal Medicines, Small Molecules of Natural Sources, and Supplements for Health Promotion in Lupus Conditions

Ardalan Pasdaran [1,2], Bahareh Hassani [3], Ali Tavakoli [4], Ekaterina Kozuharova [5] and Azadeh Hamedi [1,2,*]

1 Department of Pharmacognosy, School of Pharmacy, Shiraz University of Medical Sciences, Shiraz 7146864685, Iran; pasdaran@sums.ac.ir
2 Medicinal Plants Processing Research Center, Shiraz University of Medical Sciences, Shiraz 7146864685, Iran
3 Student Research Committee, School of Pharmacy, Shiraz University of Medical Sciences, Shiraz 7146864685, Iran; hasanibahareh@yahoo.com
4 Research Center for Traditional Medicine and History of Medicine, Department of Persian Medicine, School of Medicine, Shiraz University of Medical Sciences, Shiraz 7134845794, Iran; tavakkolia@sums.ac.ir
5 Department of Pharmacognosy, Faculty of Pharmacy, Medical University of Sofia, 1431 Sofia, Bulgaria; ina_kozuharova@yahoo.co.uk
* Correspondence: hamediaz@sums.ac.ir

Abstract: The Latin word lupus, meaning wolf, was in the medical literature prior to the 1200s to describe skin lesions that devour flesh, and the resources available to physicians to help people were limited. The present text reviews the ethnobotanical and pharmacological aspects of medicinal plants and purified molecules from natural sources with efficacy against lupus conditions. Among these molecules are artemisinin and its derivatives, antroquinonol, baicalin, curcumin, emodin, mangiferin, salvianolic acid A, triptolide, the total glycosides of paeony (TGP), and other supplements such as fatty acids and vitamins. In addition, medicinal plants, herbal remedies, mushrooms, and fungi that have been investigated for their effects on different lupus conditions through clinical trials, in vivo, in vitro, or in silico studies are reviewed. A special emphasis was placed on clinical trials, active phytochemicals, and their mechanisms of action. This review can be helpful for researchers in designing new goal-oriented studies. It can also help practitioners gain insight into recent updates on supplements that might help patients suffering from lupus conditions.

Keywords: systemic lupus erythematosus; clinical; autoimmune diseases; natural product; supplement; medicinal plant; inflammation; immunomodulatory; review; herbal medicine

Citation: Pasdaran, A.; Hassani, B.; Tavakoli, A.; Kozuharova, E.; Hamedi, A. A Review of the Potential Benefits of Herbal Medicines, Small Molecules of Natural Sources, and Supplements for Health Promotion in Lupus Conditions. *Life* **2023**, *13*, 1589. https://doi.org/10.3390/life13071589

Academic Editors: Azahara Rodríguez-Luna and Salvador González

Received: 30 May 2023
Revised: 5 July 2023
Accepted: 12 July 2023
Published: 19 July 2023

Copyright: © 2023 by the authors. Licensee MDPI, Basel, Switzerland. This article is an open access article distributed under the terms and conditions of the Creative Commons Attribution (CC BY) license (https://creativecommons.org/licenses/by/4.0/).

1. Introduction

Systemic lupus erythematosus (SLE) is an autoimmune disease involving multiple organs and clinical manifestations. SLE is more common in young women. The incidence of SLE in males compared to females is 1:5~10 [1,2]. In SLE, autoantibodies and antibody-immune complexes are produced that eventually cause damage to body tissues and induce inflammation [3,4]. SLE patients experience relapsing and remission courses [5]. In SLE, various organs can be involved, including the skin, kidneys, joints, heart, lungs, liver, and blood vessels [6,7]. Since different organs are involved in SLE, a variety of indices can be used to assess the status of diseases, such as the Systemic Lupus Erythematosus Disease Activity Index (SLEDAI), Safety of Estrogens in Lupus Erythematosus National Assessment (SELENA), or British Isles Lupus Activity Group (BILAG) index [8].

The level of anti-double-stranded DNA (anti-dsDNA) antibodies is associated with disease activity, and anti-dsDNA plays an important role in the pathogenesis of SLE. In some cases, the goal of treatment is to bring the level of anti-dsDNA antibodies back to normal (5). Mechanisms involved in kidney damage due to lupus nephritis include dysregulation of T-regulatory cells due to overactivity of B and T lymphocytes, activation of inflammatory responses, improper production of autoantibodies, and deposition of immune complexes in kidney tissue [9,10].

The current medications used to treat SLE include glucocorticoids, immunosuppressive drugs, non-steroidal anti-inflammatory drugs, anti-malarial drugs, systemic lymph node irradiation therapy, and plasma treatment. Despite this, the morbidity and mortality ratios in SLE patients are still unacceptably high [6,11]. The mentioned medications lead patients to be exposed to side effects and also reduce the patient's quality of life [12]. This encourages patients to try complementary and alternative medicines such as herbal remedies, medicinal plants, phytochemicals, vitamins and mineral supplements, acupuncture, moxibustion, and spiritual therapy such as yoga. Moreover, tremendous efforts have been made by researchers to develop safe and efficient drugs and supplements from natural molecules and their synthetic derivatives for the condition.

Some previous articles can be found on the treatment of lupus conditions that are not on natural products or supplements and mostly discussed orthodox medicine [13–18]. Some other review articles discussing natural products on a specific condition of lupus, such as cutaneous lupus [19] or lupus nephritis [20], can be found in the literature. Some discussed only a specific natural compound such as curcumin [21], omega-3 fatty acids [22], and triptolide [23]. On the other hand, some other review articles reflected a specific traditional medicine approach such as traditional Chinese medicine (TCM), traditional Iranian medicine (TIM) [24], or Ayurveda [25], which mostly focused on polyherbal formulations and plant extracts that are composed of a complex mixture of phytochemicals with a holistic approach. Some of these articles have neglected minerals, vitamins, and pure active phytoconstituents. Conversely, some reviews and meta-analyses specifically focused on fatty acids [22,26,27] or vitamins [28,29]. Moreover, some articles focused mostly on pharmacology and possible mechanisms of action of a group of natural components without scoping clinical trials or meta-analyses [30]. All the mentioned articles provided precious information for specific groups of researchers and highlighted specific aspects of treatment and drug development for lupus conditions. But a comprehensive systematic review that simultaneously gives detailed evidence-based information on the efficacy of natural products and supplements on different lupus conditions, which helps practitioners and clinicians to understand the mechanism of action and the level of evidence for any of these molecules, herbs, or supplements and researchers to design new goal-oriented experimental studies or clinical trials, seems to be lacking. The present article tries to cover the last-mentioned points in this paragraph and reviewed natural molecules, phytochemicals, medicinal plants, fungi, vitamins, minerals, and other supplements that have been reported to be beneficial for lupus conditions, with special emphasis on clinical trials and the molecules' mechanism of actions, as well as their adverse effects and toxicity.

2. Materials and Methods

A comprehensive literature search was conducted in PubMed, Cochrane Library, Web of Science, Scopus, the National Library of Medicine (NLM) catalog, and Google Scholar, from January 1970 up until January 2023. The obtained records were assessed for eligibility in accordance with the PRISMA 2000 guidelines.

Inclusion criteria: Combinations of various keywords including systemic lupus erythematosus, SLE, lupus nephritis, or lupus AND natural molecule, natural medicine, phytochemicals, herbal medicine, medicinal plants, fungi, mushrooms, minerals, vitamins, fatty acids, supplements, nutrition, meta-analysis, and clinical trial, toxicity, and side effects have been considered in the search strategy. Moreover, the compound names and scientific names of the identified plants were searched again with keywords related to lupus. No restriction was set on the language. A special focus was set on reported purified molecules from natural sources including artemisinin and its derivatives, antroquinonol, baicalin, curcumin, emodin, mangiferin, salvianolic acid A, triptolide, the total glycosides of paeony (TGP), some fatty acids, and vitamins.

Exclusion criteria: (1) Duplicate article; (2) addressed a natural compound, but not related to lupus conditions; (3) article on lupus conditions, but not related to natural compounds; (4) did not address specific natural compounds or plant extracts; (5) studies involving entirely synthetic molecules or antibodies; (6) herbal remedies from traditional medicine that had not been scientifically evaluated for lupus conditions; and (7) polyherbal formulations from traditional medicine.

3. Results

The search strategy yielded 14,300 studies. The titles or abstracts were reviewed to exclude duplicates or irrelevant ones. Excluded were 13,887 records that were identified as irrelevant, duplicate, or not reliable. As a result, 413 studies were included in the review and 74 were included in the synthesized tables.

3.1. Ethnobotany

The Latin word lupus, meaning wolf, was in the medical literature prior to the 1200s to describe skin lesions that devour flesh, and the resources available to physicians to help people were limited [31,32]. Traditional knowledge on how to deal with this condition involves the use of several medicinal plants or plant-based mixtures. Ethnobotanical and ethnopharmacological studies reveal that *Cinchona* spp. [33] and "Thanatka" made of *Hesperethusa crenulata* and *Limonia acidissima* bark [34] have dermatologic uses, specifically in the treatment of lupus erythematosus. Also, sieketroos *Arctopus* species [35], *Juniperus* species [36], *Onopordum acanthium* [37], and *Centella asiatica* [38] were documented to treat systemic lupus erythematosus. According to Iranian traditional medicine (traditional Persian medicine), infectious diseases and fever are the main reasons for nephritis, which is called "Varam-e-Kolye". Several medicinal plants have been advised to control for lupus nephritis or "Varam-e-Kolye", which are *Anethum graveolens* L., *Carum carvi* L., *Coriandrum sativum* L., *Cucurbita pepo* L., *Cydonia oblonga* Mill., *Ficus carica* L., *Linum usitatissimum* L., *Melissa officinalis* L., *Prunus amygdalus*, and *Ziziphus jujuba* Mill. Some recent research reported nephroprotective and anti-inflammatory properties of these plants [24,39,40].

As examples, *Cuminum cyminum* L. (in Persian کرویا or زیره سبز), *Carum carvi* L. (in Persian کمون کرمانی or زیره سیاه), *Lagoecia cuminoides* L. (in Persian زیره وحشی or قردمانا) [41,42], and *Bunium persicum* (Boiss.) Fedtch (in Persian Zire Kermani) [39] are other plants advised for Varam-e-Kolye and/or other kidney diseases such as "Riah-e-Gorde" [41,42]. Carvia (کرویا) is the Arabic version of the Latin word "craviya" or the Syriac word "Ceravi"; in Greek the word is "Azhamyon", in Roman "Fadroni", and in Arabic "Taghdeh", "Taghrad", and "Comone Roomi" [42]. *B. persicum* has shown antiglycation, antioxidant, anti-inflammatory, and nephroprotective (possibly due to antiglycation) effects [43–45].

According to the literature, several medicinal plants and fungi have been considered to be beneficial for conditions related to lupus. This review focused on those that are evidence-based with in vitro or in vivo studies or clinical trials. The ethnobotanical aspects of these plants are summarized in Table 1.

Table 1. Main botanical aspects and pharmacological benefits of medicinal plants, mushrooms, and fungi advised for lupus conditions.

Plant Name (Scientific Name or Species)	English Name	Endemic Region Endemic Name	Family	Main Parts Used	Main Botanical Characteristics (Description)	Preparations	Main Pharmacological Benefits/Activities/Properties	Reference
Taiwanofungus camphoratus (M.Zang and C.H.Su) Sheng H.Wu, Z.H.Yu, Y.C.Dai and C.H.Su (Basionym: *Ganoderma camphoratum*) M.Zang and C.H.Su	Stout camphor fungus	In Taiwan as "Niu-chang-chih" or "Chang-chih" or "Niu-chang-ku" or "Chang-ku"	Polyporaceae	Fruiting bodies	A parasitic fungus on the endemic rotting trees of *Cinnamomum kanehirai Hayata* causing a brown heart rot	Extract	Antioxidant Anti-inflammatory Immunomodulatory Antimicrobial Anti-diabetic Hepatoprotective Neuroprotective Vasorelaxation Antitumor Anti-hypertensive	[46–48]
Artemisia annua L.	Annual absinthe, Sweet wormwood, Sweet annie Annual mugwort, Annual wormwood, Sweet wormwood, Chinese wormwood, Sweet sagewort	-Mild climates of Asia -In China as Qinghao	Asteraceae	Leaves	An annual plant with hairless brown erect stems (a height of 50 to 100 cm) and tender leaves about 3 to 5 cm long and many small yellowish-green flowers and brown rounded seeds	Extract	Anti-inflammatory Immunomodulatory Immunosuppressive Antioxidant, Antitumor Anti-malarial, Antibacterial Antifungal, Anti-cancer Anti-obesity, Hepatoprotective, Anti-asthmatic	[49,50]
Astragalus propinquus Schischkin (Syn. *A. membranaceus* fo. *propinquus* (Schischk.)) Kitag.	Mongolian milkvetch	Huang Qi, in Chinese, Ogi, in Japanese, Milk-Vetch, in English, and Gavan, in Persian	Fabaceae (or Leguminosae)	Roots	A Perennial flowering plant 50–150 cm high, including a straight and long (up to 50 cm) cylindrical root, erect stems branched in the upper parts with small ovate-lanceolate or elliptical leaves	Extract	Anti-inflammatory Immune system boosting Antioxidant, Anti-cancer Neuroprotection, Renoprotection (reduces proteinuria and creatinine), Hepatoprotection, Hypoglycemic Anti-osteoporosis, Anti-fatigue, Cardioprotection	[51]
Kalanchoe pinnata (Lam.) Pers. Syn. *Bryophyllum pinnatum* (Lam.) Oken	Air plant Canterbury bells Cathedral bells Curtain plant Life-plant Fortune plant Good luck leaf Green mother of millions Air plant Mexican love plant	Madagascar In Brazil as "folha-da-fortuna" or "folha-de-pirarucu" or saião "coirama", In Caribbean as "bruja"	Crassulaceae	Leaves	A perennial succulent herb 0.3–1.2 m tall, including four-angled stems and fleshy simple/compound dark green leaves to 20 cm long and red pendulous flowers	Extract	Anti-inflammatory, Antinociceptive Antianaphylactic Immunomodulator, Immunosuppressive, Antitumor, Antiulcer, Anti-diabetic, Hypotensive, CNS depressant, Antimicrobial Antileishmania, Anthelmintic, Gastroprotective Hepatoprotective Anti-urolithiatic	[52–54]
Camellia sinensis (L.) Kuntze	Tea plant tea shrub Tea tree	China and Southeast Asia	Theaceae	Leaves	Evergreen shrub or small tree with strong taproot and yellow-white flowers with seven or eight petals	Extract	Antioxidant, Analgesic Anti-inflammatory, Anti-cancer, Anti-fibrotic, Antimutagenic, Cholesterol lowering, Cardiovascular protection, Hepatoprotective, Neuroprotective Anti-diabetic, Anti-obesity Analgesic	[55–57]

Table 1. *Cont.*

Plant Name (Scientific Name or Species)	English Name	Endemic Region Endemic Name	Family	Main Parts Used	Main Botanical Characteristics (Description)	Preparations	Main Pharmacological Benefits/Activities/Properties	Reference
Curcuma longa L.	Turmeric	Indian subcontinent South-east Asian Curcum in the Arab region Indian saffron Haridra (Sanskrit, Ayurvedic) Jianghuang (yellow ginger in Chinese) Kyoo or Ukon (Japanese)	Zingiberaceae	Rhizome	A perennial herbaceous plant with 1 m high with highly branched, yellow to orange, cylindrical, aromatic rhizomes and oblong, pointed leaves and bears funnel-shaped yellow flowers	Powder Extract	Anti-inflammatory Antioxidant Hepatoprotective Anticarcinogenic Anti-diabetic Antimicrobial Antidepressant Lowering cholesterol, triglyceride and low-density lipoprotein (LDL) levels	[58–60]
Ganoderma lucidum aggregate	Reishi	In China as Lingzhi or herb of spiritual potency	Ganodermataceae Or Polyporaceae	Mycelia Spores Fruit body	Polypore mushroom grows on decaying hardwood trees	Powders Dietary supple-ments Tea	Immunomodulation Anti-inflammatory, Antioxidant Radical-scavenging Analgesic, Chemo-preventive Antitumor, Chemo and radio protective, Anti-atherosclerotic, Sleep promoting Antimicrobial, Hypolipidemic, Anti-fibrotic Hepatoprotective, Anti-diabetic, Anti-androgenic Anti-angiogenic, Anti-aging, Estrogenic activity, Anti-ulcer	[61–63]
Ganoderma tsugae Murrill	Hemlock varnish shelf		Ganodermataceae Or Polyporaceae	Fruit body Mycelium	Flat polypore mushroom	Extract	High antioxidant Immunomodulatory, Antitumor, Anti-inflammatory Anti-fibrotic, Antiautoantibody formation	[64]
Gentiana burseri subsp. *villarsii* (Griseb.) Rouy (Syn. *G. macrophylla* Pall., *G. macrophylla* Bertol.)	Large leaf gentian	qin jiao in Chinese	Gentianaceae	Roots Radix Gentianae macrophyl-lae Flowers	A glabrous pale green perennial with erect, gross, 3–6 mm in diameter and 40–70 cm long stems and very large, acute, elongate-lanceolate and up to 40 cm long leaves with intensely bluish-violet and bell-shaped flowers and brown, bright and wingless seeds.	Extract	Anti-inflammatory, Wound healing, Analgesic, Antioxidant, Immunomodulation Hepatic protection, Joint protection, Cardio-protective Neuro-protective Anti-influenza	[65–68]

Table 1. *Cont.*

Plant Name (Scientific Name or Species)	English Name	Endemic Region Endemic Name	Family	Main Parts Used	Main Botanical Characteristics (Description)	Preparations	Main Pharmacological Benefits/Activities/Properties	Reference
Glycyrrhiza glabra L.	Licorice Liquorice	"Sus", "Mahak","Mehak", "Bikh-e-Mahak", in Persian ethnomedicine, "Shirin Bayan" (Persian) and "Irk-es-sus" (Arabic).	Fabaceae	Roots	An herbaceous perennial plant, growing to 1 m in height, with pinnate leaves about 7–15 cm long, with 9–17 leaflets. The flowers are 8–12 mm long, purple to pale whitish blue, produced in a loose inflorescence. The roots are stoloniferous.	Extract	Immunomodulatory Anti-inflammatory Antiapoptotic Neuroprotective Nephroprotective	[69–72]
Linum usitatissimum L.	Flax Flaxseed linseed	Cooler parts of Asia, Europe and Mediterranean regions In Persian as Bazr-e katan	Linaceae	Seeds	Annual small flowering plant with an approximate 60–100 cm length and 5 petal pale blue flowers and dark to yellow, oval, flat and pointed seed	Oil Extract Bread	Antioxidant, Immunomodulatory, Anti-inflammatory Antimicrobial, Antiprotozoal, Analgesic, Antihyperlipidemia, Antihyperglycemic, Antitumor	[73–75]
Nelumbo nucifera Gaertn. *(Syn.: Nelumbium nelumbo, N. speciosa, N. speciosum, Nymphaea nelumbo etc.)*	Lotus	sacred lotus, Laxmi lotus, Indian lotus, Egyptian bean In Chinese, seeds are called Lian Zi Xin, In India (lotus, kamala, pundarika or padma)	Nelumbonaceae	All parts of the plant (seeds, root, flowers, leaf, stem)	An aquatic species. The flowers (up to 30 cm in diameter) are on thick stems rising several centimeters above the leaves (that spread up to 80 cm in diameter). The leaf stalks grow up to 200 cm long while the peltate leaf blade or lamina can have a horizontal spread of 1 m	Extracts	Anti-inflammatory Immunomodulatory Antipyretic Antioxidant Cardioprotective Hepatoprotective	[76,77]
Ophiocordyceps sinensis (Berk.) G.H.Sung, J.M.Sung, Hywel-Jones and Spatafora (syn. *Cordyceps sinensis*)	Caterpillar fungus	In Tibet as yartsa gunbu In Chinese as "Dong Chong Xia Cao" or "winter worm summer grass" In India as keera jhar, keeda jadi, keeda ghas	Clavicipitaceae Ophiocordycip-itaceae	fruiting body	A mushroom with plant-like fruiting body fill with mycelia	Oral liquids Capsules	Immunomodulatory Anti-inflammatory Antioxidative Anti-fibrogenic Antitumor Anti-thrombotic Anti-viral Antifungal	[78,79]
Paeonia lactiflora Pall.	Chinese peony Chinese herbaceous peony Common garden peony White Peony	East Asia In Chinese as "Bai Shao Yao".	Paeoniaceae Or Ranunculaceae	Root (*Radix Paeoniae Alba)*	Herbaceous perennial flowering plant is about 60–100 cm tall with fleshy roots, annual stems and large compound leaves. The flower buds are large and round, opening into large flowers with 5–10 white, pink, or crimson petals and yellow stamen.	Dried root bark Extract	Anti-inflammatory Immunomodulatory Analgesic Antioxidative Hepatoprotective	[80]

Table 1. *Cont.*

Plant Name (Scientific Name or Species)	English Name	Endemic Region Endemic Name	Family	Main Parts Used	Main Botanical Characteristics (Description)	Preparations	Main Pharmacological Benefits/Activities/Properties	Reference
Paeonia × suffruticosa Andrews	Cortex Moutan (king of flowers)	In China as *mŭdān* or Mudanpi	Ranunculaceae or Paeoniaceae	Roots bark	Collective name of cultivated tree peonies, originated from the hybridization of multiple species of wild tree peonies, belong to P. suffruticosa complex. A bush or a tree with attractive flowers and woody stems	Dried root bark Extract	Anti-inflammatory Antioxidative Anti-allergic Antitumor Cardiovascular System Protective Anti-Diabetic Neuroprotective Hepatoprotective Analgesic Sedative	[81]
Rehmannia glutinosa (Gaertn.) Liboschitz ex Fischer and C. A. Meyer	Rehmanniae Radix	In China as Di-Huang	Orobanchaceae or Scrophulariaceae	Roots Root tuber	A perennial plant that the roots of it are thick and yellowish in color, while the stems have grayish short hairs. The leaves appear to be gathered at the roots with edges that are blunt and serrated. The flowering form is raceme and blooms purple flowers from June to July.	Extract	Anti-inflammatory Antitumor Immunomodulatory Neuroprotective Hypoglycemic Cardioprotective Antioxidant Hematopoiesis promotion Antianxiety	[82,83]
Scutellaria baicalensis Georgi	Baikal skullcap Chinese skullcap Wogonin, baicalin, and baicalein	In China as *huángqín*	Lamiaceae	Roots and rhizomes	A perennial flowering herb with thick, fleshy, elongated and branched rhizome up to 2 cm in diameter. The leaves are lanceolate to linear lanceolate, 1.5–4.5 cm in length and 0.5–1.2 cm in width. The fruits are hard, ovoid, dark brown with a tumor, 1.5 mm in height, and 1 mm in diameter	Dried Extract	Anti-inflammatory, Antioxidant, Antitumor, Antibacterial, Antiviral, Liver protection, Effects on the nervous system, Effects on the immune system	[84–86]
Tripterygium wilfordii Hook. f.	-Thunder god vine -Thunder duke vine	In China (Mandarin) as léi gōng téng	Celastraceae	Roots without bark	wood vine plant perennial vien	Extract	Anti-inflammatory Immunomodulatory Immunosuppressive Anti-cancer Anti-fertility	[87,88]
Urtica dioica L.	Common nettle Burn nettle Stinging nettle	Europe Asia Africa	*Urticaceae*	leaves Roots Seeds Aerial parts	A perennial herbaceous plant with underground rhizome and bi-arch root and erect, green and quadrangular stem with dark-green, oblong or ovate leaves and small dioecious, brown to greenish flowers. Tinging trichomes cover both stems and leaves and contain a fluid.	Extract	Anti-inflammatory Antioxidant Analgesic	[89]

3.2. Purified Molecules from Natural Sources

Different herbal remedies, medicinal plants, and mushrooms have been utilized to cure a range of medical ailments in both developing and developed communities. Additionally, it is estimated that roughly 25% of currently marketed medicines were developed from the primary or secondary metabolites of natural medicines [90]. On the other hand, the absence of a well-organized regulatory and legal framework for herbal products has caused the World Health Organization (WHO) to express worry regarding the efficacy and safety of herbal treatments [91]. Due to varying growth circumstances and harvesting times, different primary and secondary metabolites have varying concentrations in medicinal plants [92]. These problems motivate researchers to find and purify the medicinal plant's active components. Researchers have gained a greater understanding of the mechanisms of action by working with highly purified compounds. When compared to herbal extracts, pure natural molecules are more reliable at determining dosage and detecting unwanted effects or potential toxicities. Moreover, natural molecules can be considered lead compounds for developing new drugs. In the case of lupus, several natural products and their derivatives, in purified and structure-elucidated form, have been reported to exhibit considerable therapeutic potential. Although the mechanisms of action of some of these molecules have yet to be fully elucidated, more extensive research can generate new data that can be used in clinical trials. The reported data on these molecules is discussed in detail in the following (Table 2, Figure 1).

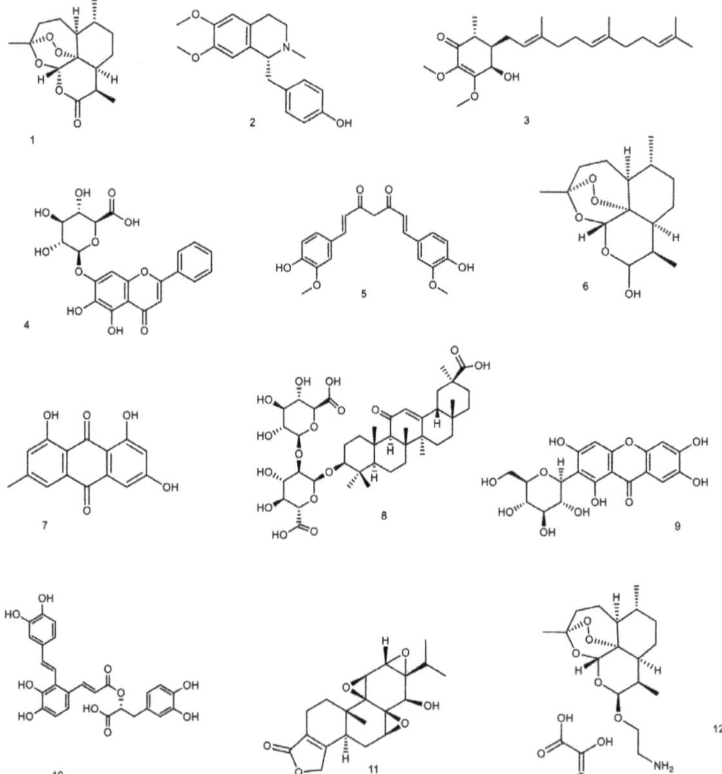

Figure 1. The structures of small molecules from natural sources and their derivatives with reported efficacy against lupus conditions. The structures are 1: (+)−artemisinin, 2: (S)−armepavine, 3: antroquinonol, 4: baicalin, 5: curcumin, 6: dihydroartemisinin, 7: emodin, 8: glycyrrhizic acid, 9: mangiferin, 10: salvianolic acid A, 11: triptolide, and 12: β-aminoarteether maleate.

Table 2. Purified molecules from natural sources and their derivatives, fatty acids, minerals, and vitamins with reported efficacy against lupus (↓: decrease in activity or level; ↑: increase in activity or level).

Compound Name	Class of Compound	Natural Source	Type of Study	Method	Result or Side Effect or Mechanism	Ref.
			Phytochemicals			
Antroquinonol	Farnesylated quinone	*Antrodia camphorata*	In vivo	NZB/NZW F1 mice were used Doses: 15 mg/kg antroquinonol orally for 5 weeks.	- ↓ hematuria, proteinuria and IL-18 production - Inhibited T cell proliferation and induced Treg cell suppression - ↓ reactive oxygen species and nitric oxide production - ↑ Nrf2 activation - Inhibited NF-κB activation	[93]
			In vivo	SLE-prone NZB/W F1 mice Doses: 100, 200, and 400 mg/kg of the mycelial extract of *A. camphorate* for 5 consecutive days per week for 12 weeks via gavage.	- ↓ proteinuria, creatinine and serum BUN levels - ↓ the thickness of the kidney glomerular basement membrane - ↓ the production of TNF-α and IL-1β	[46]
(*S*)-Armepavine	Benzylisoquinoline alkaloid	*N. nucifera*	In vivo	MRL/MpJ-lpr/lpr mice *Doses:* 5 and 10 mg/kg/day of oral (S)-armepavine, for 6 weeks.	- ↑ the life expectancy of mice - Inhibit splenocytes proliferation - prevent lymphadenopathy - Inhibit the expression of IL-2, IL-4, IL-10 and IFN-γ genes - Inhibit T cells proliferation - ↓ proteinuria and anti-dsDNA autoantibody	[94]
Artemisinin	sesquiterpene lactone	*A. annua*	Clinical trial	Treatment group: 31 patients were treated with 2–4 g/d of *C. sinensis* and 0.6 g/d of artemisinin. *C. sinensis* was taken before three main meals and artemisinin after three main meals for three years. Control group: 30 patients received tripterygiitotorum and/or Baoshenkang.	- The creatinine clearance rate did not change - The complement C3 level stabilized in the normal range - The combination of C. sinensis and artemisinin could be effective in the prevention of recurrence of lupus nephritis	[95]
			In vivo	MRL lpr/lpr mice aged 12 weeks Doses: 40 μg/kg/d of H1-A daily for 8 weeks.	- ↓ the production of anti-ds-DNA, lymphadenopathy, and proteinuria - Improved renal function - No significant changes have been observed in immune complex deposition	[96]

Table 2. *Cont.*

Phytochemicals						
Compound Name	**Class of Compound**	**Natural Source**	**Type of Study**	**Method**	**Result or Side Effect or Mechanism**	**Ref.**
Beta-Aminoarteether maleate (SM934)	Synthetic artemisinin derivative (Sesquiterpenes)	an analogue of artemisinin	In vivo, in vitro and ex vivo	In vitro: the effects of SM934 on the activation of polyclonal $CD4^-$ T cells and the differentiation of naive $CD4^+$ T cells were investigated. In vivo: preventive or therapeutic effects of SM934 were investigated in MRL/lpr mice. Ex vivo: the treatment mechanisms were investigated.	- ↓ the production of IFNγ and IL-17 by polyclonal CD4+ T cells activated by T cell receptor rearrangement - Development of naive CD4+ T cells into Th1 and Th17 cells, but not Treg cells - Attenuate the renal lesion severity - ↓ proteinuria, ↓anti-dsDNA autoantibody - ↓ the spleen sizes - ↓ the levels of serum IFN-Y and blood urea nitrogen - ↑ lifespan of mice - Inhibit the production of Th1 and Th17 while an increase in the level of Treg cells - Inhibit the complete activation of STAT-1, STAT-3, and STAT-5 proteins	[97]
Baicalin	Flavone glycoside	*Scutellaria baicalensis*	In vivo	MRL/lpr lupus-prone mice Doses: 200 mg/kg of baicalin daily for 4 weeks, intraperitoneally	- ↓ anti-ds-DNA antibody - ↓ urine protein levels - Inhibit mTOR activation - ↓ reduce mTOR agonist-mediated Tfh cell expansion - ↑ Tfr cells - ↓ IL-21 production - Inhibit Tfh cell differentiation and $Foxp3^+$ regulatory T cell differentiation	[98]
			In vivo	Adult female BALB/c mice Doses: 50 mg/kg of baicalin, orally, once in a day for 10 days.	- ↓ production of proinflammatory cytokines such as TNF-α, IL-6, IL-10, and IFN-γ - Inhibit the overproduction of IL-6 and PGE2 - Downregulate the aberrant activation of T cells	[99]

Table 2. *Cont.*

Phytochemicals						
Compound Name	**Class of Compound**	**Natural Source**	**Type of Study**	**Method**	**Result or Side Effect or Mechanism**	**Ref.**
Curcumin	Diarylheptanoid	*C. longa* (turmeric)	Clinical trial	The study was conducted on six SLE patients and six healthy individuals. $CD4^+$ of these people was collected then stimulated by Th17 differentiating factors and exposed to 0.1 and 1 μg/mL curcumin.	- ↓ Th17 percentage - ↓ IL-17a productions - ↑ Treg percentage - ↑ TGF-β1 productions - Modulate Th17/Treg balance	[100]
			In vivo and in vitro	Female MRL/lpr mice were treated with 200 mg/kg of curcumin for 8 weeks.	- ↓ proteinuria, ↓ renal inflammation and, ↓ spleen size - ↓ NLRP3 inflammasome activation - Inhibit anti-dsDNA serum induced expression of NLRP3 inflammasome in podocytes	[101]
			In vivo	Female NZBWF1 have been treated with 500 mg/kg/day curcumin by oral gavage for 14 days.	- Weight and body composition were maintained - ↓ spleen weight and renal injury (glomerulosclerosis)	[102]
			In vivo	Female BALB/c mice with lupus (0.5 cc pristane, i.p.) Doses: 0, 12.5, 50, and 200 mg/kg bw/day curcumin intragastrically daily for 16 weeks.	- ↓ arthritis score and proteinuria level - No significant alteration in body weights - ↓ Th1, Th2, and Th17 percentages - ↑ Treg percentages - ↓ serum IL-6 and IFN-α levels - ↓ the level of antinuclear antibody	[103]
Dihydroartemisinin	A derivative of artemisinin (Sesquiterpenes)	*A. annua*	In vivo	Male BXSB mice Doses: of 5 mg/kg, 25 mg/kg, or 125 mg/kg	- ↓ TNF-alpha production - ↓ NF-κB activation - ↓ p65 subunit expression - Inhibit NF-κB translocation to the nucleus - Inhibit IκB-α protein degradation	[104]

Table 2. *Cont.*

Phytochemicals						
Compound Name	**Class of Compound**	**Natural Source**	**Type of Study**	**Method**	**Result or Side Effect or Mechanism**	**Ref.**
Dihydroartemisinin	A derivative of artemisinin (Sesquiterpenes)	*A. annua*	In vivo	BALB/C female mice were infected with *Toxoplasma gondii* or *Plasmodium berghei*. Doses: 2 mg of dihydroartemisinin daily. Spleen and blood samples were collected.	- ↓ the number of B cells in the bloodstream and spleen cells - ↑ the ratio of T helper to $CD8^+$ T cells - ↓ the proinflammatory cytokines	[105]
			In vivo and In vitro	Female BALB/c mice have been used. Dose:100 mg/kg of dihydroartemisinin, the fourth group was given 5 mg/kg prednisolone, and the last group was given a combination of prednisolone and dihydroartemisinin with the same previous doses. Dihydroartemisinin and prednisolone were administered daily and orally for two months.	- Restore the Treg/Th17 balance - Inhibit inflammation - ↑ the levels of TGF-β, IL-17 and Foxp3 - ↓ transcription of RORγt - Inhibit Th17 cell differentiation - Induce Treg cell differentiation	[106]
Emodin	Anthraquinone	*Rheum palmatum*	In vivo	BXSB lupus mice Mice were treated with different doses of emodin for 30 days.	- ↓ proteinuria - ↓ the expression of ICAM 1 in the renal glomerulus	[107]
			In vivo	Lupus-prone male BXSB mice Doses: 0, 5, 10, and 20 mg/kg/day emodin for 30 days	- ↓ glomerular levels of TNF-α, ICAM-1 - ↓ fibronectin - ↓ the levels of urinary protein and serum anti-dsDNA antibody	[108]
Esculetin	Coumarin	*C. intybus* and *H. paniculate*	In vivo and In vitro	MRL/lpr mice Doses: esculetin (20 mg/kg and 40 mg/kg) from 10 to 20 weeks	- Attenuated renal impairment by reducing - ↓ BUN, serum creatinine and albuminuria - Ameliorated the glomerular hypertrophy, tubular interstitial fibrosis and mononuclear cell infiltration into interstitium. - Down-regulated complement cascade, inflammation and fibrosis pathway - Up-regulated Nrf2-related antioxidation genes - Inhibited the complement activation (both classical and alternative pathways) - Blocked the C3 convertase (C4b2a)	[109]

Table 2. *Cont.*

Phytochemicals						
Compound Name	**Class of Compound**	**Natural Source**	**Type of Study**	**Method**	**Result or Side Effect or Mechanism**	**Ref.**
Glycyrrhizin (Glycyrrhizic acid)	Triterpenoid saponin	*G. glabra*	In vivo and In vitro	Female BALB/c mice received 0.5 mg/day of glycyrrhizin for two months.	- Inhibited HMGB1 function - Caused sharp decline in serum HMGB1 levels - reduced the severity of SLE	[110]
			In vivo and In vitro	Glycyrrhizin (10 mg/kg) was injected every other three days into female BALB/c mice for 12 weeks.	- ↓ levels of anti-dsDNA antibodies - ↓ levels of inflammatory cytokines - ↓ glomerular IgG and C3 deposition - ↓ proteinuria	[111]
			In vitro		- nhibit the immunocomplex formation of 60S acidic ribosomal P proteins from porcine liver	[112]
Mangiferin	Xanthone	*A. aspheloidis*	In vivo and In vitro	lupus-prone B6/gld mice Doses: 20 or 40 mg/kg/day of mangiferin for 12 weeks, orally	- Suppressed mTOR signaling pathways - Upregulated $CD4^+$ $FoxP3^+$ Tregs - Inhibited T cell proliferation - Improved renal immunopathology - ↓ renal T cell infiltration - ↓ serum creatinine - ↓ urinary protein levels - ↑ $CD4^+$ $FoxP3^+$ Treg frequencies in the spleens, lymph nodes, and kidneys	[113]
Salvianolic acid A	A type of phenolic acid	*Salvia miltiorrhiza*	In vivo	BALB/c mice Doses: 5 mg/kg/day of salvianolic acid A for 5 months	- ↓ anti-Sm autoantibodies - Inhibit phosphorylation of IKK, IκB and NFκB in kidney tissue - ↓ pathological effects	[114]
Total glucosides of paeony (TGP)	Glucosides of paeony	*P. lactiflora*	Clinical trial	One group of 29 cases received TGP for 5 years or more, the other group of 47 cases received TGP for 1 year or more (but less than 5 years), and the third group was the control.	- Decreased the required daily dose of prednisolone and cyclophosphamide - ↓ SLEDAI score - No significant difference has been observed in urinary protein - No side effects were observed following TGP use	[115]

Table 2. *Cont.*

Phytochemicals						
Compound Name	**Class of Compound**	**Natural Source**	**Type of Study**	**Method**	**Result or Side Effect or Mechanism**	**Ref.**
Total glucosides of paeony (TGP)	Glucosides of paeony	*P. lactiflora*	Clinical trial	70 SLE patients were divided into control and treatment groups. Both groups used conventional medicine, but the treatment group also received TGP for three months.	- Decreased dose of glucocorticoids required - A combination of TGP and glucocorticoids was suggested to be beneficial	[116]
			Meta-analysis on clinical trials	- Total of 23 articles were included - 792 patients overall in the treatment group and 781 patients overall in the control group	- ↓ IL-18 and IL-6 levels - ↓ anti-dsDNA level - ↓ serum creatinine levels - ↑ Albumin and complement C3, C4 levels - ↓ the disease activity of SLE - ↓ the incidence of adverse reactions - was more effective and safer when used in combination with conventional therapeutics - improved immunoglobulins, (IgA, IgM and, IgG), ESR, CRP, 24 h urine protein, and recurrence rate	[117]
			In vivo	MRL/lpr mice were treated with 50 mg/kg/d of TGP by gavage for 4 weeks.	- ↓ ERα expression - ↑ DNMTs expression - ↓ renal damage - ↓ expression of IFN-γ, IL6 and IL12 cytokines - ↓ anti-dsDNA autoantibody	[118]
			In vivo	The effect of TGP on lupus nephritis has been investigated on MRL/lpr mice	- ↓ the urinary protein contents - ↓ the level of anti-dsDNA antibodies and ANA	[119]
			In vitro	$CD4^+$ T cells were treated with doses of 0, 62.5, 312.5, and 1562.5 mg/L of TGP for 48 h to elevate the effect of TGP on expression and DNA methylation status of ITGAL gene (CD11a) in $CD4^+$ T cells.	- Down-regulate the level of ITGAL mRNA and protein - ↑ DNA methylation of the ITGAL promoter - ↓ CD11a gene expression	[120]

Table 2. *Cont.*

Phytochemicals						
Compound Name	**Class of Compound**	**Natural Source**	**Type of Study**	**Method**	**Result or Side Effect or Mechanism**	**Ref.**
Triptolide	Diterpenoid epoxide	*T. wilfordii*	In vitro	To evaluate the effect of triptolide doses: 0, 5, 10, 30 μg/L of triptolide	- Inhibited the differentiation and maturation of DCs. - ↓ the immune function of DCs. - ↓ secretion of IFN-α, IL-6, and TNF-α	[121]
			In vivo	BALB/c-in nude mice Dose: 5 mg/kg/d of triptolide, orally blood samples were collected before treatment and 1, 3, and 6 months after treatment.	- ↓ the percentage of $CD8^+$, Tcl, Thl cells, $CD4^+/CD8^+$, Thl/Th2 and Tcl/Tc2 - ↑ the percentage of $CD4^+$, Tc2 and Th2 cells	[122]
			In vivo	(NZB×NZW) F1 mice Dose: 6 μg of triptolide or tripdiolide, orally for 15 weeks.	- ↓ levels of BUN, proteinuria, and anti-dsDNA antibody - ↓ production of cytokines such as IL-6 and TNF and monocyte chemoattractant protein 1	[123]
			In vivo	Female MRL/lpr mice Control group received 20 mg/kg/w of cyclophosphamide. treatment groups received 0.2 or 0.3 mg/kg/d of triptolide for 13 weeks.	- ↓ levels of proteinuria, serum anti-dsDNA - ↓ renal histopathologic assessment - ↑ the proportion of Treg - Induce the expression of the miR-125a-5p	[124]
(5R)-5-Hydroxytriptolide (LLDT-8)	Diterpenoid epoxide		In vivo	The effect of LLDT-8 on lupus nephritis was investigated so, female MRL/lpr mice were treated with 0.125 mg/kg/2 days of LLDT-8 for 9 weeks.	- ↓ proteinuria - ↓ serum creatinine - ↓ glomerular IgG deposits - ↓ histopathology - ↑ the lifespan of mice - ↓ the expression of inflammatory cytokines such as IFN-γ, IL-17, IL-6, TNF-α - Inhibit immune cell infiltration in the kidneys	[125]

Table 2. *Cont.*

Phytochemicals						
Compound Name	**Class of Compound**	**Natural Source**	**Type of Study**	**Method**	**Result or Side Effect or Mechanism**	**Ref.**
Fatty acids, vitamins, and minerals						
Fatty acids	Fatty acids		*meta-analysis*	A systematic review and meta-analysis study has been conducted to evaluate the effect of omega-3 fatty acids on SLE disease activity in adults.	- Omega-3 fatty acids could provide therapeutic benefit in addition to immunosuppressive regimens used for SLE - Omega-3 fatty acids: ↑lifespan and ↓autoantibody levels	[22]
			In vivo	Weanling (NZB 3 NZW) F1 female mice mice were switched to semi purified diets containing 10% corn oil as control oil and FOs enriched in either EPA or DHA	- Omega-3 fatty acids: - regulate blood pressure and proteinuria and ↓anti-dsDNA levels and ↓ TNF-α, IL-1α, IL-1β and IL-2 - DHA: - Inhibit IL-18 induction and ↑lifespan and suppress glomerulonephritis	[126]
Vit A	Vitamin		*Clinical trial*	45 female patients with SLE and 45 healthy age-matched and sex-matched patients as control group	- Regulate the balance between Th17 and Treg	[127]
			Clinical trial	Sixty-two female SLE patients and sixty-two female controls	- ↓ level of Th17 - ↑ level of Treg	[128]
Vitamin B	Vitamin		*Clinical trial*	The association of each nutrient intake with the risk of developing active disease was investigated in 216 patients who initially had inactive disease. The association with atherosclerotic vascular events was assessed in 196 women with inactive disease and no history of atherosclerotic disease at baseline.	- Vitamins B6, B12, and folate: ↓homocysteine levels and improve atherosclerosis and ↓ levels of inflammatory cytokines and CRP - Vitamin B6: ↓homocysteine and ↓ risk of active disease	[129]
			Clinical trial	Seventeen patients with SLE were used for 12 weeks	- ↓ sodium intake and maintain adequate intakes of most nutrients except B12, dietary fiber, iron, calcium, and folate. - Niacin: ↓ triglyceride and LDL-C levels	[130]

Table 2. *Cont.*

Phytochemicals						
Compound Name	**Class of Compound**	**Natural Source**	**Type of Study**	**Method**	**Result or Side Effect or Mechanism**	**Ref.**
Vitamin C	Vitamin		*Clinical trial*	279 female patients with SLE were followed over 4 years	- dietary nutrients may modify clinical course of disease in female patients with SLE. - Vitamin C intake is inversely associated with the risk of active disease. - Vitamin C intake may prevent the occurrence of active SLE disease.	[131]
Vitamin D	Vitamin		*Clinical trial*	25-OH vitamin D levels were measured in 198 consecutively recruited SLE patients.	- Inhibit dendritic cell activation and maturation	[132]
			Systematic review	1268 articles have been reviewed to determine whether supplementation with vitamin D can reduce the risk or modify the course of autoimmune diseases.	- Basic, genetic, and epidemiological studies indicate a potential role of vitamin D in the prevention of autoimmune diseases.	[133]
Vitamin E	Vitamin		*Clinical trial*	12 women among 36 outpatients received vitamin E (150 to 300 mg/day) together with prednisolone (PSL).	- ↓ the generation of autoantibodies - vitamin E can suppress autoantibody production via a mechanism independent of antioxidant activity.	[134]
Iron	Mineral		In vivo	Weanling female MRL/MPJ-lpr/lpr mice were used. Mice were fed diets with 3, 10, 35, and 250 mg Fe/kg diet	- ↑ cell damage and renal lesions - Worsen renal impairment	[135]
Selenium	Mineral		In vivo	NZB/NZW female mice Selenium supplementation was provided by adding sodium selenite to the drinking water at 0, 2, or 4 parts per million (mg/L).	- ↑ survival - ↑ levels of natural killer cell activity	[136]
			In vivo and in vitro	They have investigated the impact of Se on B cells and macrophages using in vitro Se supplementation assays and the B6.Sle1b mouse model of lupus with an oral Se or placebo supplementation regimen.	- Inhibit the activation, differentiation, and maturation of macrophages and B cells - ↓ splenomegaly and splenic cellularity compared	[137]

3.2.1. *Artemisinin* and Its Derivatives

Artemisinin is a sesquiterpene lactone with a peroxide bridge extracted from the plant *Artemisia annua* [138,139]. Several semi-synthetic derivatives of artemisinin with greater solubility or bioactivity, such as dihydroartemisinin, artemether, and arteether, artesunate have been developed and investigated in several research works [140,141].

Along with its anti-malarial effect, artemisinin and its derivatives have exhibited anti-inflammatory, immunoregulatory, and antioxidant properties [142]. Like some other conventional anti-malarial drugs, including chloroquine and hydroxychloroquine, artemisinin derivatives are assumed to have beneficial therapeutic effects on SLE [143,144]. But in contrast to chloroquine and hydroxychloroquine, which have serious side effects in some cases, no significant side effects have been associated with artemisinin except for mild side effects such as nausea and vomiting or diarrhea [145].

Studies that examined the effectiveness of artemisinin and its derivatives in patients with lupus have shown that long-term use can be effective in improving renal lesions and can prevent recurrence of lupus nephritis. They can relieve the symptoms of patients with SLE. They increased complement levels and also lowered creatinine and urinary protein levels and reduced erythrocyte sedimentation rates [144]. Artesunate increases CD3 and CD4 and increases the CD4/CD8 T lymphocytes ratio. It can regulate the immune function by increasing IL-2 activity and decreasing the level of soluble interleukin-2 receptor (sIL-2R) [144].

Artemisinin can be effective in improving kidney disorders by modulating immune-inflammatory responses. Anti-inflammatory effects of artemisinin are due to its ability to suppress nuclear factor-kB (NF-kB), phosphatidylinositol 3 kinase (PI3K)/protein kinase B (AKT) activity, signal transducer and activator of transcription (STAT), and toll-like receptors (TLRs) [142]. Following the use of artemisinin, the production of proinflammatory cytokines such as TNF-α, IL-6, IL-10, IL-17, and IL-21 is inhibited, but the production of anti-inflammatory cytokines such as IL-4 and IL-10 is increased [146].

Artesunate has suppressed the Jak2-Stat3 signaling pathway in MRL/lpr mice. It has also regulated T follicular helper cell differentiation; thus, it resulted in an increase in follicular regulatory T cells (Tfr) and a decrease in follicular T helper cells (Tfh). It has also reduced the levels of pathogenic cytokines such as IL-6, IFN-γ, and IL-21. It has reduced the level of anti-dsDNA antibodies deposited in the kidney. This means that it might be able to help lessen the symptoms of lupus nephritis [147].

Dihydroartemisinin has been shown to reduce the senescence of myeloid-derived suppressor cells (MDSCs) by regulating the Nrf2/HO-1 pathway. MDSCs are involved in exacerbating the pathogenesis of SLE [148]. Dihydroartemisinin can also restore balance in Treg/Th17 by inducing Foxp3 expression in T cells in mice model [106]. Therefore, dihydroartemisinin is assumed to be effective in improving the condition of SLE patients [148].

Toxicity and Side Effects

According to meta-analyses and large clinical studies on artemisinin and its derivatives, they did not demonstrate serious side effects. However, this group of compounds has a number of side effects that could be mentioned, such as neurotoxicity, genotoxicity, hematotoxicity, immunotoxicity, and cardiotoxicity. According to both animal and human studies, artemisinin toxicity is caused by long-term availability rather than by short-term peak concentrations. It is worth mentioning that taking artemisinin orally has a faster rate of elimination than administering it intramuscularly. Therefore, it provides a relatively safe route of administration. This explains why significant toxicities were discovered in the majority of animal research but not in those involving humans [149]. This topic is still open for further research [150].

3.2.2. Antroquinonol

Antroquinonol is a derivative of tetrahydro ubiquinone, which was found in the mycelium of *Antrodia camphorata* [151,152]. *A. camphorata* is a mushroom that grows in the inner cavity of the *Cinnamomum kanehirai* (Lauraceae) tree [153] and produces some antroquinonol drivatives, including antroquinonol, antroquinonol B, C, D, L, and M, and 4-acetyantroquinonol B [154]. Hocena is an antroquinonol capsule intended for the treatment of acute myeloid leukemia, hepatocellular carcinoma, and pancreatic cancer and has an orphan drug status from the US Food and Drug Administration [155]. Antroquinonol has been claimed to have the potential to prevent renal disorders and the worsening of lupus nephritis [156]. Inhibiting T cell activation and proliferation, lowering free radical and nitric oxide production, enhancing Nrf2 activation, and decreasing inflammation by inhibiting NF-kB function in the kidney are some of the proposed involved mechanisms [93,156].

In one study, the effect of antroquinonol on preventing the mild form of lupus nephritis from becoming severe was investigated. NZB/NZW F1 mice were used for this purpose and were treated orally with 15 mg/kg antroquinonol for 5 weeks. Eventually, *A. camphorata* reduced hematuria, proteinuria, and IL-18 production in the kidneys. T cell proliferation was also inhibited and Treg cell suppression was induced. Also, reactive oxygen species and nitric oxide production were inhibited, Nrf2 activation was increased, and NF-κB activation was inhibited. It was concluded that antroquinonol might be effective in preventing the progression of lupus nephritis [93]. In another study, antroquinonol reduced proteinuria and lowered creatinine and serum BUN levels. It also reduces the thickness of the kidney glomerular basement membrane and inhibits the production of TNF-α and IL-1β. Therefore, the use of *A. camphorata* in autoimmune diseases such as SLE can protect the kidneys [46].

Toxicity and Side Effects

In numerous research on animal toxicology, *A. camphorata* exhibited no obvious toxicity. Thus, no significant side effects or deaths were reported, and nausea, vomiting, and diarrhea were the most frequent side effects [157]. Although antroquinonol exhibits cytotoxic activities against cancer cell lines MCF-7, MDA-MB-231, Hep 3B, Hep G2, DU-145, and LNCaP with IC_{50} values ranging from 0.13 to 6.09 μM it is considered safe [158]. Antroquinonol dosages below 30 mg/kg/day do not appear to be associated with any adverse effects [159]. Overall, *A. camphorata* has revealed very little toxicity or side effects in clinical practice.

3.2.3. Baicalin

Baicalin is another compound that has a high potential to be considered as a bioactive molecule against SLE. It is a flavonoid isolated from the root of *Scutellaria baicalensis* and has anti-inflammatory and antioxidant effects [99]. Baicalin in MRL/lpr lupus-prone mice has been shown to reduce anti-ds-DNA antibody and urine protein levels. Baicalin has been able to inhibit mTOR activation and also reduce mTOR agonist-mediated Tfh cell expansion and increase Tfr cells. This molecule can inhibit IL-21 production, Tfh cell differentiation, and Foxp3^{+} regulatory T cell differentiation [98]. In a study on pristane-induced lupus in BALB/c mice, baicalin reduced the production of proinflammatory cytokines such as TNF-α, IL-6, IL-10, and IFN-γ. It also inhibited the overproduction of IL-6 and PGE2 and downregulated the aberrant activation of T cells. Thus, it was concluded that baicalin can reduce the severity of SLE and attenuate autoimmunity [99,160].

Toxicity and Side Effects

Scutellaria baicalensis has long been recognized as a safe and non-toxic herb. *S. baicalensis* oral preparation has no significant side effects; however, some patients may experience stomach discomfort, diarrhea, etc., and those with allergic constitutions may develop a blister-like medication eruption. When used in high doses of injectable preparations, *S. baicalensis* may also result in symptoms such as hypothermia, muscle discomfort, and leucopenia [161]. Some data about possible nephrotoxicity of high doses of baicalin are published, but as a whole, the safety and toxicity of this compound remain still insufficiently studied [162]. Various drug transporters and metabolic enzymes are involved in the disposition of baicalin, and they may be influenced or reciprocally influenced by co-administered medications. These factors can justify the wide herb-drug interactions between baicalin and chemical drugs. Baicalin can significantly alter the pharmacokinetics of medications that have a high protein binding affinity or share the same cytochrome P450 (CYP) enzymes. Phenacetin, theophylline, midazolam, dextromethorphan, nifedipine, and chlorzoxazone can be mentioned among drugs that can interfere with baicalin [163].

3.2.4. Curcumin

Curcumin is the major diarylheptanoid component of turmeric (*Curcuma longa*, Zingiberaceae) [21]. A variety of clinical trials assessing the curcumin effect on inflammation, skin, eye, CNS, respiratory, cardiovascular, gastrointestinal, urogenital, and metabolic disorders have been reported so far [164]. Since curcumin has shown immunomodulatory properties, it has been considered for the improvement of SLE patients. The recommended dosage for SLE ranges from 100–200 mg daily to 4.5 g/day [165]. Curcumin is found to have protective effects against aluminum toxicity and cisplatin-associated neurotoxicity and neuropathy [166,167]. Hypothetically, curcumin may help lupus induced peripheral neuropathy.

The immunomodulatory property of curcumin results from its interaction with various immune mediators, including B and T lymphocytes, macrophage and dendritic cells, cytokines, and various transcription factors such as nuclear factor kappa B (NF-κB), activator protein-1 (AP-1), and signal transducer and activator of transcription (STAT) [168–173]. It has been found that curcumin can inhibit the maturation and function of dendritic cells. This function of curcumin is achieved through reducing the expression of MHC-II and co-stimulatory molecules such as CD11c, CD40, CD54, CD80, CD83, CD86, CD252, and CD256. It can also be due to the reduction of proinflammatory cytokines such as IL-1, IL-6, IL-12, IL-12p40, IL-12p70, and TNF-α. In general, curcumin can keep dendritic cells in an immature state, and as a result, it suppresses dendritic cell-mediated stimulation of inflammatory T cells, which play a key role in the severity of symptoms observed in SLE [21].

In Vivo and In Vitro Studies

In a study that was conducted on six SLE patients and six healthy individuals, the balance between T helper 17 (Th17) and regulatory T cells (Treg) in SLE patients was investigated. The $CD4^+$ cells of these people have been collected, stimulated by Th17 differentiating factors, and exposed to 0.1 and 1 μg/mL of curcumin. Finally, it was found that curcumin can decrease Th17 percentage, decrease IL-17a production, and can increase Treg percentage and increase TGF-β1 production on $CD4^+$ T cells of SLE patients. In general, curcumin can modulate the Th17/Treg balance on $CD4^+$ T cells of SLE patients without affecting healthy subjects [100].

A study conducted on lupus-prone female MRL/lpr mice has shown that curcumin has the potential to be considered for the treatment of lupus. In this study, mice were treated with 200 mg/kg of curcumin for 8 weeks. As a result, proteinuria, renal inflammation, and spleen size have decreased following the use of curcumin, and a decrease in NLRP3 inflammasome activation was also observed. Following in vitro studies, it has also been found that curcumin can inhibit anti-dsDNA serum induced expression of NLRP3 inflammasome in podocytes [101]. In another study, the ability of oral curcumin consumption to attenuate autoimmunity and renal injury during SLE was evaluated. In order to do this, the female NZBWF1 was given 500 mg/kg/day of curcumin through an oral gavage for 14 days. Finally, it was found that following the consumption of curcumin, weight and body composition were maintained and a decrease in spleen weight and renal injury (glomerulosclerosis) were observed compared to the control group. Ultimately, it has been determined that curcumin can modulate autoimmune activity and probably reduce renal injury in female mice with SLE [102].

In a study, the immune modulation effects of curcumin on pristane-induced lupus mice have been investigated. The female BALB/c mice received an intraperitoneal injection of 0.5 mL pristane for lupus induction. Afterwards, they were treated with 0, 12.5, 50, and 200 mg/kg of bw/day curcumin intragastrically for 16 weeks. As a result, the arthritis score and proteinuria level decreased. However, no significant alteration was observed in body weight. Following 200 mg/kg bw/day curcumin consumption, Th1, Th2, and Th17 percentages decreased, Treg percentages increased slightly, serum IL-6 and IFN-α levels decreased, and antinuclear antibody levels decreased significantly. Therefore, the results have shown that curcumin could be useful as a therapeutic intervention in SLE [103].

Toxicity and Side Effects

Long-standing safety data exist for curcumin. For instance, curcumin's allowable daily intake (ADI) value is 0–3 mg/kg body weight, according to reports from the JECFA and EFSA organizations (the Joint United Nations and World Health Organization Expert Committee on Food Additives and the European Food Safety Authority, respectively) [174]. Despite its well-known safety, several unfavorable side effects have been documented. In a dose-response investigation, seven patients who received 500–12,000 mg and were monitored for 72 h reported symptoms including diarrhea, headache, rash, and yellow stools [175]. In a different study, some participants who received 0.45 to 3.6 g of curcumin per day for one to four months experienced diarrhea, nausea, and a rise in the levels of the enzymes lactate dehydrogenase and alkaline phosphatase in their serum [176].

3.2.5. Emodin

Emodin (1,3,8-trihydroxy-6-methylanthraquinone) is actually a natural anthraquinone that can be found in the barks and roots of many plants, lichens, and molds [177]. One of the main sources of emodin is *Rheum palmatum* (Polygonaceae) which is also known as Chinese rhubarb.

Emodin can reduce steroid resistance by inhibiting P-glycoprotein efflux function. Steroid therapy is part of the common treatment for SLE patients, and a decreased response to steroid therapy following overexpression of p-glycoprotein in peripheral lymphocytes has been observed in some patients [178].

An attempt was made to investigate the effect of emodin on nephritis in a study on BXSB lupus mice. Mice were treated with different doses of emodin for 30 days. As a result, it has been shown that following emodin consumption, the level of proteinuria is reduced and the expression of intercellular adhesion molecule-1 (ICAM 1) in the renal glomerulus is also reduced [107].

The effect of emodin on renal injury in lupus nephritis was investigated. Lupus-prone male BXSB mice were treated with 0, 5, 10, and 20 mg/kg/day emodin for 30 days. Finally, it was observed that following the administration of emodin, glomerular levels of TNF-α, ICAM-1, and fibronectin (FN) decreased, and the levels of urinary protein and serum anti-dsDNA antibody also decreased, and these decreases were dose-dependent. The mechanism of action of emodin is probably through inhibition of dsDNA antibody and decreased levels of TNF-α, ICAM-1, and FN in the glomeruli [108].

Toxicity and Side Effects

According to reports, emodin can reduce sperm motility in a dose-dependent manner in mice. Emodin has also been found to have dose- and time-dependent toxicity in kidney and liver cell lines. Intestinal discomfort and severe diarrhea brought on by an overdose of emodin due to its laxative properties lead to an electrolyte imbalance and dehydration [179]. Generally, it is also known to have kidney toxicity, hepatotoxicity, and reproductive toxicity, especially at high doses and long-term use [180] The extremely low bioavailability of emodin further limits its use in therapeutic applications [179].

3.2.6. Esculetin

Esculetin (also known as aesculetin, 6,7-dihydroxycoumarin, and cichorigenin) is a coumarin that has been isolated from a variety of medicinal and toxic plants such as *Cichorium intybus* (chicory) and in *Hydrangea paniculate* Siebold. In a study conducted on MRL/lpr mice, esculetin significantly attenuated renal impairment by reducing BUN, serum creatinine, and albuminuria. Esculetin could improve glomerular hypertrophy and tubular interstitial fibrosis and reduce mononuclear cell infiltration into the interstitium. It was suggested that this molecule could significantly down-regulate the complement cascade as well as the inflammation and fibrosis pathway. In addition, esculetin could up-regulate Nrf2-related antioxidation genes. The authors reported that esculetin could inhibit complement activation both in classical and alternative pathways. The molecule blocked the C3 convertase (C4b2a) to exert this inhibitory capability. Moreover, it was suggested that the antioxidation effect of esculetin was dependent on Nrf2 activation, which means that esculetin could inhibit NFκB nuclear translocation and TGFβ-smad3 profibrosis pathway [109]. Lupus nephritis is one of the important complications of lupus, and complement activation contributes to kidney injury; the inhibition of complement activation by herbal compounds might be beneficial for lupus. It was also reported that the coumarin derivates that are isolated from *H. paniculata* could improve renal injuries in cationized-BSA-induced membranous nephropathy. The suggested mechanism was the inhibition of complement activation and interleukin 10-mediated interstitial fibrosis [181].

Toxicity and Side Effects of Esculetin

Acute toxicity studies reported LD_{50} for intraperitoneal injection to mice as 1450 mg/kg and >2000 mg/kg by mouth. No reported adverse effects are known other than LD_{50} [182].

3.2.7. Mangiferin

The main source of mangiferin is *Mangifera indica,* although it is found in 96 species, 28 genera, and 19 families of angiospermic plants. *Mangifera indica* belongs to the family Anacardiaceae and is known as mango. Almost all parts of *M. indica,* such as fruits, twigs, leaves, and stem bark, contain mangiferin [183]. Mangiferin is a xanthonoid polyphenol with a variety of pharmacological effects such as anti-inflammatory, antioxidant, immunomodulatory, nephroprotective, hepatoprotective, anti-cancer, anti-diabetic, and anti-asthma [184]. According to certain research, its renal protective actions may be beneficial for those with lupus nephritis [166,167].

Mangiferin has been shown to improve lupus nephritis in lupus-prone B6/gld mice. In a study, the effect of mangiferin on lupus nephritis was investigated. Mice were treated orally with 20 or 40 mg/kg/day of mangiferin for 12 weeks. Finally, Mangiferin has been shown to be effective in treating lupus nephritis with its anti-inflammatory and immunomodulatory effects. Mangiferin was effective by suppressing mTOR signaling pathways, upregulating $CD4^+$ $FoxP3^+$ Tregs, and inhibiting T cell proliferation. Mangiferin improved renal immunopathology and reduced renal T cell infiltration. It also lowered serum creatinine and urinary protein levels and increased $CD4^+$ $FoxP3^+$ Treg frequencies in the spleens, lymph nodes, and kidneys [113].

Toxicity and Side Effects

Mangiferin is typically regarded as a non-toxic natural substance. Adults receiving 0.9 g of mangiferin orally demonstrated no toxicity. LD_{50} of the mangiferin was considered to be 400 mg/kg on mice [185]. Mangiferin was found to be safe and helpful in enhancing cellular function, according to numerous research works [186]. In a study that assessed the toxicity of mango leaf extract, which was given orally to rats for three months at a dose of 2 g/kg body weight per day, neither mortality nor toxic effects were observed [187]. The *Mangifera indica* leaf aqueous extract was not particularly mutagenic or genotoxic. Mangiferin has generally been shown to be safe in cell and animal research. In contrast, there are insufficient safety data from human research [186].

3.2.8. Salvianolic Acid A

Salvianolic acid A (or Dan phenolic acid A) is a phenolic compound extracted from *Salvia miltiorrhiza* (Lamiaceae family). The plant is also known as Chinese sage, Danshen, and red sage. Salvia species such as *S. officinalis* and *S. miltiorrhiza* have shown antioxidant, antibacterial, anti-cancer, and anti-diarrheal effects and have been used to treat lupus and autism, lower cholesterol, treat Alzheimer's, reduce sweating, and reduce menopausal hot flashes [188].

In a study performed on BALB/c mice, the effect of salvianolic acid A isolated from the root of *S. miltiorrhiza,* on lupus nephritis was investigated. Mice were treated with 5 mg/kg/day of salvianolic acid A for 5 months. As a result, it was observed that following the consumption of salvianolic acid A, anti-Sm autoantibodies decreased, phosphorylation of IKK, IκB, and NFκB in kidney tissue was inhibited, and pathological effects were reduced [114].

Toxicity and Side Effects

In an acute toxicity study, the LD_{50} of salvianolic acid A was reported as 1161.2 mg/kg in mice. In dogs' animal model, the minimum lethal dose and maximal non-lethal dose of salvianolic acid A were reported as 682 mg/kg and 455 mg/kg in dogs, respectively. Based on a 4-week repeated-dose, no observed adverse effect level was 20 mg/kg. It was suggested to examine liver and kidney function during the administration of salvianolic acid A in a clinic [189]. According to a system review of the drug's safety, the clinical use of salvianolate injection did not result in the occurrence of any common or major side effects. Blood loss and allergic reactions are the most common adverse effects of salvianolic acid injections. In general, it has been determined that salvianolic acid is well tolerated in the general population. Rash, erythemas, pruritus, palpitations, headaches, dizziness, elevated blood bilirubin, elevated transaminases, elevated blood creatinine, positive fecal occult blood, and abnormal platelet count are among the most common adverse effects that were reported for salvianolic acids [190].

3.2.9. Triptolide

Triptolide is a diterpene triepoxide isolated from *Tripterygium wilfordii* Hook F (*T. wilfordii*). The molecule has immunosuppressive and anti-inflammatory effects and has been shown to have therapeutic effects on autoimmune and inflammatory diseases such as lupus nephritis, arthritis, neurodegenerative disorders, and asthma [23,139]. Despite the beneficial effects of Triptolide in the treatment of various inflammatory disorders, it should be noted that the use of *T. wilfordii* can cause severe toxicity and side effects. This may limit the clinical use of this plant [191]. Triptolide's anti-inflammatory and immunosuppressive effects are due to its ability to inhibit the proliferation of immune cells and inflammation-related cells and reduce cytokines and proinflammatory mediators [192].

The effects of triptolide on SLE are assumed to be through induction of miR-125a-5p and an increase in the proportion of Treg [124]. Triptolide has been shown to reduce the expression of transforming growth factor-beta (TGF-β) and vascular cell adhesion molecule (VCAM-1) [193]. It can also reduce the expression of C3 and CD40, so it generally has immunosuppressive and anti-inflammatory effects and is useful in renal disorders [194]. Another way triptolide can be immunosuppressive and anti-inflammatory is through changing signaling pathways. Triptolide has been reported to inhibit nuclear factor-κB (NF-κB) signaling pathway [195], lower the IL-17 level, and suppress IL-6/signal transducer and transcription 3 (STAT3) signaling pathway [196].

(5R)-5-Hydroxytriptolide (LLDT-8) is a triptolide analogue. It has strong anti-inflam matory and immunosuppressive activity [196]. LLDT-8 improves anti-GBM glomerulonephritis because it can regulate Fcγ signaling pathway [197]. It can also improve lupus nephritis and reduce the infiltration of kidney immune cells because it inhibits the expression of renal chemokines [125].

Toxicity and Side Effects

The subject of triptolide's safety in clinical applications has been brought up because of its broad usage. Despite the valuable pharmacological effects of triptolide, its application requires particular caution because it is well known to have hepatotoxicity, nephrotoxicity, reproductive toxicity, etc. [198]. Hepatic cytochrome P450s are involved in the metabolism of triptolide, and triptolide toxicity and CPY3A also have a close relationship. Clinical case reports have shown through research that triptolide exposure can be involved in damaging a variety of organs, including the kidney, liver, heart, ovary, and testicles. Additionally, it has been shown that triptolide has a variety of harmful effects on cells, including damage to membranes, oxidative stress, endoplasmic reticulum stress, metabolism dysfunction, mitochondrial dysfunction, apoptosis, and autophagy [199].

3.2.10. Total Glycosides of Paeony (TGP)

Total glycosides of paeony (TGP) are extracted from the root of *Paeonia lactiflora*. TGP has long been used to treat autoimmune diseases [200]. The beneficial effects of TGP on lupus patients are dependent on its anti-inflammatory and immunosuppressive effects [201]. The effects of TGP on the production of proinflammatory cytokines, antibody production, apoptosis of lymphocytes, and lymphocyte proliferation are dual and dose-dependent [202]. TGP increased the mouse splenocytes' proliferation at low doses (0.05~0.4 mg/L), while it decreased it at high doses (0.4~1.6 mg/L) [202–204]. The ratio of T helper cells to T suppressor cells (Th/Ts) increases at low doses of TGP (0.2 mg/L) and decreases at high doses of TGP (6.0 mg/L) [204]. IL-1 production increases at low doses of TGP (0.5~12.5 mg/L) and decreases at high doses of TGP (12.5~312.5 mg/L) [203]. IgM-antibody production increases at low doses of TGP (0.1~0.4 mg/L) and decreases at high doses of TGP (0.4~3.2 mg/L) [205,206]. Therefore, it is assumed that the immunomodulatory effects of TGP are dose-dependent, and the dose should be adjusted for best results.

The beneficial effect of TGP on SLE has been discussed in several studies. It was reported that the anti-inflammatory effect of TGP is due to its ability to inhibit the production of nitric oxide, leukotriene B4, and prostaglandin E2 [202]. TGP reduces the SLEDAI score in SLE patients and also reduces the average daily dose of prednisolone [115]. A decrease in renal pathology has been observed following the consumption of TGP by MRL/lpr mice. TGP has also reduced the levels of anti-dsDNA antibodies and antinuclear antibodies (ANA). It could also reduce urinary protein levels. Consequently, it was concluded that TGP consumption in patients with lupus nephritis can have therapeutic effects [119]. TGP causes down-regulated Foxp3 promoter methylation levels, thus increasing the expression of Foxp3 in lupus $CD4^+$ T cells. TGP increased the number and percentage of Treg cells in lupus $CD4^+$ T cells and increased IFN-γ and IL-2 expression [200]. TGP increases DNA methylation of ITGAL promoter in $CD4^+$ T cells, thereby reducing CD11a gene expression [120].

Toxicity and Side Effects

In general, TGP is considered a safe and effective compound that is tolerable and does not cause any serious side effects. According to studies, the likelihood of developing diarrhea after consuming TGP may rise. TGP can accelerate the gastrointestinal tract's peristalsis, which may be the cause of the diarrhea. The majority of patients only experience moderate and acceptable symptoms, and the gastrointestinal system is not organically harmed. Drowsiness, dry mouth, dizziness, and weakness are some additional side effects that have been reported [207].

3.3. *Fatty Acids, Vitamins, and Minerals*

Certain nutrients and vitamins as dietary supplements have been consumed to improve lupus [208]. The efficacy of some of them have been investigated and discussed in several studies (Table 2). Safety and side effects of fatty acids, vitamins, and minerals is well studied by numerous publications and are available even on the indications of these over-the-counter (OTC) medicines.

3.3.1. Fatty Acids

Unsaturated oils play an important role in the immune system regulation. In human nutrition and/or healthcare, seed oils have long been utilized as a daily supplement, a food ingredient, or a therapeutic cure. Long chain fatty acids (LCFAs) are fatty acids with more than 14 carbons and make up the majority of vegetable oils. They are necessary for the human body's ongoing regular cell growth and development. Among these, polyunsaturated fatty acids (PUFA) like n-3 and n-6 fatty acids are crucial for the prevention and treatment of many chronic diseases, including diabetes, coronary artery disease, inflammatory and autoimmune disorders, and many other ailments. Some significant fatty acids, such as linoleic acid (an n-6 fatty acid) found in the majority of vegetable oils and plant seeds, are regarded as essential fatty acids (EFAs). Arachidonic acid, which can be further elongated and desaturated to form prostaglandins, thromboxanes, and leukotrienes, is one of these EFAs. A different class of EFA is the n-3 fatty acids, which include linolenic acid and are present in soy, linseed, and flaxseed oils. According to some evidence, n-3 fatty acids have protective effects on eicosanoid metabolism. Docosahexaenoic acid (DHA), a crucial component of cellular membranes and another significant n-3 fatty acid, has a favorable impact on coronary heart disease, inflammatory disease, atherosclerosis, and disorders of the nervous system [209]. Dietary lipids are also involved in autoimmune phenomena by affecting the balance between Th1 and Th2 cells [210,211].

Dysregulation of PUFAs induces a wide range of neurological and developmental disorders. Linoleic acid and linolenic acid are required as part of the immune cell membrane [212]. α-linolenic acid and γ-linolenic acid are among the omega-3 acids that have beneficial effects following the reduction of TNF-α and IL-2 in SLE patients. Omega-3 fatty acid supplementation has shown potential benefit on SLE disease activity as demonstrated by Systemic Lupus Activity Measure-Revised (SLAM-R), SLE Disease Activity Index (SLEDAI), and British Isles Lupus Assessment Group (BILAG) scores as well as plasma membrane arachidonic acid composition and urinary 8-isoprostane levels, with minimal adverse effects [213].

Finding the optimal ratio of ω-6/ω-3 PUFAs is essential in therapeutic interventions. As an example, linoleic/alpha-linolenic of 1:3 is the optimal ratio for enhancing both the proliferation and differentiation of cells such as neural stem cells [214]. Wei et al. concluded in a meta-analysis that low-ratio n-6/n-3 PUFA supplementation could significantly reduce serum TNF-α and IL-6 concentrations but not CRP concentrations [215].

In the NZB $\times$ NZW mice animal models, essential fatty acid deficient diets can reduce arachidonic acid levels, thus reducing proinflammatory prostaglandins and leukotrienes, and also reduce nephritis by inhibiting autoantibody production [212]. Studies have shown that the lifespan increased and autoantibody levels decreased in animal models of SLE following a diet rich in omega-3 fatty acids [22].

The presence of omega-3 PUFA in the diet of SLE patients can regulate blood pressure and proteinuria and also reduce anti-dsDNA levels, as well as TNF-α, IL-1α, IL-1β, and IL-2 [126,213].

A meta-analysis conducted in 2020 found that omega-3 fatty acids could reduce SLE activity. In this study, 136 patients in the comparison group and 138 in the treatment group were used, and the mean age of patients was 43 years. The follow-up time of the trial varied between 12 and 52 weeks. This study showed that the use of omega-3 fatty acids is more effective than placebo in reducing disease activity in SLE [22].

Eicosapentaenoic (EPA) and docosahexaenoic (DHA) are some of the unsaturated fatty acids that exert their anti-inflammatory effects by lowering the level of C reactive protein (CRP) and other inflammatory mediators [126,216,217]. The most widely available dietary source of EPA and DHA is cold-water oily fish, such as salmon, herring, mackerel, anchovies, and sardines.

EPA and DHA can affect the immune system through various mechanisms. They can inhibit the enzyme lipoxygenase and subsequently reduce the inflammatory factors derived from arachidonic acid. DHA can inhibit nuclear factor κB (NF-κB) and TNF-α [218].

DHA has increased the lifespan of and suppressed glomerulonephritis in NZB × NZW mice with systemic lupus erythematosus, possibly due to inhibition of IL-18 induction [126]. DHA has also reduced IL-18 levels, lowered serum levels of anti-dsDNA, and regulated IgG renal deposition in mice [126,219].

3.3.2. Vitamin A

Some studies have been conducted to investigate the effectiveness of vitamin A in lupus. Retinoic acid is a metabolite of vitamin A. Vitamin A deficiency in lupus patients has been shown to have a negative effect on the prognosis of the disease. Consumption of retinoic acid and vitamin A regulates the balance between Th17 and Treg. It was reported that following the intake of vitamin A by lupus patients, the level of Th17 decreased and the level of Treg increased [127,128].

3.3.3. Vitamin B

Vitamins B6, B12, and folate reduce homocysteine levels, so they can be helpful in improving atherosclerosis in SLE patients. They can also lower levels of inflammatory cytokines and C-reactive protein (CRP). Vitamin B6 can also reduce the risk of active disease by lowering homocysteine [129]. Following the use of niacin, a decrease in triglyceride and LDL-C levels was observed, with no significant effect on HDL-C levels [130]. In general, it was suggested that taking supplements of the vitamin B complex could be beneficial for people with SLE.

3.3.4. Vitamin C

Vitamin C has an antioxidant effect. It can release inflammatory mediators and modulate immune function. It also lowers anti-dsDNA levels and IgG. Vitamin C can prevent active SLE [131]. Concomitant use of 500 mg of vitamin C and 800 mg of vitamin E daily for 3 months has shown a slight decrease in lipid peroxidation. In SLE patients with high doses of vitamin C, ascorbate is found in the urine, so the maximum dose of vitamin C is 1000 mg/day [220].

3.3.5. Vitamin D

It has been shown that there is a link between vitamin D deficiency and the severity of SLE. Higher SLEDAI scores have been reported in patients with low levels of vitamin D. Supplementation with vitamin D in SLE patients inhibits dendritic cell activation and maturation [132]. Calcitriol is the active form of vitamin D and acts on autoimmune diseases such as SLE by regulating the response of T and B cells and boosting the innate immune response [133]. SLE patients are photosensitive and should use sunscreen when exposed to the sun. On the other hand, sunlight is needed to produce vitamin D, so it can be assumed that taking vitamin D supplements might be useful for SLE patients [221,222].

3.3.6. Vitamin E

Vitamin E has antioxidant and anti-inflammatory effects and, because of its anti-inflamm atory effect, seeks to reduce IL-2, IL-4, and TNF-α, which can be effective in lupus [223]. Furthermore, vitamin E consumption by SLE patients reduces the generation of autoantibodies [134].

3.3.7. Calcium

In some SLE patients, a decrease in bone mineral density has been observed, which may or may not be dependent on corticosteroid use. On the other hand, these patients are mostly deficient in vitamin D and avoid exposure to sunlight. Therefore, adequate calcium intake is important for SLE patients [223].

3.3.8. Iron

There should be a balance of iron intake in SLE patients. Iron supplementation to MRL/MPJ-lpr/lpr mice resulted in cell damage, renal lesions, and worsened renal impairment in an in vivo study. Iron chelators have also been shown to be beneficial in autoimmune diseases. In contrast, iron deficiency increases the symptoms of the disease, so iron should be used in SLE patients who have anemia [135,224].

3.3.9. Selenium

Selenium supplementation has been considered in the treatment of lupus because of its antioxidant and anti-inflammatory effects. A study on NZB/NZW female mice found that survival increased following selenium supplementation, which may be due to increased levels of natural killer cell activity [136]. In an in vitro study performed on the B6.Sle1b mouse model of lupus, an attempt was made to find the mechanism of the effect of selenium on lupus. It has been shown that selenium can inhibit the activation, differentiation, and maturation of macrophages and B cells. Therefore, its use can be useful in patients with lupus [137]. Reduced selenium levels have been observed in patients with autoimmune diseases, which may be considered a risk factor for the onset of autoimmunity and inflammation. Due to the anti-inflammatory effect of selenium, it has been suggested that consuming a certain amount of selenium in patients with autoimmune diseases can lead to better management of disease complications [225].

3.3.10. Zinc

It was shown that a zinc-restricted diet can increase serum levels of corticosteroids and subsequently reduce the symptoms of autoimmune diseases such as SLE, so it can be useful in controlling SLE [226]. A study of NZB/NZW mice showed a decrease in autoantibody production. In MRL/lpr mice, the use of zinc-restricted diets decreased the level of anti-dsDNA, decreased lymphoproliferation, and recovered glomerulonephritis [210]. On the other hand, it should be noted that a study conducted on humans has shown that zinc deficiency causes immune dysfunction by acting on Th cells and can lead to neurosensory disorders and reduced body mass [227].

3.4. *Herbal Medicines, Medicinal Plants, Mushrooms, and Fungi and Their Crude Extracts*

A variety of medicinal plants and mushrooms have been reported to exhibit efficacy against lupus conditions (Table 3). In some traditional remedies, they have been used in the form of dried powdered plant material or fungi. In some others, a crude extract of the plant or fungus was prepared using solvents such as water or ethanol or a mixture of both (hydroalcoholic extract). Crude extracts are a complex mixture of molecules with the same range of polarity but different concentrations. Sometimes, several molecules with a variety of mechanisms work synergistically to produce a specific effect. Although the crude extracts of herbal medicine can reflect the efficacy of a special herb or fungi, due to the variation of compounds in the natural sources, special attention should be given to the standardization and authentication of them in order to have repeatable and reliable effects. Compared to purified bioactive molecules, crude extracts usually exhibit milder efficacy and side effects, and introducing supplements from herbal medicine to the market is much easier.

The use of herbal medicines has long been used to treat various diseases, such as SLE. In this article, an attempt has been made to review effective herbs for improving lupus. The use of traditional medicine along with Western medicine can reduce the dose of Western medicine drugs, reduce their side effects, and ultimately improve the quality of life of SLE patients.

Table 3. Medicinal plants, mushrooms, and fungi with reported efficacy against lupus.

Plant Name	Other Names	Type of Study	Method	Result or Side Effect or Mechanism	Ref.
Antrodia camphorata	Stout camphor fungus	In vivo	SLE-prone NZB/W F1 mice took 100, 200, and 400 mg/kg of *A. camphorata* extract, orally on 5 consecutive days per week for 12 weeks.	- ↓ urine protein - ↓creatinine - ↓ serum BUN levels - ↓ the kidney glomerular basement membrane's thickness	[46]
Artemisia annua Pall.	Annual wormwood, sweet wormwood, Chinese wormwood, sweet sagewort sweet Annie	In vivo and in vitro	Six groups of female ICR mice were used and there were 5 mice in each group. Splenocytes of immunized mice were isolated and exposed to different concentrations, and finally the number of specific antibodies was counted by indirect ELISA.	- ↓ the levels of a series of antibodies	[228]
Astragalus propinquus Schischkin (*syn. Astragalus membranaceus* (Fisch.) Bunge)	Mongolian milkvetch	In vitro	NK activity of peripheral blood mononuclear cells (PBMC) from 28 patients with SLE was measured using enzyme-release assay.	- ↓ NK activity - ↓ disease activity	[229]
Bryophyllum pinnatum (Lam.) *Oken*	Kalanchoe pinnata, Zakham-e-hyat, life plant, cathedral bells	In vivo and in vitro	BALB/c mice were treated with different doses of ethanolic extract of *B. pinnatum* leaves Doses: 10.5, 21, and 42 mg/kg/day for 12 weeks.	- ↓ TNF-α, IL-17, IL-12, CRP, and matured B cells - ↑ complement C3 and C4 and TGF-β - No specific side effects have been reported	[230, 231]
		In vitro	The effect of ethanolic extract of *B. pinnatum* leaves on spleen cells of BALB/c mice Doses: 0, 0.02, 0.1, or 0.5 μg/mL	- ↓ B cells maturation - ↑ B cells apoptosis - ↓ NF-κB p65 expression	[232]
		In vivo and in silico	The effect of aqueous extract of *B. pinnatum* leaf on lupus nephritis in female Balb/c mice Doses: 200, 400, or 600 mg/kg/day, orally for 21 days	- ↓ proteinuria levels - ↓ glomerular inflammation - bryophyllin A is probably the active compound of *B. pinnatum* for its anti-inflammatory effect	[233]

Table 3. *Cont.*

Plant Name	Other Names	Type of Study	Method	Result or Side Effect or Mechanism	Ref.
Camellia sinensis (L.) Kuntze	Tea	Clinical trial	68 SLE patients were divided into control and study groups. The study group was treated with 1000 mg of *C. sinensis* extract daily for 12 weeks.	- ↑ the quality of life of patients - ↑ general health - ↓ disease activity	[234]
Curcuma longa L.	Turmeric	Clinical trial	24 patients with lupus nephritis divided into study and control groups. The study group was treated with 1500 mg of turmeric daily for 3 months.	- ↓ proteinuria and hematuria - ↓ systolic blood pressure - No side effects have been observed	[235]
Ganoderma lucidum and Ganoderma tsugae	Lingzhi Reishi	In vivo	Female MRL/lpr mice with mild, moderate, and severe lupus Doses: Initially, 500 mg/kg/day was administered orally for 7 days, and then 50 mg/kg/day was injected intraperitoneally for 7 days.	- Significant reduction in anti-ds-DNA - ↑ the percentages of IL-10, $CD4^+$, $CD25^+$, $Foxp3^+$ and Treg cells - ↑ the concentrations of IL-2 and IL-12P70 - ↓ the concentrations of IL-21, IL-10, and IL-17A	[236]
		In vivo	All groups of NZB/NZW F1 mice (two months of age) were given standard laboratory chow feeding. The first study group was given 0.1 cm^3 of oral ganoderma extract, and the second study group was given 0.2 cm^3 of oral ganoderma extract, daily. The third study group was given 0.5 mg/kg/day of prednisolone.	- ↑ life expectancy - ↓ anti-dsDNA autoantibody, proteinuria, parenchyma and perivascular mononuclear cell infiltration	[237]

Table 3. *Cont.*

Plant Name	Other Names	Type of Study	Method	Result or Side Effect or Mechanism	Ref.
Gentiana macrophylla Pall.	Qinjiao	Clinical trial	Treatment group: 62 patients with SLE were treated with *G. macrophylla* complex tablets (10 tablets BID or 5 tablets TID) with 10 to 30 mg of prednisolone daily. Control group: 19 SLE patients were treated with prednisolone alone.	- The recovery rate was significantly higher in the treated group - Improvement of nephropathy, erythema, arthralgia and restoration of ESR, LE cells, C3 and CH50 - No significant complication was observed	[238]
		In vivo and In vitro	Female NZB/W F1 mice divided into the control group, the cholesterol consuming group and the cholesterol and *G. macrophylla* consuming group. (a 12-week study, heart tissues were examined)	- ↓ cholesterol-aggravated apoptosis - ↑ IGF-1 survival signal - ↑ anti-apoptotic proteins	[67]
		In vivo and In vitro	Female NZB/W F1 mice—an 8-week study	- reduced liver inflammation	[239]
Linum usitatissimum L.	Flaxseed, linseed	Clinical trial	23 patients with lupus nephritis (15 volunteers remained) Doses: 30 g/day A two-year nonplacebo-controlled crossover study	- Viscosity of serum and plasma lipids remained unchanged. - ↓ serum creatinine - ↓ microalbumin	[240]
		Clinical trial	8 patients with lupus nephritis Doses: 15, 30, and 45 mg of flaxseed daily for four weeks, with a washout of 5 weeks between doses.	- ↓ blood viscosity and LDL - Inhibit AF-induced platelet aggregation - ↓ serum creatinine and proteinuria - ↑ complement C3 - ↓ expression of CD11b	[241]

Table 3. *Cont.*

Plant Name	Other Names	Type of Study	Method	Result or Side Effect or Mechanism	Ref.
Ophiocordyceps sinensis (syn. *Cordyceps sinensis*)	Yartsa gumba caterpillar fungus	meta-analysis study	This meta-analysis study was conducted on a total of 14 studies comprising 1301 participants.	- Consumption of *O. sinensis* mycelium for lupus nephritis is more effective than not using it. - There was no significant difference between the Bailing group and the control group in anti-ds-DNAIgM levels and complement C3 levels. - Improvement of other indicators of the disease, such as SLEDAI score, Alb, 24 h urinary protein, serum creatinine, and the number of effective treatments and complications.	[142]
		In vivo	Lupus-prone (NZB/NZW) F1 hybrid mice with different ages (three, six, and eight months) Doses: 2.4 mg/g/day of cultured mycelia of *C. sinensis* orally	- In groups who started taking it at the ages of 3 and 6 months, survival increased, proteinuria decreased, and titers of anti-double-stranded DNA antibody decreased. - ↓ the percentage of $CD4^+$ T cells in PBMC and ↑ the percentage of $CD8^+$ T cells - Early administration of *C. sinensis* reduces the severity of lupus disease	[242]

Table 3. *Cont.*

Plant Name	Other Names	Type of Study	Method	Result or Side Effect or Mechanism	Ref.
Ophiocordyceps sinensis (syn. *Cordyceps sinensis*)	Yartsa gumba caterpillar fungus	In vivo	MRL lpr/lpr mice aged 12 weeks Doses: 40 µg/kg/d of H1-A daily for 8 weeks.	- ↓ the production of anti-ds-DNA, lymphadenopathy, and proteinuria - Improved renal function - No significant changes have been observed in immune complex deposition.	[96]
		In vivo	The plants studied in this article include: *Cordyceps sinensis, Anqelica sinensis, Atractylodes ovata, Codonopsis pilosula, Ligustrum lucidum,* and *Homo sapiens*. 144 NZB/NZW F1 mice at one month of age were divided into 19 groups	- *C. sinensis* was found to inhibit anti-ds-DNA production and increase the lifespan of mice. - *A. sinensis* does not inhibit anti-ds-DNA production, but it has been able to increase the lifespan of mice.	[243]
Paeonia lactiflora Pall.	Chinese peony common garden peony shaoyao	In vivo and In vitro	MRL/lpr lupus mice Received Radix Paeoniae Rubra for 12 weeks	- ↓renal pathological damage - ↓urinary protein levels - ↓ the expression of ICAM-1, VCAM-1 and PECAM-1	[244]
Paeonia × *suffruticosa* Andrews	mǔdān Moutan Cortex	Clinical trial	Moutan Cortex extract group: 84 patients control group: 84 patients	- ↓ the percentage of Th17 cell, - ↑ the percentage of Th1 cell - ↓ IL-6 level, -↓ ESR and SLEDAI score - ↓ side effects	[245]

Table 3. *Cont.*

Plant Name	Other Names	Type of Study	Method	Result or Side Effect or Mechanism	Ref.
Rehmannia glutinosa (Gaertn.) DC.	*Rehmanniae Radix* (Di Huang)	Clinical trial	72 patients were divided into control and treated groups. The control group was given prednisolone and cyclophosphamide. In the treatment group *Radix Rehmanniae* and *Radix Astragali* was added to the treatment regime for 6 months.	- The dose reduction of prednisolone was greater in the treated group. - Patients who had to increase the dose of prednisolone due to aggravating disease were fewer in the treated group. - Infection, cardiovascular anomalies, hot flush, insomnia, and Cushing's syndrome were less common in the treated group. - There was no difference in blood immunoglobulin G and blood complement 3 in the two groups. - The protein in the 24 h urine was lower in the treated group.	[246]
		In vitro	The effect of fresh Rehmanniae radix methanol extract has been investigated in adult female BALB/c mice. Mouse splenocytes were investigated.	- ↓ inflammatory cytokines such as IL-2, IFN-γ, IL-6 and IL-10	[247]
Scutellaria baicalensis Georgi	Baikal skullcap Chinese skullcap	In vitro	In vitro study on splenocytes of pristane-induced lupus BALB/c mice.	- Downregulated the production of proinflammatory cytokines such as TNF-α, IL-6, IL-10 and IFN-γ. - The expression of $CD69^+$ $CD4^+$ T cells and $CD4^+$ T cells decreased but $CD8^+$ did not decrease.	[248]

Table 3. *Cont.*

Plant Name	Other Names	Type of Study	Method	Result or Side Effect or Mechanism	Ref.
Tripterygium wilfordii Hook. F.	Thunder duke vine Thunder god vine Léi gōng téng	Clinical trial	There were 23 cases of lupus erythematosus, of which 15 had SLE and 8 had DLE. They were given 45 mg/day of crude extract of *T. wilfordii*. The control group consisted of 19 cases of SLE treated with prednisolone.	- The rate of improvement was almost the same in the two groups, and no significant difference was observed. - Recovery from erythematosus rash and arthralgia following use of *T. wilfordii*.	[249]
		Clinical trial	26 DLE cases treated with *T. wilfordii*	- Progress with varying degrees was observed in 24 cases.	[250]
Urtica dioica L.	stinging nettle grande ortie anonhasquara	Case report	One female lupus with status renal allograft and elevated serum creatinine was consuming immunosuppressants. She consumed an herbal remedy, consisting of a combination of *Agropyron repens* rhizome and *U. dioica* seed extracts (1:3, 5 mL, three times a day for 46 days), and then she received *U. dioica* seed extract as a monotherapy for three months.	- ↓ serum creatinine after consuming mixed herbal remedy - Serum creatinine levels were normalized to acceptable levels after monotherapy of U. *dioica* seed extract for three months.	[251]
		In vivo	MRL lpr/lpr mice Dose: 100 µg, inj., every 2 weeks for 4.5 months, blood samples were every 3	- Inhibited the development of overt clinical signs of lupus and nephritis. - Pathogenic T cell clones are thus found among the V8.3+ T cell population, which also contains an enlarged T cell clone. - Affected autoantibody production in a sex-dependent manner.	[252]

3.4.1. *Tripterygium wilfordii* Hook. F.

Tripterygium wilfordii Hook. F. (Celastraceae) has the common names of "thunder duke vine" and "thunder god vine" and is also known as "léi gōng téng" in Mandarin. The plant has long been used in traditional Chinese medicine (TCM) and it is widespread in southern and eastern China [88,253]. Its root extract is used, but its bark must be removed because of its toxicity. There are various compounds with a range of biological effects in the root extract, and the procedures and methods of extraction play a role in overcoming the plants' toxicity [254,255]. In TCM, the *T. wilfordii* preparations have long been used for some health conditions, including sores and swelling, inflammations, ankylosing spondylitis, hepatitis, nephropathy, allergic skin diseases, inflammatory lesions of leprosy, and cancer [253,254,256–259]. The effect of *T. wilfordii* on autoimmune diseases such as SLE, rheumatoid arthritis (RA), BehCet's disease, psoriasis, etc., has been investigated in several studies [253,260]. Its effect on kidney transplantation, nephrotic syndrome, and diabetic nephropathy has also been investigated [261–263].

Clinical Trials

A number of studies have examined the efficacy of *T. wilfordii* in lupus erythematosus. A clinical trial study compared the effect of *T. wilfordii* with prednisolone. In this study, there were 23 cases of lupus erythematosus, of which 15 had SLE and 8 had discoid lupus erythematosus (DLE). They had been given 45 mg/day of crude extract of *T. wilfordii*. The control group consisted of 19 cases of SLE treated with prednisolone. The rate of improvement was almost the same in the two groups, and no significant difference was observed. However, there are benefits to using *T. wilfordii*, including the recovery from erythematosus rash and arthralgia [249]. The immunomodulatory and anti-inflammatory effects of *T. wilfordii* have also been reported in a clinical study on 26 DLE cases [250].

Despite the benefits of this plant, side effects have also been reported. Headaches, gastrointestinal complications, nausea, diarrhea, infertility, etc., are among these adverse effects [249,250] One of the side effects reported due to long-term use of *T. wilfordii* in women is decreased bone mineral density [121]. Due to some toxic side effects of *T. wilfordii* such as kidney damage, its use is not recommended in SLE patients who also have nephropathy [249]. A systematic review has discussed cardiovascular, hematologic, and skin complications as well as infertility and gastrointestinal complications of this plant [122]). Side effects are more common with high concentrations. Consuming the right amounts and rational treatment will help control the side effects and obtain therapeutic effects [121].

Active Compounds and Possible Mechanisms of Efficacy

A variety of compounds have been isolated from *T. wilfordii* such as triptonide, tripodine, triptolide (TPT), tripdiolide (TPO), etc. [123]. Among the phytochemicals of *T. wilfordii* root extract, celastrol (tripterine) and triptolide (diterpenoid triepoxide) are the most investigated [124,258]. Triptolide's anti-inflammatory and immunosuppressive effects are mediated by inhibition of T cells and inhibition of IL-17 and STAT3 transcription [125]. It has been observed that NF-κB activity is significantly reduced in patients with SLE after consuming *T. wilfordii*. It was suggested that *T. wilfordii* may exert an immunosuppressive effect on SLE patients by inhibiting NF-κB expression [123].

An in vitro study on dendritic cells (DCs) of SLE patients has shown that triptolide can inhibit the differentiation and maturation of DCs and reduce the immune function of DCs. In this study, doses of 0, 5, 10, and 30 μg/L of triptolide were used. Triptolide also reduced secretion of IFN-α, IL-6, and TNF-α [121]. In another study, to evaluate the effect of triptolide, BALB/c-un nude mice were used. Mice were treated orally with 5 mg/kg/d of triptolide, and their blood samples were collected before treatment and 1, 3, and 6 months after treatment. Finally, it was found that with the use of triptolide, a decrease in the percentage of $CD8^+$, Tcl, Thl cells, $CD4^+/CD8^+$, Thl/Th2, and Tcl/Tc2 and an increase in the percentage of $CD4^+$, Tc2, and Th2 cells was observed [122].

The effect of triptolide and tripdiolide on lupus nephritis in (NZB × NZW) F1 mice has been investigated. Mice were treated orally with 6 μg of triptolide or tripdiolide for 15 weeks. Finally, triptolide and tripdiolide have been shown to reduce BUN, proteinuria, and anti-dsDNA antibody levels, as well as the production of cytokines such as IL-6 and TNF and monocyte chemoattractant protein 1. Therefore, triptolide and tripdiolide have therapeutic effects in lupus nephritis [123].

In a study, the effect of triptolide on SLE was investigated in female MRL/lpr mice treated with 0.2 or 0.3 mg/kg/d of triptolide for 13 weeks. Compared to the control (vehicle) group, triptolide significantly reduced proteinuria, serum anti-dsDNA, and renal histopathologic assessment. The effect was comparable to that of cyclophosphamide (20 mg/kg/w). Triptolide also increased the proportion of Treg and induced expression of miR-125a-5p [124].

(5R)-5-Hydroxytriptolide (LLDT-8) is a triptolide derivative that has strong anti-inflammatory and immunosuppressive effects and low toxicity. In a study, the effect of LLDT-8 on lupus nephritis was investigated. Female MRL/lpr mice were treated with 0.125 mg/kg/2 days of LLDT-8 for 9 weeks. Finally, LLDT-8 has been shown to reduce proteinuria, serum creatinine, and glomerular IgG deposits, and it could also ameliorate histopathology and increase the lifespan of mice. LLDT-8 reduced the expression of inflammatory cytokines such as IFN-γ, IL-17, IL-6, and TNF-α and inhibited immune cell infiltration in the kidneys. It was suggested that LLDT-8 could have therapeutic effects on lupus nephritis [154].

Toxicity and Side Effects

Refer to the section on toxicity and side effects of triptolide in this text.

3.4.2. *Ophiocordyceps sinensis* (syn. *Cordyceps sinensis*)

Ophiocordyceps sinensis belongs to the Ophiocordycipitaceae family, is an entomogenous fungus used in TCM. It is also known as Yartsa gumba or caterpillar fungus [79,264].

Clinical Trials

Several clinical trials have evaluated the efficacy of the dry powder of *O. sinensis* mycelium (Bailing capsules) as a supplement in conjunction with prednisolone, cyclophosphamide, tacrolimus, or leflunomide on lupus nephritis. They have reported controversial results, which might be due to the small sample size or differences in control groups and study design [142]. A meta-analysis study was conducted on a total of 14 studies comprising 1301 participants, which were combined for analysis in the present study. In general, this study showed that consumption of *O. sinensis* mycelium (Bailing capsule) for lupus nephritis is more effective than not using it. Although there was no significant difference between the Bailing group and the control group in anti-ds-DNAIgM levels and complement C3 levels (which is associated with the existence of some immune diseases like SLE), other indicators of the disease, such as SLEDAI score, Alb, 24 h urinary protein, serum creatinine, and the number of effective treatments and complications, improved. It was concluded that *O. sinensis* could be beneficial in the treatment of lupus nephritis [142].

In a study of 61 lupus nephritis patients, the prevention of the recurrence of lupus nephritis by *artemisinin* and *C. sinensis* was evaluated. A total of 30 patients were in the control group, and 31 patients were treated with 2–4 g/d of *C. sinensis* and 0.6 g/d of artemisinin. *C. sinensis* was taken before three main meals and artemisinin after three main meals for three years. The control group took tripterygiitotorum and/or Baoshenkang. Finally, the creatinine clearance rate did not change between before and after treatment, and the complement C3 level stabilized in the normal range. It was concluded that the combination of *C. sinensis* and artemisinin could be effective in the prevention of recurrence of lupus nephritis [95].

Active Compounds and Possible Mechanisms of Efficacy

A variety of bioactive phytochemicals with biological activities have been reported from *O. sinensis*, such as cordycepin, cordycepic acid, polysaccharides, ergosterol, nucleosides, fatty acids, proteins, minerals, etc. The immunomodulatory effects of *O. sinensis* were mostly attributed to its polysaccharides, which affect both humoral and cellular immune responses in vivo and improve the serum levels of ovalbumin-specific IgG, IgG1, and IgG2b levels. This fungus's mycelia polysaccharides have been shown to enhance the proliferation and phagocytosis of macrophages and stimulate macrophages [265]. Intracellular polysaccharides isolated from submerged cultures of *O. sinensis* have been reported to exhibit strong immunomodulatory effects on RAW264.7 macrophage cells via the MAPK and PI3K/Akt signaling pathways. It enhanced the phagocytic activity of RAW264.7 cells and increased cytokine production. This immunomodulatory response was mediated by the secretion of both proinflammatory cytokines (TNF-α, IL-6, and IL-1β) and anti-inflammatory substances (TGF-β1 and IL-10), producing NO and promoting the expression of iNOS [266].

Some of *O. sinensis*' nucleotide contents have also been reported to exhibit immunomodulatory properties by lowering NO and increasing IL-1β and TNF-α release from macrophages [265]. Deoxynucleic acids from *O. sinensis* activated mouse bone marrow-derived DCs via a toll-like receptor 9-dependent pathway [265]. Aside from the total extracts of *O. sinensis*, specific components, including 1-(5-Hydroxymethyl-2-furyl)-β-carboline, and cordymin (a purified peptide from *O. sinensis*), displayed significant anti-inflammatory properties [79,267–269].

Hypothetically, besides immunomodulatory and anti-inflammatory effects, the antioxidant [270], cardiovascular [271], and kidney protective [272] properties of *O. sinensis* may also have a role in the beneficial effects of this fungus on SLE.

A number of studies have been performed to investigate the effect of *O. sinensis* on lupus. A study was performed on lupus-prone (NZB/NZW) F1 hybrid mice. The mice were divided into four groups of different ages (three, six, and eight months) and were given 2.4 mg/g/day of cultured mycelia of *C. sinensis* orally. The fourth group was also used as a control. The results showed that in groups who started taking it at the ages of 3 and 6 months, survival increased, proteinuria decreased, and titers of anti-double-stranded DNA antibodies decreased. The percentage of $CD4^+$ T cells in peripheral blood mononuclear cells (PBMC) decreased significantly, while the percentage of $CD8^+$ T cells increased. Eventually, the results of this study showed that early administration of C. sinensis reduces the severity of lupus disease [242]. In another study on MRL lpr/lpr mice, a triterpenoid, component H1-A, was extracted from *C. sinensis*. Administration of 40 μg/kg/d of H1-A daily for 8 weeks to mice aged 12 weeks resulted in a reduction in the production of anti-dsDNA, lymphadenopathy, and proteinuria. Renal function has improved, and no significant changes have been observed in immune complex deposition. In general, H1-A intake has increased the survival of mice with lupus [96].

In another study, the effects of Chinese herbs on SLE were investigated in NZB/NZW F1 mice at one month of age. *C. sinensis* was found to inhibit anti-ds-DNA production and increase the lifespan of mice. Although *A. sinensis* does not inhibit anti-ds-DNA production, it has been able to increase the lifespan of mice [243].

Toxicity and Side Effects

In addition to the claimed therapeutic or positive effects for the chemicals derived from the Cordyceps fungi, cytotoxicity and/or neurological toxicity adverse effects have also been described for these compounds. After daily intake of Cordyceps fruiting bodies or associated products, reports of nausea, diarrhea, and even significant post-extraction bleeding have been documented. There have also been a few rare reports of dry mouth, nausea, and diarrhea [273].

3.4.3. *Ganoderma Lucidum* and *Ganoderma Tsugae*

Ganoderma lucidum belongs to the family Ganodermataceae or Polyporaceae and is also known as Lingzhi or Reishi. This fungus has been observed to have anti-inflammatory, antioxidant, and analgesic effects, and it has been used in TMC since the ancient era. This fungus is used for various diseases that, in most cases, have an inflammatory basis, such as arthritis, hepatitis, bronchitis, acute colitis, etc., as well as hypertension and malignancy [236].

G. lucidum has shown an immunomodulatory effect on PMNC (peripheral mononuclear cells). It has exhibited suppressive effects on tumor necrosis factor-α (TNF-α), IL-1β, IL-12, and IL-6, which are pathogenic cytokines associated with SLE [274,275].

Clinical Trials

The authors of the present study could not find a clinical trial on the efficacy of the *G. lucidum* on SLE, but there are several trials investigating the efficacy of this mushroom on some other conditions such as rheumatoid arthritis [276], fibromyalgia [111,277], neurasthenia [278], cancers [279], cardiovascular risk factors of metabolic syndrome [280], lower urinary tract symptoms (LUTS) [281], etc. Moreover, in a clinical trial, β-glucans of *G. lucidum* were reported as safe and well-tolerated immunomodulator supplements for children [282].

Active Compounds and Possible Mechanisms of Efficacy

A variety of natural polysaccharides have shown immunomodulatory, anti-inflamma tory, and wound-healing properties [283,284]. *G. lucidum* polysaccharides (GLPs), such as ganoderan and β-glucans, have been extensively studied for their biological activities, which include antioxidant, antitumor, anti-inflammatory, anti-diabetes, and immunomodulatory properties [285]. GLPs can affect different immune effector cells, including lymphocytes and myeloid cells. They can modulate innate immunity, cellular immunity, and humoral immunity [286]. Also, some triterpenoids of *G. lucidum* (GLTs), such as ganosidone A and its derivatives [287], as well as ganoderic acid D [288], and 3-oxo-5α -lanosta-8, 24-dien-21-oic acid [289] have shown anti-inflammatory effects.

In a study, the combination of *G. lucidum* and *San-Miao-San* (SMS) has been used to evaluate the anti-inflammatory effect on SLE. SMS is a Chinese herbal medicine that consists of a combination of three herbs that include Phellodendri Cortex (Huangbai), Atractylodes Rhizome (Cangzhu), and Radix achyranthis bidentatae (Niuxi). The control group consisted of female Balb/c mice (at the age of 20–24 weeks). The study group included three groups of female MRL/lpr mice that had mild, moderate, and severe lupus. Initially, 500 mg/kg/day was administered orally for 7 days, and then 50 mg/kg/day was injected intraperitoneally for 7 days. Finally, a significant reduction in anti-ds-DNA in the study group with moderate and severe SLEs was observed. In the study group, the percentages of IL-10, $CD4^+$, $CD25^+$, $Foxp3^+$ and Treg cells increased significantly, the concentrations of IL-2 and IL-12P70 increased significantly, and the concentrations of IL-21, IL-10, and IL-17A decreased significantly [236].

In a study, the effectiveness of *Ganoderma tsugae* on increasing the lifespan of NZB/NZW F1 mice was investigated. All groups of mice (two months of age) were given standard laboratory chow feeding. The first study group was given 0.1 cm^3 of oral ganoderma extract, and the second study group was given 0.2 cm^3 of oral ganoderma extract daily. The third study group was given 0.5 mg/kg/day of prednisolone. *G. tsugae* increased the life expectancy in mice with lupus and reduced anti-dsDNA autoantibody and proteinuria, as well as parenchyma and perivascular mononuclear cell infiltration [237].

Toxicity and Side Effects

Patients from various nations have reported developing human sensitivity to the *G. lucidum* antigen. Since *G. lucidum* has an anticoagulant effect and extending the prothrombin time has an additional effect on clotting factors, patients who were taking anticoagulants or antiplatelets should take caution. Hypoglycemic people should also take caution because it reduces blood sugar levels. *G. lucidum* is an anti-hypertensive agent, according to numerous research works. Before using it, persons with cardiac issues should visit a physician [290]. In a study, it has been determined that sub-chronic toxicity of the liver occurs when rats are given more than 1.2 mg/kg body weight of *G. lucidum* extract [291].

3.4.4. *Urtica Dioica* L.

Urtica dioica, also known as stinging nettle, grande ortie, or anonhasquara, belongs to the Urticaceae family. Different parts of the plant, such as roots, leaves, seeds, and aerial parts, have different therapeutic effects upon extraction by different methods. Following animal studies, various therapeutic effects without the appearance of serious side effects have been reported. This plant has shown anti-inflammatory, antioxidant, antimicrobial, antifungal, etc., effects. In traditional medicine and ethnomedicine, different parts of the plant have been used in diseases such as systemic lupus erythematosus and rheumatoid arthritis, diabetes, prostate cancer, breast cancer, atherosclerosis, cardiovascular diseases, etc. Non-aqueous extraction of the root of this plant has been found to be effective in SLE [292,293].

Clinical Trials

In a case report, one female lupus patient (24 years old) with a renal allograft status and elevated serum creatinine was featured. The patient was consuming immunosuppressants (prednisone, CellCept™ and Prograf™). After consuming an herbal remedy, consisting of a combination of *Agropyron repens* rhizome and *U. dioica* seed extracts (1:3, 5 mL, three times a day), the serum creatinine started to decline. After 46 days, U. *dioica* seed extract was used as a monotherapy for 3 months. Serum creatinine levels were normalized to acceptable levels [251].

Active Compounds and Possible Mechanisms of Efficacy

A variety of phytochemicals, including phenylpropanoids, flavonoids (such as chlorogenic acid, rutin and isoquercitrin, quercetin-3-O-rutinoside, kaempherol-3-O-rutinoside, and isorhamnetin3-O-glucoside), lignans (such as secoisolariciresinol), and coumarin (such as scopoletin), have been reported from nettle extracts. These extracts have shown anti-inflammatory and immunomodulatory effects with different selectivity toward the COX and LOX branches of the eicosanoid pathway [294,295]. Moreover, several plant sterols, such as sitosterol and its derivatives, have also been reported from the nettle root extract [296]. The root and leave extracts of *U. dioica* have shown immunomodulatory effects through different mechanisms, including lowering thromboxane production in human platelets and inhibiting the 12-LOX pathway. Different parts of the plant have shown antioxidant effects [294,297].

In a study, the effect of long-term injection of the Vβ8.3-specific superantigenic lectin *U. dioica agglutinin* (UDA) was investigated in MRL lpr/lpr mice (7 weeks of age). In contrast to the control group, injection of UDA (100 μg every two weeks for 4.5 months), inhibited the development of overt clinical signs of lupus and nephritis. Pathogenic T cell clones are thus found among the V8.3+ T cell population, which also contains an enlarged T cell clone. UDA affected autoantibody production in a sex-dependent manner [252].

Toxicity and Side Effects

Sweating and gastric discomfort are some of the side effects that have been reported to be associated with using stinging nettle. It should be noted that touching stinging nettle typically can cause skin irritation. Patients with renal conditions have been documented to experience hypersensitivity following consumption of this plant. Additionally, this plant has been shown to improve the effects of CNS depressive drugs. Consumption of stinging nettle concurrently with sedatives, such as lorazepam, phenobarbital, clonazepam, zolpidem, and others, may cause drowsiness and sleepiness [298].

3.4.5. *Nelumbo nucifera* Gaertn.

Nelumbo nucifera belongs to the Nelumbonaceae family and is also known as the sacred lotus and water lily. In addition to being used as a vegetable and food, this plant has had many therapeutic uses from the past to the present. All parts of the plants (fruits, leaves, flowers, seeds, roots, rhizomes, buds, stems, anthers, stalks, plumules, and stamens) have been used in traditional medicine. The plant has been comprehensively studied to investigate its medicinal benefits such as anti-obesity, anti-diabetic, antioxidant, anti-amnesic, anti-thrombotic, anticarcinogenic, anti-inflammatory, immunomodulatory activity, anti-neurodegenerative, antiproliferation, cardiovascular activity, etc.

The effectiveness of lotus on SLE is due to the presence of procyanadins, polyphenols, and polysaccharides in its seeds [299].

The effectiveness of *N. nucifera* on SLE has been investigated in a study. In this study, 12-week-old MRL/MpJ-lpr/lpr mice were used. *N. nucifera* seeds extracted with ethanol contain (S)-armepavine ($C_{19}H_{23}O_3N$). One group was given corn oil orally as a control; another group was given 5 and 10 mg/kg/day of oral (S)-armepavine; and the other group was given

20 mg/kg/day of oral cyclosporine. Mice were treated for 6 weeks. Finally, it was observed that taking (S)-Armepavine increased the life expectancy of mice and inhibited splenocytes proliferation and prevented lymphadenopathy. It also inhibited the expression of IL-2, IL-4, IL-10, and IFN-γ genes and inhibited T cells proliferation. Consumption of (S)-armepavine reduced proteinuria and anti-dsDNA autoantibody [94].

Toxicity and Side Effects

Despite the *N. nucifera*'s long history of therapeutic use, research on its possible toxicity and safety is required. Numerous investigations up to this point have partially supported the safety of *N. nucifera* [300]. The toxicity and safety profile of *N. nucifera* and its components have been examined in several research. A number of in vitro studies have been conducted to evaluate the toxicity of *N. nucifera* using normal cell lines. *N. nucifera* has not been found to significantly affect cell viability or to have any toxic effects, according to the findings of a number of in vitro investigations. The safety profile of *N. nucifera* has also been examined in numerous in vivo studies. In general, side effects from in vivo investigations include an increase in lymphocytes, a decrease in basophils, and a drop in creatinine, cholesterol, and hematocrit [301].

3.4.6. *Artemisia annua* Pall.

Artemisia annua belongs to the Asteraceae family. *A. annua* has not only been used in TCM for various ailments but has also been identified as a medicinal plant in the United States, Europe, Australia, and Asia. This plant used to be found in western Asia and southeastern Europe. Today, it has spread all over the world and can be found in Australia, North and South America, and many parts of Asia and Europe. Since this plant is found in different parts of the world, it is known by many names, such as annual wormwood, sweet wormwood, Chinese wormwood, sweet sagewort, and sweet Annie. Antitumor, analgesic, anti-inflammatory, and antioxidant effects of *A. annua* have been discovered in studies on this species. In traditional medicine, this plant has been used in viral and bacterial diseases, jaundice, and autoimmune diseases such as rheumatoid arthritis and SLE, bacterial dysentery, hemorrhoids, and wound healing. It has also been used as an antipyretic in the treatment of tuberculosis and malaria. The 2015 Nobel Prize in Medicine was awarded for the discovery of sesquiterpene lactone artemisinin and its effects on the treatment of malaria [302]. Recently, the possibility of *A. annua* being useful in the treatment of COVID-19 has been investigated and satisfactory results have been reported [303–305].

Clinical Trial

In a clinical trial performed on 73 patients, including 36 patients with SLE and 37 patients with DLE, 60 or 80 mg of dihydroartemisinin was given orally for 9 weeks. Finally, dihydroartemisinin has been shown to be effective in most patients, and no serious side effects have been reported [306].

Phytochemicals and Possible Mechanisms of Action

The components identified from different parts of *A. annua* are classified as sesquiterpene lactones, coumarins, saponins, flavonoids, tannins, essential oils, polyalkenes, phenolic acids, fatty acids, proteins, and phytosterols [140,307,308]. Artemisinin, a sesquiterpene lactone in glandular hairs on the leaves and flowers of *A. annua*, is one of the important components of this plant, which is assumed to be one of the potential compounds against lupus. Apart from *A. annua*, other plants containing artemisinin, such as *Artemisia apiacea*, can also be effective in treating lupus [309].

Several semi-synthetic derivatives of artemisinin such as dihydroartemisinin, artemether, arteether, and artesunate have been investigated [140,302]. The effects of dihydroarteannuin (also known as dihydroqinghaosu, artenimol, or DHA) is one of the semi-synthetic derivatives of artemisinin. In a study, the effect of dihydroarteannuin and its mechanism on lupus in BXSB mice has been investigated. In this study, male BXSB mice were given dihydroarteannuin daily for ten days. One group was considered a control and the other three groups were given doses of 5 mg/kg, 25 mg/kg, or 125 mg/kg. As a result, dihydroarteannuin consumption was found to reduce TNF-alpha production, as well as nuclear factor-κB (NF-κB) activation and p65 subunit expression. Dihydroarteannuin inhibits NF-κB translocation to the nucleus and also inhibits IκB-α protein degradation [104]. In general, the level of TNF-alpha in the serum of SLE patients is higher than normal people. In addition, the expression of TNF-alpha receptors in peripheral blood lymphocytes of SLE patients is higher. NF-κB is a transcription factor located in the inactive cytoplasmic complex. It has two subunits, p50 and p65, and is attached to the IκB family, which are inhibitory proteins. Stimulation separates NF-κB from IκB and eventually translocates NF-κB to the nucleus, where it binds to DNA, inducing the expression of genes involved in the pathology of SLE progression [104].

In a study, the effect of *A. annua* was studied in female ICR mice, and it was found that, due to its suppressive effect on the immune system, it can be effective in diseases such as SLE and rheumatoid arthritis. Splenocytes of immunized mice were isolated and exposed to different concentrations, and finally, the number of specific antibodies was counted by indirect ELISA. The results indicated that the levels of a series of antibodies decreased following the consumption of ethanolic extract of *A. annua*. It was suggested that the plant might be useful in autoimmune diseases such as SLE and rheumatoid arthritis [228].

SM934 is another derivative of artemisinin that has more water solubility, bioavailability, and bioactivity but less toxicity. A study on MRL/lpr mice showed improvement in lupus syndrome. SM934 reduced the production of IFNγ and IL-17 by polyclonal $CD4^+$ T cells activated by T cell receptor rearrangement in vitro, as well as the development of naive $CD4^+$ T cells into Th1 and Th17 cells, but not Treg cells. In the in vivo study, administration of SM934 to mice for 4 weeks attenuated the renal lesion severity, proteinuria, and anti-dsDNA autoantibody. It also decreased the spleen size and the levels of serum IFN-Y and blood urea nitrogen. Following 8 weeks of consumption, the lifespan of mice increased. Ex vivo studies have shown an inhibition in the production of Th1 and Th17 while the level of Treg cells increased. In splenocytes, SM934 inhibited the complete activation of STAT-1, STAT-3, and STAT-5 proteins [97].

The effect of oral dihydroartemisinin (2 mg, daily, three days post-infection) on the immune system of BALB/C female mice infected with *Toxoplasma gondii* or *Plasmodium berghei* has been evaluated. Following the consumption of dihydroartemisinin, the number of B cells in the bloodstream and spleen cells decreased, and the ratio of T helper to $CD8^+$ T cells increased. Dihydroartemisinin also decreased the proinflammatory cytokines. It is assumed that due to immunomodulating properties, dihydroartemisinin might be useful in autoimmune diseases such as SLE [105].

One of the factors associated with the progression of SLE is the imbalance between Treg/Th17. In a study, the effect of dihydroartemisinin alone or in combination with prednisolone on Treg/Th17 balance has been investigated. For this purpose, female BALB/c mice have been used. One group was considered the control, and one group was the SLE model group. The third group was given 100 mg/kg of dihydroartemisinin, the fourth group was given 5 mg/kg prednisolone, and the last group was given a combination of prednisolone and dihydroartemisinin with the same previous doses. Dihydroartemisinin and prednisolone were administered daily and orally for two months. As a result, the Treg/Th17 balance was restored, and the inflammation was inhibited. The effect of dihydroartemisinin has also been studied in vitro by isolating mouse spleen lymphocytes and exposing them to dihydroartemisinin or a combination of dihydroartemisinin and prednisolone. As a result, it was observed that the levels of TGF-β, IL-17, and Foxp3 increased, while transcription of RORγt decreased. Th17 cell differentiation is inhibited, but Treg cell differentiation is induced. Finally, dihydroartemisinin has been shown to have a synergistic effect on prednisolone [106]. There is also further evidence and studies to investigate the effect of *A. annua* on lupus [310–312].

Toxicity and Side Effects

Refer to the section on toxicity and side effects of artemisinin in the present text.

3.4.7. *Linum usitatissimum* L.

The plant belongs to the Linaceae family and is also known as flaxseed and linseed.

Clinical Trials

In a two-year nonplacebo-controlled crossover study, 23 patients suffering from lupus nephritis were divided into two groups. One group received ground flaxseed (30 g/day for one year), and the second group was considered the control. At the end of the one-year period, the two groups were switched after a 12-week period of washout. Fifteen volunteers remained until the end of the trial. Eventually, it was observed that the viscosity of serum and plasma lipids remained unchanged, but serum creatinine decreased. Although microalbumin decreased during both flaxseed consumption and control times, a further decrease was observed during flaxseed treatment. It was concluded that in lupus nephritis, flaxseed appears to be renoprotective, but this interpretation is hampered by underpowering due to poor adherence and possible Hawthorne effects [240].

In another study, the effect of flaxseed on lupus nephritis was investigated. Nine patients with lupus nephritis enrolled while eight of them accomplished the study. Patients consumed 15, 30, and 45 mg of flaxseed daily for four weeks, with a washout of five weeks between doses. Finally, it was observed that blood viscosity and LDL decreased significantly, and there was a further decrease following an increase of the dose. Inhibition of AF-induced platelet aggregation, a decrease in serum creatinine, a decrease in proteinuria, an increase in complement C3, and decreased expression of CD11b have been observed. It was concluded that a daily consumption of 30 mg of flaxseed is tolerable and useful for patients with lupus nephritis due to its anti-inflammatory effect and beneficial effects on the kidney and atherogenic mechanism [241].

Phytochemicals and Possible Mechanisms of Action

Flaxseed oil, or linseed oil, is rich in omega-3 fatty acids. It also contains linolenic acid, linoleic acid, secoisolariciresinol diglucoside (SDG), lignans, and cyclic peptides [313,314]. Flaxseed contains polysaccharides with immunomodulatory properties such as FP-1. In a study, this polysaccharide stimulated immune responses by inducing mRNA expression of TNF-α, NO, IL-6, and IL-12 in murine macrophages [315].

Polyphenols, polysaccharides, and lignans of flaxseed have been reported to have antioxidant and anti-angiogenic properties [316–318]. Flaxseed oil has been reported to possess anti-inflammatory and immunomodulatory properties [319,320]. Flaxseed oil is rich in phytosterols such as campesterol, brassicasterol, stigmasterol, β-sitosterol, and Δ5-avenasterol [321]. These phytosterols have been repeatedly reported as anti-inflammatory agents [322]. Adding flaxseed oil to the co-culture of 3T3-L1 adipocytes, RAW 264.7 macrophages, showed a dose-dependent shift in cytokines toward IL-4 but a decrease in TNF-α. In the in vivo model (C57bl/6 mice), oral flaxseed oil (4 weeks) increased IL-4 cytokine, serum anti-ova IgG1, and IgE levels. Anti-ova IgG2a, IgG2b, and IgG3 levels were also reduced [319]. Some polyunsaturated fatty acids (PUFA), such as α-linolenic acid, have a variety of biological activities including neural stem cell proliferative, anti-atherosclerotic, and anti-inflammatory effects [209,215,323]. They might exhibit cardiovascular protective properties and might be useful in preventing the progression of nephritis in patients with lupus [240]. The beneficial effect of flaxseed on the kidneys has been reported in a number of studies [324–327].

In general, flaxseed reduces inflammatory responses and lowers blood pressure and vascular disease due to its PUFA content, such as omega-3. Since blood pressure is a risk factor for chronic kidney disease and flaxseed has an anti-inflammatory effect on the kidneys, flaxseed improves kidney health [313].

Toxicity and Side Effects

The flaxseed contains certain chemicals that have been recognized as potentially harmful, including cyanogenic glycosides and linatine, even though no toxicity has been documented in clinical investigations with dietary supplementation of flaxseed. Intestinal β-glycosidase changes the glycoside into cyanohydrin, which subsequently breaks down into hydrogen cyanide. Acute cyanide poisoning from hydrogen cyanide could put the nervous and respiratory systems at risk. However, consuming 15–100 g of flaxseed has not been reported to cause any rise in plasma cyanide levels above the baseline. Theoretically, 1–2 tablespoons of flaxseed will result in the production of between 5 and 10 milligrams of hydrogen cyanide when consumed. Toxic effects are quite unlikely to result from this. It is crucial to note that the idea that dietary flaxseed is toxic due to any of these components has not been proven scientifically [328].

3.4.8. *Rehmannia glutinosa* (Gaertn.) DC.

Rehmanniae Radix (Di Huang) is the root of *Rehmannia glutinosa* (Gaertn.) DC. and belongs to the Orobanchaceae family. *Rehmanniae Radix* has long been used in TCM for medicinal purposes. Studies on *Rehmanniae Radix* have been shown to have antioxidant and anti-inflammatory effects and can lower blood sugar, activate the autonomic nervous system, and improve cognitive function. It has been used for treating dermatitis, cervical cancer, and nephrotic hypertension and improving liver damage. The effect of Rehmanniae Radix on lupus has also been investigated [302,329].

Clinical Trial

In a clinical trial in 72 patients with lupus, the combination of *Radix Rehmanniae* and *Radix Astragali* with glucocorticoid drugs was investigated. Patients were divided into control and treated groups. The control group was given prednisolone and cyclophosphamide, and the treated group was given a combination of Radix Rehmanniae and Radix Astragali on the basis of the control group. The duration of treatment in both groups was 6 months. The dose of prednisolone was reduced following the improvement in the patient's condition. The dose reduction of prednisolone was greater in the treated group. In the treated group, there were fewer patients who had to increase the dose of prednisolone due to an aggravation of the disease. Infection, cardiovascular anomalies, hot flushes, insomnia, and Cushing's syndrome were less common in the treated group. Although there was no difference in blood immunoglobulin G and blood complement 3 in the two groups, the protein in the 24 h urine was lower in the treated group. Therefore, following the use of *Radix Rehmanniae* and *Radix Astragal* with conventional Western medicine, fewer side effects were observed, and it was more convenient to withdraw corticosteroids [246]. A similar clinical trial was performed for patients with lupus nephritis, and it can be said that similar results were seen in general [330].

Phytochemicals and Possible Mechanisms of Action

In an in vitro study, the effect of fresh Rehmanniae radix Methanol extract has been investigated in adult female BALB/c mice. Therefore, mouse splenocytes were used. Finally, it has been observed that inflammatory cytokines such as IL-2, IFN-γ, IL-6 and IL-10 are reduced in the mouse splenocytes [247].

More than 100 chemical compounds have been reported from *R. glutinosa* that are mostly classified as iridoids, ionones, phenylethanoid glycosides, lignans, polysaccharides, and phenylpropanoids [331]. The polysaccharides of this plant have been reported to exhibit anti-inflammatory activity through suppression of IL-6 and TGFβ production on bacterial LPS-induced macrophages.

The anti-inflammatory effect of RG-B9, a polysaccharide isolated after processing of Rehmanniae Radix, involved AKT/ERK/JNK signaling pathway [332]. Two other polysaccharides, including SDH-WA and SDH-0.2A, have also been reported to have immunomodulatory properties. They increased lysozyme activity and TNF-α and IL-6 production by RAW264.7 cells. They did, however, reduce the secretion of lysozymes, TNF-α, IL-6, IL-1β, and nitric oxide by LPS-induced RAW264.7 cells [333]. In LPS-stimulated BV2 microglia, catalpol, an iridoid glucoside from *R. glutinosa* significantly suppressed LPS-induced secretion of proinflammatory mediators, NO and prostaglandin E2. Catalpol downregulated NO synthase and cyclooxygenase-2 expression. It also inhibited the TNF-α and IL-1β secretion. Moreover, catalpol downregulated the NF-κB signaling pathway, suppressed the expression of toll-like receptor 4 (TLR4), and lowered LPS-induced generation of ROS [334].

Toxicity and Side Effects

Although *R. glutinosa* is considered one of the safe plants, it can cause side effects such as dizziness, vertigo, headache, heart palpitations, nausea, diarrhea, allergies, and fatigue. It should be used with caution in patients with liver, digestive, and immune system problems [335].

3.4.9. *Paeonia* × *suffruticosa* Andrews

Paeonia suffruticosa belongs to the family Paeoniaceae and is also known as mŭdān and Moutan Cortex. Due to its anti-inflammatory, antibacterial, sedative, anti-diabetic, and analgesic effects, it is used for inflammatory diseases, menstrual problems, cardiovascular diseases, and atherosclerosis. Many benefits of this plant are due to the large number of monoterpenoids glucosides in it [336]. The root bark of this plant has been used in traditional Chinese medicine for lupus nephritis [337].

Clinical Trials

The effect of Moutan Cortex on improving the condition of SLE patients in a clinical trial has been investigated. A total of 84 patients in the study group were given Moutan Cortex extract, and 84 patients in the control group were given common drugs. Finally, it was observed that following the use of Moutan Cortex extract, the percentage of Th17 cells decreased and the percentage of Th1 cells increased. Decreases were also observed in the IL-6 level, ESR, and SLEDAI scores. Compared to the control group, fewer side effects were observed with the use of Moutan Cortex. Therefore, taking Moutan Cortex improves the condition of SLE patients [245].

Toxicity and Side Effects

Moutan Cortex can be considered a safe raw material, and no research has been done to support its toxic effects. Although, benzoic acid, a component of the extracts that are thought to be toxic, is present in small amounts in this species. It is important to consider how easily heavy metals from soil, pesticides, air dust, irrigation water, vehicle and industrial exhaust gases, and fertilizers can contaminate raw materials. Exogenous elements like heavy metals, pesticide residues, or an excessive amount of sulfur from sulfur fumigation can contaminate Moutan Cortex. To assure the high quality of the raw material, it is crucial to determine the trace elements present in Moutan Cortex [338].

3.4.10. *Paeonia lactiflora* Pall.

Paeonia lactiflora belongs to the family Ranunculaceae and the genus *Paeonia* and is also known as the Chinese peony, common garden peony, and shaoyao. *P. lactiflora* has a long history of use in TCM. Radix Paeoniae Alba and Radix Paeoniae Rubra are both derived from Paeonia roots, but in terms of the percentage of components and pharmacological actions, they are different. Radix Paeoniae Rubra, which is also known as chishao, RPR, and red peony root, is the dried root of *P. lactiflora* and has a cardiovascular and hepatoprotective effect. Radix Paeoniae Alba, which is known as baishao, RPA, and white peony root, is also the dried root of *P. lactiflora,* but there is also boiling and peeling in its production process. It affects the immune and nervous systems. Radix Paeoniae Alba and Radix Paeoniae Rubra both have antitumor and anti-inflammatory effects [339].

Clinical Trials

Total glycoside of paeony (TGP) in the hydroalcoholic extract of *Radix Paeoniae Alba,* which consists of more than 15 components. TGP has shown a direct anti-inflammatory effect by inhibiting the production of nitric oxide, leukotriene B4, and prostaglandin E2 [202].

A study investigated the effect of TGP in SLE patients. In this clinical study, one group of 29 cases received TGP for 5 years or more, the other group of 47 cases received TGP for one year or more (but less than 5 years), and the third group was the control. The results showed that following the daily intake of TGP, the required daily dose of prednisolone and cyclophosphamide decreased, and a decrease in SLEDAI score was observed, but no significant difference was observed in urinary protein. No side effects were observed following TGP use [115].

In another clinical trial, the effect of TGP on SLE patients was investigated on 70 SLE patients, who were divided into control and treatment groups. Both groups used conventional medicine, but the treatment group also received TGP for three months. Finally, it was found that the dose of glucocorticoids required by the patient decreased following the use of TGP. It was concluded that a combination of TGP and glucocorticoids might be beneficial for these patients. The side effects, including gastrointestinal effects, following the use of TGP were tolerable [116].

In a clinical trial, the effect of TGP on lupus nephritis was investigated. Forty patients were in the control group and were treated with prednisolone and cyclophosphamide. Forty patients were in the treatment group and were treated with TGP in addition to prednisolone and cyclophosphamide. As a result, following TGP intake, IL-18 and IL-6 levels decreased more than in the control group, and anti-dsDNA and serum creatinine levels decreased. Albumin and complement C3 levels increased following TGP intake. A meta-analysis on TGP, concluded that TGP is more efficient and safer when used in combination with conventional treatments. It could reduce the disease activity of SLE and the incidence of adverse reactions. Moreover, TGP could improve other outcomes related to SLE disease activity, including complement proteins (C3 and C4), immunoglobulins (IgA, IgM and, IgG), ESR, CRP, 24 h urine protein, and recurrence rate [117].

Phytochemicals and Possible Mechanisms of Action

P. lactiflora has therapeutic applications in lupus due to its anti-inflammatory and immuno modulatory effects. It has been many years since the decoction of the root of *P. lactiflora* has been used to treat SLE, rheumatoid arthritis, etc.

In a study, the effect of 12 weeks of treatment with Radix Paeoniae Rubra on lupus nephritis was investigated. MRL/lpr lupus mice were divided into three groups: control, prednisolone, and Radix Paeoniae Rubra. The results showed a decrease in renal pathological damage and urinary protein levels. Also, the expression of intercellular cell adhesion molecule-1 (ICAM-1), vascular cell adhesion molecule-1 (VCAM-1), and platelet endothelial cell adhesion molecule-1 (PECAM-1) decreased. Moreover, following the use of Radix Paeoniae Rubra, the required dose of prednisolone was reduced [244].

In another study on female MRL/lpr mice, the effect of oral administration of 50 mg/kg/d of TGP was investigated. As a result, ERα expression decreased, and DNA methyltransferases (DNMTs) expression increased. Besides a decrease in renal damage and the expression of IFN-γ, IL6, and IL12 cytokines, the serum dsDNA levels were inhibited [118]. In another study, following the consumption of TGP by MRL/lpr mice, the urinary protein content and the levels of anti-dsDNA antibodies and antinuclear antibodies (ANA) decreased [119].

In an in vitro study, the effect of TGP was investigated on the expression and DNA methylation status of the ITGAL gene (CD11a) in CD4$^+$ T cells isolated from patients with SLE. TGP led to down-regulation of ITGAL mRNA and protein levels. In addition, DNA methylation of the ITGAL promoter was increased, which can result in the repression of the CD11a gene expression [120].

Among the components of TGP, paeoniflorin (Pae), hydroxyl-paeoniflorin, paeonin, albiflorin, and benzoylpaeoniflorin can be named. Paeoniflorin, accounting for about 40% of TGP, is the major active component of TGP. This compound has been shown to modify immune cell activities, reduce inflammatory medium formation, and correct aberrant signal pathways. Paeoniflorin has been found to regulate a variety of signaling pathways, including GPCR pathway, MAPKs/NF-κB pathway, PI3K/Akt/mTOR pathway, JAK2/STAT3 pathway, TGFβ/Smads, etc. [201].

Toxicity and Side Effects

Although there is not a lot of information about *P. lactiflora*'s toxicity, peonies are typically regarded as non-toxic in texts on traditional medicine. According to the research, the roots of paeony species are not toxic, but elevated doses of some of their components, such as pyrethrin

I and phenol, can be harmful [340]. In the section of this article on purified molecules from natural sources, the side effects of this plant and TGP compound have been discussed.

3.4.11. *Scutellaria baicalensis* Georgi

Scutellaria baicalensis belongs to the family Lamiaceae and is also known as Baikal skullcap or Chinese skullcap. Wogonin, baicalin, and baicalein are some of the components of *S. baicalensis*. [248] Baicalin is a flavonoid isolated from the root of *S. baicalensis* and its anti-inflammatory, anti-cancer, and antioxidant effects have been studied [99].

Phytochemicals and Possible Mechanisms of Action

An in vitro study on pristane-induced lupus BALB/c mice found that the use of *S. baicalensis* downregulated the production of proinflammatory cytokines such as TNF-α, IL-6, IL-10, and IFN-γ. Also, the expression of $CD69^{+}$ $CD4^{+}$ T cells and $CD4^{+}$ T cells decreased, but $CD8^{+}$ did not decrease [248].

In another in vivo study, BALB/c mice were divided into three groups: healthy control mice, lupus control mice, and baicalin-treated mice. Mice were treated with 50 mg/kg of baicalin for 10 days. Finally, it was shown that following the use of baicalin, abnormal activation of T cells was downregulated and overproduction of IL-6 and PGE2 was inhibited [99,160].

To evaluate the effect of baicalin, a study was performed on lupus-prone MRL/lpr mice in which 200 mg/kg of baicalin was used peritoneally daily for 4 weeks. It was observed that following the use of baicalin, urine protein decreased, anti-dsDNA antibody titers were inhibited, and lupus nephritis was attenuated. Baicalin inhibits IL-21 production and Tfh cell differentiation and induces $Foxp3^{+}$ regulatory T cell differentiation [98].

Toxicity and Side Effects

Refer to the section on toxicity and side effects of baicalin in this text.

3.4.12. *Gentiana macrophylla* Pall.

Gentiana macrophylla belongs to the Gentianaceae family, which is also known as Qinjiao. The Gentiana genus has protective effects on the cardiovascular, gastrointestinal, and reproductive systems. They have long been recognized for their immunomodulatory properties, as well as their beneficial effects on the skin, liver, and joints. *Gentiana* spp. have been used to control visceral pain [65,66,341]. *Gentiana macrophylla* root has been used in TCM to treat inflammation and pain and systemic lupus erythematosus [239].

Clinical Trial

In a clinical trial, the effect of *G. macrophylla* was compared with that of prednisolone. Sixty-two patients with SLE were treated with *G. macrophylla* complex tablets in the form of 10 tablets twice a day or 5 tablets three times a day with 10 to 30 mg of prednisolone daily. In the control group, 19 SLE patients were treated with prednisolone alone. As a result, it was observed that the recovery rate was significantly higher in the Gentiana group. Improvement of nephropathy, erythema, and arthralgia and restoration of ESR, LE cells, C3, and CH50 following *G. macrophylla* consumption was higher in than the control group, and no significant complication was observed [238].

Animal Studies

In a study, the cardiac protective effect of *G. macrophylla* on lupus mice was investigated. Thirty female NZB/W F1 mice were used, which were divided into three groups that included the control group, the cholesterol-consuming group, and the cholesterol- and *G. macrophylla*-consuming group. After 12 weeks, the mice's heart tissue was used for further experiments. The results showed that following the use of *G. macrophylla*, cholesterol-aggravated apoptosis decreased, IGF-1 survival signal increased, and anti-apoptotic proteins increased. Therefore, *G. macrophylla* protects the heart against cholesterol-aggravated apoptosis in mice with lupus, so it can be useful in the treatment of CVD in patients with SLE [67].

In another study, the hepatoprotective effect of *G. macrophylla* on lupus mice was investigated. The female NZB/W F1 mice were divided into four groups and participated in the experiment for 8 weeks. The livers of mice were used for further experiments. Finally, they found that taking *G. macrophylla* could be effective in reducing liver inflammation in SLE patients [239].

Toxicity and Side Effects

In vivo studies on the Gentiana genus have not revealed any significant toxic effects when given elevated doses [65]. In Kunming mice treated with 500 mg/kg of *G. macrophylla* root extract, no abnormal performance or mortality was observed [342].

3.4.13. *Glycyrrhiza glabra* L.

Glycyrrhiza glabra belongs to the Fabaceae family and is also known as liquorice, sweet wood, and mulaithi [343]. *G. glabra* root contains glycyrrhizin and its derivatives, which are in the class of saponins [344]. *G. glabra* has long been used to treat a wide range of diseases. Studies have shown that it has anti-inflammatory, antioxidant, immunostimulatory, anticoagulant, hepatoprotective, and neuroprotective properties, as well as antibacterial, antiviral, antifungal, and anti-malarial effects [69,345,346].

Phytochemicals and Possible Mechanisms of Action

Licorice root is rich in triterpenoids and flavonoids, which are known for their antioxidant and anti-inflammatory properties, mostly by decreasing TNF, MMPs, PGE2, and free radicals [347].

G. glabra has been reported to be beneficial in improving SLE. High mobility group box 1 (HMGB1) has proinflammatory effects and an immune-stimulatory function and plays a role in the pathogenesis of inflammatory and autoimmune diseases such as SLE. Glycyrrhizin has the blocking effect of HMGB1 [110,348]. Elevated HMGB1 levels have been shown to be associated with exacerbation of SLE and elevated levels of proinflammatory cytokines such as IL-6 and TNF-α. The results of an animal study on female BALB/c mice that received glycyrrhizin (0.5 mg/day for two months) demonstrated that glycyrrhizin's inhibition of HMGB1 function caused a sharp decline in serum HMGB1 levels, which in turn decreased the severity of SLE [110]. Glycyrrhizin directly binds to HMGB1, reducing both the extracellular release of HMGB1 and its cytokine actions [348].

In another study based on HMGB1, glycyrrhizin was used as an HMGB1 blocker. Glycyrrhizin (10 mg/kg) was injected every other three days into female BALB/c mice for 12 weeks. Finally, an improvement in lupus nephritis was achieved following the use of glycyrrhizin. Decreased levels of anti-dsDNA antibodies and decreased levels of inflammatory cytokines, as well as decreased glomerular IgG and C3 deposition and reduction of proteinuria, were observed [349].

In an in vitro study, glycyrrhizin was shown to inhibit the immunocomplex formation of 60S acidic ribosomal P proteins from porcine liver when combined with patient serum for SLE. It was concluded that a relatively high dose of glycyrrhizin could prevent the immunocomplex formation of 60S acidic ribosomal P proteins with their specific antibodies in the sera of SLE patients [112].

No clinical trial was found on the efficacy of licorice on SLE. Considering the results of preclinical studies and the application of licorice as an anti-inflammatory agent in a variety of diseases, further studies might lead researchers to find effective remedies.

Toxicity and Side Effects

The most significant adverse effects of glycyrrhizin and licorice are secondary diseases brought on by hypokalemia and hypertension. Additionally, it may result in fatal arrhythmias and cardiomyopathy. Hypokalemia, hypertension, anorexia nervosa, prolonged gastrointestinal, advanced age, and being of female sex all enhance the risk of adverse consequences from licorice. It should be mentioned that the positive effects of beta blockers and angiotensin-converting enzyme inhibitors (ACEIs) can be countered by the hypertensive impact of licorice [350,351].

3.4.14. *Antrodia camphorata*

Antrodia camphorata is a fungal parasite on *Cinnamomum kanehirai.* It is also known as "stout camphor fungus". *Antrodia camphorata* has long been used in traditional medicine in China and Taiwan and is used to treat a wide range of diseases. Crude extracts of *A. camphorata* have been shown to have anti-cancer, antioxidant, anti-inflammatory, immunomodulatory, hepatoprotective, neuroprotective, anti-hypertensive, and vasorelaxant effects [153].

A study investigated the effect of *A. camphorata* on nephritis in SLE-prone NZB/W F1 mice. For this purpose, for 12 weeks, 100, 200, and 400 mg/kg of *A. camphorata* extract were administered orally on 5 consecutive days per week. As a result, the kidney glomerular basement membrane's thickness was reduced and urine protein and serum BUN levels were markedly controlled by the extract of *A. camphorate* (400 mg/kg) [46]. Antroquinonol was reported to be the main active ingredient (refer to the antroquinonol section in this text).

Toxicity and Side Effects

Refer to the section on toxicity and side effects of antroquinonol in this text.

3.4.15. *Astragalus propinquus* Schischkin (syn. *Astragalus membranaceus* (Fisch.) Bunge)

Astragalus membranaceus belongs to the Fabaceae family and is also known as Mongolian milkvetch. It consists of the components astragaloside, astragalus flavonoids, and astragalus polysaccharide. It has an anti-inflammatory effect and reduces proteinuria and creatinine. It has been used to treat kidney disease [352]. *A. membranaceus* is used in lupus nephritis and its effects have been reviewed in articles. In a study, the effect of *A. membranaceus* and *Tripterygium hypoglancum* on SLE patients was investigated, and it was observed that NK activity decreased and therefore disease activity decreased [229]. A bioinformatics study has attempted to investigate the mechanism of the effect of *A. membranaceus* on lupus nephritis [352].

Toxicity and Side Effects

A. membranaceus has an LD_{50} of about 40 g/kg, which makes it safe and non-toxic. However, it has been found through in vivo research that a dose of 1 mg/kg of this plant in the form of AS-IV can have side effects such as fetal toxicity and reproductive toxicity. As a result, it should be used carefully throughout pregnancy and the postpartum period. However, astragalus extract is generally safe and has no significant side effects [353].

3.4.16. *Bryophyllum pinnatum* (Lam.) Oken

Bryophyllum pinnatum belongs to the Crassulaceae family and is also known as Kalanchoe pinnata, air plant, Zakham-e-hyat, life plant, and cathedral bells. The constituents of *B. pinnatum* include alkaloids, glycosides, triterpenes, cardienolides, flavonoids, steroids, lipids, and buffadienolides. It is used for a wide range of diseases and has been found to have antibacterial, antileishmaniasis, and antimutagenic effects, as well as hepatoprotective and nephroprotective effects. *B. Pinnatum* can be beneficial in the treatment of SLE due to its anti-inflammatory and immunosuppressive effects [354]. The effect of *B. pinnatum* on SLE has been investigated.

In a study on BALB/c mice, mice were treated with different doses of ethanolic extract of *B. pinnatum* leaves for 12 weeks. For this purpose, 4 groups including control and those with doses of 10.5, 21, and 42 mg/kg/day were used. As a result, following the consumption of *B. pinnatum*, a decrease in TNF-α, IL-17, IL-12, CRP, and matured B cells was observed. On the other hand, there was an increase in complement C3 and C4 and TGF-β. No specific side effects were reported with *B. pinnatum* [230,231].

In an in vitro study performed with the help of spleen cells of BALB/c mice, the effect of ethanolic extract of *B. pinnatum* leaves at doses of 0, 0.02, 0.1, or 0.5 μg/mL was investigated. As a result, *B. pinnatum* reduces B cell maturation and increases B cell apoptosis and decreases

NF-κB p65 expression [232]. The effect of *B. pinnatum* on B cells has also been investigated in silico [355,356].

In vivo, the effect of aqueous extract of *B. pinnatum* leaf on lupus nephritis in female Balb/c mice was performed. Mice in the treatment group received 200, 400, or 600 mg/kg/day aqueous extract of *B. pinnatum* orally for 21 days. Eventually, it was found that proteinuria levels and glomerular inflammation were reduced. On the other hand, with the help of in silico studies, an attempt has been made to find the flavonoid composition that binds to the glucocorticoid receptor. It has been shown that bryophyllin A is probably the active compound of *B. pinnatum* that has an anti-inflammatory effect [233]. There is also further evidence and studies that investigate the effect of *B. pinnatum* on SLE and lupus nephritis [357–360].

Toxicity and Side Effects

Toxicological tests have been carried out mostly on leaf extracts to determine the safety of *B. pinnatum*. Even though the majority of research works have demonstrated low toxicity and acceptable safety, some have noted its cytotoxicity. For this plant, abnormalities in the animal's testicles have also been observed. As a result, it's seemed that *B. pinnatum* can be used safely in acute situations, but further research is required to determine its chronic toxicity [361].

3.4.17. *Anemarrhenae aspheloidis*

Anemarrhenae aspheloidis belongs to the family Asparagaceae and is also known as Zhi Mu. Anemarrhenae rhizoma has been used in TCM for many years to treat various ailments. One of the compounds of anemarrhenae rhizoma is mangiferin, which has antioxidant, anti-inflammatory, and immunomodulatory effects. The effect of mangiferin extracted from anemarrhenae rhizomaand *Mangifera indica* on lupus nephritis has been investigated (see the section on mangiferin in this text) [362].

Toxicity and Side Effects

Refer to the section on toxicity and side effects of mangiferin in this text.

3.4.18. *Camellia sinensis* (L.) Kuntze

Camellia sinensis belongs to the family Theaceae and is generally known as tea. The compounds in *C. sinensis* depend on various factors such as geographical environment, growing season, etc., but generally contain flavanols, flavonols, polyphenolic acids, and flavonol glycosides. *C. sinensis* has anti-inflammatory, antioxidant, and anti-arthritis effects [363]. The effect of *C. sinensis* on lupus has been studied. In an in silco study, the potency of green tea phytoconstituents as immunomodulators, anti-apoptosis agents, and anti-pyroptosis agents in SLE was investigated. The result of molecular docking can explain the mechanism of the active compound as anti-apoptosis and anti-pyroptosis. The docking results suggested theaflavin as one of the most active constituents [364].

Clinical Trial

The effect of *C. sinensis* on the improvement of SLE has been investigated due to its anti-inflammatory and immunomodulatory effects. In this clinical trial, 68 SLE patients were divided into control and study groups. The study group was treated with 1000 mg of *C. sinensis* extract daily for 12 weeks. Finally, it was observed that following the consumption of *C. sinensis*, the quality of life of patients and their general health increased and their disease activity decreased [234].

Toxicity and Side Effects

Although *C. sinensis* is one of the most commonly used and safe plants, a number of side effects have been reported for its excessive consumption, including diuresis, tremors, irritability during the day, heart irregularities, nervousness, anxiety, headache, and hypotension. In patients with anxiety, poor cardiovascular systems, renal disorders, and hyperthyroidism, it has been suggested that tea consumption be restricted [365].

3.4.19. *Curcuma longa* L.

Curcuma longa belongs to the Zingiberaceae family and is also known as turmeric. *C. longa* is widely used in Asian traditional medicine due to its medicinal properties. Since curcumin can interact with different molecular and cellular targets, it exhibits a wide range of pharmacological effects including anti-inflammatory, antioxidant, antimicrobial and chemotherapeutic activity [21]. It also has hepatoprotective effects and is useful in gastrointestinal disorders [366]. The effect of *C. longa* on lupus has been studied. One of the most active constituents is curcumin (see the section on curcumin in this text).

Clinical Trials

In a clinical trial, the effect of turmeric on the improvement of lupus nephritis has been investigated. For this purpose, 24 patients with lupus nephritis were divided into study and control groups. The study group was treated with 1500 mg of turmeric daily for 3 months. As a result, following turmeric consumption, proteinuria and hematuria decreased, and systolic blood pressure also decreased. No side effects were observed following short-term use of turmeric in this clinical trial [235].

Toxicity and Side Effects

Refer to the section on toxicity and side effects of curcumin in this text.

3.5. *Other Plants with Lower Evidence for Lupus Conditions*

In addition to the herbs listed so far, other herbs that have anti-inflammatory, antioxidant, or immunomodulatory effects can help improve SLE. By using these herbs with conventional treatment, the patient might experience fewer adverse effects, and a lower dosage of conventional medicine might be needed.

Cinchona officinalis L. belongs to the family Rubiaceae and is known as cinchona and Peruvian bark. Important components in *C. officinalis* include alkaloids such as Quinine, Quinidine, Chichonine, and Cinchonidine. It has anti-malarial, anti-inflammatory, antioxidant, antimicrobial, and anti-cancer effects and is used to treat lupus, malaria, etc. [367,368]. The plant and its phytoconstituents were applied to treat COVID-19 due to its anti-inflamm atory and immunomodulatory properties [369,370].

Bupleurum falcatum L. (Apiaceae) is known as Chinese thoroughwax and sickle-leaf hare's ear. Saikosaponins extracted from *B. falcatum* have anti-inflammatory, immunoregulatory, and anti-cancer effects. Investigating the potency of Saikosaponins to attenuate symptoms in autoimmune diseases such as lupus can be beneficial [371].

Centella asiatica (L.) Urb. (Apiceae family) is known as *Hydrocotyle asiatica, Indischer Wassernabel,* and Indian pennywort. [372] It contains triterpene saponins, madecassic acid, asiatic acid, asiaticoside, and madecassoside. It is mostly used to heal wounds, but it has also been used for skin conditions caused by lupus, leprosy, eczema, psoriasis, varicose ulcers, etc. [373].

Tinospora sinensis (Lour.) Merr. (syn. *Tinospora cordifolia*) belongs to the family Menispermaceae and is also known as guduchi, heart-leaved moonseed, and gurjo. This plant can be used in SLE due to its immunomodulatory effect. It has been reported for its anti-inflammatory, immunomodulatory, antioxidant, antibacterial, antifungal, and anti-diabetic effects. It has been used in some inflammatory and autoimmune diseases such as SLE, rheumatoid arthritis, psoriasis, etc. [374].

Acacia farnesiana (L.) Willd. (Fabaceae) is known as sweet acacia, needle bush, and huisache. *A. farnesiana,* due to its proteins, lectin, and α-amyrin, β-amyrin, and lupeol have anti-inflamma tory and antioxidant effects and can down-regulate proinflammatory mediators [375–377].

Morinda citrifolia L. belongs to the family Rubiaceae and is also known as noni. In addition to its nutritional value, *M. citrifolia* has several biological activities, including, antibacterial, antifungal, anti-inflammatory, antioxidant, and antituberculosis effects, which have been reported for this plant and its phytochemicals [378]. *M. citrifolia* has also been claimed to be beneficial in treating lupus [379,380].

Cornus officinalis Siebold and Zucc. belongs to the Cornaceae family and is known as Asiatic Dogwood, Japanese Cornel Dogwood, and Shan Zhu Yu. Its ripe and dried fruit is called Corni Fructus. Corni Fructus has been used in traditional Chinese medicine (TCM) for a variety of conditions. It has exhibited antioxidant, anti-inflammatory, nephroprotective, hepatoprotective, neuroprotective, hypoglycemic, and anti-cancer effects [381]. The combination of Corni Fructus with other plants of traditional Chinese medicine has been used to treat lupus, although we could not find a study that has specifically examined the effect of Corni Fructus on SLE. It has been assumed that, owing to its anti-inflammatory and nephroprotective effects, it can be beneficial for patients with lupus. There are studies on the mechanism of its anti-inflammatory and kidney protective effect [382–384].

Allium sativum L. (Alliaceae) is known as garlic. Alliin and allicin are the most important sulfur components in *A. sativum*. It has exhibited a wide range of therapeutic properties, including anti-inflammatory, antioxidant, antimicrobial, antifungal, immunomodulatory, antiatherosclerotic, anti-hypertensive effects, etc. [385] *A. sativum* has also been suggested to be useful in improving lupus patients [386].

Wolfiporia extensa, or *Poria cocos*, is a fungus that belongs to the family Polyporaceae and grows on the roots of the pine tree. *Poria cocos* has long been used extensively in traditional Chinese medicine and is also known as Fuling, poria, and hoelen. Due to the presence of triterpenoid and ergosterol compounds, it has anti-inflammatory effects and also affects the immune system, so it is used in diseases such as rheumatoid arthritis and SLE. In addition to anti-inflammation and immunomodulation effects, this fungus also has antitumor, anti-diabetic, anti-aging, and antioxidant effects [387–389].

Evening primrose oil (EPO) can also be helpful in the healing process of lupus and is obtained from the seeds of *Oenothera biennis* L., which belongs to the Onagraceae family. The plant is also known as evening primrose, evening star, and sundrop. Evening primrose oil (EPO) contains a high concentration of γ-linolenic acid (GLA), which has anti-inflammatory, antioxidant, radical scavenging, and immunomodulatory properties [390,391]. Some clinical trials have shown that EPO can be beneficial in improving some inflammatory conditions such as atopic eczema and atopic dermatitis. There are studies on the effect of EPO on arthritis [391,392]. Due to the presence of γ-linolenic acid and prostaglandin E1, its use can be effective in SLE patients.

Andrographis paniculata (Burm.f.) Nees belongs to the family Acanthaceae and is also known as green chiretta and creat and Chuan Xin Lian. The aerial parts of this plant have been used in traditional Chinese medicine to treat inflammation, pain, and detoxification. *A. paniculata* contains flavonoids, polyphenols, and diterpenoids [393]. Immunomodulatory effects have been reported for the aqueous extract of *A. paniculata* leaves on rats [394].

Phyllanthus emblic G.L.Webster belongs to the family Euphorbiaceae. The genus *Phyllanthus* has long been used extensively in immune-related diseases such as SLE and has been shown to have an immunomodulatory effect [395].

Coriandrum sativum L. (Apiaceae): The essential oils in the seeds and leaves of this plant, such as linalool, are responsible for many of the plant's benefits. *C. sativum* has anti-inflammatory, antioxidant, antibacterial, anti-cancer, immunostimulatory effects, etc. [396]. Consumption of this plant could be effective in lupus patients due to its anti-inflammatory and immunomodulatory effects [30].

Boswellia spp. (Burseraceae): The four main species of Boswellia, *B. sacra*, *B. frereana*, *B. papyrifera*, and *B. serrata* produce frankincense (also known as olibanum). Among the chemical compounds of frankincense 3-O-acetyl-11-keto-β boswellic acid, α- and β-boswellic acids, 11-keto-β-boswellic acid and other boswellic acids, lupeolic acids, incensole, cembrenes, triterpenediol, tirucallic acids, and olibanumols can be named. Frankincense exhibits anti-inflamm atory effects through a variety of mechanisms including inhibition of leukotriene synthesis, cyclooxygenase 1/2 and 5-lipoxygenase, and oxidative stress and by regulation of immune cells from the innate and acquired immune systems. Additionally, it modifies signal transduction, which is in charge of cell cycle arrest as well as the suppression of proliferation, angiogenesis, invasion, and metastasis [397]. The major components of frankincense are boswellic acids, among which the most important and abundant is 3-O-acetyl-11-keto-β-boswellic acid (AKBA). This compound is a strong inhibitor of 5-lipoxygenase with anti-inflammatory and anti-arthritic properties [398]. Another suggested mechanism for the anti-inflammatory effects of boswellic acid is potently inhibiting cathepsin G [399].

In clinical trials, frankincense and its phytochemicals were found to be effective in treating a range of inflammatory disorders, such as osteoarthritis, multiple sclerosis, asthma, psoriasis and erythematous dermatitis, plaque-induced gingivitis, and pain [397,400,401]. We could not find a clinical trial investigating the efficacy of frankincense on SLE, but due to its anti-inflammatory and immunomodulatory effects, it might be poetically effective.

Dioscorea polystachya Turcz. (syn. *Dioscorea batatas* Decne.) belongs to the Dioscoreaceae family and is also known as Chinese yam or cinnamon-vine. In addition to being used as food in China, it has also been used in traditional Chinese medicine to treat various diseases such as asthma, diabetes, etc. It has anti-inflammatory and antioxidant effects [402,403] and has also been used to treat lupus [402–404].

Ocimum gratissimum L. belongs to the Lamiaceae family and is also known as Ram Tulshi, clove basil, and African basil. In addition to being used in food in some countries, this plant is also used in the treatment of some diseases. *O. gratisimum* has anti-malarial, anti-inflammatory, and antioxidant effects [405]. Owing to its immunomodulatory and anti-inflammatory effects, it can be effective in treating SLE patients [30,406].

Uncaria tomentosa (Willd. ex Schult.) DC. belongs to the Rubiaceae family and is also known as cat's claw. The roots and bark of this plant have been used in traditional medicine to treat various diseases such as inflammation, viral infections, urinary tract infections, asthma, etc. *U. tomentosa* has anti-inflammatory, antioxidant, antimicrobial, and immunom odulatory effects and can be effective in treating lupus [407,408]. There is a case report that an SLE patient has ended up with acute renal failure following daily use of *U. tomentosa* [409].

Clerodendrum trichotomum Thunb. (Verbenaceae) is known as Chou Wu Tong and Kusagi. *C. trichotomum* has antioxidant, anti-inflammatory, analgesic, and sedative effects [410,411]. Due to the presence of taraxerol, friedelin, lupeol, and betulinic acid, it has an immunomodulatory effect [30].

Scrophularia ningpoensis Hemsl. belongs to the family Scrophulariaceae and is also known as Xuanshen. In traditional Chinese medicine, it has been used for many years to treat various diseases. Among the components of *S. ningpoensis*, phenylpropanoid glycosides and iridoid glycosides can be named to possess anti-inflammatory effects. The plant has been used to treat liver disease, cardiovascular disease, and diabetes and has antioxidant, anti-inflammatory, and anticarcinogenic effects [412]. Its effectiveness in lupus has also been considered [413].

4. Conclusions

Overall, a variety of natural molecules and their derivatives, in purified and structurally elucidated form, have been reported to exhibit beneficial effects in lupus conditions. Among these molecules, artemisinin and its derivatives, antroquinonol, baicalin, curcumin, emodin, mangiferin, salvianolic acid A, triptolide, and the total glycosides of paeony (TGP) have potential to be considered for further drug development studies. They showed their efficacy through interaction with various immune mediators, cytokines, and transcription factors such as nuclear factor kappa B (NF-κB), inhibition of anti-dsDNA, etc. Additionally, some omega-6 and omega-3 PUFAs, such as EPA, DHA, α-linolenic acid, and γ-linolenic acid, have been shown to be beneficial by lowering anti-dsDNA, TNF-α, IL-1, IL-1, IL-2, and/or CRP levels.

Some minerals (calcium, iron, selenium, and zinc) and vitamins (vitamins A, B, C, D, and E) have shown potential to exhibit degrees of beneficial effects. Considering the reported clinical trials on medicinal plants and fungi, *T. wilfordii*, *O. sinensis*, *G. lucidum*, *A. annua*, *U. dioica*, *L. usitatissimum*, *R. glutinosa*, *P.* × *suffruticosa*, *P. lactiflora*, *G. macrophylla*, and *C. longa* have exhibited efficacy against lupus conditions. Due to the small number of clinical trials and the small number of patents in each trial, it is not possible to do a meta-analysis on any of the phytochemicals or herbal medicines listed above. However, since many of these herbal medicines, minerals, or vitamins have a history of human consumption, until more studies are done, the use of these products in the form of complementary products can be beneficial for patients.

Author Contributions: Conceptualization, A.H., A.P., and B.H.; methodology, A.H., A.P., and B.H.; validation, A.H., A.P., and B.H.; investigation, A.H., A.P., A.T., E.K., and B.H.; writing—original draft preparation, A.H., A.P., A.T., E.K., and B.H.; writing—review and editing, A.H.; project administration, A.H. All authors have read and agreed to the published version of the manuscript.

Funding: The authors received no financial support for the research, authorship, and/or publication of this article, except Ekaterina Kozuharova, who is grateful for the financial support for the research to European Union—NextGenerationEU—through the National Recovery and Resilience Plan of the Republic of Bulgaria, Project No. BG-RRP-2.004-0004-C01.

Institutional Review Board Statement: Not applicable.

Informed Consent Statement: Not applicable.

Data Availability Statement: Not applicable.

Conflicts of Interest: The authors declare no conflict of interest.

References

1. Dörner, T.; Furie, R. Novel paradigms in systemic lupus erythematosus. *Lancet* **2019**, *393*, 2344–2358. [CrossRef] [PubMed]
2. Veeranki, S.; Choubey, D. Systemic lupus erythematosus and increased risk to develop B cell malignancies: Role of the p200-family proteins. *Immunol. Lett.* **2010**, *133*, 1–5. [CrossRef] [PubMed]
3. Lightfoot, Y.L.; Blanco, L.P.; Kaplan, M.J. Metabolic abnormalities and oxidative stress in lupus. *Curr. Opin. Rheumatol.* **2017**, *29*, 442. [CrossRef] [PubMed]
4. Bertsias, G.K.; Pamfil, C.; Fanouriakis, A.; Boumpas, D.T. Diagnostic criteria for systemic lupus erythematosus: Has the time come? *Nat. Rev. Rheumatol.* **2013**, *9*, 687–694. [CrossRef] [PubMed]
5. Bootsma, H.; Spronk, P.; de Boer, G.; Limburg, P.; Kallenberg, C.; Derksen, R.; Wolters-Dicke, J.; Gmelig-Meyling, F.; Kater, L.; Hermans, J. Prevention of relapses in systemic lupus erythematosus. *Lancet* **1995**, *345*, 1595–1599. [CrossRef]
6. Loram, L.C.; Culp, M.E.; Connolly-Strong, E.C.; Sturgill-Koszycki, S. Melanocortin peptides: Potential targets in systemic lupus erythematosus. *Inflammation* **2015**, *38*, 260–271. [CrossRef]
7. Fanouriakis, A.; Tziolos, N.; Bertsias, G.; Boumpas, D.T. Update on the diagnosis and management of systemic lupus erythematosus. *Ann. Rheum. Dis.* **2021**, *80*, 14–25. [CrossRef]
8. Mikdashi, J.; Nived, O. Measuring disease activity in adults with systemic lupus erythematosus: The challenges of administrative burden and responsiveness to patient concerns in clinical research. *Arthritis Res. Ther.* **2015**, *17*, 1–10. [CrossRef]
9. Almaani, S.; Meara, A.; Rovin, B.H. Update on lupus nephritis. *Clin. J. Am. Soc. Nephrol.* **2017**, *12*, 825–835. [CrossRef]
10. Hoover, P.J.; Costenbader, K.H. Insights into the epidemiology and management of lupus nephritis from the US rheumatologist's perspective. *Kidney Int.* **2016**, *90*, 487–492. [CrossRef]
11. Wallace, D.J. The evolution of drug discovery in systemic lupus erythematosus. *Nat. Rev. Rheumatol.* **2015**, *11*, 616–620. [CrossRef] [PubMed]
12. Basta, F.; Fasola, F.; Triantafyllias, K.; Schwarting, A. Systemic lupus erythematosus (SLE) therapy: The old and the new. *Rheumatol. Ther.* **2020**, *7*, 433–446. [CrossRef] [PubMed]
13. Kuhn, A.; Bonsmann, G.; Anders, H.-J.; Herzer, P.; Tenbrock, K.; Schneider, M. The diagnosis and treatment of systemic lupus erythematosus. *Dtsch. Ärzteblatt Int.* **2015**, *112*, 423. [CrossRef] [PubMed]
14. Gurevitz, S.; Snyder, J.; Wessel, E.; Frey, J.; Williamson, B. Systemic lupus erythematosus: A review of the disease and treatment options. *Consult. Pharm.* **2013**, *28*, 110–121. [CrossRef]
15. Tanaka, Y. State-of-the-art treatment of systemic lupus erythematosus. *Int. J. Rheum. Dis.* **2020**, *23*, 465–471. [CrossRef] [PubMed]
16. Liossis, S.N.; Staveri, C. What's new in the treatment of systemic lupus erythematosus. *Front. Med.* **2021**, *8*, 655100. [CrossRef] [PubMed]
17. Verdelli, A.; Corrà, A.; Mariotti, E.B.; Aimo, C.; Ruffo di Calabria, V.; Volpi, W.; Quintarelli, L.; Caproni, M. An update on the management of refractory cutaneous lupus erythematosus. *Front. Med.* **2022**, *9*, 941003. [CrossRef]
18. Schwartz, N.; Stock, A.D.; Putterman, C. Neuropsychiatric lupus: New mechanistic insights and future treatment directions. *Nat. Rev. Rheumatol.* **2019**, *15*, 137–152. [CrossRef]
19. Lubov, J.E.; Jamison, A.S.; Baltich Nelson, B.; Amudzi, A.A.; Haas, K.N.; Richmond, J.M. Medicinal plant extracts and natural compounds for the treatment of cutaneous lupus erythematosus: A systematic review. *Front. Pharmacol.* **2022**, *13*, 802624. [CrossRef]
20. Cao, F.; Cheng, M.-H.; Hu, L.-Q.; Shen, H.-H.; Tao, J.-H.; Li, X.-M.; Pan, H.-F.; Gao, J. Natural products action on pathogenic cues in autoimmunity: Efficacy in systemic lupus erythematosus and rheumatoid arthritis as compared to classical treatments. *Pharmacol. Res.* **2020**, *160*, 105054. [CrossRef]
21. Momtazi-Borojeni, A.A.; Haftcheshmeh, S.M.; Esmaeili, S.-A.; Johnston, T.P.; Abdollahi, E.; Sahebkar, A. Curcumin: A natural modulator of immune cells in systemic lupus erythematosus. *Autoimmun. Rev.* **2018**, *17*, 125–135. [CrossRef] [PubMed]

22. Duarte-Garcia, A.; Myasoedova, E.; Karmacharya, P.; Hocaoğlu, M.; Murad, M.H.; Warrington, K.J.; Crowson, C.S. Effect of omega-3 fatty acids on systemic lupus erythematosus disease activity: A systematic review and meta-analysis. *Autoimmun. Rev.* **2020**, *19*, 102688. [CrossRef] [PubMed]
23. Yuan, K.; Li, X.; Lu, Q.; Zhu, Q.; Jiang, H.; Wang, T.; Huang, G.; Xu, A. Application and mechanisms of triptolide in the treatment of inflammatory diseases—A review. *Front. Pharmacol.* **2019**, *10*, 1469. [CrossRef] [PubMed]
24. Vahedi-Mazdabadi, Y.; Saeedi, M. Treatment of Lupus Nephritis from Iranian Traditional Medicine and Modern Medicine Points of View: A Comparative Study. *Evid.-Based Complement. Altern. Med.* **2021**, *2021*, 6645319. [CrossRef]
25. Gururaja, D.; Hegde, V. Ayurvedic management of systemic lupus erythematosus overlap vasculitis. *J. Ayurveda Integr. Med.* **2019**, *10*, 294–298. [CrossRef]
26. Salek, M.; Hosseini Hooshiar, S.; Salek, M.; Poorebrahimi, M.; Jafarnejad, S. Omega-3 fatty acids: Current insights into mechanisms of action in systemic lupus erythematosus. *Lupus* **2023**, *32*, 7–22. [CrossRef]
27. Huerta, M.D.R.; Trujillo-Martin, M.M.; Rúa-Figueroa, Í.; Cuellar-Pompa, L.; Quiros-Lopez, R.; Serrano-Aguilar, P.; Spanish, S. Healthy lifestyle habits for patients with systemic lupus erythematosus: A systemic review. *Semin. Arthritis Rheum.* **2016**, *45*, 463–470. [CrossRef]
28. Zheng, R.; Gonzalez, A.; Yue, J.; Wu, X.; Qiu, M.; Gui, L.; Zhu, S.; Huang, L. Efficacy and safety of vitamin D supplementation in patients with systemic lupus erythematosus: A meta-analysis of randomized controlled trials. *Am. J. Med. Sci.* **2019**, *358*, 104–114. [CrossRef]
29. Islam, M.A.; Khandker, S.S.; Alam, S.S.; Kotyla, P.; Hassan, R. Vitamin D status in patients with systemic lupus erythematosus (SLE): A systematic review and meta-analysis. *Autoimmun. Rev.* **2019**, *18*, 102392. [CrossRef]
30. Balkrishna, A.; Thakur, P.; Singh, S.; Chandra Dev, S.N.; Varshney, A. Mechanistic paradigms of natural plant metabolites as remedial candidates for systemic lupus erythromatosus. *Cells* **2020**, *9*, 1049. [CrossRef]
31. Thomas, D.E. *The Lupus Encyclopedia: A Comprehensive Guide for Patients and Families*; JHU Press: Baltimore, MD, USA, 2014.
32. Fu, S.M.; Gaskin, F. History of systemic lupus erythematosus with an emphasis on certain recent major issues. In *Systemic Lupus Erythematosus*; Elsevier: Amsterdam, The Netherlands, 2021; pp. 3–8.
33. Tene, V.; Malagon, O.; Finzi, P.V.; Vidari, G.; Armijos, C.; Zaragoza, T. An ethnobotanical survey of medicinal plants used in Loja and Zamora-Chinchipe, Ecuador. *J. Ethnopharmacol.* **2007**, *111*, 63–81. [CrossRef] [PubMed]
34. Kyaw, M.S.; Aye, M.M.; Grinnell, M.; Rabach, M. Traditional and ethnobotanical dermatology practices in Myanmar. *Clin. Dermatol.* **2018**, *36*, 320–324. [CrossRef] [PubMed]
35. Magee, A.; Van Wyk, B.-E.; Van Vuuren, S. Ethnobotany and antimicrobial activity of sieketroos (*Arctopus species*). *S. Afr. J. Bot.* **2007**, *73*, 159–162. [CrossRef]
36. Orhan, N.; Akkol, E.; Ergun, F. Evaluation of antiinflammatory and antinociceptive effects of some Juniperus species growing in Turkey. *Turk. J. Biol.* **2012**, *36*, 719–726. [CrossRef]
37. Gilca, M.; Tiplica, G.S.; Salavastru, C.M. Traditional and ethnobotanical dermatology practices in Romania and other Eastern European countries. *Clin. Dermatol.* **2018**, *36*, 338–352. [CrossRef]
38. Verma, R.; Singh, H.; Thakur, A.; Kohli, S. Ethnobotanical survey of medicinal and aromatic plants of Bhagalpur Region. *Int. J. Appl. Sci. Biotechnol.* **2020**, *8*, 216–222. [CrossRef]
39. Sharififar, F.; Yassa, N.; Mozaffarian, V. Bioactivity of major components from the seeds of *Bunium persicum* (Boiss.) Fedtch. *Pak. J. Pharm. Sci.* **2010**, *23*, 300–304.
40. Abou El-Soud, N.H.; El-Lithy, N.A.; El-Saeed, G.; Wahby, M.S.; Khalil, M.Y.; Morsy, F.; Shaffie, N. Renoprotective effects of caraway (*Carum carvi* L.) essential oil in streptozotocin induced diabetic rats. *J. Appl. Pharm. Sci.* **2014**, *4*, 027–033. [CrossRef]
41. Aghili, M.H. *Makhzan-al-Advia*; Tehran University of Medical Sciences: Tehran, Iran, 2009; pp. 227–228. (In Persian)
42. Ardakani, M.R.S.; Farjadmand, F.; Rahimi, R. Makhzan al adviyeh and pointing to the scientific names of medicinal plants for the first time in a persian book. *Tradit. Integr. Med.* **2018**, *3*, 186–195.
43. Seri, A.; Khorsand, M.; Rezaei, Z.; Hamedi, A.; Takhshid, M.A. Inhibitory effect of bunium persicum hydroalcoholic extract on glucose-induced albumin glycation, oxidation, and aggregation in vitro. *Iran. J. Med. Sci.* **2017**, *42*, 369.
44. Mehrabadi, M.M.; Zarshenas, M.M. A Concise Overview of Phytochemistry, Pharmacology and Clinical Aspects of Persian Cumin; *Bunium persicum* (Boiss.) B. Fedtsch. *Curr. Drug Discov. Technol.* **2021**, *18*, 485–491. [CrossRef] [PubMed]
45. Bansal, S.; Sharma, K.; Gautam, V.; Lone, A.A.; Malhotra, E.V.; Kumar, S.; Singh, R. A comprehensive review of *Bunium persicum*: A valuable medicinal spice. *Food Rev. Int.* **2021**, *39*, 1184–1202. [CrossRef]
46. Chang, J.-M.; Lee, Y.-R.; Hung, L.-M.; Liu, S.-Y.; Kuo, M.-T.; Wen, W.-C.; Chen, P. An extract of *Antrodia camphorata* mycelia attenuates the progression of nephritis in systemic lupus erythematosus-prone NZB/W F1 mice. *Evid.-Based Complement. Altern. Med.* **2011**, *2011*, 465894. [CrossRef] [PubMed]
47. Yang, Y.; Han, C.; Sheng, Y.; Wang, J.; Li, W.; Zhou, X.; Ruan, S. *Antrodia camphorata* polysaccharide improves inflammatory response in liver injury via the ROS/TLR4/NF-κB signal. *J. Cell. Mol. Med.* **2022**, *26*, 2706–2716. [CrossRef]
48. Kushairi, N.; Tarmizi, N.A.K.A.; Phan, C.W.; Macreadie, I.; Sabaratnam, V.; Naidu, M.; David, P. Modulation of neuroinflammatory pathways by medicinal mushrooms, with particular relevance to Alzheimer's disease. *Trends Food Sci. Technol.* **2020**, *104*, 153–162. [CrossRef]
49. Septembre-Malaterre, A.; Lalarizo Rakoto, M.; Marodon, C.; Bedoui, Y.; Nakab, J.; Simon, E.; Hoarau, L.; Savriama, S.; Strasberg, D.; Guiraud, P.; et al. *Artemisia annua*, a Traditional Plant Brought to Light. *Int. J. Mol. Sci.* **2020**, *21*, 4986. [CrossRef]
50. Alesaeidi, S.; Miraj, S. A Systematic Review of Anti-malarial Properties, Immunosuppressive Properties, Anti-inflammatory Properties, and Anti-cancer Properties of *Artemisia Annua*. *Electron. Physician* **2016**, *8*, 3150–3155. [CrossRef]

51. Durazzo, A.; Nazhand, A.; Lucarini, M.; Silva, A.M.; Souto, S.B.; Guerra, F.; Severino, P.; Zaccardelli, M.; Souto, E.B.; Santini, A. *Astragalus (Astragalus membranaceus Bunge)*: Botanical, geographical, and historical aspects to pharmaceutical components and beneficial role. *Rend. Lincei. Sci. Fis. Nat.* **2021**, *32*, 625–642. [CrossRef]
52. Chibli, L.A.; Rodrigues, K.C.; Gasparetto, C.M.; Pinto, N.C.; Fabri, R.L.; Scio, E.; Alves, M.S.; Del-Vechio-Vieira, G.; Sousa, O.V. Anti-inflammatory effects of *Bryophyllum pinnatum* (Lam.) Oken ethanol extract in acute and chronic cutaneous inflammation. *J. Ethnopharmacol.* **2014**, *154*, 330–338. [CrossRef]
53. Smith, G.F.; Figueiredo, E.; van Wyk, A.E. Chapter 12—Taxonomic Treatment. In *Kalanchoe (Crassulaceae) in Southern Africa*; Smith, G.F., Figueiredo, E., van Wyk, A.E., Eds.; Academic Press: Cambridge, MA, USA, 2019; pp. 131–303.
54. Mule, P.; Upadhye, M.; Taru, P.; Dhole, S. A Review on *Bryophyllum pinnatum* (Lam.) Oken. *Res. J. Pharmacogn. Phytochem.* **2020**, *12*, 111. [CrossRef]
55. Sánchez, M.; González-Burgos, E.; Iglesias, I.; Lozano, R.; Gómez-Serranillos, M.P. The Pharmacological Activity of *Camellia sinensis* (L.) Kuntze on Metabolic and Endocrine Disorders: A Systematic Review. *Biomolecules* **2020**, *10*, 603. [CrossRef]
56. Oliveira, A.; Guimarães, A.; Oliveira Junior, R.; Quintans, J.; Medeiros, F.; Barbosa Filho, J.; Quintans-Júnior, L.; Almeida, J.R. *Camellia sinensis* (L.) Kuntze: A Review of Chemical and Nutraceutical Properties. *Nat. Prod. Res. Rev.* **2016**, *4*, 21–62.
57. Umesh, C.V. Chapter 13—*Camellia sinensis*. In *Herbs, Spices and Their Roles in Nutraceuticals and Functional Foods*; Amalraj, A., Kuttappan, S., Varma A.C, K., Matharu, A., Eds.; Academic Press: Cambridge, MA, USA, 2023; pp. 219–231.
58. Sahoo, J.P.; Behera, L.; Praveena, J.; Sawant, S.; Mishra, A.; Sharma, S.S.; Ghosh, L.; Mishra, A.P.; Sahoo, A.R.; Pradhan, P. The golden spice turmeric (*Curcuma longa*) and its feasible benefits in prospering human health—A review. *Am. J. Plant Sci.* **2021**, *12*, 455–475. [CrossRef]
59. Varma, A.C.K.; Jude, S.; Varghese, B.A.; Kuttappan, S.; Amalraj, A. Chapter 2—*Curcuma longa*. In *Herbs, Spices and Their Roles in Nutraceuticals and Functional Foods*; Amalraj, A., Kuttappan, S., Varma A.C, K., Matharu, A., Eds.; Academic Press: Cambridge, MA, USA, 2023; pp. 15–30.
60. Jyotirmayee, B.; Mahalik, G. A review on selected pharmacological activities of *Curcuma longa* L. *Int. J. Food Prop.* **2022**, *25*, 1377–1398. [CrossRef]
61. Wachtel-Galor, S.; Yuen, J.; Buswell, J.A.; Benzie, I.F.F. *Ganoderma lucidum* (*Lingzhi* or *Reishi*): A Medicinal Mushroom. In *Herbal Medicine: Biomolecular and Clinical Aspects*, 2nd ed.; CRC Press/Taylor & Francis: Boca Raton, FL, USA, 2011.
62. Sanodiya, B.S.; Thakur, G.S.; Baghel, R.K.; Prasad, G.B.; Bisen, P.S. *Ganoderma lucidum*: A potent pharmacological macrofungus. *Curr. Pharm. Biotechnol.* **2009**, *10*, 717–742. [CrossRef]
63. Ahmad, R.; Riaz, M.; Khan, A.; Aljamea, A.; Algheryafi, M.; Sewaket, D.; Alqathama, A. *Ganoderma lucidum* (Reishi) an edible mushroom; a comprehensive and critical review of its nutritional, cosmeceutical, mycochemical, pharmacological, clinical, and toxicological properties. *Phytother. Res. PTR* **2021**, *35*, 6030–6062. [CrossRef]
64. Yu, Y.H.; Kuo, H.P.; Hsieh, H.H.; Li, J.W.; Hsu, W.H.; Chen, S.J.; Su, M.H.; Liu, S.H.; Cheng, Y.C.; Chen, C.Y.; et al. *Ganoderma tsugae* Induces S Phase Arrest and Apoptosis in Doxorubicin-Resistant Lung Adenocarcinoma H23/0.3 Cells via Modulation of the PI3K/Akt Signaling Pathway. *Evid.-Based Complement. Altern. Med.* **2012**, *2012*, 371286. [CrossRef] [PubMed]
65. Jiang, M.; Cui, B.-W.; Wu, Y.-L.; Nan, J.-X.; Lian, L.-H. *Genus Gentiana*: A review on phytochemistry, pharmacology and molecular mechanism. *J. Ethnopharmacol.* **2021**, *264*, 113391. [CrossRef]
66. Zhang, X.; Zhan, G.; Jin, M.; Zhang, H.; Dang, J.; Zhang, Y.; Guo, Z.; Ito, Y. Botany, traditional use, phytochemistry, pharmacology, quality control, and authentication of *Radix Gentianae Macrophyllae*-A traditional medicine: A review. *Phytomedicine* **2018**, *46*, 142–163. [CrossRef] [PubMed]
67. Huang, C.-Y.; Hsu, T.-C.; Kuo, W.-W.; Liou, Y.-F.; Lee, S.-D.; Ju, D.-T.; Kuo, C.-H.; Tzang, B.-S. The root extract of *Gentiana macrophylla* Pall. Alleviates cardiac apoptosis in lupus prone mice. *PLoS ONE* **2015**, *10*, e0127440. [CrossRef]
68. Pasdaran, A.; Naychov, Z.; Batovska, D.; Kerr, P.; Favre, A.; Dimitrov, V.; Aneva, I.; Hamedi, A.; Kozuharova, E. Some European Gentiana Species Are Used Traditionally to Cure Wounds: Bioactivity and Conservation Issues. *Diversity* **2023**, *15*, 467. [CrossRef]
69. Ravanfar, P.; Namazi, G.; Atigh, M.; Zafarmand, S.; Hamedi, A.; Salehi, A.; Izadi, S.; Borhani-Haghighi, A. Efficacy of whole extract of licorice in neurological improvement of patients after acute ischemic stroke. *J. Herb. Med.* **2016**, *6*, 12–17. [CrossRef]
70. Petramfar, P.; Hajari, F.; Yousefi, G.; Azadi, S.; Hamedi, A. Efficacy of oral administration of licorice as an adjunct therapy on improving the symptoms of patients with Parkinson's disease, A randomized double blinded clinical trial. *J. Ethnopharmacol.* **2020**, *247*, 112226. [CrossRef] [PubMed]
71. Ayeka, P.A.; Bian, Y.; Githaiga, P.M.; Zhao, Y. The immunomodulatory activities of licorice polysaccharides (*Glycyrrhiza uralensis* Fisch.) in CT 26 tumor-bearing mice. *BMC Complement. Altern. Med.* **2017**, *17*, 536. [CrossRef]
72. Jiang, M.; Zhao, S.; Yang, S.; Lin, X.; He, X.; Wei, X.; Song, Q.; Li, R.; Fu, C.; Zhang, J. An "essential herbal medicine"—Licorice: A review of phytochemicals and its effects in combination preparations. *J. Ethnopharmacol.* **2020**, *249*, 112439. [CrossRef]
73. Ansari, R.; Zarshenas, M.M.; Dadbakhsh, A. A Review on Pharmacological and Clinical Aspects of *Linum usitatissimum* L. *Curr. Drug Discov. Technol.* **2018**, *15*, 148–158. [CrossRef]
74. Ganguly, S.; Panjagari, N.R.; Raman, R.K. Flaxseed (*Linum usitatissimum*). In *Oilseeds: Health Attributes and Food Applications*; Springer: Berlin/Heidelberg, Germany, 2021; pp. 253–283.
75. Heydarirad, G.; Tavakoli, A.; Cooley, K.; Pasalar, M. A review on medical plants advised for neuralgia from the perspective of "canon of medicine". *Adv. Integr. Med.* **2021**, *8*, 230–235. [CrossRef]

76. Mehta, N.; Patel, E.P.; Pragnesh, B.S.V.P.; Shah, B. *Nelumbo nucifera* (Lotus): A review on ethanobotany, phytochemistry and pharmacology. *Indian J. Pharm. Biol. Res.* **2013**, *1*, 152–167. [CrossRef]
77. Sharma, B.R.; Gautam, L.N.S.; Adhikari, D.; Karki, R. A comprehensive review on chemical profiling of *Nelumbo nucifera*: Potential for drug development. *Phytother. Res.* **2017**, *31*, 3–26. [CrossRef] [PubMed]
78. Luo, Y.; Yang, S.K.; Zhou, X.; Wang, M.; Tang, D.; Liu, F.Y.; Sun, L.; Xiao, L. Use of Ophiocordyceps sinensis (syn. *Cordyceps sinensis*) combined with angiotensin-converting enzyme inhibitors (ACEI)/angiotensin receptor blockers (ARB) versus ACEI/ARB alone in the treatment of diabetic kidney disease: A meta-analysis. *Ren. Fail.* **2015**, *37*, 614–634. [CrossRef]
79. Lo, H.-C.; Hsieh, C.; Lin, F.-Y.; Hsu, T.-H. A systematic review of the mysterious caterpillar fungus *Ophiocordyceps sinensis* in DongChongXiaCao (冬蟲夏草 Dōng Chóng Xià Cǎo) and related bioactive ingredients. *J. Tradit. Complement. Med.* **2013**, *3*, 16–32. [CrossRef] [PubMed]
80. He, D.Y.; Dai, S.M. Anti-inflammatory and immunomodulatory effects of paeonia lactiflora pall., a traditional chinese herbal medicine. *Front. Pharmacol.* **2011**, *2*, 10. [CrossRef]
81. Wang, Z.; He, C.; Peng, Y.; Chen, F.; Xiao, P. Origins, Phytochemistry, Pharmacology, Analytical Methods and Safety of Cortex Moutan (*Paeonia suffruticosa* Andrew): A Systematic Review. *Molecules* **2017**, *22*, 946. [CrossRef]
82. Kim, Y.G.; Komakech, R.; Jeong, D.H.; Park, Y.M.; Lee, T.K.; Kim, K.H.; Lee, A.Y.; Moon, B.C.; Kang, Y. Verification of the Field Productivity of *Rehmannia glutinosa* (Gaertn.) DC. Developed Through Optimized In Vitro Culture Method. *Plants* **2020**, *9*, 317. [CrossRef] [PubMed]
83. Bian, Z.; Zhang, R.; Zhang, X.; Zhang, J.; Xu, L.; Zhu, L.; Ma, Y.; Liu, Y. Extraction, structure and bioactivities of polysaccharides from *Rehmannia glutinosa*: A review. *J. Ethnopharmacol.* **2023**, *305*, 116132. [CrossRef] [PubMed]
84. Zhao, T.; Tang, H.; Xie, L.; Zheng, Y.; Ma, Z.; Sun, Q.; Li, X. *Scutellaria baicalensis* Georgi. (Lamiaceae): A review of its traditional uses, botany, phytochemistry, pharmacology and toxicology. *J. Pharm. Pharmacol.* **2019**, *71*, 1353–1369. [CrossRef]
85. Song, J.-W.; Long, J.-Y.; Xie, L.; Zhang, L.-L.; Xie, Q.-X.; Chen, H.-J.; Deng, M.; Li, X.-F. Applications, phytochemistry, pharmacological effects, pharmacokinetics, toxicity of Scutellaria baicalensis Georgi. and its probably potential therapeutic effects on COVID-19: A review. *Chin. Med.* **2020**, *15*, 102. [CrossRef]
86. Billah, M.M.; Khan, F.; Niaz, K. Chapter 3.38—*Scutellaria baicalensis* Georgi. In *Nonvitamin and Nonmineral Nutritional Supplements*; Nabavi, S.M., Silva, A.S., Eds.; Academic Press: Cambridge, MA, USA, 2019; pp. 403–408.
87. Song, C.Y.; Xu, Y.G.; Lu, Y.Q. Use of *Tripterygium wilfordii* Hook F for immune-mediated inflammatory diseases: Progress and future prospects. *J. Zhejiang Univ. Sci. B* **2020**, *21*, 280–290. [CrossRef]
88. Brinker, A.M.; Ma, J.; Lipsky, P.E.; Raskin, I. Medicinal chemistry and pharmacology of genus *Tripterygium* (Celastraceae). *Phytochemistry* **2007**, *68*, 732–766. [CrossRef]
89. Devkota, H.P.; Paudel, K.R.; Khanal, S.; Baral, A.; Panth, N.; Adhikari-Devkota, A.; Jha, N.K.; Das, N.; Singh, S.K.; Chellappan, D.K.; et al. Stinging Nettle (*Urtica dioica* L.): Nutritional Composition, Bioactive Compounds, and Food Functional Properties. *Molecules* **2022**, *27*, 5219. [CrossRef]
90. Hamedi, A.; Sakhteman, A.; Moheimani, S.M. An in silico approach towards investigation of possible effects of essential oils constituents on receptors involved in cardiovascular diseases (CVD) and associated risk factors (Diabetes Mellitus and Hyperlipidemia). *Cardiovasc. Hematol. Agents Med. Chem. (Former. Curr. Med. Chem.-Cardiovasc. Hematol. Agents)* **2021**, *19*, 32–42. [CrossRef]
91. Ebrahimi-Najafabadi, H.; Kazemeini, S.S.; Pasdaran, A.; Hamedi, A. A novel similarity search approach for high-performance thin-layer chromatography (HPTLC) fingerprinting of medicinal plants. *Phytochem. Anal.* **2019**, *30*, 405–414. [CrossRef] [PubMed]
92. Mojab, F.; Hamedi, A.; Nickavar, B.; Javidnia, K. Hydrodistilled volatile constituents of the leaves of *Daucus carota* L. subsp. sativus (Hoffman.) Arcang. (Apiaceae) from Iran. *J. Essent. Oil Bear. Plants* **2008**, *11*, 271–277. [CrossRef]
93. Tsai, P.Y.; Ka, S.M.; Chang, J.M.; Lai, J.H.; Dai, M.S.; Jheng, H.L.; Kuo, M.T.; Chen, P.; Chen, A. Antroquinonol differentially modulates T cell activity and reduces interleukin-18 production, but enhances Nrf2 activation, in murine accelerated severe lupus nephritis. *Arthritis Rheum.* **2012**, *64*, 232–242. [CrossRef]
94. Liu, C.-P.; Tsai, W.-J.; Shen, C.-C.; Lin, Y.-L.; Liao, J.-F.; Chen, C.-F.; Kuo, Y.-C. Inhibition of (S)-armepavine from Nelumbo nucifera on autoimmune disease of MRL/MpJ-lpr/lpr mice. *Eur. J. Pharmacol.* **2006**, *531*, 270–279. [CrossRef]
95. Lan, L. Study on effect of *Cordyceps sinensis* and artemisinin in preventing recurrence of lupus nephritis. *Chin. J. Integr. Tradit. West. Med.* **2002**, *8*, 89. [CrossRef]
96. Yang, L.-Y.; Chen, A.; Kuo, Y.-C.; Lin, C.-Y. Efficacy of a pure compound H1-A extracted from *Cordyceps sinensis* on autoimmune disease of MRL lpr/lpr mice. *J. Lab. Clin. Med.* **1999**, *134*, 492–500. [CrossRef] [PubMed]
97. Hou, L.F.; He, S.J.; Li, X.; Yang, Y.; He, P.L.; Zhou, Y.; Zhu, F.H.; Yang, Y.F.; Li, Y.; Tang, W. Oral administration of artemisinin analog SM934 ameliorates lupus syndromes in MRL/lpr mice by inhibiting Th1 and Th17 cell responses. *Arthritis Rheum.* **2011**, *63*, 2445–2455. [CrossRef]
98. Yang, J.; Yang, X.; Yang, J.; Li, M. Baicalin ameliorates lupus autoimmunity by inhibiting differentiation of Tfh cells and inducing expansion of Tfr cells. *Cell Death Dis.* **2019**, *10*, 140. [CrossRef]
99. Chae, B.-S. Baicalin Ameliorates Dysimmunoregulation in Pristane-Induced Lupus Mice: Production of IL-6 and PGE_2 and Activation of T cells. *Nat. Prod. Sci.* **2011**, *17*, 354–362.
100. Handono, K.; Pratama, M.Z.; Endharti, A.T.; Kalim, H. Treatment of low doses curcumin could modulate Th17/Treg balance specifically on CD4+ T cell cultures of systemic lupus erythematosus patients. *Cent. Eur. J. Immunol.* **2015**, *40*, 461–469. [CrossRef] [PubMed]

101. Zhao, J.; Wang, J.; Zhou, M.; Li, M.; Li, M.; Tan, H. Curcumin attenuates murine lupus via inhibiting NLRP3 inflammasome. *Int. Immunopharmacol.* **2019**, *69*, 213–216. [CrossRef] [PubMed]
102. Dent, E.L.; Taylor, E.B.; Turbeville, H.R.; Ryan, M.J. Curcumin attenuates autoimmunity and renal injury in an experimental model of systemic lupus erythematosus. *Physiol. Rep.* **2020**, *8*, e14501. [CrossRef]
103. Kalim, H.; Handono, K.; Khalasha, T.; Pratama, M.Z.; Dantara, T.W.I.; Wulandari, A.P.; Albinsaid, F.; Fitria, S.N.; Mahardika, M.V. Immune modulation effects of curcumin in pristane-induced lupus mice. *Indian J. Rheumatol.* **2017**, *12*, 86. [CrossRef]
104. Li, W.-D.; Dong, Y.-J.; Tu, Y.-Y.; Lin, Z.-B. *Dihydroarteannuin ameliorates* lupus symptom of BXSB mice by inhibiting production of TNF-alpha and blocking the signaling pathway NF-kappa B translocation. *Int. Immunopharmacol.* **2006**, *6*, 1243–1250. [CrossRef]
105. Zhang, T.; Zhang, Y.; Jiang, N.; Zhao, X.; Sang, X.; Yang, N.; Feng, Y.; Chen, R.; Chen, Q. Dihydroartemisinin regulates the immune system by promotion of $CD8^+$ T lymphocytes and suppression of B cell responses. *Sci. China Life Sci.* **2020**, *63*, 737–749. [CrossRef] [PubMed]
106. Chen, Y.; Tao, T.; Wang, W.; Yang, B.; Cha, X. Dihydroartemisinin attenuated the symptoms of mice model of systemic lupus erythematosus by restoring the Treg/Th17 balance. *Clin. Exp. Pharmacol. Physiol.* **2021**, *48*, 626–633. [CrossRef]
107. Xia, Y.; Xu, S. Effect of emodin on nephritis of BXSB lupus mice and its pharmacological mechanism. *Cent. China Med. J.* **2003**, *27*, 63–64.
108. Yuan, X.; Dai, B.; Yang, L.; Lin, B.; Lin, E.; Pan, Y. Emodin ameliorates renal injury in BXSB mice by modulating TNF-α/ICAM-1. *Biosci. Rep.* **2020**, *40*, BSR20202551. [CrossRef]
109. Zhang, Y.; Li, Z.; Wu, H.; Wang, J.; Zhang, S. Esculetin alleviates murine lupus nephritis by inhibiting complement activation and enhancing Nrf2 signaling pathway. *J. Ethnopharmacol.* **2022**, *288*, 115004. [CrossRef] [PubMed]
110. Lu, M.; Yu, S.; Xu, W.; Gao, B.; Xiong, S. HMGB1 promotes systemic lupus erythematosus by enhancing macrophage inflammatory response. *J. Immunol. Res.* **2015**, *2015*, 946748. [CrossRef]
111. Garcia-Gordillo, M.A.; Collado-Mateo, D.; Hernández-Mocholi, M.A.; Pazzi, F.; Gusi, N.; Dominguez-Munoz, F.J.; Adsuar, J.C. Cost-Utility Analysis of a Six-Weeks Ganoderma Lucidum-Based Treatment for Women with Fibromyalgia: A Randomized Double-Blind, Active Placebo-Controlled Trial. *Myopain* **2015**, *23*, 188–194. [CrossRef]
112. Maekawa, T.; Kosuge, S.; Karino, A.; Okano, T.; Ito, J.; Munakata, H.; Ohtsuki, K. Biochemical characterization of 60S acidic ribosomal P proteins from porcine liver and the inhibition of their immunocomplex formation with sera from systemic lupus erythematosus (SLE) patients by glycyrrhizin in vitro. *Biol. Pharm. Bull.* **2000**, *23*, 27–32. [CrossRef]
113. Liang, C.-L.; Lu, W.; Zhou, J.-Y.; Chen, Y.; Zhang, Q.; Liu, H.; Qiu, F.; Dai, Z. Mangiferin attenuates murine lupus nephritis by inducing $CD4^+$ $Foxp3^+$ regulatory T cells via suppression of mTOR signaling. *Cell. Physiol. Biochem.* **2018**, *50*, 1560–1573. [CrossRef] [PubMed]
114. Lin, Y.; Yan, Y.; Zhang, H.; Chen, Y.; He, Y.; Wang, S.; Fang, L.; Lv, Y.; Du, G. Salvianolic acid A alleviates renal injury in systemic lupus erythematosus induced by pristane in BALB/c mice. *Acta Pharm. Sin. B* **2017**, *7*, 159–166. [CrossRef] [PubMed]
115. Zhang, H.; Xiao, W.; Hou, P. Clinical study of total glucosides of paeony in patients with systemic lupus erythematosus. *Zhongguo Zhong Xi Yi Jie He Za Zhi Zhongguo Zhongxiyi Jiehe Zazhi = Chin. J. Integr. Tradit. West. Med.* **2011**, *31*, 476–479.
116. Shuai, Z.-w.; Xu, J.-h.; Liu, S. Clinical study on effect of Total Glucosides of Peony in treating systemic lupus erythematosus as adjuvant treatment. *Chin. J. Integr. Tradit. West. Med.* **2003**, *23*, 188–191.
117. Gong, X.; Li, H.; Guo, H.; Wu, S.; Lu, C.; Chen, Y.; Li, S. Efficacy and safety of total glucosides of paeony in the treatment of systemic lupus erythematosus: A systematic review and meta-analysis. *Front. Pharmacol.* **2022**, *13*, 932874. [CrossRef] [PubMed]
118. Li, M.; Jiang, A. DNA methylation was involved in total glucosides of paeony regulating ERα for the treatment of female systemic lupus erythematosus mice. *J. Pharmacol. Sci.* **2019**, *140*, 187–192. [CrossRef]
119. Ding, Z.-X.; Yang, S.-F.; Wu, Q.-F.; Lu, Y.; Chen, Y.-Y.; Nie, X.-L.; Jie, H.-Y.; Qi, J.-M.; Wang, F.-S. Therapeutic effect of total glucosides of paeony on lupus nephritis in MRL/lpr mice. *J. South. Med. Univ.* **2011**, *31*, 656–660.
120. Zhao, M.; Liang, G.; Luo, S.; Lu, Q. Effect of total glucosides of peony on expression and DNA methylation status of ITGAL gene in CD4 (+) T cells of systemic lupus erythematosus. *Zhong Nan Da Xue Xue Bao. Yi Xue Ban = J. Cent. South Univ. Med. Sci.* **2012**, *37*, 463–468.
121. Liu, L.; Jiao, W.; Zhang, X.; Zhang, Y.; Zhao, G.; Sun, Y.; Shuhua, L. Regulation on function and maturation of dendritic cells from systemic lupus erythematosus patients by triptolide. *Chin. J. Microbiol. Immunol.* **2011**, *31*, 824–829.
122. Liu, Y.-F.; He, H.-Q.; Ding, Y.-L.; Wu, S.-Y.; Chen, D.-S.; CL, E. Effects of Triptolide on Tc and Th Cell Excursion in Peripheral Blood of Nude Mice with Systemic Lupus Erythematosus BALB/c-un. *Zhongguo Shi Yan Xue Ye Xue Za Zhi* **2019**, *27*, 1691–1695. [PubMed]
123. Tao, X.; Fan, F.; Hoffmann, V.; Gao, C.Y.; Longo, N.S.; Zerfas, P.; Lipsky, P.E. Effective therapy for nephritis in (NZB $\times$ NZW) F_1 mice with triptolide and tripdiolide, the principal active components of the Chinese herbal remedy *Tripterygium wilfordii* Hook F. *Arthritis Rheum. Off. J. Am. Coll. Rheumatol.* **2008**, *58*, 1774–1783. [CrossRef] [PubMed]
124. Zhao, X.; Tang, X.; Yan, Q.; Song, H.; Li, Z.; Wang, D.; Chen, H.; Sun, L. *Triptolide ameliorates* lupus via the induction of miR-125a-5p mediating Treg upregulation. *Int. Immunopharmacol.* **2019**, *71*, 14–21. [CrossRef] [PubMed]
125. Zhang, L.-Y.; Li, H.; Wu, Y.-W.; Cheng, L.; Yan, Y.-X.; Yang, X.-Q.; Zhu, F.-H.; He, S.-J.; Tang, W.; Zuo, J.-P. (5R)-5-hydroxytriptolide ameliorates lupus nephritis in MRL/lpr mice by preventing infiltration of immune cells. *Am. J. Physiol.-Ren. Physiol.* **2017**, *312*, F769–F777. [CrossRef] [PubMed]
126. Halade, G.V.; Rahman, M.M.; Bhattacharya, A.; Barnes, J.L.; Chandrasekar, B.; Fernandes, G. Docosahexaenoic acid-enriched fish oil attenuates kidney disease and prolongs median and maximal life span of autoimmune lupus-prone mice. *J. Immunol.* **2010**, *184*, 5280–5286. [CrossRef]
127. Fettouh, D.S.; Saif, D.S.; El Gazzar, S.F.; Sonbol, A.A. Study the relationship between vitamin A deficiency, T helper 17, regulatory T cells, and disease activity in patients with systemic lupus erythematosus. *Egypt. Rheumatol. Rehabil.* **2019**, *46*, 244–250. [CrossRef]

128. Handono, K.; Firdausi, S.N.; Pratama, M.Z.; Endharti, A.T.; Kalim, H. Vitamin A improve Th17 and Treg regulation in systemic lupus erythematosus. *Clin. Rheumatol.* **2016**, *35*, 631–638. [CrossRef] [PubMed]
129. Minami, Y.; Hirabayashi, Y.; Nagata, C.; Ishii, T.; Harigae, H.; Sasaki, T. Intakes of vitamin B6 and dietary fiber and clinical course of systemic lupus erythematosus: A prospective study of Japanese female patients. *J. Epidemiol.* **2011**, *21*, 246–254. [CrossRef]
130. Shah, M.; Adams-Huet, B.; Kavanaugh, A.; Coyle, Y.; Lipsky, P. Nutrient intake and diet quality in patients with systemic lupus erythematosus on a culturally sensitive cholesterol lowering dietary program. *J. Rheumatol.* **2004**, *31*, 71–75.
131. Minami, Y.; Sasaki, T.; Arai, Y.; Kurisu, Y.; Hisamichi, S. Diet and systemic lupus erythematosus: A 4 year prospective study of Japanese patients. *J. Rheumatol.* **2003**, *30*, 747–754. [PubMed]
132. Ben-Zvi, I.; Aranow, C.; Mackay, M.; Stanevsky, A.; Kamen, D.L.; Marinescu, L.M.; Collins, C.E.; Gilkeson, G.S.; Diamond, B.; Hardin, J.A. The impact of vitamin D on dendritic cell function in patients with systemic lupus erythematosus. *PLoS ONE* **2010**, *5*, e9193. [CrossRef] [PubMed]
133. Antico, A.; Tampoia, M.; Tozzoli, R.; Bizzaro, N. Can supplementation with vitamin D reduce the risk or modify the course of autoimmune diseases? A systematic review of the literature. *Autoimmun. Rev.* **2012**, *12*, 127–136. [CrossRef] [PubMed]
134. Maeshima, E.; Liang, X.-M.; Goda, M.; Otani, H.; Mune, M. The efficacy of vitamin E against oxidative damage and autoantibody production in systemic lupus erythematosus: A preliminary study. *Clin. Rheumatol.* **2007**, *26*, 401–404. [CrossRef]
135. Leiter, L.M.; Reuhl, K.R.; Racis Jr, S.P.; Sherman, A.R. Iron status alters murine systemic lupus erythematosus. *J. Nutr.* **1995**, *125*, 474–484.
136. O'Dell, J.R.; McGivern, J.P.; Kay, H.; Klassen, L.W. Improved survival in murine lupus as the result of selenium supplementation. *Clin. Exp. Immunol.* **1988**, *73*, 322.
137. Soni, C.; Sinha, I.; Fasnacht, M.J.; Olsen, N.J.; Rahman, Z.S.; Sinha, R. Selenium supplementation suppresses immunological and serological features of lupus in B6. Sle1b mice. *Autoimmunity* **2019**, *52*, 57–68. [CrossRef]
138. D'Cunha, N.M.; Peterson, G.; Baby, K.; Thomas, J. Impetigo: A need for new therapies in a world of increasing antimicrobial resistance. *J. Clin. Pharm. Ther.* **2017**, *43*, 150–153. [CrossRef]
139. Corson, T.W.; Crews, C.M. Molecular understanding and modern application of traditional medicines: Triumphs and trials. *Cell* **2007**, *130*, 769–774. [CrossRef]
140. Garcia, L.C. A Review of *Artemisia annua* L.: Its genetics, biochemical characteristics, and anti-malarial efficacy. *Int. J. Sci. Technol.* **2015**, *5*, 38–46.
141. Krishna, S.; Bustamante, L.; Haynes, R.K.; Staines, H.M. Artemisinins: Their growing importance in medicine. *Trends Pharmacol. Sci.* **2008**, *29*, 520–527. [CrossRef] [PubMed]
142. Li, Y.; Xu, T.; Qiu, X.; Tian, B.; Bi, C.; Yao, L. Effectiveness of Bailing capsules in the treatment of lupus nephritis: A meta-analysis. *Mol. Med. Rep.* **2020**, *22*, 2132–2140. [CrossRef] [PubMed]
143. Wang, C.; Fortin, P.; Li, Y.; Panaritis, T.; Gans, M.; Esdaile, J. Discontinuation of antimalarial drugs in systemic lupus erythematosus. *J. Rheumatol.* **1999**, *26*, 808–815. [PubMed]
144. Mu, X.; Wang, C. Artemisinins—A promising new treatment for systemic lupus erythematosus: A descriptive review. *Curr. Rheumatol. Rep.* **2018**, *20*, 1–10. [CrossRef] [PubMed]
145. Golenser, J.; Waknine, J.H.; Krugliak, M.; Hunt, N.H.; Grau, G.E. Current perspectives on the mechanism of action of artemisinins. *Int. J. Parasitol.* **2006**, *36*, 1427–1441. [CrossRef] [PubMed]
146. Efferth, T.; Oesch, F. The immunosuppressive activity of artemisinin-type drugs towards inflammatory and autoimmune diseases. *Med. Res. Rev.* **2021**, *41*, 3023–3061. [CrossRef] [PubMed]
147. Dang, W.-Z.; Li, H.; Jiang, B.; Nandakumar, K.S.; Liu, K.-F.; Liu, L.-X.; Yu, X.-C.; Tan, H.-J.; Zhou, C. Therapeutic effects of artesunate on lupus-prone MRL/lpr mice are dependent on T follicular helper cell differentiation and activation of JAK2–STAT3 signaling pathway. *Phytomedicine* **2019**, *62*, 152965. [CrossRef]
148. Li, D.; Qi, J.; Wang, J.; Pan, Y.; Li, J.; Xia, X.; Dou, H.; Hou, Y. Protective effect of dihydroartemisinin in inhibiting senescence of myeloid-derived suppressor cells from lupus mice via Nrf2/HO-1 pathway. *Free Radic. Biol. Med.* **2019**, *143*, 260–274. [CrossRef]
149. Efferth, T.; Kaina, B. Toxicity of the antimalarial artemisinin and its dervatives. *Crit. Rev. Toxicol.* **2010**, *40*, 405–421. [CrossRef]
150. Yin, J.; Wang, H.; Ding, R. Artemisinin and its derivatives: Progress in toxicology. *Chin. J. Pharmacol. Toxicol.* **2014**, *6*, 309–314.
151. Lee, W.-T.; Lee, T.-H.; Cheng, C.-H.; Chen, K.-C.; Chen, Y.-C.; Lin, C.-W. Antroquinonol from *Antrodia Camphorata* suppresses breast tumor migration/invasion through inhibiting ERK-AP-1-and AKT-NF-κB-dependent MMP-9 and epithelial-mesenchymal transition expressions. *Food Chem. Toxicol.* **2015**, *78*, 33–41. [CrossRef] [PubMed]
152. Tsai, P.-Y.; Ka, S.-M.; Chao, T.-K.; Chang, J.-M.; Lin, S.-H.; Li, C.-Y.; Kuo, M.-T.; Chen, P.; Chen, A. Antroquinonol reduces oxidative stress by enhancing the Nrf2 signaling pathway and inhibits inflammation and sclerosis in focal segmental glomerulosclerosis mice. *Free Radic. Biol. Med.* **2011**, *50*, 1503–1516. [CrossRef] [PubMed]
153. Geethangili, M.; Tzeng, Y.-M. Review of pharmacological effects of Antrodia camphorata and its bioactive compounds. *Evid.-Based Complement. Altern. Med.* **2011**, *2011*, 212641. [CrossRef]
154. Zhang, B.-B.; Hu, P.-F.; Huang, J.; Hu, Y.-D.; Chen, L.; Xu, G.-R. Current advances on the structure, bioactivity, synthesis, and metabolic regulation of novel ubiquinone derivatives in the edible and medicinal mushroom *Antrodia cinnamomea*. *J. Agric. Food Chem.* **2017**, *65*, 10395–10405. [CrossRef]
155. Villaume, M.T.; Sella, E.; Saul, G.; Borzilleri, R.M.; Fargnoli, J.; Johnston, K.A.; Zhang, H.; Fereshteh, M.P.; Dhar, T.M.; Baran, P.S. Antroquinonol a: Scalable synthesis and preclinical biology of a phase 2 drug candidate. *ACS Cent. Sci.* **2016**, *2*, 27–31. [CrossRef]

156. Angamuthu, V.; Shanmugavadivu, M.; Nagarajan, G.; Velmurugan, B.K. Pharmacological activities of antroquinonol-Mini review. *Chem.-Biol. Interact.* **2019**, *297*, 8–15. [CrossRef]
157. Kuang, Y.; Li, B.; Wang, Z.; Qiao, X.; Ye, M. Terpenoids from the medicinal mushroom *Antrodia camphorata*: Chemistry and medicinal potential. *Nat. Prod. Rep.* **2021**, *38*, 83–102. [CrossRef]
158. Huang, C.-Y.; Ju, D.-T.; Chang, C.-F.; Reddy, P.M.; Velmurugan, B.K. A review on the effects of current chemotherapy drugs and natural agents in treating non–small cell lung cancer. *Biomedicine* **2017**, *7*, 23. [CrossRef]
159. Chang, W.-H.; Chen, M.C.; Cheng, I.H. Antroquinonol lowers brain amyloid-β levels and improves spatial learning and memory in a transgenic mouse model of Alzheimer's disease. *Sci. Rep.* **2015**, *5*, 15067. [CrossRef]
160. Chae, B.S. Effect of Baicalin on the Ex vivo Production of Cytokines in Pristane-Induced Lupus Mice. *YAKHAK HOEJI* **2016**, *60*, 21–28. [CrossRef]
161. Delerue, T.; Barroso, M.F.; Dias-Teixeira, M.; Figueiredo-González, M.; Delerue-Matos, C.; Grosso, C. Interactions between *Ginkgo biloba* L. and *Scutellaria baicalensis* Georgi in multicomponent mixtures towards cholinesterase inhibition and ROS scavenging. *Food Res. Int.* **2021**, *140*, 109857. [CrossRef] [PubMed]
162. Cai, Y.; Ma, W.; Xiao, Y.; Wu, B.; Li, X.; Liu, F.; Qiu, J.; Zhang, G. High doses of baicalin induces kidney injury and fibrosis through regulating TGF-β/Smad signaling pathway. *Toxicol. Appl. Pharmacol.* **2017**, *333*, 1–9. [CrossRef] [PubMed]
163. Huang, T.; Liu, Y.; Zhang, C. Pharmacokinetics and bioavailability enhancement of baicalin: A review. *Eur. J. Drug Metab. Pharmacokinet.* **2019**, *44*, 159–168. [CrossRef] [PubMed]
164. Salehi, B.; Stojanović-Radić, Z.; Matejić, J.; Sharifi-Rad, M.; Anil Kumar, N.V.; Martins, N.; Sharifi-Rad, J. The therapeutic potential of curcumin: A review of clinical trials. *Eur. J. Med. Chem.* **2019**, *163*, 527–545. [CrossRef] [PubMed]
165. Constantin, M.M.; Nita, I.E.; Olteanu, R.; Constantin, T.; Bucur, S.; Matei, C.; Raducan, A. Significance and impact of dietary factors on systemic lupus erythematosus pathogenesis. *Exp. Ther. Med.* **2019**, *17*, 1085–1090. [CrossRef]
166. Kumar, A.; Dogra, S.; Prakash, A. Protective effect of curcumin (*Curcuma longa*), against aluminium toxicity: Possible behavioral and biochemical alterations in rats. *Behav. Brain Res.* **2009**, *205*, 384–390. [CrossRef]
167. Rezaee, R.; Momtazi, A.A.; Monemi, A.; Sahebkar, A. Curcumin: A potentially powerful tool to reverse cisplatin-induced toxicity. *Pharmacol. Res.* **2017**, *117*, 218–227. [CrossRef]
168. Soetikno, V.; Sari, F.R.; Veeraveedu, P.T.; Thandavarayan, R.A.; Harima, M.; Sukumaran, V.; Lakshmanan, A.P.; Suzuki, K.; Kawachi, H.; Watanabe, K. Curcumin ameliorates macrophage infiltration by inhibiting NF-κB activation and proinflammatory cytokines in streptozotocin induced-diabetic nephropathy. *Nutr. Metab.* **2011**, *8*, 1–11. [CrossRef]
169. Shishodia, S.; Sethi, G.; Aggarwal, B.B. Curcumin: Getting back to the roots. *Ann. N. Y. Acad. Sci.* **2005**, *1056*, 206–217. [CrossRef]
170. Shishodia, S.; Singh, T.; Chaturvedi, M.M. Modulation of transcription factors by curcumin. *Mol. Targets Ther. Uses Curcumin Health Dis.* **2007**, *595*, 127–148.
171. Gonzales, A.M.; Orlando, R.A. Curcumin and resveratrol inhibit nuclear factor-kappaB-mediated cytokine expression in adipocytes. *Nutr. Metab.* **2008**, *5*, 17. [CrossRef] [PubMed]
172. JIANG, C.-X. Curcumin analog exhibited anti-inflammatory activity through inhibiting ERK/JNK and NF-κB signaling pathway. *Chin. Tradit. Herb. Drugs* **2016**, *24*, 2871–2876.
173. Han, S.-S.; Keum, Y.-S.; Seo, H.-J.; Surh, Y.-J. Curcumin suppresses activation of NF-κB and AP-1 induced by phorbol ester in cultured human promyelocytic leukemia cells. *BMB Rep.* **2002**, *35*, 337–342. [CrossRef]
174. Kocaadam, B.; Şanlier, N. Curcumin, an active component of turmeric (*Curcuma longa*), and its effects on health. *Crit. Rev. Food Sci. Nutr.* **2017**, *57*, 2889–2895. [CrossRef] [PubMed]
175. Lao, C.D.; Ruffin, M.T.; Normolle, D.; Heath, D.D.; Murray, S.I.; Bailey, J.M.; Boggs, M.E.; Crowell, J.; Rock, C.L.; Brenner, D.E. Dose escalation of a curcuminoid formulation. *BMC Complement. Altern. Med.* **2006**, *6*, 1–4. [CrossRef]
176. Sharma, R.A.; Euden, S.A.; Platton, S.L.; Cooke, D.N.; Shafayat, A.; Hewitt, H.R.; Marczylo, T.H.; Morgan, B.; Hemingway, D.; Plummer, S.M. Phase I clinical trial of oral curcumin: Biomarkers of systemic activity and compliance. *Clin. Cancer Res.* **2004**, *10*, 6847–6854. [CrossRef]
177. Srinivas, G.; Babykutty, S.; Sathiadevan, P.P.; Srinivas, P. Molecular mechanism of emodin action: Transition from laxative ingredient to an antitumor agent. *Med. Res. Rev.* **2007**, *27*, 591–608. [CrossRef]
178. Zhang, B.; Shi, Y.; Lei, T.C. Detection of active P-glycoprotein in systemic lupus erythematosus patients with poor disease control. *Exp. Ther. Med.* **2012**, *4*, 705–710. [CrossRef]
179. Sharifi-Rad, J.; Herrera-Bravo, J.; Kamiloglu, S.; Petroni, K.; Mishra, A.P.; Monserrat-Mesquida, M.; Sureda, A.; Martorell, M.; Aidarbekovna, D.S.; Yessimsiitova, Z. Recent advances in the therapeutic potential of emodin for human health. *Biomed. Pharmacother.* **2022**, *154*, 113555. [CrossRef]
180. Akkol, E.K.; Tatlı, I.I.; Karatoprak, G.Ş.; Ağar, O.T.; Yücel, Ç.; Sobarzo-Sánchez, E.; Capasso, R. Is emodin with anticancer effects completely innocent? Two sides of the coin. *Cancers* **2021**, *13*, 2733. [CrossRef] [PubMed]
181. Wang, W.; Sheng, L.; Chen, Y.; Li, Z.; Wu, H.; Ma, J.; Zhang, D.; Chen, X.; Zhang, S. Total coumarin derivates from *Hydrangea paniculata* attenuate renal injuries in cationized-BSA induced membranous nephropathy by inhibiting complement activation and interleukin 10-mediated interstitial fibrosis. *Phytomedicine* **2022**, *96*, 153886. [CrossRef]
182. Tubaro, A.; Del Negro, P.; Ragazzi, E.; Zampiron, S.; Della Loggia, R. Anti-inflammatory and peripheral analgesic activity of esculetin in vivo. *Pharmacol. Res. Commun.* **1988**, *20*, 83–85. [CrossRef] [PubMed]

183. Lum, P.T.; Sekar, M.; Gan, S.H.; Jeyabalan, S.; Bonam, S.R.; Rani, N.N.I.M.; Ku-Mahdzir, K.-M.; Seow, L.J.; Wu, Y.S.; Subramaniyan, V. Therapeutic potential of mangiferin against kidney disorders and its mechanism of action: A review. *Saudi J. Biol. Sci.* **2022**, *29*, 1530–1542. [CrossRef] [PubMed]
184. Jangra, A.; Arora, M.K.; Kisku, A.; Sharma, S. The multifaceted role of mangiferin in health and diseases: A review. *Adv. Tradit. Med.* **2021**, *21*, 619–643. [CrossRef]
185. Morozkina, S.N.; Nhung Vu, T.H.; Generalova, Y.E.; Snetkov, P.P.; Uspenskaya, M.V. Mangiferin as new potential anti-cancer agent and mangiferin-integrated polymer systems—A novel research direction. *Biomolecules* **2021**, *11*, 79. [CrossRef]
186. Mei, S.; Ma, H.; Chen, X. Anticancer and anti-inflammatory properties of mangiferin: A review of its molecular mechanisms. *Food Chem. Toxicol.* **2021**, *149*, 111997. [CrossRef]
187. Reddeman, R.A.; Glávits, R.; Endres, J.R.; Clewell, A.E.; Hirka, G.; Vértesi, A.; Béres, E.; Szakonyiné, I.P. A toxicological evaluation of mango leaf extract (*Mangifera indica*) containing 60% mangiferin. *J. Toxicol.* **2019**, *2019*, 4763015. [CrossRef] [PubMed]
188. Hamidpour, M.; Hamidpour, R.; Hamidpour, S.; Shahlari, M. Chemistry, pharmacology, and medicinal property of sage (*Salvia*) to prevent and cure illnesses such as obesity, diabetes, depression, dementia, lupus, autism, heart disease, and cancer. *J. Tradit. Complement. Med.* **2014**, *4*, 82–88. [CrossRef] [PubMed]
189. Yang, M.-Y.; Song, Z.-Y.; Gan, H.-L.; Zheng, M.-H.; Liu, Q.; Meng, X.-T.; Pan, T.; Li, Z.-Y.; Peng, R.-X.; Liu, K. Non-clinical safety evaluation of salvianolic acid A: Acute, 4-week intravenous toxicities and genotoxicity evaluations. *BMC Pharmacol. Toxicol.* **2022**, *23*, 83. [CrossRef]
190. Du, G.; Song, J.; Du, L.; Zhang, L.; Qiang, G.; Wang, S.; Yang, X.; Fang, L. Chemical and pharmacological research on the polyphenol acids isolated from Danshen: A review of salvianolic acids. *Adv. Pharmacol.* **2020**, *87*, 1–41. [PubMed]
191. Wang, Q.; Meng, J.; Dong, A.; Yu, J.-z.; Zhang, G.-X.; Ma, C.-G. The pharmacological effects and mechanism of *Tripterygium wilfordii* Hook F in central nervous system autoimmunity. *J. Altern. Complement. Med.* **2016**, *22*, 496–502. [CrossRef] [PubMed]
192. Zhou, Z.-L.; Yang, Y.-X.; Ding, J.; Li, Y.-C.; Miao, Z.-H. Triptolide: Structural modifications, structure–activity relationships, bioactivities, clinical development and mechanisms. *Nat. Prod. Rep.* **2012**, *29*, 457–475. [CrossRef] [PubMed]
193. Crews, G.; Erickson, L.; Pan, F.; Fisniku, O.; Jang, M.-S.; Wynn, C.; Benediktsson, H.; Kobayashi, M.; Jiang, H. Down-regulation of TGF-β and VCAM-1 is associated with successful treatment of chronic rejection in rats. *Transplant. Proc.* **2005**, 1926–1928. [CrossRef]
194. Hong, Y.; Zhou, W.; Li, K.; Sacks, S.H. Triptolide is a potent suppressant of C3, CD40 and B7h expression in activated human proximal tubular epithelial cells. *Kidney Int.* **2002**, *62*, 1291–1300. [CrossRef]
195. Zhou, Y.; Hong, Y.; Huang, H. Triptolide attenuates inflammatory response in membranous glomerulo-nephritis rat via downregulation of NF-κB signaling pathway. *Kidney Blood Press. Res.* **2016**, *41*, 901–910. [CrossRef]
196. Li, Y.; Yu, C.; Zhu, W.-M.; Xie, Y.; Qi, X.; Li, N.; Li, J.-S. Triptolide ameliorates IL-10-deficient mice colitis by mechanisms involving suppression of IL-6/STAT3 signaling pathway and down-regulation of IL-17. *Mol. Immunol.* **2010**, *47*, 2467–2474. [CrossRef]
197. Qi, Q.; Li, H.; Lin, Z.-M.; Yang, X.-Q.; Zhu, F.-H.; Liu, Y.-T.; Shao, M.-J.; Zhang, L.-Y.; Xu, Y.-S.; Yan, Y.-X. (5 R)-5-hydroxytriptolide ameliorates anti-glomerular basement membrane glomerulonephritis in NZW mice by regulating Fcγ receptor signaling. *Acta Pharmacol. Sin.* **2018**, *39*, 107–116. [CrossRef]
198. Fan, D.; Guo, Q.; Shen, J.; Zheng, K.; Lu, C.; Zhang, G.; Lu, A.; He, X. The effect of triptolide in rheumatoid arthritis: From basic research towards clinical translation. *Int. J. Mol. Sci.* **2018**, *19*, 376. [CrossRef]
199. Xi, C.; Peng, S.; Wu, Z.; Zhou, Q.; Zhou, J. Toxicity of triptolide and the molecular mechanisms involved. *Biomed. Pharmacother.* **2017**, *90*, 531–541. [CrossRef]
200. Zhao, M.; Liang, G.-P.; Tang, M.-N.; Luo, S.-Y.; Zhang, J.; Cheng, W.-J.; Chan, T.-M.; Lu, Q.-J. Total glucosides of paeony induces regulatory $CD4^+$ $CD25^+$ T cells by increasing Foxp3 demethylation in lupus $CD4^+$ T cells. *Clin. Immunol.* **2012**, *143*, 180–187. [CrossRef] [PubMed]
201. Zhang, L.; Wei, W. Anti-inflammatory and immunoregulatory effects of paeoniflorin and total glucosides of paeony. *Pharmacol. Ther.* **2020**, *207*, 107452. [CrossRef] [PubMed]
202. Chen, Y.; Wang, L.; Cao, Y.; Li, N. Total glucosides of *Paeonia lactiflora* for safely reducing disease activity in systemic lupus erythematosus: A systematic review and meta-analysis. *Front. Pharmacol.* **2022**, *13*, 834947. [CrossRef] [PubMed]
203. Jun, L.; Junshan, L.; Aiwu, Z.; Minzhu, C.; Shuyun, X. Modulatory effects of total glucosides of paeony on B lymphocyte proliferation and interleukin 1 production. *Chin. J. Pharmacol. Toxicol.* **1994**, *8*, 53–55.
204. Wang, X.W.; Chen, M.Z.; Xu, S.Y. The effects of total glucosides of paeony (TGP) on T lymphocyte subsets. *Chin. Pharmacol. Bull.* **1992**, *8*, 340–343.
205. Wang, X.; Cheng, M.; Xu, S. Effects of total glucosides of paeony on immune system. *Zhongguo Bing Li Sheng Li Za Zhi* **1991**, *7*, 609–611.
206. Wang, X.; Chen, M.; Xu, S. The effects of total glucosides' of paeony (TGP) on T lymphocyte subsets. *Zhongguo Yao Li Xue Tong Bao* **1992**, *8*, 340–343.
207. Li, M.; Li, Y.; Xiang, L.; Li, L. Efficacy and safety of total glucosides of paeony as an add-on treatment in adolescents and adults with chronic urticaria: A systematic review and meta-analysis. *Front. Pharmacol.* **2022**, *13*, 961371. [CrossRef]
208. Goh, E.; Tan, L.; Chow, S.; Teh, L.; Yeap, S. Use of complementary medicine in systemic lupus erythematosus patients in Malaysia. *APLAR J. Rheumatol.* **2003**, *6*, 21–25. [CrossRef]
209. Hamedi, A.; Sohrabpour, M.; Zarshenas, M.M.; Pasdaran, A. Phytochemical investigation and quantitative analysis of the fatty acids and sterol compounds of seven pharmaceutical valuable seeds. *Curr. Pharm. Anal.* **2018**, *14*, 475–482. [CrossRef]
210. Leiba, A.; Amital, H.; Gershwin, M.E.; Shoenfeld, Y. Diet and lupus. *Lupus* **2001**, *10*, 246–248. [CrossRef] [PubMed]

211. Maki, P.A.; Newberne, P.M. Dietary lipids and immune function. *J. Nutr.* **1992**, *122*, 610–614. [CrossRef] [PubMed]
212. Harbige, L.S. Fatty acids, the immune response, and autoimmunity: A question of n − 6 essentiality and the balance between n − 6 and n − 3. *Lipids* **2003**, *38*, 323–341. [CrossRef] [PubMed]
213. Ramessar, N.; Borad, A.; Schlesinger, N. The effect of Omega-3 fatty acid supplementation in systemic lupus erythematosus patients: A systematic review. *Lupus* **2022**, *31*, 287–296. [CrossRef] [PubMed]
214. Hejr, H.; Ghareghani, M.; Zibara, K.; Ghafari, M.; Sadri, F.; Salehpour, Z.; Hamedi, A.; Negintaji, K.; Azari, H.; Ghanbari, A. The ratio of 1/3 linoleic acid to alpha linolenic acid is optimal for oligodendrogenesis of embryonic neural stem cells. *Neurosci. Lett.* **2017**, *651*, 216–225. [CrossRef]
215. Wei, Y.; Meng, Y.; Li, N.; Wang, Q.; Chen, L. The effects of low-ratio n-6/n-3 PUFA on biomarkers of inflammation: A systematic review and meta-analysis. *Food Funct.* **2021**, *12*, 30–40. [CrossRef]
216. Borges, M.C.; Santos, F.d.M.M.; Telles, R.W.; Correia, M.I.T.D.; Lanna, C.C.D. Polyunsaturated omega-3 fatty acids and systemic lupus erythematosus: What do we know? *Rev. Bras. Reumatol.* **2014**, *54*, 459–466. [CrossRef]
217. MacLean, C.H.; Mojica, W.A.; Morton, S.C.; Pencharz, J.; Garland, R.H.; Tu, W.; Newberry, S.J.; Jungvig, L.K.; Grossman, J.; Khanna, P. Effects of omega-3 fatty acids on lipids and glycemic control in type II diabetes and the metabolic syndrome and on inflammatory bowel disease, rheumatoid arthritis, renal disease, systemic lupus erythematosus, and osteoporosis: Summary. In *AHRQ Evidence Report Summaries*; Agency for Healthcare Research and Quality: Rockville, MD, USA, 2004.
218. Pestka, J.J. n-3 polyunsaturated fatty acids and autoimmune-mediated glomerulonephritis. *Prostaglandins Leukot. Essent. Fat. Acids (PLEFA)* **2010**, *82*, 251–258. [CrossRef]
219. Fassett, R.G.; Gobe, G.C.; Peake, J.M.; Coombes, J.S. Omega-3 polyunsaturated fatty acids in the treatment of kidney disease. *Am. J. Kidney Dis.* **2010**, *56*, 728–742. [CrossRef]
220. Tam, L.S.; Li, E.K.; Leung, V.Y.; Griffith, J.F.; Benzie, I.F.; Lim, P.L.; Whitney, B.; Lee, V.W.; Lee, K.K.; Thomas, G.N. Effects of vitamins C and E on oxidative stress markers and endothelial function in patients with systemic lupus erythematosus: A double blind, placebo controlled pilot study. *J. Rheumatol.* **2005**, *32*, 275–282.
221. Kamen, D.L.; Cooper, G.S.; Bouali, H.; Shaftman, S.R.; Hollis, B.W.; Gilkeson, G.S. Vitamin D deficiency in systemic lupus erythematosus. *Autoimmun. Rev.* **2006**, *5*, 114–117. [CrossRef] [PubMed]
222. Solovastru, L.G.; Vâta, D.; Statescu, L.; Constantin, M.M.; Andrese, E. Skin cancer between myth and reality, yet ethically constrained. *Rev. Romana Bioet.* **2014**, *12*, 47–52.
223. Klack, K.; Bonfa, E.; Borba Neto, E.F. Diet and nutritional aspects in systemic lupus erythematosus. *Rev. Bras. Reumatol.* **2012**, *52*, 395–408. [CrossRef]
224. Falcão, S.; Barros, R.; Mateus, M.; Nero, P.; De Matos, A.A.; Pimentão, J.B.; Ribeiro, I.; Weigert, A.; Branco, J. Lúpus eritematoso sistiémico e anemia. *Acta Reumatol. Port.* **2007**, *32*, 73–79.
225. Sahebari, M.; Rezaieyazdi, Z.; Khodashahi, M. Selenium and autoimmune diseases: A review article. *Curr. Rheumatol. Rev.* **2019**, *15*, 123–134. [CrossRef] [PubMed]
226. Brown, A.C. Lupus erythematosus and nutrition: A review of the literature. *J. Ren. Nutr.* **2000**, *10*, 170–183. [CrossRef] [PubMed]
227. Selmi, C.; Tsuneyama, K. Nutrition, geoepidemiology, and autoimmunity. *Autoimmun. Rev.* **2010**, *9*, A267–A270. [CrossRef] [PubMed]
228. Zhang, Y.-X.; Sun, H.-X. Immunosuppressive effect of ethanol extract of *Artemisia annua* on specific antibody and cellular responses of mice against ovalbumin. *Immunopharmacol. Immunotoxicol.* **2009**, *31*, 625–630. [CrossRef] [PubMed]
229. Zhao, X. Effects of *Astragalus membranaceus* and *Tripterygium hypoglancum* on natural killer cell activity of peripheral blood mononuclear in systemic lupus erythematosus. *Zhongguo Zhong Xi Yi Jie He Za Zhi Zhongguo Zhongxiyi Jiehe Zazhi = Chin. J. Integr. Tradit. West. Med.* **1992**, *12*, 669–671, 645.
230. Nurdiana; Dantara, T.W.I.; Syaban, M.F.R.; Mustafa, S.A.; Ikhsani, H.; Syafitri, F.E.; Hapsari, N.K.; Khoirunnisa, A. Efficacy and side effects studies of *Bryophyllum pinnatum* leaves ethanol extract in pristane-induced SLE BALB/c mice model. *AIP Conf. Proc.* **2019**, *2108*, 020016. [CrossRef]
231. Dantara, T.W.I.; Syaban, M.F.R.; Mustafa, S.A.; Ikhsani, H.; Syafitri, F.E.; Hapsari, N.K.; Khoirunnisa, A. Effect of *Bryophyllum pinnatum* leaves Ethanol extract in TNF-α and TGF-β as candidate therapy of SLE in pristane-induced SLE BALB/c mice model. *Res. J. Pharm. Technol.* **2021**, *14*, 1069–1072.
232. Handono, K.; Dantara, T.W.; Dewi, E.S.; Pratama, M.Z.; Nurdiana, N. *Bryophyllum pinnatum* leaves ethanol extract inhibit maturation and promote apoptosis of systemic lupus erythematosus BALB/c mice B cells. *Med. J. Indones.* **2017**, *26*, 253–260. [CrossRef]
233. Indriyanti, N.; Garmana, A.N.; Setiawan, F. Repairing effects of aqueous extract of *Kalanchoe pinnata* (Lmk) Pers. on lupus nephritis mice. *Pharmacogn. J.* **2018**, *10*, 548–552. [CrossRef]
234. Shamekhi, Z.; Amani, R.; Habibagahi, Z.; Namjoyan, F.; Ghadiri, A.; Saki Malehi, A. A Randomized, Double-blind, Placebo-controlled Clinical Trial Examining the Effects of Green Tea Extract on Systemic Lupus Erythematosus Disease Activity and Quality of Life. *Phytother. Res.* **2017**, *31*, 1063–1071. [CrossRef] [PubMed]
235. Khajehdehi, P.; Zanjaninejad, B.; Aflaki, E.; Nazarinia, M.; Azad, F.; Malekmakan, L.; Dehghanzadeh, G.-R. Oral supplementation of turmeric decreases proteinuria, hematuria, and systolic blood pressure in patients suffering from relapsing or refractory lupus nephritis: A randomized and placebo-controlled study. *J. Ren. Nutr.* **2012**, *22*, 50–57. [CrossRef]
236. Cai, Z.; Wong, C.K.; Dong, J.; Jiao, D.; Chu, M.; Leung, P.C.; San Lau, C.B.; Lau, C.P.; Tam, L.S.; Lam, C.W.K. Anti-inflammatory activities of *Ganoderma lucidum* (*Lingzhi*) and San-Miao-San supplements in MRL/lpr mice for the treatment of systemic lupus erythematosus. *Chin. Med.* **2016**, *11*, 23. [CrossRef] [PubMed]

237. Lai, N.; Lin, R.; Lai, R.; Kun, U.; Leu, S. Prevention of autoantibody formation and prolonged survival in New Zealand Black/New Zealand White F1 mice with an ancient Chinese herb, *Ganoderma tsugae*. *Lupus* **2001**, *10*, 461–465. [CrossRef]
238. Yuan, Z.; Feng, J. Observation on the treatment of systemic lupus erythematous with a *Gentiana macrophylla* complex tablet and a minimal dose of prednisone. *Zhong Xi Yi Jie He Za Zhi = Chin. J. Mod. Dev. Tradit. Med.* **1989**, *9*, 156–157, 133.
239. Sheu, M.-J.; Chiu, C.-C.; Yang, D.-J.; Hsu, T.-C.; Tzang, B.-S. The root extract of *Gentiana macrophylla* Pall. alleviates B19-NS1-exacerbated liver injuries in NZB/W F1 mice. *J. Med. Food* **2017**, *20*, 56–64. [CrossRef]
240. Clark, W.F.; Kortas, C.; Heidenheim, A.P.; Garland, J.; Spanner, E.; Parbtani, A. Flaxseed in lupus nephritis: A two-year nonplacebo-controlled crossover study. *J. Am. Coll. Nutr.* **2001**, *20*, 143–148. [CrossRef] [PubMed]
241. Clark, W.F.; Parbtani, A.; Huff, M.W.; Spanner, E.; de Salis, H.; Chin-Yee, I.; Philbrick, D.J.; Holub, B.J. Flaxseed: A potential treatment for lupus nephritis. *Kidney Int.* **1995**, *48*, 475–480. [CrossRef] [PubMed]
242. Chen, J.-L.; Chen, Y.-C.; Yang, S.-H.; Ko, Y.-F.; Chen, S.-Y. Immunological alterations in lupus-prone autoimmune (NZB/NZW) F1 mice by mycelia Chinese medicinal fungus *Cordyceps sinensis*-induced redistributions of peripheral mononuclear T lymphocytes. *Clin. Exp. Med.* **2009**, *9*, 277–284. [CrossRef] [PubMed]
243. Chen, J.-R.; Yen, J.-H.; Lin, C.-C.; Tsai, W.-J.; Liu, W.-J.; Tsai, J.-J.; Lin, S.-F.; Liu, H.-W. The effects of Chinese herbs on improving survival and inhibiting anti-ds DNA antibody production in lupus mice. *Am. J. Chin. Med.* **1993**, *21*, 257–262. [CrossRef] [PubMed]
244. Wang, W.; Cao, L.; Wang, X.; Fan, Y. *Radix Paeoniae Rubra* Ameliorates Lupus Nephritis in Lupus-Like Symptoms of Mrl Mice by Reducing Intercellular Cell Adhesion Molecule-1, Vascular Cell Adhesion Molecule-1, and Platelet Endothelial Cell Adhesion Molecule-1 Expression. *Comb. Chem. High Throughput Screen.* **2020**, *23*, 675–683. [CrossRef] [PubMed]
245. Yine, H.; Dai, S.; Wang, B.; Wei, Q. Effect of moutan cortex extract on Th17 cells of patients with systemic lupus erythematosus. *Chin. J. Biochem. Pharm.* **2015**, *6*, 119–121.
246. Ming, L.; Jingjing, M.; Zhao, X.; Zhu, Y. Clinical study of *Radix Astragali*, *Radix Rehmanniae* combined with glucocorticoid in treating systemic lupus erythematosus. *Int. J. Tradit. Chin. Med.* **2012**, *34*, 203–206.
247. Chae, B.-S.; Yang, J.-H. Regulatory effect of fresh *rehmanniae radix* extract on the in vitro production of proinflammatory cytokines in pristane-induced lupus mice. *Nat. Prod. Sci.* **2007**, *13*, 322–327.
248. Shin, T.-Y.; Oh, C.-H.; Kim, D.-K.; Eun, J.-S.; Jeon, H.; Park, J.-S.; Kim, M.-S.; Yang, J.-H.; Chae, B.-S. Regulatory effect of *Scutellariae radix* on the proinflammatory cytokine production and abnormal T-cell activation in vitro in pristane-induced lupus mice. *Nat. Prod. Sci.* **2007**, *13*, 207–213.
249. Wang, B.; Yuan, Z. A tablet of *Tripterygium wilfordii* in treating lupus erythematosus. *Zhong Xi Yi Jie He Za Zhi = Chin. J. Mod. Dev. Tradit. Med.* **1989**, *9*, 407–408, 389.
250. Qin, W.Z.; Zhu, G.D.; Yang, S.M.; Han, K.Y.; Wang, J. Clinical observations on *Tripterygium wilfordii* in the treatment of 26 cases of discoid lupus erythematosus. *J. Trad. Chin. Med.* **1983**, *3*, 131–132.
251. Treasure, J. Urtica semen reduces serum creatinine levels. *J. Am. Herb. Guild* **2003**, *4*, 22–25.
252. Musette, P.; Galelli, A.; Chabre, H.; Callard, P.; Peumans, W.; Truffa-Bachi, P.; Kourilsky, P.; Gachelin, G. Urtica dioica agglutinin, a Vβ8. 3-specific superantigen, prevents the development of the systemic lupus erythematosus-like pathology of MRL lpr/lpr mice. *Eur. J. Immunol.* **1996**, *26*, 1707–1711. [CrossRef]
253. Tao, X.; Davis, L.S.; Lipsky, P.E. Effect of an extract of the Chinese herbal remedy *Tripterygium wilfordii* Hook F on human immune responsiveness. *Arthritis Rheum. Off. J. Am. Coll. Rheumatol.* **1991**, *34*, 1274–1281. [CrossRef] [PubMed]
254. Tao, X.; Lipsky, P.E. The Chinese anti-inflammatory and immunosuppressive herbal remedy *Tripterygium wilfordii* Hook F. *Rheum. Dis. Clin. North Am.* **2000**, *26*, 29–50. [CrossRef] [PubMed]
255. Setty, A.R.; Sigal, L.H. Herbal medications commonly used in the practice of rheumatology: Mechanisms of action, efficacy, and side effects. *Semin. Arthritis Rheum.* **2005**, 773–784. [CrossRef] [PubMed]
256. Li, X.-Y. Anti-inflammatory and immunosuppressive components of *Tripterygium wilfordii* Hook F. *Int. J. Immunother.* **1993**, *9*, 181–187.
257. Takaishi, Y.; Shishido, K.; Wariishi, N.; Shibuya, M.; Goto, K.; Kido, M.; Taka, M.; Ono, Y. Triptoquinone A and B novel interleukin-1 inhibitors from *Tripterygium wilfordii* var Regeli. *Tetrahedron Lett.* **1992**, *33*, 7177–7180. [CrossRef]
258. Juling, G.; Shixiang, Y.; Xichun, W.; Shixi, X.; Deda, L. *Tripterygium wilfordii* Hook, f. in rheumatoid arthritis and ankylosing spondylitis. *Zhonghua Yixue Zazhi* **1981**, *94*, 405–412.
259. Wong, K.-F.; Yuan, Y.; Luk, J.M. *Tripterygium wilfordii* bioactive compounds as anticancer and anti-inflammatory agents. *Clin. Exp. Pharmacol. Physiol.* **2012**, *39*, 311–320. [CrossRef]
260. Qin, W.; Wanzhang, Q.; Chenghuang, L.; Shumei, Y.; Guangdou, Z. *Tripterygium wilfordii* Hook F in systemic lupus erythematosus: Report of 103 cases. *Chin. Med. J.* **1981**, *94*, 827–834.
261. Xu, G.; Tu, W.; Jiang, D.; Xu, C. *Tripterygium wilfordii* Hook F treatment for idiopathic refractory nephrotic syndrome in adults: A meta-analysis. *Nephron Clin. Pract.* **2009**, *111*, c223–c228. [CrossRef] [PubMed]
262. Ge, Y.; Xie, H.; Li, S.; Jin, B.; Hou, J.; Zhang, H.; Shi, M.; Liu, Z. Treatment of diabetic nephropathy with *Tripterygium wilfordii* Hook F extract: A prospective, randomized, controlled clinical trial. *J. Transl. Med.* **2013**, *11*, 1–9. [CrossRef] [PubMed]
263. Ji, S.-M.; Wang, Q.-W.; Chen, J.-S.; Sha, G.-Z.; Liu, Z.-H.; Li, L.-S. Clinical trial of *Tripterygium wilfordii* Hook F. in human kidney transplantation in China. *Transplant. Proc.* **2006**, 1274–1279. [CrossRef] [PubMed]
264. Baral, B.; Shrestha, B.; da Silva, J.T. A review of Chinese Cordyceps with special reference to Nepal, focusing on conservation. *Environ. Exp. Biol.* **2015**, *13*, 61–73.

265. Xu, J.; Huang, Y.; Chen, X.X.; Zheng, S.C.; Chen, P.; Mo, M.H. The mechanisms of pharmacological activities of Ophiocordyceps sinensis fungi. *Phytother. Res.* **2016**, *30*, 1572–1583. [CrossRef] [PubMed]
266. Liu, Y.; Li, Q.-Z.; Li, L.-D.-J.; Zhou, X.-W. Immunostimulatory effects of the intracellular polysaccharides isolated from liquid culture of *Ophiocordyceps sinensis* (Ascomycetes) on RAW264.7 cells via the MAPK and PI3K/Akt signaling pathways. *J. Ethnopharmacol.* **2021**, *275*, 114130. [CrossRef] [PubMed]
267. Qian, G.-M.; Pan, G.-F.; Guo, J.-Y. Anti-inflammatory and antinociceptive effects of cordymin, a peptide purified from the medicinal mushroom Cordyceps sinensis. *Nat. Prod. Res.* **2012**, *26*, 2358–2362. [CrossRef] [PubMed]
268. Ng, T.B.; Wong, J.H.; Fang, E.F. Recent Research on Pharmacological Activities of the Medicinal Fungus Cordyceps sinensis. In *Antitumor Potential and other Emerging Medicinal Properties of Natural Compounds*; Fang, E.F., Ng, T.B., Eds.; Springer: Dordrecht, The Netherlands, 2013; pp. 303–314.
269. Yang, M.-L.; Kuo, P.-C.; Hwang, T.-L.; Wu, T.-S. Anti-inflammatory principles from Cordyceps sinensis. *J. Nat. Prod.* **2011**, *74*, 1996–2000. [CrossRef]
270. Tong, X.; Guo, J. High Throughput Identification of the Potential Antioxidant Peptides in Ophiocordyceps sinensis. *Molecules* **2022**, *27*, 438. [CrossRef]
271. Xiang, F.; Lin, L.; Hu, M.; Qi, X. Therapeutic efficacy of a polysaccharide isolated from Cordyceps sinensis on hypertensive rats. *Int. J. Biol. Macromol.* **2016**, *82*, 308–314. [CrossRef]
272. Ding, C.; Tian, P.; Xue, W.; Ding, X.; Yan, H.; Pan, X.; Feng, X.; Xiang, H.; Hou, J.; Tian, X. Efficacy of Cordyceps sinensis in long term treatment of renal transplant patients. *Front. Biosci.-Elite* **2011**, *3*, 301–307.
273. Chen, B.; Sun, Y.; Luo, F.; Wang, C. Bioactive metabolites and potential mycotoxins produced by Cordyceps fungi: A review of safety. *Toxins* **2020**, *12*, 410. [CrossRef] [PubMed]
274. Maczynska, I.; Millo, B.; Ratajczak-Stefańska, V.; Maleszka, R.; Szych, Z.; Kurpisz, M.; Giedrys-Kalemba, S. Proinflammatory cytokine (IL-1β, IL-6, IL-12, IL-18 and TNF-α) levels in sera of patients with subacute cutaneous lupus erythematosus (SCLE). *Immunol. Lett.* **2006**, *102*, 79–82. [CrossRef]
275. Umare, V.; Pradhan, V.; Nadkar, M.; Rajadhyaksha, A.; Patwardhan, M.; Ghosh, K.K.; Nadkarni, A.H. Effect of proinflammatory cytokines (IL-6, TNF-α, and IL-1β) on clinical manifestations in Indian SLE patients. *Mediat. Inflamm.* **2014**, *2014*, 385297. [CrossRef] [PubMed]
276. Bao, Y.X.; Wong, K.C.; Li, E.K.M.; Tam, S.L.; Leung, C.P.; Yin, B.Y.; Lam, C.W.K. Immunomodulatory effects of lingzhi and san-miao-san supplementation on patients with rheumatoid arthritis. *Immunopharmacol. Immunotoxicol.* **2006**, *28*, 197–200. [CrossRef]
277. Pazzi, F.; Fraile-Fabero, R. Effects of *Ganoderma Lucidum* on Pain in Women with Fibromyalgia. *Fibrom Open Access* **2017**, *2*, 2.
278. Tang, W.; Gao, Y.; Chen, G.; Gao, H.; Dai, X.; Ye, J.; Chan, E.; Huang, M.; Zhou, S. A randomized, double-blind and placebo-controlled study of a *Ganoderma lucidum* polysaccharide extract in neurasthenia. *J. Med. Food* **2005**, *8*, 53–58. [CrossRef]
279. Zhong, L.; Yan, P.; Lam, W.C.; Yao, L.; Bian, Z. Coriolus versicolor and *Ganoderma lucidum* related natural products as an adjunct therapy for cancers: A systematic review and meta-analysis of randomized controlled trials. *Front. Pharmacol.* **2019**, *10*, 703. [CrossRef]
280. Klupp, N.L.; Kiat, H.; Bensoussan, A.; Steiner, G.Z.; Chang, D.H. A double-blind, randomised, placebo-controlled trial of *Ganoderma lucidum* for the treatment of cardiovascular risk factors of metabolic syndrome. *Sci. Rep.* **2016**, *6*, 1–10. [CrossRef]
281. Noguchi, M.; Kakuma, T.; Tomiyasu, K.; Yamada, A.; Itoh, K.; Konishi, F.; Kumamoto, S.; Shimizu, K.; Kondo, R.; Matsuoka, K. Randomized clinical trial of an ethanol extract of *Ganoderma lucidum* in men with lower urinary tract symptoms. *Asian J. Androl.* **2008**, *10*, 777–785. [CrossRef]
282. Henao, S.L.D.; Urrego, S.A.; Cano, A.M.; Higuita, E.A. Randomized Clinical Trial for the Evaluation of Immune Modulation by Yogurt Enriched with β-Glucans from *Lingzhi* or *Reishi* Medicinal Mushroom, *Ganoderma lucidum* (Agaricomycetes), in Children from Medellin, Colombia. *Int. J. Med. Mushrooms* **2018**, *20*, 705–716. [CrossRef] [PubMed]
283. Hamedi, A.; Rezaei, H.; Azarpira, N.; Jafarpour, M.; Ahmadi, F. Effects of *Malva sylvestris* and its isolated polysaccharide on experimental ulcerative colitis in rats. *J. Evid.-Based Complement. Altern. Med.* **2016**, *21*, 14–22. [CrossRef] [PubMed]
284. Yu, Y.; Shen, M.; Song, Q.; Xie, J. Biological activities and pharmaceutical applications of polysaccharide from natural resources: A review. *Carbohydr. Polym.* **2018**, *183*, 91–101. [CrossRef] [PubMed]
285. Lu, J.; He, R.; Sun, P.; Zhang, F.; Linhardt, R.J.; Zhang, A. Molecular mechanisms of bioactive polysaccharides from *Ganoderma lucidum* (*Lingzhi*), a review. *Int. J. Biol. Macromol.* **2020**, *150*, 765–774. [CrossRef]
286. Ren, L.; Zhang, J.; Zhang, T. Immunomodulatory activities of polysaccharides from *Ganoderma* on immune effector cells. *Food Chem.* **2021**, *340*, 127933. [CrossRef] [PubMed]
287. Koo, M.H.; Chae, H.-J.; Lee, J.H.; Suh, S.-S.; Youn, U.J. Antiinflammatory lanostane triterpenoids from *Ganoderma lucidum*. *Nat. Prod. Res.* **2021**, *35*, 4295–4302. [CrossRef]
288. Liang, C.; Tian, D.; Liu, Y.; Li, H.; Zhu, J.; Li, M.; Xin, M.; Xia, J. Review of the molecular mechanisms of *Ganoderma lucidum* triterpenoids: Ganoderic acids A, C2, D, F, DM, X and Y. *Eur. J. Med. Chem.* **2019**, *174*, 130–141. [CrossRef] [PubMed]
289. Sharma, C.; Bhardwaj, N.; Sharma, A.; Tuli, H.S.; Batra, P.; Beniwal, V.; Gupta, G.K.; Sharma, A.K. Bioactive metabolites of *Ganoderma lucidum*: Factors, mechanism and broad spectrum therapeutic potential. *J. Herb. Med.* **2019**, *17*, 100268. [CrossRef]
290. Ahmad, M.F. *Ganoderma lucidum*: Persuasive biologically active constituents and their health endorsement. *Biomed. Pharmacother.* **2018**, *107*, 507–519. [CrossRef]

291. Dewi, S.C.; Sargowo, D.; Widodo, M.A.; Wihastuti, T.A.; Heriansyah, T.; Hartanto, M.A.T.; Pambayun, I.D.A.; Bakhri, S.; A'ini, N.Q.; Sahudi, D.P. *Ganoderma lucidum* subchronic toxicity on the liver as anti-oxidant and anti-inflamatory agent for cardivascular disease. *J. Hypertens.* **2015**, *33*, e30. [CrossRef]
292. Fattahi, S.; Golpour, M.; Akhavan Niaki, H. *Urtica Dioica*, An emerauld in the medical Kingdom. *Int. Biol. Biomed. J.* **2016**, *2*, 1–10.
293. Yarnell, E.; Abascal, K. Lupus erythematosus and herbal medicine. *Altern. Complement. Ther.* **2008**, *14*, 9–12. [CrossRef]
294. Francišković, M.; Gonzalez-Pérez, R.; Orčić, D.; Sanchez de Medina, F.; Martínez-Augustin, O.; Svirčev, E.; Simin, N.; Mimica-Dukić, N. Chemical Composition and Immuno-Modulatory Effects of *Urtica dioica* L. (Stinging Nettle) Extracts. *Phytother. Res.* **2017**, *31*, 1183–1191. [CrossRef] [PubMed]
295. Akbay, P.; Basaran, A.A.; Undeger, U.; Basaran, N. In vitro immunomodulatory activity of flavonoid glycosides from *Urtica dioica* L. *Phytother. Res. Int. J. Devoted Pharmacol. Toxicol. Eval. Nat. Prod. Deriv.* **2003**, *17*, 34–37. [CrossRef] [PubMed]
296. Semalty, M.; Adhikari, L.; Semwal, D.; Chauhan, A.; Mishra, A.; Kotiyal, R.; Semalty, A. A comprehensive review on phytochemistry and pharmacological effects of stinging nettle (*Urtica dioica*). *Curr. Tradit. Med.* **2017**, *3*, 156–167. [CrossRef]
297. Grauso, L.; de Falco, B.; Lanzotti, V.; Motti, R. Stinging nettle, *Urtica dioica* L.: Botanical, phytochemical and pharmacological overview. *Phytochem. Rev.* **2020**, *19*, 1341–1377. [CrossRef]
298. Taheri, Y.; Quispe, C.; Herrera-Bravo, J.; Sharifi-Rad, J.; Ezzat, S.M.; Merghany, R.M.; Shaheen, S.; Azmi, L.; Prakash Mishra, A.; Sener, B. *Urtica* dioica-derived phytochemicals for pharmacological and therapeutic applications. *Evid.-Based Complement. Altern. Med.* **2022**, *2022*, 4024331. [CrossRef]
299. Showkat, Q.A.; Rather, J.A.; Jabeen, A.; Dar, B.; Makroo, H.A.; Majid, D. Bioactive components, physicochemical and starch characteristics of different parts of lotus (*Nelumbo nucifera* Gaertn.) plant: A review. *Int. J. Food Sci. Technol.* **2021**, *56*, 2205–2214. [CrossRef]
300. Limwachiranon, J.; Huang, H.; Shi, Z.; Li, L.; Luo, Z. Lotus flavonoids and phenolic acids: Health promotion and safe consumption dosages. *Compr. Rev. Food Sci. Food Saf.* **2018**, *17*, 458–471. [CrossRef]
301. Bishayee, A.; Patel, P.A.; Sharma, P.; Thoutireddy, S.; Das, N. Lotus (*Nelumbo nucifera* Gaertn.) and its bioactive phytocompounds: A tribute to cancer prevention and intervention. *Cancers* **2022**, *14*, 529. [CrossRef]
302. Ekiert, H.; Świątkowska, J.; Klin, P.; Rzepiela, A.; Szopa, A. *Artemisia annua*–Importance in Traditional Medicine and Current State of Knowledge on the Chemistry, Biological Activity and Possible Applications. *Planta Med.* **2021**, *87*, 584–599. [CrossRef] [PubMed]
303. Rolta, R.; Salaria, D.; Sharma, P.; Sharma, B.; Kumar, V.; Rathi, B.; Verma, M.; Sourirajan, A.; Baumler, D.J.; Dev, K. Phytocompounds of *Rheum emodi*, *Thymus serpyllum*, and *Artemisia annua* Inhibit Spike Protein of SARS-CoV-2 Binding to ACE2 Receptor: In Silico Approach. *Curr. Pharmacol. Rep.* **2021**, *7*, 135–149. [CrossRef]
304. Law, S.; Leung, A.W.; Xu, C. Is the traditional Chinese herb "*Artemisia annua*" possible to fight against COVID-19? *Integr. Med. Res.* **2020**, *9*, 100474. [CrossRef]
305. Fuzimoto, A.D. An overview of the anti-SARS-CoV-2 properties of *Artemisia annua*, its antiviral action, protein-associated mechanisms, and repurposing for COVID-19 treatment. *J. Integr. Med.* **2021**, *19*, 375–388. [CrossRef] [PubMed]
306. Yang, L. 杨岚; Zhang, D. 张东. 双氢青蒿素及其红斑狼疮新适应症研究概述. 科学通报 **2017**, *62*, 2007–2012.
307. Willcox, M.; Bodeker, G.; Rasoanaivo, P.; Addae-Kyereme, J. *Traditional Medicinal Plants and Malaria*; CRC Press: Boca Raton, FL, USA, 2004; Volume 4.
308. Ashok, P.K.; Upadhyaya, K. Preliminary Phytochemical Screening and Physico-Chemical Parameters of *Artemisia absinthium* and *Artemisia annua*. *J. Pharmacogn. Phytochem.* **2013**, *1*, 229–235.
309. Zhuang, G. Clinical study on the treatment of lupus erythematosus with *Artemisia apiacea* Hce. *Zhonghua Yi Xue Za Zhi* **1982**, *62*, 365–367.
310. Guokang, Z. Discoid Lupus Erythematosus Treated by *Artemisia Annua*. *Altern. Complement. Ther.* **2008**, *14*, 9–12.
311. Jun, D.Y.; Dong, L.W.; You, T.Y.; Na, Z.H.; Zhong, Z.W.; Lan, Y.; Bin, L.Z. The effects of DQHS on the pathologic changes in BXSB mice lupus nephritis and the effect mechanism. *Chin. Pharmacol. Bull.* **2003**, *19*, 1125–1128.
312. Dong, Y.; Li, W.; Tu, Y. Effect of dihydro-qinghaosu on auto-antibody production, TNF alpha secretion and pathologic change of lupus nephritis in BXSB mice. *Zhongguo Zhong Xi Yi Jie He Za Zhi Zhongguo Zhongxiyi Jiehe Zazhi = Chin. J. Integr. Tradit. West. Med.* **2003**, *23*, 676–679.
313. Yang, J.; Wen, C.; Duan, Y.; Deng, Q.; Peng, D.; Zhang, H.; Ma, H. The composition, extraction, analysis, bioactivities, bioavailability and applications in food system of flaxseed (*Linum usitatissimum* L.) oil: A review. *Trends Food Sci. Technol.* **2021**, *118*, 252–260. [CrossRef]
314. Akter, Y.; Junaid, M.; Afrose, S.S.; Nahrin, A.; Alam, M.S.; Sharmin, T.; Akter, R.; Hosen, S. A Comprehensive Review on *Linum usitatissimum* Medicinal Plant: Its Phytochemistry, Pharmacology, and Ethnomedicinal Uses. *Mini Rev. Med. Chem.* **2021**, *21*, 2801–2834. [CrossRef] [PubMed]
315. Liang, S.; Li, X.; Ma, X.; Li, A.; Wang, Y.; Reaney, M.J.; Shim, Y.Y. A flaxseed heteropolysaccharide stimulates immune responses and inhibits hepatitis B virus. *Int. J. Biol. Macromol.* **2019**, *136*, 230–240. [CrossRef] [PubMed]
316. Ibrahim, I.; Shalaby, A.; Abdallah, H.; El-Zohairy, N.; Bahr, H. Immunomodulatory and anti-angiogenic properties of *Linum usitatissium* (flaxseed) seeds ethanolic extract in atherogenic diet treated rats. *Adv. Anim. Vet. Sci* **2020**, *8*, 18–25. [CrossRef]
317. Touré, A.; Xu, X. Flaxseed lignans: Source, biosynthesis, metabolism, antioxidant activity, bio-active components, and health benefits. *Compr. Rev. Food Sci. Food Saf.* **2010**, *9*, 261–269. [CrossRef]
318. Yuan, B.; Han, J.; Cheng, Y.; Cheng, S.; Huang, D.; Mcclements, D.J.; Cao, C. Identification and characterization of antioxidant and immune-stimulatory polysaccharides in flaxseed hull. *Food Chem.* **2020**, *315*, 126266.
319. Bashir, S.; Ali, S.; Khan, F. Partial reversal of obesity-induced insulin resistance owing to anti-inflammatory immunomodulatory potential of flaxseed oil. *Immunol. Investig.* **2015**, *44*, 451–469. [CrossRef]

320. Chytilová, M.; Mudroňová, D.; Nemcová, R.; Gancarčíková, S.; Buleca, V.; Koščová, J.; Tkáčiková, Ľ. Anti-inflammatory and immunoregulatory effects of flax-seed oil and *Lactobacillus plantarum*—Biocenol™ LP96 in gnotobiotic pigs challenged with enterotoxigenic *Escherichia coli*. *Res. Vet. Sci.* **2013**, *95*, 103–109. [CrossRef]
321. Herchi, W.; Harrabi, S.; Sebei, K.; Rochut, S.; Boukhchina, S.; Pepe, C.; Kallel, H. Phytosterols accumulation in the seeds of *Linum usitatissimum* L. *Plant Physiol. Biochem.* **2009**, *47*, 880–885. [CrossRef]
322. Vilahur, G.; Ben-Aicha, S.; Diaz-Riera, E.; Badimon, L.; Padró, T. Phytosterols and inflammation. *Curr. Med. Chem.* **2019**, *26*, 6724–6734. [CrossRef]
323. Hamedi, A.; Ghanbari, A.; Razavipour, R.; Saeidi, V.; Zarshenas, M.M.; Sohrabpour, M.; Azari, H. *Alyssum homolocarpum* seeds: Phytochemical analysis and effects of the seed oil on neural stem cell proliferation and differentiation. *J. Nat. Med.* **2015**, *69*, 387–396. [CrossRef] [PubMed]
324. Ogborn, M.R.; Nitschmann, E.; Bankovic-Calic, N.; Weiler, H.A.; Aukema, H. Dietary flax oil reduces renal injury, oxidized LDL content, and tissue n− 6/n− 3 FA ratio in experimental polycystic kidney disease. *Lipids* **2002**, *37*, 1059–1065. [CrossRef] [PubMed]
325. Naqshbandi, A.; Rizwan, S.; Khan, F. Dietary supplementation of flaxseed oil ameliorates the effect of cisplatin on rat kidney. *J. Funct. Foods* **2013**, *5*, 316–326. [CrossRef]
326. Al Za'abi, M.; Ali, H.; Ali, B.H. Effect of flaxseed on systemic inflammation and oxidative stress in diabetic rats with or without chronic kidney disease. *PLoS ONE* **2021**, *16*, e0258800. [CrossRef]
327. Ogborn, M.R. Flaxseed and flaxseed products in kidney disease. In *Flaxseed in Human Nutrition*; AOCS Press: Champaign, IL, USA, 2003; pp. 301–318.
328. Parikh, M.; Maddaford, T.G.; Austria, J.A.; Aliani, M.; Netticadan, T.; Pierce, G.N. Dietary flaxseed as a strategy for improving human health. *Nutrients* **2019**, *11*, 1171. [CrossRef] [PubMed]
329. Kim, S.-H.; Yook, T.-H.; Kim, J.-U. *Rehmanniae radix*, an effective treatment for patients with various inflammatory and metabolic diseases: Results from a review of Korean publications. *J. Pharmacopunct.* **2017**, *20*, 81.
330. Li, M.; Ma, J.-J.; Zhao, X.-L.; Zhu, Y. Treating lupus nephritis by a drug pair of *radix astragali* and *rehmanniae radix* combined with glucocorticoid: A preliminary clinical study. *Zhongguo Zhong Xi Yi Jie He Za Zhi Zhongguo Zhongxiyi Jiehe Zazhi = Chin. J. Integr. Tradit. West. Med.* **2014**, *34*, 956–959.
331. Li, M.; Jiang, H.; Hao, Y.; Du, K.; Du, H.; Ma, C.; Tu, H.; He, Y. A systematic review on botany, processing, application, phytochemistry and pharmacological action of *Radix Rehmnniae*. *J. Ethnopharmacol.* **2022**, *285*, 114820. [CrossRef]
332. Lu, M.-K.; Chang, C.-C.; Chao, C.-H.; Hsu, Y.-C. Structural changes, and anti-inflammatory, anti-cancer potential of polysaccharides from multiple processing of *Rehmannia glutinosa*. *Int. J. Biol. Macromol.* **2022**, *206*, 621–632. [CrossRef]
333. Zhou, Y.; Wang, S.; Feng, W.; Zhang, Z.; Li, H. Structural characterization and immunomodulatory activities of two polysaccharides from *Rehmanniae Radix* Praeparata. *Int. J. Biol. Macromol.* **2021**, *186*, 385–395. [CrossRef]
334. Choi, Y.H. Catalpol attenuates lipopolysaccharide-induced inflammatory responses in BV2 microglia through inhibiting the TLR4-mediated NF-κB pathway. *Gen. Physiol. Biophys.* **2019**, *38*, 111–122. [CrossRef] [PubMed]
335. Chi, X.; Wang, S.; Baloch, Z.; Zhang, H.; Li, X.; Zhang, Z.; Zhang, H.; Dong, Z.; Lu, Y.; Yu, H. Research progress on classical traditional Chinese medicine formula Lily Bulb and Rehmannia Decoction in the treatment of depression. *Biomed. Pharmacother.* **2019**, *112*, 108616. [CrossRef] [PubMed]
336. Song, W.-H.; Cheng, Z.-H.; Chen, D.-F. Anticomplement monoterpenoid glucosides from the root bark of *Paeonia suffruticosa*. *J. Nat. Prod.* **2014**, *77*, 42–48. [CrossRef] [PubMed]
337. Wang, Y.-J.; Li, Y.-X.; Li, S.; He, W.; Wang, Z.-R.; Zhan, T.-P.; Lv, C.-Y.; Liu, Y.-P.; Yang, Y.; Zeng, X.-X. Progress in traditional Chinese medicine and natural extracts for the treatment of lupus nephritis. *Biomed. Pharmacother.* **2022**, *149*, 112799. [CrossRef] [PubMed]
338. Ekiert, H.; Klimek-Szczykutowicz, M.; Szopa, A. *Paeonia* × *suffruticosa* (Moutan Peony)—A review of the chemical composition, traditional and professional use in medicine, position in cosmetics industries, and biotechnological studies. *Plants* **2022**, *11*, 3379. [CrossRef] [PubMed]
339. Tan, Y.-Q.; Chen, H.-W.; Li, J.; Wu, Q.-J. Efficacy, chemical constituents, and pharmacological actions of *Radix Paeoniae rubra* and *Radix Paeoniae alba*. *Front. Pharmacol.* **2020**, *11*, 1054. [CrossRef]
340. Parker, S.; May, B.; Zhang, C.; Zhang, A.L.; Lu, C.; Xue, C.C. A pharmacological review of bioactive constituents of *Paeonia lactiflora* Pallas and *Paeonia veitchii* Lynch. *Phytother. Res.* **2016**, *30*, 1445–1473. [CrossRef]
341. Pasdaran, A.; Butovska, D.; Kerr, P.; Naychov, Z.; Aneva, I.; Kozuharova, E. Gentians, natural remedies for future of visceral pain control; an ethnopharmacological review with an in silico approach. *Biol. Futur.* **2022**, *73*, 219–227. [CrossRef]
342. Cui, Y.; Jiang, L.; Shao, Y.; Mei, L.; Tao, Y. Anti-alcohol liver disease effect of *Gentianae macrophyllae* extract through MAPK/JNK/p38 pathway. *J. Pharm. Pharmacol.* **2019**, *71*, 240–250. [CrossRef]
343. Lohar, A.V.; Wankhade, A.M.; Faisal, M.; Jagtap, A. A review on *Glycyrrhiza glabra* linn (liquorice)—An excellent medicinal plant. *Eur. J. Biomed.* **2020**, *7*, 330–334.
344. Pandey, S.; Verma, B.; Arya, P. A review on constituents, pharmacological activities and medicinal uses of *Glycyrrhiza glabra*. *Pharm. Res.* **2017**, *2*, 26–31. [CrossRef]
345. Damle, M. *Glycyrrhiza glabra* (Liquorice)—A potent medicinal herb. *Int. J. Herb. Med.* **2014**, *2*, 132–136.
346. Kwon, Y.-J.; Son, D.-H.; Chung, T.-H.; Lee, Y.-J. A review of the pharmacological efficacy and safety of licorice root from corroborative clinical trial findings. *J. Med. Food* **2020**, *23*, 12–20. [CrossRef] [PubMed]

347. Yang, R.; Yuan, B.-C.; Ma, Y.-S.; Zhou, S.; Liu, Y. The anti-inflammatory activity of licorice, a widely used Chinese herb. *Pharm. Biol.* **2017**, *55*, 5–18. [CrossRef] [PubMed]
348. Mollica, L.; De Marchis, F.; Spitaleri, A.; Dallacosta, C.; Pennacchini, D.; Zamai, M.; Agresti, A.; Trisciuoglio, L.; Musco, G.; Bianchi, M.E. Glycyrrhizin binds to high-mobility group box 1 protein and inhibits its cytokine activities. *Chem. Biol.* **2007**, *14*, 431–441. [CrossRef]
349. Li, X.; Yue, Y.; Zhu, Y.; Xiong, S. Extracellular, but not intracellular HMGB1, facilitates self-DNA induced macrophage activation via promoting DNA accumulation in endosomes and contributes to the pathogenesis of lupus nephritis. *Mol. Immunol.* **2015**, *65*, 177–188. [CrossRef]
350. Nazari, S.; Rameshrad, M.; Hosseinzadeh, H. Toxicological effects of *Glycyrrhiza glabra* (licorice): A review. *Phytother. Res.* **2017**, *31*, 1635–1650. [CrossRef] [PubMed]
351. Wahab, S.; Ahmad, I.; Irfan, S.; Siddiqua, A.; Usmani, S.; Ahmad, M.P. Pharmacological Efficacy and Safety of *Glycyrrhiza glabra* in the treatment of respiratory tract infections. *Mini Rev. Med. Chem.* **2022**, *22*, 1476–1494. [CrossRef]
352. Liu, T.; Qiao, X.; Kong, K.; Wang, X.; Li, R.; Zhang, X. Bioinformatics-based Identification of the Mechanism Whereby Astragalus Membranaceus Inhibits Inflammation and Autophagy in Lupus Nephritis. *preprint* **2021**. [CrossRef]
353. Sheik, A.; Kim, K.; Varaprasad, G.L.; Lee, H.; Kim, S.; Kim, E.; Shin, J.-Y.; Oh, S.Y.; Huh, Y.S. The anti-cancerous activity of adaptogenic herb *Astragalus membranaceus*. *Phytomedicine* **2021**, *91*, 153698. [CrossRef]
354. Kamboj, A.; Saluja, A. *Bryophyllum pinnatum* (*Lam.*) Kurz.: Phytochemical and pharmacological profile: A review. *Pharmacogn. Rev.* **2009**, *3*, 364.
355. Handono, K.; Dantara, T.; Dewi, E.; Pratama, M.; Nurdiana, N. 94 Development of therapy using bryophyllum pinnatum to decrease maturation and increase apoptotic b cells from balb/c lupus mice: In silico and in vitro study approach. *Lupus Sci. Med.* **2017**, *4*. [CrossRef]
356. Kalsum, U.; Nurdiana, N.; Pratama, M.; Kalim, H.; Handono, K. 91 Potential novel natural b cell depleting and immunosuppression agent in lupus treapment using *bryophyllum pinnatum*. in silico and in pristane induced lupus mice. *Lupus Sci. Med.* **2017**, *4*. [CrossRef]
357. Indriyanti, N.; Soeroso, J.; Khotib, J. T-cell activation controlling effects of ethyl acetate fraction of *Kalanchoe pinnata* (lmk) pers on tmpd-treated lupus mice. *Int. J. Pharm. Sci. Res.* **2018**, *9*, 475–482.
358. Indriyanti, N.; Soeroso, J.; Khotib, J. The benefits of active compounds in *Kalanchoe pinnata* (LMK) pers ethyl acetate fraction on lupus arthritis mice. *Asian J. Pharm. Clin. Res.* **2017**, *10*, 1–5. [CrossRef]
359. Indriyanti, N.; Soeroso, J.; Khotib, J. Positive Impact of Ethyl Acetate Fraction of *Kalanchoe pinnata* on Anti-Smith Antibody and T Reg in Lupus Mice. *J. Ilmu Kefarmasian Indones.* **2017**, *15*, 57–62.
360. Indriyanti, N.; Garmana, A.N. Ekstrak Daun Cocor Bebek (*Kalanchoe pinnata*) Untuk Terapi Preventif Lupus pada Mencit yang Diinduksi dengan 2, 6, 10, 14 Tetramethylpentadecane. *J. Trop. Pharm. Chem.* **2011**, *1*, 221–226. [CrossRef]
361. Fernandes, J.M.; Cunha, L.M.; Azevedo, E.P.; Lourenço, E.M.; Fernandes-Pedrosa, M.F.; Zucolotto, S.M. *Kalanchoe laciniata* and *Bryophyllum pinnatum*: An updated review about ethnopharmacology, phytochemistry, pharmacology and toxicology. *Rev. Bras. Farmacogn.* **2019**, *29*, 529–558. [CrossRef]
362. Liu, T.; An, X.-N.; Liu, D.-L.; Wei, Y.-J. A comparison of several second-order algorithms for simultaneous determination of neomangiferin and mangiferin with severe spectral overlapping in Anemarrhenae Rhizoma. *Spectrochim. Acta Part A Mol. Biomol. Spectrosc.* **2019**, *208*, 172–178. [CrossRef]
363. Singh, K.G.; Sonia, S.; Konsoor, N. In-vitro and ex-vivo studies on the antioxidant, anti-inflammatory and antiarthritic properties of *Camellia sinensis, Hibiscus rosa sinensis, Matricaria chamomilla, Rosa* SP., *Zingiber officinale* tea extracts. *Inflammation* **2018**, *49*, 50.
364. Mawarti, H.; Nugraha, J.; Purwanto, D.A.; Soeroso, J. Identifying and Revealing Active Compound from Green Tea (*Camellia sinensis*) for Curing Systemic Lupus Erythematosus by Acting as CASPASE 1 Inhibitor. *Med. -Leg. Update* **2020**, *20*, 323–329.
365. Bedrood, Z.; Rameshrad, M.; Hosseinzadeh, H. Toxicological effects of *Camellia sinensis* (green tea): A review. *Phytother. Res.* **2018**, *32*, 1163–1180. [CrossRef] [PubMed]
366. Labban, L. Medicinal and pharmacological properties of Turmeric (*Curcuma longa*): A review. *Int. J. Pharm. Biomed. Sci.* **2014**, *5*, 17–23.
367. Gurung, P.; De, P. Spectrum of biological properties of *Cinchona alkaloids*: A brief review. *J. Pharmacogn. Phytochem.* **2017**, *6*, 162–166.
368. Kinsley-Scott, T.R.; Norton, S.A. Useful plants of dermatology. VII: Cinchona and antimalarials. *J. Am. Acad. Dermatol.* **2003**, *49*, 499–502. [CrossRef]
369. Weng, J.-K. Plant solutions for the COVID-19 pandemic and beyond: Historical reflections and future perspectives. *Mol. Plant* **2020**, *13*, 803–807. [CrossRef]
370. Shah, R.R. Chloroquine and hydroxychloroquine for COVID-19: Perspectives on their failure in repurposing. *J. Clin. Pharm. Ther.* **2021**, *46*, 17–27. [CrossRef]
371. Wu, G.-C.; Wu, H.; Fan, L.-Y.; Pan, H.-F. Saikosaponins: A potential treatment option for systemic lupus erythematosus. *Ir. J. Med. Sci.* **2011**, *180*, 259–261. [CrossRef]
372. Brinkhaus, B.; Lindner, M.; Schuppan, D.; Hahn, E. Chemical, pharmacological and clinical profile of the East Asian medical plant *Centella aslatica*. *Phytomedicine* **2000**, *7*, 427–448. [CrossRef]
373. Kundu, P.; Sharma, P.; Mahato, R.; Saha, M.; Das, S.; Ghosh, P. A brief review for the development of bio-nanoparticles using some important Indian ethnomedicinal plants. *J. Med. Plants* **2020**, *8*, 26–33. [CrossRef]
374. Islam, M. A Review on *Tinospora cordifolia* and its Immunomodulatory Activities. Ph.D. Thesis, Brac University, Dhaka, Bangladesh, 2020.

375. Abrantes, V.E.F.; Matias da Rocha, B.A.; Batista da Nóbrega, R.; Silva-Filho, J.C.; Teixeira, C.S.; Cavada, B.S.; de Almeida Gadelha, C.A.; Ferreira, S.H.; Figueiredo, J.G.; Santi-Gadelha, T. Molecular modeling of Lectin-like protein from *Acacia farnesiana* reveals a possible anti-inflammatory mechanism in carrageenan-induced inflammation. *BioMed Res. Int.* **2013**, *2013*, 253483. [CrossRef]
376. Delgadillo Puga, C.; Cuchillo Hilario, M.; Navarro Ocaña, A.; Medina Campos, O.N.; Nieto Camacho, A.; Ramírez Apan, T.; López Tecpoyotl, Z.G.; Díaz Martínez, M.; Álvarez Izazaga, M.A.; Cruz Martínez, Y.R. Phenolic Compounds in Organic and Aqueous Extracts from *Acacia farnesiana* Pods Analyzed by ULPS-ESI-Q-oa/TOF-MS. In Vitro Antioxidant Activity and Anti-Inflammatory Response in CD-1 Mice. *Molecules* **2018**, *23*, 2386.
377. Leal, L.S.S.; Silva, R.O.; Araújo, T.d.S.L.; Silva, V.G.d.; Barbosa, A.L.d.R.; Medeiros, J.V.R.; Oliveira, J.S.; Ventura, C.A. The anti-inflammatory and antinociceptive effects of proteins extracted from *Acacia farnesiana* seeds. *Rev. Bras. Plantas Med.* **2016**, *18*, 38–47. [CrossRef]
378. Almeida, É.S.; de Oliveira, D.; Hotza, D. Properties and applications of *Morinda citrifolia* (noni): A review. *Compr. Rev. Food Sci. Food Saf.* **2019**, *18*, 883–909. [CrossRef] [PubMed]
379. Dixon, A.R.; McMillen, H.; Etkin, N.L. Ferment this: The transformation of Noni, a traditional Polynesian medicine (*Morinda citrifolia, Rubiaceae*). *Econ. Bot.* **1999**, *53*, 51–68. [CrossRef]
380. McClatchey, W. From Polynesian healers to health food stores: Changing perspectives of Morinda citrifolia (*Rubiaceae*). *Integr. Cancer Ther.* **2002**, *1*, 110–120. [CrossRef]
381. Dong, Y.; Feng, Z.-L.; Chen, H.-B.; Wang, F.-S.; Lu, J.-H. *Corni Fructus*: A review of chemical constituents and pharmacological activities. *Chin. Med.* **2018**, *13*, 1–20. [CrossRef]
382. Sung, Y.-H.; Chang, H.-K.; Kim, S.-E.; Kim, Y.-M.; Seo, J.-H.; Shin, M.-C.; Shin, M.-S.; Yi, J.-W.; Shin, D.-H.; Kim, H. Anti-inflammatory and analgesic effects of the aqueous extract of corni fructus in murine RAW 264.7 macrophage cells. *J. Med. Food* **2009**, *12*, 788–795. [CrossRef]
383. Park, C.H.; Tanaka, T.; Yokozawa, T. Anti-diabetic action of 7-O-galloyl-D-sedoheptulose, a polyphenol from *Corni Fructus*, through ameliorating inflammation and inflammation-related oxidative stress in the pancreas of type 2 diabetics. *Biol. Pharm. Bull.* **2013**, *36*, 723–732. [CrossRef]
384. Tao, J.-H.; Zhao, M.; Jiang, S.; Pu, X.-L.; Wei, X.-Y. Comparative metabolism of two major compounds in *Fructus Corni* extracts by gut microflora from normal and chronic nephropathy rats in vitro by UPLC-Q-TOF/MS. *J. Chromatogr. B* **2018**, *1073*, 170–176. [CrossRef]
385. Londhe, V.; Gavasane, A.; Nipate, S.; Bandawane, D.; Chaudhari, P. Role of garlic (*Allium sativum*) in various diseases: An overview. *Angiogenesis* **2011**, *12*, 13.
386. Abdou, I.; Abou-Zeid, A.; El-Sherbeeny, M.; Abou-El-Gheat, Z. Antimicrobial activities of *Allium sativum, Allium cepa, Raphanus sativus, Capsicum frutescens, Eruca sativa, Allium kurrat* on bacteria. *Qual. Plant. Mater. Veg.* **1972**, *22*, 29–35. [CrossRef]
387. Zou, Y.-T.; Zhou, J.; Wu, C.-Y.; Zhang, W.; Shen, H.; Xu, J.-D.; Zhang, Y.-Q.; Long, F.; Li, S.-L. Protective effects of *Poria cocos* and its components against cisplatin-induced intestinal injury. *J. Ethnopharmacol.* **2021**, *269*, 113722. [CrossRef] [PubMed]
388. Li, X.; He, Y.; Zeng, P.; Liu, Y.; Zhang, M.; Hao, C.; Wang, H.; Lv, Z.; Zhang, L. Molecular basis for *Poria cocos* mushroom polysaccharide used as an antitumour drug in China. *J. Cell. Mol. Med.* **2019**, *23*, 4–20. [CrossRef]
389. Bae, M.-J.; See, H.-J.; Choi, G.; Kang, C.-Y.; Shon, D.-H.; Shin, H.S. Regulatory T cell induced by *Poria cocos* bark exert therapeutic effects in murine models of atopic dermatitis and food allergy. *Mediat. Inflamm.* **2016**, *2016*, 3472608. [CrossRef]
390. Knorr, R.; Hamburger, M. Quantitative analysis of anti-inflammatory and radical scavenging triterpenoid esters in evening primrose oil. *J. Agric. Food Chem.* **2004**, *52*, 3319–3324. [CrossRef]
391. Ismail, M.F.; El-Maraghy, S.A.; Sadik, N.A. Study of the immunomodulatory and anti-inflammatory effects of evening primrose oil in adjuvant arthritis. *Afr. J. Biochem. Res.* **2008**, *2*, 074–080.
392. Belch, J.J.; Hill, A. Evening primrose oil and borage oil in rheumatologic conditions. *Am. J. Clin. Nutr.* **2000**, *71*, 352s–356s. [CrossRef]
393. Chao, W.-W.; Lin, B.-F. Isolation and identification of bioactive compounds in *Andrographis paniculata* (*Chuanxinlian*). *Chin. Med.* **2010**, *5*, 17. [CrossRef]
394. Bukoye, O.; Musbau, A. Immune modulation potentials of aqueous extract of *Andrographis paniculata* leaves in male rat. *Researcher* **2011**, *3*, 48–57.
395. Jantan, I.; Haque, M.; Ilangkovan, M.; Arshad, L. An insight into the modulatory effects and mechanisms of action of phyllanthus species and their bioactive metabolites on the immune system. *Front. Pharmacol.* **2019**, *10*, 878. [CrossRef] [PubMed]
396. Ishida, M.; Nishi, K.; Kunihiro, N.; Onda, H.; Nishimoto, S.; Sugahara, T. Immunostimulatory effect of aqueous extract of *Coriandrum sativum* L. seed on macrophages. *J. Sci. Food Agric.* **2017**, *97*, 4727–4736. [CrossRef]
397. Efferth, T.; Oesch, F. Anti-inflammatory and anti-cancer activities of frankincense: Targets, treatments and toxicities. *Semin. Cancer Biol.* **2022**, *80*, 39–57. [CrossRef] [PubMed]
398. Maksimović, Z. On Frankincense. *Arch. Pharm.* **2021**, *71*, 1–21.
399. WERZ, O. Gender Medicine and Frankincense: Novel Findings in Inflammation Research. *Sci. Pharm.* **2009**, *77*, 163. [CrossRef]
400. Soni, A.; Bohra, N. *Boswellia serrata*-Propogation and uses—A Review. *Int. J. Adv. Res. Biol. Sci.* **2021**, *8*, 35–46.
401. Stürner, K.H.; Stellmann, J.-P.; Dörr, J.; Paul, F.; Friede, T.; Schammler, S.; Reinhardt, S.; Gellissen, S.; Weissflog, G.; Faizy, T.D. A standardised frankincense extract reduces disease activity in relapsing-remitting multiple sclerosis (the SABA phase IIa trial). *J. Neurol. Neurosurg. Psychiatry* **2018**, *89*, 330–338. [CrossRef]
402. Lim, J.S.; Hahn, D.; Gu, M.J.; Oh, J.; Lee, J.S.; Kim, J.-S. Anti-inflammatory and antioxidant effects of 2, 7-dihydroxy-4, 6-dimethoxy phenanthrene isolated from Dioscorea batatas Decne. *Appl. Biol. Chem.* **2019**, *62*, 1–9.

403. Chen, Y.-F.; Zhu, Q.; Wu, S. Preparation of oligosaccharides from Chinese yam and their antioxidant activity. *Food Chem.* **2015**, *173*, 1107–1110. [CrossRef]
404. Jiao, B.; Gao, J. Intensive research on the prospective use of complementary and alternative medicine to treat systemic lupus erythematosus. *Drug Discov. Ther.* **2013**, *7*, 167–171. [CrossRef]
405. Alabi, Q.K.; Akomolafe, R.O.; Akomolafe, J.G.; Aturamu, A.; Ige, M.S.; Kayode, O.O.; Kajewole, D.I. Polyphenol-rich extract of *Ocimum gratissimum* leaves Prevented Toxic Effects of Cyclophosphamide on the kidney Function of Wistar rats. *BMC Complement. Med. Ther.* **2021**, *21*, 274. [CrossRef] [PubMed]
406. Mahapatra, S.K.; Chakraborty, S.P.; Roy, S. Immunomodulatory role of *Ocimum gratissimum* and ascorbic acid against nicotine-induced murine peritoneal macrophages in vitro. *Oxidative Med. Cell. Longev.* **2011**, *2011*, 734319. [CrossRef] [PubMed]
407. Batiha, G.E.-S.; Magdy Beshbishy, A.; Wasef, L.; Elewa, Y.H.; El-Hack, A.; Mohamed, E.; Taha, A.E.; Al-Sagheer, A.A.; Devkota, H.P.; Tufarelli, V. *Uncaria tomentosa* (Willd. ex Schult.) DC.: A review on chemical constituents and biological activities. *Appl. Sci.* **2020**, *10*, 2668. [CrossRef]
408. Kemper, K.J. *Cat's Claw (Uncaria tomentosa)*; The Center for Holistic Pediatric Education and Research: Boston, MA, USA, 1999.
409. Hilepo, J.N.; Bellucci, A.G.; Mossey, R.T. Acute renal failure caused by'cat's claw'herbal remedy in a patient with systemic lupus erythematosus. *Nephron* **1997**, *77*, 361. [CrossRef] [PubMed]
410. Kim, K.H.; Kim, S.; Jung, M.Y.; Ham, I.H.; Whang, W.K. Anti-inflammatory phenylpropanoid glycosides from *Clerodendron trichotomum* leaves. *Arch. Pharmacal Res.* **2009**, *32*, 7–13. [CrossRef]
411. Chae, S.; Kim, J.S.; Kang, K.A.; Bu, H.D.; Lee, Y.; Hyun, J.W.; Kang, S.S. Antioxidant activity of jionoside D from *Clerodendron trichotomum*. *Biol. Pharm. Bull.* **2004**, *27*, 1504–1508. [CrossRef]
412. Ren, D.; Shen, Z.-Y.; Qin, L.-P.; Zhu, B. Pharmacology, phytochemistry, and traditional uses of *Scrophularia ningpoensis* Hemsl. *J. Ethnopharmacol.* **2020**, *269*, 113688. [CrossRef]
413. Chang, C.-M.; Chu, H.-T.; Wei, Y.-H.; Chen, F.-P.; Wang, S.; Wu, P.-C.; Yen, H.-R.; Chen, T.-J.; Chang, H.-H. The core pattern analysis on chinese herbal medicine for Sjögren's syndrome: A nationwide population-based study. *Sci. Rep.* **2015**, *5*, 1–11. [CrossRef]

Disclaimer/Publisher's Note: The statements, opinions and data contained in all publications are solely those of the individual author(s) and contributor(s) and not of MDPI and/or the editor(s). MDPI and/or the editor(s) disclaim responsibility for any injury to people or property resulting from any ideas, methods, instructions or products referred to in the content.

 life

MDPI

Review

Molecular Insights into Royal Jelly Anti-Inflammatory Properties and Related Diseases

Lilla Bagameri, Sara Botezan, Otilia Bobis *, Victorita Bonta and Daniel Severus Dezmirean

Department of Apiculture and Sericulture, Faculty of Animal Sciences and Biotechnologies, University of Agricultural Sciences and Veterinary Medicine Cluj-Napoca, 400372 Cluj-Napoca, Romania; lilla.bagameri@usamvcluj.ro (L.B.); sara.botezan@student.usamvcluj.ro (S.B.); victorita.bonta@usamvcluj.ro (V.B.); ddezmirean@usamvcluj.ro (D.S.D.)

* Correspondence: obobis@usamvcluj.ro

Abstract: Royal jelly (RJ), a highly nutritious natural product, has gained recognition for its remarkable health-promoting properties, leading to its widespread use in the pharmaceutical, food, and cosmetic industries. Extensive investigations have revealed that RJ possesses a broad spectrum of therapeutic effects, including anti-inflammatory, antioxidant, antitumor, anti-aging, and antibacterial activities. Distinctive among bee products, RJ exhibits a significantly higher water and relatively lower sugar content. It is characterized by its substantial protein content, making it a valuable source of this essential macronutrient. Moreover, RJ contains a diverse array of bioactive substances, such as lipids, phenolic compounds, flavonoids, organic acids, minerals, vitamins, enzymes, and hormones. This review aims to provide an overview of current research on the bioactive components present in RJ and their associated health-promoting qualities. According to existing literature, these bioactive substances hold great potential as alternative approaches to enhancing human health. Notably, this review emphasizes the anti-inflammatory properties of RJ, particularly in relation to inflammatory diseases, such as multiple sclerosis (MS), rheumatoid arthritis (RA), and inflammatory bowel diseases (IBD). Furthermore, we delve into the antitumor and antioxidant activities of RJ, aiming to deepen our understanding of its biological functions. By shedding light on the multifaceted benefits of RJ, this review seeks to encourage its utilization and inspire further investigation in this field.

Keywords: royal jelly; anti-inflammatory activity; antioxidant activity; natural products; anticancer potential; biological activity

Citation: Bagameri, L.; Botezan, S.; Bobis, O.; Bonta, V.; Dezmirean, D.S. Molecular Insights into Royal Jelly Anti-Inflammatory Properties and Related Diseases. *Life* **2023**, *13*, 1573. https://doi.org/10.3390/life13071573

Academic Editors: Azahara Rodríguez-Luna and Salvador González

Received: 31 May 2023
Revised: 3 July 2023
Accepted: 13 July 2023
Published: 17 July 2023

Copyright: © 2023 by the authors. Licensee MDPI, Basel, Switzerland. This article is an open access article distributed under the terms and conditions of the Creative Commons Attribution (CC BY) license (https://creativecommons.org/licenses/by/4.0/).

1. Introduction

Infections, chemical toxins, mechanical injuries, and many other factors can cause inflammation, a natural defense mechanism. When bacteria invade the organism, host cells are driven to secrete pro-inflammatory cytokines. Employing natural remedies as a treatment has received special attention recently because taking anti-inflammatory pharmaceuticals may have a variety of negative side effects [1].

Royal jelly (RJ) is a yellowish-white, creamy substance secreted by the hypopharyngeal and mandibular glands of worker bees. It serves as the primary nourishment for developing larvae during their initial three days and continues to be the exclusive food source for honeybee queens throughout their lives. Recognized for its exceptional nutritional value, RJ has gained the reputation of a "superfood" with numerous potential health benefits for human consumption. It is characterized by its creamy texture and has a pH range of 3.6–4.2. It comprises a diverse range of essential components, including proteins, lipids, carbohydrates, minerals, amino acids, vitamins, enzymes, and hormones. These constituents contribute to its remarkable nutritional profile and may play a role in its physiological effects [2,3].

With an estimated yearly output of more than 4000 tons of RJ, China is regarded as the world's greatest producer and exporter, accounting for over 90% of the total amount collected globally [4,5]. Several countries have created their own national standards for the quality requirements of RJ [6].

Due to its health benefits, RJ has been used as an alternative treatment since ancient times and was especially widespread in Asia and Ancient Egypt [6]. In recent years, there has been a significant rise in interest in the food sector, with the significance of RJ and its distinct pharmacological and therapeutic qualities being highlighted. RJ meets functional requirements for dietary supplements by possessing anti-inflammatory, anticancer, antioxidant, hypotensive, anti-aging, and anti-microbial activities [5]. Considering these therapeutic properties of RJ, the present review aims to highlight recent findings on its activity against inflammatory diseases.

2. Chemical Composition

RJ's rich composition not only holds potential for pharmaceutical applications but also for use in dietary supplements and functional foods, aiding in overall health and well-being [5]. As the critical sustenance for queen bees, RJ necessitates the inclusion of all life-supporting nutrients, encompassing sugars, proteins, lipids, and water, harmoniously proportioned. RJ stands as the bee product with the most significant water content, accounting for 60–70%, while exhibiting a lower sugar ratio, 7–16%, relative to other bee derivatives. It also manifests a high protein content, spanning 10–18%, and lipids ranging between 3–8%. The composition extends to phenolic compounds, flavonoids, organic acids, minerals, vitamins, enzymes, and hormones [3,5,7–10].

2.1. *Proteins and Amino Acids*

Proteins are a major component of RJ, representing more than 50% of its dry matter [6]. Major royal jelly proteins (MRJPs) and specific peptides, namely jeleines, royalsin, royalactin, apidaecin, and defensin-1, constitute the protein composition of RJ [11]. There have been nine significant royal jelly proteins (MRJP1-9) distinguished by several researchers [3,12,13], each with molecular weights spanning between 49 to 87 kDa. Among these, MRJP1 and MRJP3 contribute the most significant proportions, accounting for 31% and 26%, respectively. They are closely followed by MRJP2 and MRJP5 [14]. Expanding our knowledge of protein dynamics, a groundbreaking study by Buttstedt [15] brought to light the unique role of 10-Hydroxy-Δ2-decenoic acid (10-HDA) in manipulating the structure of a particular protein ensemble found in honeybee (*Apis mellifera*) RJ. This complex consists of three key elements: MRJP1, apisimin, and 24-methylenecholesterol. Buttstedt's research illustrates how the presence of 10-HDA influences this complex, encouraging the formation of fibrils—a process with potential implications for the functionality of RJ. The revelations from this research open new avenues in the world of protein studies and provide valuable insights for fields such as entomology and apiculture [15]. In their 2019 research, Mureşan and Buttstedt [16] investigated the stability of MRJPs in the context of varying pH environments. They found a direct correlation between the pH levels and the stability of these proteins, influencing their digestion process. The study further revealed that the beneficial aspects of these proteins, such as their anti-inflammatory properties, are reliant on this pH-dependent stability [16].

In addition to these proteins and peptides, RJ also encompasses a collection of free amino acids, which includes lysine, proline, cystine, aspartic acid, valine, glutamic acid, serine, glycine, cysteine, threonine, alanine, tyrosine, phenylalanine, leucine, isoleucine, and glutamine [17,18].

2.2. *Carbohydrates*

RJ is recognized for its complex and diverse nutrient content, which encompasses an array of carbohydrates. These include glucose, fructose, sucrose, maltose, turanose, trehalose, and isomaltose [19]. The presence and distribution of these carbohydrates in RJ can be a crucial factor in determining its authenticity [11]. The carbohydrate content in RJ appears to be influenced by various factors such as the harvesting season, the species of bees, and the geographical origin of the product [20]. Consequently, these variables could potentially serve as distinctive markers in authenticity tests for RJ products, providing valuable insights for both food science researchers and the apiculture industry [11].

2.3. Lipids and Fatty Acids

RJ is a remarkable substance, rich in a range of lipids. Its lipid composition is predominantly composed of medium-chain fatty acids, among which 10-HDA is a major constituent [10,21]. This fatty acid alone accounts for a substantial portion of RJ's lipidic profile. It is regarded as a freshness and quality marker [10]. However, other fatty acids, such as sebacic acid and 9-hydroxy-2-decenoic acid [5,10], are also present and contribute to the overall lipid matrix of RJ. Beyond fatty acids, RJ also comprises other lipid-soluble substances, including waxes, sterols, and phospholipids [22]. Due to the wide variety and complexity of lipids found in RJ, a recent study [23] utilized sophisticated techniques, such as gas chromatography (GC) and ultra-high performance liquid chromatography paired with ion mobility-quadrupole-time-of-flight-mass spectrometry (UHPLC-IM-Q-TOF-MS). These methods offer high precision, sensitivity, and accuracy, making them highly effective in analyzing intricate mixtures of lipids.

The investigation was successful in identifying and quantifying nine distinct classes of lipids in RJ. These included phosphatidylcholine (PC), diacylglycerol (DG), lyso-PC (LPC), sphingomyelin (SM), triglycerides (TG), phosphatidylethanolamine (PE), ceramide (Cer), cardiolipin (CL), and lyso-PE (LPE). This research marks a significant step forward in understanding the intricate lipid composition of RJ, which, in turn, sheds light on its various biological activities and potential health benefits. Future investigations in this area could provide even more insights into the functional properties of RJ, contributing to its potential use in various health and nutritional applications.

Studies about the anti-inflammatory effects of fatty acids will be discussed in the "RJ as an adjuvant in anti-inflammatory diseases" section.

2.4. Other Constituents

RJ is characterized by its phenolic compounds, comprising substances such as pinobanksin, hesperetin, naringenin, isosakuranetin, chrysin, acacetin, luteolin, apigenin, kaempferol, isorhamnetin, formononetin, along with various glycosides. These compounds contribute to the biological activities and potential health benefits associated with RJ [20,24].

The vitamin profile of RJ is also noteworthy. It includes essential vitamins, such as biotin, riboflavin, thiamin, pantothenic acid, inositol, niacin, pyridoxine, and vitamin E. These vitamins are key to multiple biological functions, further enhancing the nutritional value and health-promoting potential of RJ [25]. This bee product also incorporates a range of minerals, such as potassium (K), phosphorus (P), calcium (Ca), magnesium (Mg), sodium (Na), zinc (Zn), chromium (Cr), and cadmium (Cd). The presence and concentration of these minerals in RJ are affected by a wide range of factors. These include environmental conditions, the time of the harvest season, and the unique biological traits of the bees. Therefore, these factors should be taken into account when studying the composition and properties of RJ [11,26,27].

A variety of factors significantly influence the production and chemical composition of RJ, as emphasized in the study conducted by Al-Kahtani and Taha [28]. They highlighted the pivotal role of the post-grafting period, demonstrating that it profoundly affects the output and macro and trace elements of this bee product. A precise post-grafting duration is hence critical for optimizing both quantity and quality. Harvest timing also wields substantial influence, with suboptimal timing potentially impacting both yield and chemical constitution. Al-Kahtani and Taha [29] sought to determine the optimal timing for harvesting RJ based on its yield and nutritional composition, examining three timeframes: 24, 48, and 72 h post-grafting. The results indicated that RJ collected after 72 h of grafting produced the highest yield per queen cell, contrasting with the minimum yield seen at the 24 h mark. The nutritional constitution of this bee product also exhibited a temporal dynamic. The 72 h harvest exhibited the lowest moisture levels and highest amounts of crude protein, ash, fructose, and glucose. RJ harvested at 24 h, however, showed the highest concentration of lipids. Furthermore, a trend towards decreasing pH and increasing acidity was observed over time. The study's conclusions suggest a 72 h grafting period to maximize RJ yield and optimize nutritional content. Yet, it is important to recognize that the changing nutritional characteristics over time mean that RJ properties may vary depending on when it is harvested post-grafting [29]. Additionally, the dietary intake of the bee colonies directly affects the volume and biochemical composition of the RJ produced. Seasonal variations introduce another layer of complexity, resulting in oscillations in its production and chemical attributes. These findings highlight the multifactorial aspect of RJ production and emphasize the need for a comprehensive understanding to optimize beekeeping practices [28].

3. RJ as an Adjuvant in Inflammatory Diseases

When the organism is challenged by pathogens, harmful toxins, irradiation, or microbial infections, inflammation occurs as a natural defense mechanism. Numerous physiological and immunological pathways are activated by this complex process; thus, when inflammation becomes dysregulated as a result of certain circumstances, it can harm nearby tissue and induce a wide range of diseases [30].

Characteristic inflammatory symptoms, including pain, redness, and heat, involve immune cells, such as B and T lymphocytes, monocytes, macrophages, neutrophils, and basophils, as well as mast and dendritic cells. The inflammatory response is largely influenced by the type of stimulus, all sharing similar mechanisms, such as the identification of pattern-recognition receptors on foreign pathogen's cell surfaces or intracellular signaling caused by harmed cells or tissues [31].

Nowadays, a wide range of anti-inflammatory medications on the market work by preventing the production of prostaglandins, but they have substantial adverse effects as well [31]. As a result, several studies investigating natural products with the potential to enhance anti-inflammatory effects without having negative side effects have been carried out, and various natural substances with the potential to enhance immune function have been found. It is worth noting that these natural products have a beneficial effect on both the human body and animals. As an example, a natural product called RJ has been traditionally used to enhance immune system performance [32]. Numerous biological and chemical factors, including cytokines, pro-inflammatory enzymes, as well as the enzymatic degradation of tissues, all contribute to the inflammatory process. Interleukin-1 (IL-1), interleukin-6 (IL-6), and tumor necrosis factor-alpha (TNF-α) production are effectively inhibited by RJ treatment in a dose-dependent manner without having any deleterious effects on macrophages in vitro. In autoimmune disorders, such as rheumatoid arthritis (RA) and inflammatory bowel diseases (IBD), RJ may play a crucial role in enhancing life quality [5]. Figure 1 represents a summary of the information described in the text regarding three important inflammatory-related diseases taken into discussion.

3.1. RJ in Multiple Sclerosis

Multiple sclerosis (MS) is an inflammatory condition that affects the nervous system and is thought to be immunologically mediated and persistent. This disease comes with a damaged blood-brain barrier, lymphocytes, microglia, and macrophages being attracted to the lesion sites. It is distinguished by the destruction of fatty myelin sheaths around the brain and spinal cord's axons, which results in demyelination along with a wide range of symptoms. Although genetic and environmental variables have been demonstrated to play an essential role, the fundamental etiology of MS remains unclear [33].

A descriptive analysis performed by Podbielska et al. [34] indicated that lipids could potentially play a key role in the immunopathogenesis of MS while also being involved in the progression and remission of the disease.

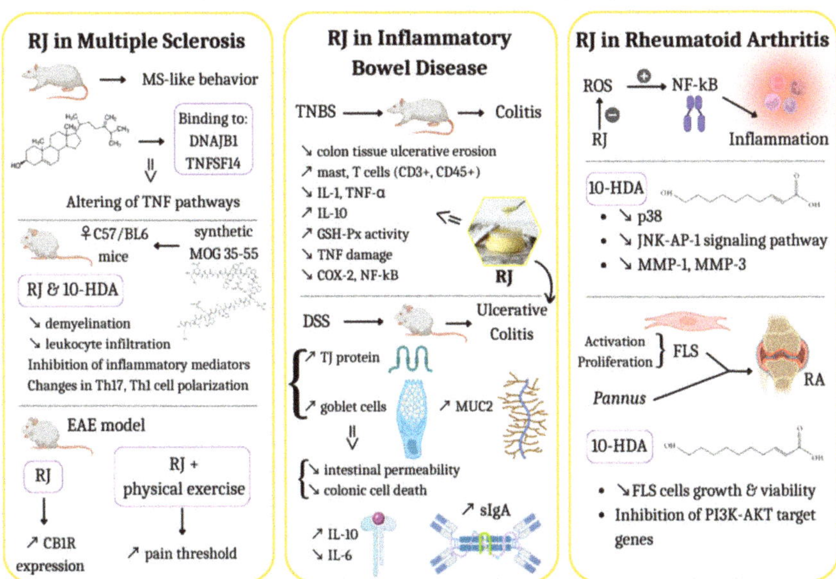

Figure 1. RJ—Mechanisms of action in inflammatory-related diseases. First panel: RJ in Multiple Sclerosis (MS); Second panel: RJ in Inflammatory Bowel Disease (IBD); Third panel: RJ in Rheumatoid Arthritis (RA). ↗ increase, ↘ decrease (Created with canva.com, accessed on 22 April 2023).

Natural product-based therapies have shown promising outcomes in the treatment of MS symptoms and the disease's progression. Due to its antioxidant, anti-inflammatory, and anti-apoptotic properties, RJ is used to treat a variety of diseases, including neurological disorders. RJ has been linked to relieving chronic pain, neuroplasticity, and altering neurotransmitters involved in anxiety and depression conditions [35,36].

Lohrasbi et al. [35] pointed out that rats with MS-like behaviors, supplemented with 100 mg/kg/day of RJ, could show an improvement in terms of mobility, pro-inflammatory cytokine functions, and demyelination. Researchers highlighted 10 bioactive RJ compounds with suitable binding affinities, including Vitamin A, Vitamin B2, Vitamin E, hesperetin, kaempferol, naringenin, formononetin, genistein, isosakuranetin, and 24-methylenecholesterol. Based on the artificial intelligence analysis (AI) study, they discovered that these bioactive substances might bind to DNAJB1 and TNFSF14 protein-coding genes. According to the molecular docking analysis of bioactive chemicals, RJ consumption may alter important molecular signaling pathways, improving life quality and increasing muscular strength in patients with MS. Based on their new findings, researchers revealed that 24-methylenecholesterol, a low-energy molecule, is bound to the surfaces of the TNFSF14 and DNAJB1. These results suggest that TNF pathways were altered because of consuming this natural bee product [35].

As Cannabinoid-1-receptors (CB1R) are therapeutic targets for the treatment of various associated symptoms of MS, Kheirdah et al. [36] aimed to assess the impact of aerobic exercise and two doses of RJ on hippocampus CB1R and pain threshold (PT) in an EAE model. RJ was administered intraperitoneal at dosages of 50 and 100 mg/kg/day throughout a five-week period of exercise training (ET), which was completed four times per week at different speeds ranging between 11 to 15 m/min for 30 min. In rats with EAE, endurance training had no remarkable impact on PT or hippocampus CB1R. In comparison to the EAE group, the CB1R gene expression levels in the RJ100 group were increased. Additionally, a higher PT level was observed in the ETRJ50 and ETRJ100 groups compared to the EAE group. PT and CB1R were more significantly affected by the combination of ET and RJ50 than by either of them alone. These findings indicate that RJ at 50 or 100 mg/kg doses should be consumed, and this supplementation should be combined with physical exercise for better results [36].

Another study investigated the effects of RJ on female C57/BL6 mice with induced experimental autoimmune encephalomyelitis (EAE), a model for MS. The mice received daily assessments for 25 days after receiving synthetic MOG35-55. Histological analysis using techniques such as H&E and LFB staining, BrdU incorporation, ELISA, and Real-time PCR were employed to evaluate demyelination, proliferation, lymphocyte infiltration, cytokine profiles, and gene expression levels. The results indicated that RJ and 10-HDA prevented the development of EAE. The treated groups exhibited reduced demyelination, decreased leukocyte infiltration in the central nervous system, and dose-dependent inhibition of inflammatory mediators. RJ and 10-HDA were found to modulate the immune response by primarily affecting the polarization of Th17 and Th1 cells. These findings highlight the potential therapeutic benefits of RJ and 10-HDA in MS and support their use in the treatment of inflammatory diseases [37].

A randomized clinical trial conducted by Oshvandi et al. [38] aimed to assess the impact of RJ supplementation on the quality of life in patients diagnosed with MS. The study involved 100 MS patients, divided into an experimental group and a control group. The experimental group received a daily capsule of 500 mg of RJ for 90 days, while the control group received a placebo. The researchers utilized the MS-specific quality of life questionnaire and the Barthel Index of Daily Living Activities to evaluate patients' quality of life and daily activities before and after the intervention. The results demonstrated that the experimental group experienced a significantly higher average quality of life score after the intervention compared to the control group, even after accounting for potential confounding factors. Additionally, the experimental group exhibited a significant improvement in daily activity status scores compared to the control group. These findings suggest that incorporating RJ supplements into the daily routine can have a positive impact on the quality of life for MS patients [38].

In summary, the studies mentioned above provide insights into the potential benefits of RJ in the context of MS. The in vitro and in vivo experiments demonstrated improvements in mobility, pro-inflammatory cytokine functions, demyelination, and immune response modulation. The molecular docking analysis indicated the interaction of bioactive compounds with specific genes and molecular pathways related to MS. Furthermore, clinical data highlighted the positive effects of RJ supplementation on the quality of life and daily activities of MS patients. These findings support the exploration of this bee product as a potential therapeutic option for managing MS symptoms and disease progression (Figure 1, First panel).

3.2. *RJ in Inflammatory Bowel Disease*

Both Crohn's disease (CD) and ulcerative colitis (UC) are chronic inflammatory bowel diseases (IBD), which share many of the same symptoms and cause inflammation in the digestive tract. Its incidence is ascribed to a number of circumstances, some of which include geographic location, low-quality food, genetics, and decreased immunological response [39].

Weight loss, fever, rectal bleeding, diarrhea, and stomach discomfort are some of the symptoms of CD and UC, inflammation being the fundamental characteristic of them. These disorders can affect men and women equally. The terminal ileum, cecum, perianal region, and colon are often affected by CD, although it can also affect any part of the intestine. In contrast, UC affects the rectum and can spread continuously across the entire colon or only a portion of it. The inflammation in UC is only present in the mucosa and submucosa with cryptitis and crypt abscesses, in contrast to the thickened submucosa, transmural inflammation, fissuring ulceration, and granulomas that are histologically present in CD [40].

In vitro studies have provided valuable insights into the anti-inflammatory properties of RJ and its components in the context of colitis. For example, Yang et al. [41] demonstrated in vitro that 10-HDA, a component isolated from RJ, can prevent the secretion of pro-inflammatory cytokines TNF-α, IL-1β, and IL-8 in WiDr adenocarcinoma cells. Additionally, they found that this fatty acid effectively increased the production of IL-1Ra (IL-1 receptor antagonist), which acted as a constraint on IL-1β production.

Moving to in vivo animal studies, the effects of RJ have been investigated in 2, 4, 6-trinitrobenzene sulfonic acid (TNBS)-induced colitis model. It was discovered that RJ supplementation reduced colon tissue ulcerative erosion and increased the presence of mast cells, CD3+, and CD45+ T cells. Furthermore, pre-treatment with RJ decreased the production of pro-inflammatory cytokines, such as IL-1 and TNF-α, while enhancing the production of the anti-inflammatory cytokine IL-10 in the colon. RJ was also found to improve plasma glutathione peroxidase (GSH-Px) activity, reduce TNF-α damage, and suppress the production of major inflammatory mediators, including COX-2 and NF-kB [3,42].

In another animal study conducted by Guo et al. [43], the protective effects of RJ were investigated in a DSS-induced UC model in mice. RJ supplementation in these mice increased the levels of tight-junction proteins, goblet cells, and mucin secretion (MUC2), leading to alleviation of symptoms, reduced intestinal permeability, and colonic cell death. Moreover, RJ supplementation resulted in increased expression of the anti-inflammatory cytokines IL-10 and IgA while decreasing the expression of the pro-inflammatory cytokine IL-6. The relative abundance of gut microbiota was also affected by DSS, with specific changes observed. Treatment with RJ increased the relative abundance of certain beneficial gut microbiota. These findings demonstrated that RJ could mitigate DSS-induced colitis by strengthening the intestinal mucosal barrier and modulating gut microbiota composition.

In conclusion, in vitro and in vivo animal studies collectively suggest the anti-inflammatory effects of RJ in colitis (Figure 1, Second panel). In vitro studies demonstrated the ability of RJ components, such as 10-HDA, to suppress the secretion of pro-inflammatory cytokines and enhance the production of anti-inflammatory cytokines. Animal studies showed that RJ supplementation reduced tissue damage, modulated immune response, strengthened the intestinal mucosal barrier, and influenced gut microbiota composition. These findings highlight the potential of RJ as a therapeutic option for managing colitis, but further research and clinical trials are necessary to fully understand the underlying mechanisms and optimize its clinical application.

3.3. *RJ in Rheumatoid Arthritis*

Rheumatoid arthritis (RA) is a chronic, autoimmune condition that seriously threatens healthspan, evolving gradually, affecting primarily elderly people [20]. It is frequently characterized by pain, edema, stiffness, inflammation, and joint damage. Chronic synovial membrane inflammation causes bone and cartilage deterioration, resulting in the degeneration of the afflicted joints. Inflammation of the blood vessels, internal organs, and tendon sheaths is among the most common symptoms, along with discomfort, exhaustion, and movement restrictions. Early recognition and the beginning of treatment as soon as possible are crucial as the immune process can be considerably altered if this condition is identified and treated within three to six months of the onset of symptoms [44]. Results of RA therapy (steroids and anti-TNF drugs) in the elderly are unsatisfactory due to age-related loss in organ function, comorbidities, and body composition changes [45].

Clinical investigations have indicated that oxidative stress plays a significant role in the genesis of RA, resulting in high levels of oxidative stress biomarkers and low antioxidant status. NF-kB is activated by rROS, which might lead to inflammatory reactions in RA. Thus, supplementation with RJ as an antioxidant agent may contribute to symptom relief and life quality enhancement. Nowadays, pharmacologic treatments for RA symptoms include non-steroidal anti-inflammatory drugs (NSAIDs), corticosteroids, and disease-modifying anti-rheumatic drugs (DMARDs), although these treatments come with certain adverse effects. As a result, complementary therapies, particularly nutritional supplements, have gained more attention [46].

In vitro studies were conducted to investigate the effects of 10-HDA, a compound found in RJ, on key signaling pathways and enzymes associated with RA. The researchers observed that this fatty acid significantly reduced the activity of p38, c-Jun N-terminal kinases-activating protein-1 (JNK-AP-1) signaling pathway, and matrix metalloproteinases (MMP-1, MMP-3) [5,47]. These findings suggested that 10-HDA may have a protective effect against the adverse effects of RA treatment.

Researchers sought to explore the impact of 10-HDA on fibroblast-like synoviocytes (FLS) cells, which play a crucial role in the pathogenesis of RA. Increased activation and proliferation of FLS cells, along with the development of pannus that invades nearby bone and cartilage, are characteristic features of RA. Inhibition of FLS cell growth and control of pannus formation are important therapeutic goals in RA treatment. Previous research indicated that 10-HDA could potentially reduce FLS cell growth [44]. However, recent findings highlighted the inverse relationship between dosage and time with regard to the viability and histone deacetylase (HDAC) activity of FLS cells. This discovery opened up new possibilities for the development of histone deacetylase inhibitors (HDACI) as a potential treatment option for RA. Consequently, 10-HDA was implicated in inhibiting the target genes of the phosphoinositide-3-kinase-protein kinase B/AKT (PI3K-AKT) pathway, thereby suggesting its potential as an alternative treatment option for RA [32] (Figure 1, Third panel).

Finally, clinical data is needed to validate the findings from in vitro studies. Clinical trials involving human participants with RA would be required to evaluate the efficacy and safety of 10-HDA as a therapeutic agent. These trials would assess the impact of this fatty acid on disease progression, symptoms, and other relevant factors. If the clinical data support the earlier findings, 10-HDA could potentially be considered as a complementary treatment option for RA.

4. Additional Anti-Inflammatory Effects of RJ

A group of researchers developed a personalized treatment based on the protease enzyme technique to hydrolyze RJ, aiming to compare the anti-inflammatory and immune-boosting properties of enzyme-treated RJ (ERJ) on macrophages and mice. They discovered that ERJ could affect macrophage proliferation and offer protection against LPS-induced stress. The mice selected for this study that were given ERJ for 4 weeks and stimulated with LPS presented considerably lower levels of TNF-α, IL-1, IL-6, IL-10, and IL-12, as well as IFN-γ. Moreover, B-lymphocyte and T-lymphocyte proliferation, as well as the activity of naturally occurring natural killer cells, were all markedly and dose-dependently boosted by ERJ. These findings show that ERJ possesses substantial anti-inflammatory and immune-promoting properties, making it a promising dietary ingredient for the treatment of inflammatory diseases [48].

The three main fatty acids found in RJ, namely 10-HDA, 10-hydroxydecanoic acid (10-HDAA), and sebacic acid (SEA), were examined for their anti-inflammatory properties by Chen et al. [49]. Whereas all of them exhibited significant, dose-dependent inhibitory effects on the release of the key inflammatory mediators (nitric oxide and IL-10), findings showed that only SEA had the ability to reduce the production of TNF-α. These RJ fatty acids have also been shown to affect a number of important inflammatory genes, with 10-HDA exhibiting different modulatory actions in comparison to the other two fatty acids. The authors also discovered that all of them possessed a regulatory effect on a number of proteins involved in the mitogen-activated protein kinase (MAPK) and nuclear factor kappa-light-chain-enhancer of activated B cells (NF-κB) signaling pathways [50]. Chen et al. [50] examined the in vitro anti-inflammatory effects of 10-HDA and noticed a reduction in the expression of crucial inflammatory genes, such as IL-1, IL-6, COX-2, and MCP-1. Mice were enrolled to examine the effects of 10-HDA on lung damage brought on by LTA. In accordance with the findings, 100 mg/kg of 10-HDA can have protective effects by limiting the production of inflammatory cytokines, including IL-10, MCP-1, and TNF-α. Even though the inhibitory effects of this fatty acid were dose-dependent, it was speculated to possess anti-inflammatory effects. Further investigations to comprehend this fatty acid's accurate action mechanisms were suggested by the authors.

RJ's anti-inflammatory effects on the BV-2 murine microglial cell line after exposure to LPS were investigated by You et al. [51]. RJ was discovered to have a protective effect by reducing the inflammatory response; the underlying processes may be connected to pro-inflammatory cytokines production inhibition. RJ has the potential to drastically reduce the expression of the pro-inflammatory protein cyclooxygenase-2 (COX-2). Moreover, RJ can decrease inflammatory mediator levels by inhibiting the nuclear factor kappa B (NF-kB) and c-Jun N-terminal kinase (JNK) pathways [51]. Studies involving phytohaemagglutinin-activated peripheral blood mononuclear cells (PBMCs) as an in vitro model have shown that RJ fatty acids, including 3, 10-dihydroxy-decanoic acid (3,10-DDA) and 10-HDA at the concentration of 500 μM have an inhibitory effect on PBMC proliferation. Moreover, it suppressed Th1 and Th2 immune responses, along with modulating TNF-α and IL-1β production. While TNF-α and IL-1β levels were unaffected by 3,10-DDA at a concentration of 500 μM, the production of these cytokines by stimulated PBMCs was inhibited by the same amount of 10-HDA. The researchers concluded that RJ fatty acids had a substantial, dose-dependent immunomodulatory impact in vitro [52].

It was shown that RJ can stimulate the proliferation of healthy lymphocytes and has a stimulatory effect on interferon-gamma (IFN-γ) production. RJ treatment shifted the Th1/Th2 cytokine ratio in favor of the Th1 in patients with autoimmune Basedow Graves' illness. This indicates that RJ may be beneficial in the treatment of Graves' disease as an immunomodulatory agent and an antithyroid medication [53].

Arzi et al. [54] also examined the anti-inflammatory effects of RJ in formalin-induced rat paw edema. The researchers showed that RJ inhibited inflammation in a dose-dependent manner, considerably reducing edema at doses of 50 and 100 mg/kg. It was also demonstrated that there was no discernible difference in the inhibition of formalin-induced edema between groups treated with dosages of 50 and 100 mg/kg and the positive control group treated with aspirin 300 mg/kg.

Minegaki et al. [55] aimed to investigate the anti-inflammatory effects of MRJP3 and its derived peptides both in vitro and in vivo since in vitro testing revealed that MRJP3 had anti-inflammatory activities. The addition of MRJP3 or its C-terminal tandem pentapeptide repeats (TPRs) sequence was shown to inhibit the expression of both TNF-α and IL-6 mRNAs in LPS-stimulated THP-1 cells. TPRs were injected into mice that suffered from herpes stromal keratitis (HSK) caused by the herpes simplex virus type 1 (HSV-1), the main objective consisting in the decrease of both disease scores and levels of TNF-α and IL-6 expression. Additionally, in both in vivo and in vitro models, the expression of TNF-α and IL-6 was decreased by TPRs derived from synthetic pentapeptides.

The study conducted by Fatmawati et al. [56] evaluated endogenous antioxidant expression of nuclear-related factor 2 (Nrf2), transcription factor (Nf-kB), and pro-inflammatory cytokine TNF-α to determine the effectiveness of RJ as a UV radiation protector. Wistar rats were exposed for 2 h daily to 40 Watt UV-B lamps for a period of two weeks. RJ cream was applied in different concentrations, including 2.5%, 5%, and 10%, the highest dose inducing an increased Nrf2 and a decreased Nf-kB expression level. Moreover, TNF-α expression was significantly reduced with an increased RJ dose. As a result, the application of RJ cream shielded the skin from UV rays by reducing inflammatory reactions and strengthening cellular antioxidants.

Nanoparticles, such as nano-silver (NS), have the potential to cause inflammation by activating immune cells, which causes the release of pro-inflammatory cytokines. The skin, cardiovascular system, gastrointestinal and respiratory tract can all be negatively affected by NS. These adverse effects, including DNA oxidation and cell cytotoxicity, can be alleviated by using tissue-protecting compounds, such as RJ, along with NS. Thus, Pourmobini et al. [57] aimed to investigate RJ's protective effect against NS's inflammatory effect. Pro-inflammatory cytokines IL-1β, IL-2, IL-6, and IL-33 levels were measured in the kidney and liver of rats treated with NS, RJ, or a combination of NS and RJ. This study highlights the possibility that when combined, RJ and NS might have anti-inflammatory effects and affect immune cell activity. Therefore, combining RJ with NS may alter NS's impact on the production of important pro-inflammatory cytokines, which indicates the need for more research on this topic in the future.

The goal of the Salashoor et al. [58] study was to evaluate the anti-inflammatory and protective properties of RJ against ischemia/reperfusion (I/R)-induced renal diseases. Forty male rats were randomly assigned to one of four groups: placebo (0.9% saline), I/R, RJ (treated for 15 days with 300 mg/kg/day RJ), and I/R + RJ (pretreated for 15 days with 300 mg/kg/day RJ). Their study demonstrated that RJ has a protective effect against damage caused by I/R, perhaps due to its antioxidant and anti-inflammatory activities.

In the research conducted by Lin et al. [59], the focus lies in exploring the variations in the wound-healing properties of RJ produced by *Apis mellifera* L. during different blossom seasons of various floral sources. Their primary objective was to provide guidelines for the future rational application of RJ in cutaneous wounds based on the findings. Additionally, the study seeks to contribute to the further discovery of substances that promote wound repair. In this study, RJ samples collected during the flowering seasons of *Castanea mollissima* BL. (chestnut) and *Brassica napus* L. (rapeseed) in South China were investigated. Hydrophilic and lipophilic fractions were extracted from these samples. The wound-healing potential of the RJ samples was assessed in vivo using Wistar rats with excisional full-thickness wounds. The mechanisms of action were further explored through in vitro assays using human epidermal keratinocytes and macrophages stimulated with lipopolysaccharide (LPS). The results revealed distinct wound-healing properties among the different RJ samples. *Castanea mollissima* BL. RJ demonstrated higher potency, significantly accelerating wound closure between day 2 and day 4 with a rate of 0.25 cm^2/day ($p < 0.05$). It also enhanced the proliferative and migratory capabilities of keratinocytes by 50.9% ($p < 0.001$) and 14.9% ($p < 0.001$), respectively. Furthermore, it modulated inflammation by inhibiting nitric oxide (NO) production by 46.2% ($p < 0.001$) and promoting cell growth through increased secretion of transforming growth factor-β (TGF-β1) by 44.7% ($p < 0.001$). On the other hand, *Brassica napus* L. RJ exhibited anti-inflammatory effects by reducing tumor necrosis factor-α levels by 21.3% ($p < 0.001$). Based on these findings, the study highlights the potential of *Castanea mollissima* BL. RJ for treating challenging wounds [59]. A randomized, double-blind trial conducted by Petelin et al. [60] aimed to find out how RJ supplementation affects oxidative and inflammatory indicators, as well as the metabolic profile of asymptomatic overweight people. Total cholesterol and inflammatory marker C-reactive protein both showed statistically significant declines after RJ supplementation, whereas anti-inflammatory marker adiponectin showed an increase. This investigation has pointed out that the daily use of 666 mg of lyophilized RJ has favorable effects on overweight people's lipid profiles, inflammation, and antioxidant capacity.

The purpose of the research conducted by Chansuwan et al. [61] was to explore the potential anti-inflammatory and anti-allergic properties of RJ-derived proteins and their enzymatic hydrolysates in mammalian cell lines. The bioactivity of protein hydrolysates depends on different factors, including peptide sequences and amino acids. Alcalase, Flavourzyme, and Protamex, three commercial proteases used to hydrolyze RJ, all possess anti-inflammatory and anti-allergic effects without altering the survival of macrophages and mast cells. Findings suggested that the protease type and hydrolysis duration may have an impact on their capacity to inhibit NO generation and -HEX release; Flavourzyme hydrolysates display the highest activity for both characteristics. Their anti-oxidative qualities and DNA damage-protecting action may be related to the anti-inflammatory impact. These findings indicate that RJ hydrolysates may be further developed into functional food and components for anti-inflammatory and anti-allergic activities.

The inhibition of RAW264.7 cell expression of the inflammatory gene was analyzed by Uthaibutra et al. [1]. RT-PCR technique was used to investigate inducible nitric oxide synthase (iNOS), COX-2, and IL-6 cells. Total RNA was isolated from the cells, and following a three-hour RJ treatment, they were stimulated with LPS. The outcomes demonstrated that LPS might activate inflammatory genes, and the expression of the iNOS gene could be suppressed at doses of 5 and 10 mg/mL RJ-LP1 and RJ-CM1.

5. Other Bioactive Effects of RJ

RJ is noteworthy for having a large number of bioactive substances, such as MRJPs and 10-HDA, which are considered to be the basis of all the valuable effects of bee-derived secretion [30].

Figure 2 presents other biological effects of RJ that this review takes into discussion, other than its anti-inflammatory one. These six properties of the food of the queens will be briefly discussed in the following paragraphs and table.

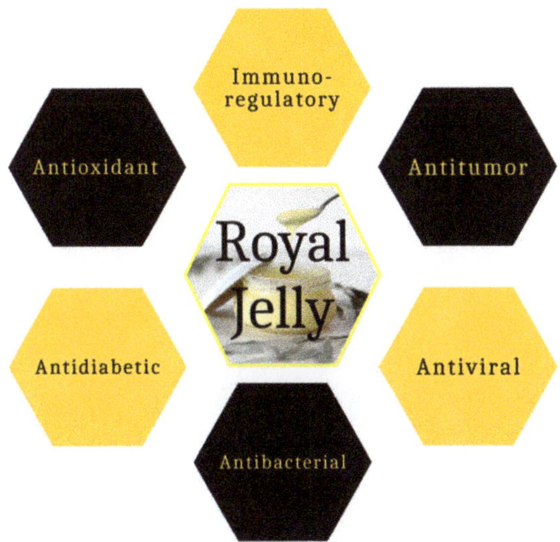

Figure 2. Other bioactive properties of RJ taken into discussion. (Created with canva.com, accessed on 9 April 2023).

Some recent investigations suggest that RJ has potent antibacterial properties [6,57]. Antibiotic-resistant microorganisms are threatening the health of people across the world. Finding and creating modern antibacterial drugs to stop their spread is therefore becoming increasingly popular. Bee-derived products, such as RJ, honey, and propolis, are regarded as a natural alternative to synthetic antibiotics since they are rich in bioactive chemicals. RJ is renowned for combating deadly illnesses caused by a series of pathogenic agents in both humans and animals [27].

The antibacterial property of RJ has been tested on various bacterial strains, such as *Escherichia coli*, *Staphylococcus aureus*, *Pseudomonas aeruginosa*, *Listeria monocytogenes*, *Prevotella intermedia*, *Porphyromonas gingivalis*, *Aggregatibacter actinomycetemcomitans*, *Fusobacterium nucleatum* [27,62] and many more. Some authors pointed out that this antimicrobial effect appears due to RJ's most important bioactive compounds: proteins, peptides, and 10-HDA [62]. Moreover, due to its minimal adverse effects, RJ can be viewed as a potential substitute for synthetic antibiotics [63]. Thus, RJ contains bioactive substances, including proteins, peptides, and fatty acids, which provide significant antibacterial effects against a variety of infections [27]. In other words, RJ can be regarded as a natural antibiotic with additional health benefits due to its unique chemical profile.

A wide range of studies have shed light on the antioxidant property of RJ, among its therapeutic benefits. Biomolecules essential for a healthy life can be degraded as a result of oxidative stress caused by the increased synthesis of ROS. This process is linked to the emergence of several pathogenic processes and chronic disorders. Researchers have emphasized the many benefits of employing antioxidants to treat and help prevent these conditions [30], and there has been an increase in RJ use in this sense in recent years. Thus, RJ may very well be able to play the part of a free-radical scavenger, according to a growing body of research [21]. In all of the investigations, including RJ compounds, their antioxidant property was expressed by favorable outcomes in a number of representative criteria [21,30]. Furthermore, according to available data, the radical-scavenging activity of RJ was proven by both in vitro and in vivo experiments, which confirmed the value of hive products in alleviating the negative consequences of oxidative stress and the illnesses caused by it (e.g., diabetes, atherosclerosis, neurodegenerative disorders, cancer) [21].

It has also been reported that bee products, such as RJ and bee bread, have anticancer properties [64]. Scientists and medical experts have worked together to develop a number of approaches that are still in use today in the battle against cancer, such as chemotherapy or radiotherapy. Nevertheless, the majority of these treatment strategies have severe drawbacks. It is well documented that several bee-derived compounds can aid in the suppression of cancer. More precisely, processes such as cancer cell proliferation, metastasis, and tumorigenesis inhibition have been discovered and linked to the antitumor effect of bee secretion [30]. Thus, there are studies that confirm RJ's antitumor property and the fact that this bee product is a contender as far as complementary therapies for cancer go [65].

Moreover, RJ influences the production of various chemokines and growth factors, as well as the expression of cancer-related molecules in patients with malignancies, particularly in those treated with cytostatic drugs. It also reduces cell growth and activates cell death in malignant cells. RJ is hence believed to have anti-cancer effects on tumor development and to have protective qualities against toxic medication side effects [2]. For example, via the control of several cancer-related pathways, RJ and one of its key constituents, 10-HDA, can reduce tumor development and the migration of malignant cells [2].

In general, the research on the antitumor effects of RJ showcases this hive product as a potential source of cytotoxic chemicals with a variety of antitumor effects and mechanisms [64]. In addition, when RJ or its components are used alongside anticancer medication, their effects on the disease are synergistic, increasing the drug's efficacy and even minimizing its negative effects. Recent investigations have provided strong evidence that RJ has an anticancer impact, whether used alone or in conjunction with other traditional cytostatic medications [66].

Furthermore, RJ presents various other health-boosting effects, including immunoregulatory, antiviral, and antidiabetic effects [30]. Studies that have explored the three aforementioned properties of RJ or some of its key components are briefly presented in Table 1.

Table 1. Mechanisms underlying the immunoregulatory, antiviral, and antidiabetic effects of RJ.

Effect	Key Players	Mechanism	Type of Study	References
Immuno-regulatory	RJ	↓ total proteins and immunoglobulins, ↑ plaque-forming splenocytes, ↑ antibody production, and immunocompetent cell proliferation	In vivo (CBA mice)	[67]
	MRJP3	Suppression of IL-4 production, inhibition of serum anti-OVA IgE and IgG1 levels	In vivo (Mice)	[68]
	RJ	Innate immunity modulation through IIS/DAF-16, p38 MAPK, and Wnt signaling pathways	In vivo (*C. elegans*)	[69]
	RJ	↓ inflammation, ↑ cell regeneration	In vivo (Rats)	[70]
	MRJPs	Positive effects on immunoglobulin content, immune factor level, and proliferation of spleen lymphocytes	In vivo (Mice)	[71]
	RJ	Modulation of immune responses via downregulation of NLRP1	Clinical (human patients)	[72]
	RJ	Induction of antibody production, maturation of immune cells, stimulation of the innate and adaptive immune responses	Review	[73]

Table 1. *Cont.*

Effect	Key Players	Mechanism	Type of Study	References
Antiviral	RJ	Inhibition of HSV-1	In vitro (Vero cells)	[74]
	MRJP2 and MRJP2 isoform X1	Sialic acid hydrolysis, attachment prevention (high binding affinity to viral receptor-binding sites), inhibition of SARS-CoV-2 enzymes, prevention of hemoglobin attack	In vitro (WI-38 lung cells)	[75]
	Erlose, Kaempferol glucoside, Iridin, Luteolin glucoside (Cynaroside)	Antiviral affinity (binding to COVID-19 binding sites through H-bonding), blocking of SARS-CoV-2 protease (through hydrogen bond and π-π T-shaped interactions)	In vivo and in vitro tests	[76]
Antidiabetic	RJ-propolis	Hypoglycemic and antioxidant activity	In vivo (Mice)	[77]
	RJ-honey	↓ plasma VLDL-C and TG, normalization of glycemic control indices	In vivo (Rats)	[78]
	RJ	↓ FBS, ↓ HbA1c, ↓ HOM A-I R, improvement in serum levels of triglycerides, cholesterol, HDL, LDL, VLDL, and Apo-A1, ↓ oxidative stress, ↑ antioxidant enzymes	Review	[78]
	RJ	Glycemic regulation (fasting blood glucose and glucose clearance as the most affected parameters)	Review	[78]

↓ decreasing effect; ↑ increasing effect.

6. Conclusions

Recent research articles have highlighted that RJ represents a valuable resource for treating inflammatory diseases, owing to its distinct pharmacological and therapeutic properties. The anti-inflammatory effects of RJ have been extensively investigated and substantiated by various experimental models and clinical studies. RJ has been shown to modulate key inflammatory mediators, including cytokines, chemokines, and adhesion molecules, thereby mitigating the inflammatory response. It exerts its anti-inflammatory actions through diverse mechanisms, such as inhibition of pro-inflammatory cytokine production, suppression of NF-κB signaling pathway, and modulation of immune cell function. These findings highlight the potential of RJ as a natural remedy for managing inflammation and related disorders. Although further research, including well-designed clinical trials, is warranted to establish optimal dosage, treatment regimens, and long-term safety, the existing scientific evidence underscores the potential of RJ as a valuable natural remedy for inflammatory diseases and warrants further exploration of its clinical applications.

Author Contributions: Conceptualization, L.B. and S.B.; resources, L.B., S.B., V.B. and O.B.; writing—original draft preparation, L.B. and S.B.; writing—review and editing, L.B., S.B. and O.B.; supervision, D.S.D. and O.B.; funding acquisition, O.B. All authors have read and agreed to the published version of the manuscript.

Funding: This research received no external funding.

Institutional Review Board Statement: Not applicable.

Informed Consent Statement: Not applicable.

Data Availability Statement: Not applicable.

Conflicts of Interest: The authors declare no conflict of interest.

References

1. Uthaibutra, V.; Kaewkod, T.; Prapawilai, P.; Pandith, H.; Tragoolpua, Y. Inhibition of Skin Pathogenic Bacteria, Antioxidant and Anti-Inflammatory Activity of Royal Jelly from Northern Thailand. *Molecules.* **2023**, *28*, 996. [CrossRef]
2. Salama, S.; Shou, Q.; Abd El-Wahed, A.A.; Elias, N.; Xiao, J.; Swillam, A.; Umair, M.; Guo, Z.; Daglia, M.; Wang, K.; et al. Royal Jelly: Beneficial Properties and Synergistic Effects with Chemotherapeutic Drugs with Particular Emphasis in Anticancer Strategies. *Nutrients* **2022**, *14*, 4166. [CrossRef]
3. Guo, J.; Wang, Z.; Chen, Y.; Cao, J.; Tian, W.; Ma, B.; Dong, Y. Active Components and Biological Functions of Royal Jelly. *J. Funct. Food* **2021**, *82*, 104514. [CrossRef]

4. Clarke, M.; Peter, M. Australian Royal Jelly Market Opportunity Assessment Based on Production That Uses New Labour Saving Technology. 2017. Available online: https://www.agrifutures.com.au/wp-content/uploads/publications/17-017.pdf (accessed on 3 July 2023).
5. Ahmad, S.; Campos, M.G.; Fratini, F.; Altaye, S.Z.; Li, J. New Insights into the Biological and Pharmaceutical Properties of Royal Jelly. *Mol. Sci.* **2020**, *21*, 382. [CrossRef] [PubMed]
6. Collazo, N.; Carpena, M.; Nuñez-Estevez, B.; Otero, P.; Simal-Gandara, J.; Prieto, M.A. Health Promoting Properties of Bee Royal Jelly: Food of the Queens. *Nutrients* **2021**, *13*, 543. [CrossRef] [PubMed]
7. Sesta, G. Determination of sugars in royal jelly by HPLC. *Apidologie* **2006**, *37*, 84–90. [CrossRef]
8. Sabatini, A.G.; Marcazzan, G.L.; Caboni, M.F.; Bogdanov, S.; Almeida-Muradian, L. Quality and standardization of royal jelly. *J. ApiProd. ApiMed. Sci.* **2009**, *1*, 1–6. [CrossRef]
9. Terada, Y.; Narukawa, M.; Watanabe, T. Specific hydroxy fatty acids in royal jelly activate TRPA1. *J. Agric. Food Chem.* **2011**, *59*, 2627–2635. [CrossRef]
10. Kolayli, S.; Sahin, H.; Can, Z.; Yildiz, O.; Malkoc, M.; Asadov, A. A member of complementary medicinal food: Anatolian royal jellies, their chemical compositions, and antioxidant properties. *J. Evid. Based Complement. Alt. Med.* **2016**, *21*, NP43–NP48. [CrossRef]
11. Fratini, F.; Cilia, G.; Mancini, S.; Felicioli, A. Royal Jelly: An ancient remedy with remarkable antibacterial properties. *Microbiol. Res.* **2016**, *192*, 130–141. [CrossRef]
12. Lin, N.; Chen, S.; Zhang, H.; Li, J.; Fu, L. Quantification of Major Royal Jelly Protein 1 in Fresh Royal Jelly by Ultraperformance Liquid Chromatography–Tandem Mass Spectrometry. *J. Agric. Food Chem.* **2018**, *66*, 1270–1278. [CrossRef] [PubMed]
13. Maghsoudlou, A.; Mahoonak, A.S.; Mohebodini, H.; Toldra, F. Royal Jelly: Chemistry, Storage and Bioactivities. *J. Apic. Sci.* **2019**, *63*, 17–40. [CrossRef]
14. Schmitzová, J.; Klaudiny, J.; Albert, Š.; Schroder, W.; Schreckengost, W.; Hanes, J.; Judova, J.; Simuth, J. A family of major royal jelly proteins of the honeybee *Apis mellifera* L. *CMLS Cell Mol. Life Sci.* **1998**, *54*, 1020–1030. [CrossRef]
15. Buttstedt, A. 10-Hydroxy-Δ2-decenoic acid's role in the fibril formation of the major royal jelly protein 1/apisimin/24-methylenecholesterol complex isolated from honeybee (*Apis mellifera*) royal jelly. *Eur. J. Entomol.* **2022**, *119*, 448–453. [CrossRef]
16. Mureşan, C.I.; Buttstedt, A. pH-dependent stability of honey bee (*Apis mellifera*) major royal jelly proteins. *Sci. Rep.* **2019**, *9*, 9014. [CrossRef]
17. Ozcan, S.; Senyuva, H.Z. Improved and simplified liquid chromatography/ atmospheric pressure chemical ionization mass spectrometry method for the analysis of underivatized free amino acids in various foods. *J. Chromatogr. A* **2006**, *1135*, 179–185. [CrossRef]
18. Jie, H.; Li, P.M.; Zhao, G.J.; Feng, X.L.; Zeng, D.J.; Zhang, C.L.; Lei, M.Y.; Yu, M.; Chen, Q. Amino acid composition of royal jelly harvested at different times after larval transfer. *Genet. Mol. Res.* **2016**, *15*, 15038306. [CrossRef]
19. Mărgăoan, R.; Mărghitaş, L.A.; Dezmirean, D.S.; Bobiş, O.; Bonta, V.; Cătană, C.; Urcan, A.; Mureşan, C.I.; Margin, M.G. Comparative study on quality parameters of royal jelly, apilarnil and queen bee larvae triturate. *Bull. UASVM Cluj-Napoca* **2017**, *74*, 51–58. [CrossRef]
20. Kunugi, H.; Ali, A.M. Royal Jelly and Its Components Promote Healthy Aging and Longevity: From Animal Models to Humans. *Int. J. Mol. Sci.* **2019**, *20*, 4662. [CrossRef]
21. Kocot, J.; Kiełczykowska, M.; Luchowska-Kocot, D.; Kurzepa, J.; Musik, I. Antioxidant potential of propolis, bee pollen, and royal jelly: Possible medical application. *Oxid. Med. Cell Longev.* **2018**, *2018*, 7074209. [CrossRef]
22. Melliou, E.; Chinou, I. Chemistry and bioactivities of royal jelly. In *Studies in Natural Products Chemistry*; Elsevier: Amsterdam, The Netherlands, 2014; Volume 43, pp. 261–290.
23. Yan, S.; Wang, X.; Sun, M.; Wang, W.; Wu, L.; Xue, X. Investigation of the lipidomic profile of royal jelly from different botanical origins using UHPLC-IM-Q-TOF-MS and GC-MS. *LWT* **2022**, *169*, 113894. [CrossRef]
24. López-Gutiérrez, N.; Aguilera-Luiz, M.D.M.; Romero-González, R.; Vidal, J.L.M.; Garrido Frenich, A. Fast analysis of polyphenols in royal jelly products using automated TurboFlow™-liquid chromatography-Orbitrap high resolution mass spectrometry. *J. Chromatogr. B* **2014**, *973*, 17–28. [CrossRef] [PubMed]
25. Kanbur, M.; Eraslan, G.; Beyaz, L.; Silici, S.; Liman, B.C.; Altinordulu, Ş.; Aasever, A. The effects of royal jelly on liver damage induced by paracetamol in mice. *Exp. Toxicol. Pathol.* **2009**, *61*, 123–132. [CrossRef] [PubMed]
26. Adaškevičiūtė, V.; Kaškonienė, V.; Kaškonas, P.; Barčauskaitė, K.; Maruška, A. Comparison of physicochemical properties of bee pollen with other bee products. *Biomolecules* **2019**, *9*, 819. [CrossRef] [PubMed]
27. Bagameri, L.; Baci, G.M.; Dezmirean, D.S. Royal jelly as a nutraceutical natural product with focus on its antibacterial activity. *Pharmaceutics* **2022**, *14*, 1142. [CrossRef] [PubMed]
28. Al-Kahtani, S.N.; Taha, E.-K.A. Post grafting time significantly influences royal jelly yield and content of macro and trace elements. *PLoS ONE* **2020**, *15*, e0238751. [CrossRef]
29. Al-Kahtani, S.N.; Taha, E.-K.A. Effect of harvest time on royal jelly yield and chemical composition. *J. Kans. Entomol. Soc.* **2020**, *93*, 132–139. [CrossRef]

30. Botezan, S.; Baci, G.-M.; Bagameri, L.; Pașca, C.; Dezmirean, D.S. Current Status of the Bioactive Properties of Royal Jelly: A Comprehensive Review with a Focus on Its Anticancer, Anti-Inflammatory, and Antioxidant Effects. *Molecules* **2023**, *28*, 1510. [CrossRef]
31. El-Seedi, H.R.; Eid, N.; Abd El-Wahed, A.A.; Rateb, M.E.; Afifi, H.S.; Algethami, A.F.; Zhao, C.; Al Naggar, Y.; Alsharif, S.M.; Tahir, H.E.; et al. HoneyBee Products: Preclinical and Clinical Studies of Their Anti-inflammatory and Immunomodulatory Properties. *Front. Nutr.* **2022**, *8*, 761267. [CrossRef] [PubMed]
32. Gu, H.; Song, I.B.; Han, H.J.; Lee, N.Y.; Cha, J.Y.; Son, Y.K.; Kwon, J. Anti-inflammatory and immune-enhancing effects of enzyme-treated royal jelly. *Appl. Biol. Chem.* **2018**, *61*, 227–233. [CrossRef]
33. Hegazi, A.G.; Al-Menabbawy, K.; Abd El Rahman, E.; Helal, S.I. Novel Therapeutic Modality Employing Apitherapy for Controlling of Multiple Sclerosis. *J. Clin. Cell Immunol.* **2015**, *6*, 299. [CrossRef]
34. Podbielska, M.; O'Keeffe, J.; Pokryszko-Dragan, A. New Insights into Multiple Sclerosis Mechanisms: Lipid on the Track to Control Inflammation and Neurodegeneration. *Int. J. Mol. Sci.* **2021**, *22*, 7319. [CrossRef] [PubMed]
35. Lohrasbi, M.; Taghian, F.; Jalali Dehkordi, K.; Hosseini, S.A. The functional mechanisms of synchronizing royal jelly consumption and physical activity on rat with multiple sclerosis-like behaviors hallmarks based on bioinformatics analysis, and experimental survey. *BMC Neurosci.* **2022**, *23*, 34. [CrossRef]
36. Kheirdeh, M.; Koushkie Jahromi, M.; Brühl, A.B.; Brand, S. The Effect of Exercise Training and Royal Jelly on Hippocampal Cannabinoid-1-Receptors and Pain Threshold in Experimental Autoimmune Encephalomyelitis in Rats as Animal Model of Multiple Sclerosis. *Nutrients.* **2022**, *14*, 4119. [CrossRef]
37. Oshvandi, K.; Aghamohammadi, M.; Kazemi, F.; Masoumi, S.Z.; Mazdeh, M.; Molavi Vardanjani, M. Effect of royal jelly capsule on quality of life of patients with multiple sclerosis: A double-blind randomized controlled clinical trial. *Iran. Red Crescent Med. J.* **2020**, *22*, e74. [CrossRef]
38. Seyedian, S.S.; Nokhostin, F.; Malamir, M.D. A review of the diagnosis, prevention, and treatment methods of inflammatory bowel disease. *J. Med. Life* **2019**, *12*, 113–122. [CrossRef]
39. Guan, Q. A Comprehensive Review and Update on the Pathogenesis of Inflammatory Bowel Disease. *J. Immunol. Res.* **2019**, *2019*, 7247238. [CrossRef]
40. Yang, Y.; Chou, W.; Widowati, D.A.; Lin, I.; Peng, C. 10-Hydroxy-2-Decenoic Acid of Royal Jelly Exhibits Bactericide and Anti-Inflammatory Activity in Human Colon Cancer Cells. *BMC Complement. Altern. Med.* **2018**, *18*, 202. [CrossRef]
41. Karaca, T.; Uz, Y.H.; Demirtas, S.; Karaboga, I.; Can, G. Protective effect of royal jelly in 2,4,6 trinitrobenzene sulfonic acid-induced colitis in rats. *Iran. J. Basic Med. Sci.* **2015**, *18*, 370–379. [PubMed]
42. Guo, J.; Ma, B.; Wang, Z.; Chen, Y.; Tian, W.; Dong, Y. Royal Jelly Protected against Dextran-Sulfate-Sodium-Induced Colitis by Improving the Colonic Mucosal Barrier and Gut Microbiota. *Nutrients* **2022**, *14*, 2069. [CrossRef] [PubMed]
43. Münstedt, K. Rheumatoid arthritis—Is there a role for apitherapy? *J. Aphyter. Nat.* **2022**, *5*, 103–118. [CrossRef]
44. Dalal, D.S.; Durán, J.; Brar, T.; Alqadi, R.; Halladay, C.W.; Lakhani, A.; Rudolph, J.L. Efficacy and safety of biological agents in the older rheumatoid arthritis patients compared to Young: A systematic review and meta-analysis. *Semin. Arthritis Rheum.* **2019**, *48*, 799–807. [CrossRef] [PubMed]
45. Nattagh-Eshtivani, E.; Jokar, M.H.; Tabesh, H.; Nematy, M.; Safarian, M.; Pahlavani, N.; Maddahi, M.; Khosravi, M. The effect of propolis supplementation on inflammatory factors and oxidative status in women with rheumatoid arthritis: Design and research protocol of a double-blind, randomized controlled. *Contemp. Clin. Trials Comm.* **2021**, *23*, 100807. [CrossRef]
46. Yang, X.Y.; Yang, D.S.; Wei, Z.; Wang, J.M.; Li, C.Y.; Hui, Y.; Lei, K.F.; Chen, X.F.; Shen, N.H.; Jin, L.Q.; et al. 10-Hydroxy-2-decenoic acid from Royal jelly: A potential medicine for RA. *J. Ethnopharmacol.* **2010**, *128*, 314–321. [CrossRef] [PubMed]
47. Wang, J.; Zhang, W.; Zou, H.; Lin, Y.; Lin, K.; Zhou, Z.; Qiang, J.; Lin, J.; Chuka, C.M.; Ge, R.; et al. 10-Hydroxy-2-decenoic acid inhibiting the proliferation of fibroblast-like synoviocytes by PI3K–AKT pathway. *Int. Immunopharmacol.* **2015**, *28*, 97–104. [CrossRef] [PubMed]
48. Chen, Y.F.; Wang, K.; Zhang, Y.Z.; Zheng, Y.F.; Hu, F.L. In Vitro Anti-Inflammatory Effects of Three Fatty Acids from Royal Jelly. *Mediat. Inflamm.* **2016**, *2016*, 3583684. [CrossRef]
49. Chen, Y.-F.; You, M.-M.; Liu, Y.-C.; Shi, Y.-Z.; Wang, K.; Lu, Y.-Y.; Hu, F.-L. Potential protective effect of Trans-10-hydroxy-2-decenoic acid on the inflammation induced by Lipoteichoic acid. *J. Funct. Foods* **2018**, *45*, 491–498. [CrossRef]
50. You, M.M.; Chen, Y.F.; Pan, Y.M.; Liu, Y.C.; Tu, J.; Wang, K.; Hu, F.L. Royal Jelly Attenuates LPS-Induced Inflammation in BV-2 Microglial Cells through Modulating NF-κB and p38/JNK Signaling Pathways. *Mediat. Inflamm.* **2018**, *2018*, 7834381. [CrossRef]
51. Mihajlovic, D.; Vucevic, D.; Chinou, I.; Colic, M. Royal jelly fatty acids modulate proliferation and cytokine production by human peripheral blood mononuclear cells. *Eur. Food Res. Technol.* **2014**, *238*, 881–887. [CrossRef]
52. Erem, C.; Deger, O.; Ovali, E.; Barlak, Y. The effects of royal jelly on autoimmunity in Graves' disease. *Endocrine* **2006**, *30*, 175–183. [CrossRef]
53. Arzi, A.; Olapour, S.; Yaghooti, H.; Karampour, N.S. Effect of Royal Jelly on Formalin Induced-Inflammation in Rat Hind Paw. *Jundishapur J. Nat. Pharm. Prod.* **2015**, *10*, e22466. [CrossRef] [PubMed]
54. Minegaki, N.; Koshizuka, T.; Hatasa, K.; Kondo, H.; Kato, H.; Tannaka, M.; Takahashi, K.; Tsuji, M.; Inoue, N. The C-Terminal Penta-Peptide Repeats of Major Royal Jelly Protein 3 Ameliorate the Progression of Inflammation *"in Vivo"* and *"in Vitro"*. *Biol. Pharmaceutic. Bull.* **2022**, *45*, 583–589. [CrossRef] [PubMed]

55. Fatmawati, F.; Erizka, E.; Hidayat, R. Royal Jelly (Bee Product) Decreases Inflammatory Response in Wistar Rats Induced with Ultraviolet Radiation. *Open Access Maced. J. Med. Sci.* **2019**, *7*, 2723–2727. [CrossRef] [PubMed]
56. Pourmobini, H.; Kazemi Arababadi, M.; Salahshoor, M.R.; Roshankhah, S.; Taghavi, M.M.; Taghipour, Z.; Shabanizadeh, A. The effect of royal jelly and silver nanoparticles on liver and kidney inflammation. *Avicenna J. Phytomed.* **2021**, *11*, 218–223. [PubMed]
57. Salahshoor, M.R.; Jalili, C.; Roshankhah, S. Can royal jelly protect against renal ischemia/reperfusion injury in rats? *Chin. J. Physiol.* **2019**, *62*, 131. [CrossRef] [PubMed]
58. Lin, Y.; Zhang, R.; Wang, L.; Lin, T.; Wang, G.; Peng, J.; Su, S. The in vitro and in vivo wound-healing effects of royal jelly derived from *Apis mellifera* L. during blossom seasons of *Castanea mollissima* Bl. and *Brassica napus* L. in South China exhibited distinct patterns. *BMC Complement. Med. Ther.* **2020**, *20*, 357. [CrossRef]
59. Petelin, A.; Lah, T.T.; Kopinc, R.; Deželak, M.; Bizjak, M.; Pražnikar, Z.J. Effects of Royal Jelly Administration on Lipid Profile, Satiety, Inflammation, and Antioxidant Capacity in Asymptomatic Overweight Adults. *Evid.-Based Complement. Altern. Med.* **2019**, *2019*, 4969720. [CrossRef]
60. Chansuwan, W.; Khamhae, M.; Yang, Z.; Sirinupong, N. Hydrolase-treated royal jelly attenuates LPS-induced inflammation and IgE-antigen-mediated allergic reaction. *Funct. Foods Health Dis.* **2020**, *10*, 127–142. [CrossRef]
61. Coutinho, D.; Karibasappa, S.N.; Mehta, D.S. Royal Jelly Antimicrobial Activity against Periodontopathic Bacteria. *J. Interdiscip. Dent.* **2018**, *8*, 17–22.
62. Khosla, A.; Gupta, S.J.; Jain, A.; Shetty, D.C.; Sharma, N. Evaluation and Comparison of the Antimicrobial Activity of Royal Jelly-A Holistic Healer against Periodontopathic Bacteria. *J. Indian Soc. Periodontol.* **2020**, *24*, 221–226. [CrossRef]
63. Nainu, F.; Masyita, A.; Bahar, M.A.; Raihan, M.; Prova, S.R.; Mitra, S.; Emran, T.B.; Simal-Gandara, J. Pharmaceutical Prospects of Bee Products: Special Focus on Anticancer, Antibacterial, Antiviral, and Antiparasitic Properties. *Antibiotics* **2021**, *10*, 822. [CrossRef] [PubMed]
64. Bălan, A.; Moga, M.A.; Dima, L.; Toma, S.; Neculau, E.A.; Anastasiu, C.V. Royal Jelly—A Traditional and Natural Remedy for Postmenopausal Symptoms and Aging-Related Pathologies. *Molecules* **2020**, *25*, 3291. [CrossRef] [PubMed]
65. Miyata, Y.; Sakai, H. Anti-Cancer and Protective Effects of Royal Jelly for Therapy-Induced Toxicities in Malignancies. *Int. J. Mol. Sci.* **2018**, *19*, 3270. [CrossRef] [PubMed]
66. Sver, L.; Orsolić, N.; Tadić, Z.; Njari, B.; Valpotić, I.; Basić, I. A royal jelly as a new potential immunomodulator in rats and mice. *Comp. Immunol. Microbiol. Infect. Dis.* **1996**, *19*, 31–38. [CrossRef] [PubMed]
67. Okamoto, I.; Taniguchi, Y.; Kunikata, T.; Kohno, K.; Iwaki, K.; Ikeda, M.; Kurimoto, M. Major royal jelly protein 3 modulates immune responses in vitro and in vivo. *Life Sci.* **2003**, *73*, 2029–2045. [CrossRef] [PubMed]
68. Natarajan, O.; Angeloni, J.T.; Bilodeau, M.F.; Russi, K.E.; Dong, Y.; Cao, M. The Immunomodulatory Effects of Royal Jelly on Defending Against Bacterial Infections in the *Caenorhabditis elegans* Model. *J. Med. Food.* **2021**, 358–369. [CrossRef] [PubMed]
69. Karaca, T.; Simşek, N.; Uslu, S.; Kalkan, Y.; Can, I.; Kara, A.; Yoruk, M. The effect of royal jelly on $CD3^{+}$, $CD5^{+}$, $CD45^{+}$, T-cell and $CD68^{+}$ cell distribution in the colon of rats with acetic acid-induced colitis. *Allergol. Immunopathol.* **2012**, *40*, 357–361. [CrossRef]
70. Wang, W.; Li, X.; Li, D.; Pan, F.; Fang, X.; Peng, W.; Tian, W. Effects of Major Royal Jelly Proteins on the Immune Response and Gut Microbiota Composition in Cyclophosphamide-Treated Mice. *Nutrients* **2023**, *15*, 974. [CrossRef]
71. Bahaaldin-beygi, M.; Kariminik, A.; Kazemi Arababadi, M. Royal jelly significantly alters inflammatiom pathways in patients with chronic hepatitis B. *Indian J. Exp. Biol.* **2022**, *50*, 875–879. [CrossRef]
72. Lima, W.G.; Brito, J.C.M.; da Cruz Nizer, W.S. Bee products as a source of promising therapeutic and chemoprophylaxis strategies against COVID-19 (SARS-CoV-2). *Phytother. Res.* **2021**, *35*, 743–750. [CrossRef]
73. Habashy, N.H.; Abu-Serie, M.M. The potential antiviral effect of major royal jelly protein2 and its isoform X1 against severe acute respiratory syndrome coronavirus 2 (SARS-CoV-2): Insight on their sialidase activity and molecular docking. *J. Funct. Foods* **2020**, *75*, 104282. [CrossRef] [PubMed]
74. Amirkhani, R.; Zarei, A.; Gholampour, M.; Tavakoli, H.; Ramazani, A. Honey and royal jelly natural products as possible antiviral nominations to combat SARS-CoV-2 main protease. *Euras. Chem. Comm.* **2022**, *4*, 567–579. [CrossRef]
75. Dania, F.; Bazelidze, N.; Chinou, I.; Melliou, E.; Rallis, M.; Papaioannou, G. In vivo antidiabetic activity of Greek propolis and royal jelly. *Planta Med.* **2008**, *74*, PH43. [CrossRef]
76. Nohair, S.F.A. Antidiabetic efficacy of a honey-royal jelly mixture: Biochemical study in rats. *Int. J. Health Sci.* **2021**, *15*, 4–9.
77. Maleki, V.; Jafari-Vayghan, V.; Saleh-Ghadimi, S.; Adibian, M.; Kheirouri, S.; Alizadeh, M. Effects of Royal jelly on metabolic variables in *diabetes mellitus*: A systematic review. *Complement. Ther. Med.* **2019**, *43*, 20–27. [CrossRef] [PubMed]
78. Omer, K.; Gelkopf, M.J.; Newton, G. Effectiveness of royal jelly supplementation in glycemic regulation: A systematic review. *World J. Diabetes* **2019**, *10*, 96–113. [CrossRef] [PubMed]

Disclaimer/Publisher's Note: The statements, opinions and data contained in all publications are solely those of the individual author(s) and contributor(s) and not of MDPI and/or the editor(s). MDPI and/or the editor(s) disclaim responsibility for any injury to people or property resulting from any ideas, methods, instructions or products referred to in the content.

 life

Article

Comparative Studies on the Anti-Inflammatory and Apoptotic Activities of Four Greek Essential Oils: Involvement in the Regulation of NF-κB and Steroid Receptor Signaling

Achilleas Georgantopoulos, Athanasios Vougioukas, Foteini D. Kalousi, Ioannis Tsialtas and Anna-Maria G. Psarra *

Department of Biochemistry and Biotechnology, University of Thessaly, Biopolis, 41500 Larissa, Greece; ageorgant@uth.gr (A.G.); atvougioukas@uth.gr (A.V.); fokalous@uth.gr (F.D.K.); tsialtasj@gmail.com (I.T.)
* Correspondence: ampsarra@uth.gr; Tel.: +30-2410-565221

Abstract: Essential oils (EOs) are well-known for their anti-fungal, anti-microbial, anti-inflammatory and relaxing activities. Steroid hormones, especially glucocorticoids, are also well-known for their anti-inflammatory activities and control of the hypothalamus–pituitary–adrenal (HPA) axis and glucose homeostasis. The biological activities of glucocorticoids render them the most widely prescribed anti-inflammatory drugs, despite their adverse side effects. In this study, comparative studies of the anti-inflammatory activities and interference with glucocorticoids receptor (GR) and estrogen receptor (ER) signaling of EOs from Greek Oregano, *Melissa officinalis*, Lavender and from the Chios Mastic, produced from the Greek endemic mastic tree, were performed in Human Embryonic Kidney 293 (HEK-293) cells. Chios Mastic (Mastiha) and oregano EOs exhibited the highest anti-inflammatory activities. The former showed a reduction in both NF-κB activity and protein levels. Mastic essential oil also caused a reduction in GR protein levels that may compensate for its boosting effect on dexamethasone (DEX)-induced GR transcriptional activation, ending up in no induction of the gluconeogenic phoshoenolpyruvate carboxykinase (PEPCK) protein levels that constitute the GR target. Oregano, *Melissa officinalis* and lavender EOs caused the suppression of the transcriptional activation of GR. Furthermore, the most active EO, that taken from *Melissa officinalis*, showed a reduction in both GR and PEPCK protein levels. Thus, the anti-inflammatory and anti-gluconeogenic activities of the EOs were uncovered, possibly via the regulation of GR signaling. Moreover, cytotoxic actions of *Melissa officinalis* and lavender EOs via the induction of mitochondrial-dependent apoptosis were revealed. Our results highlight these essentials oils' anti-inflammatory and apoptotic actions in relation to their implication on the regulation of steroid hormones' actions, uncovering their potential use in steroid therapy, with many applications in pharmaceutical and health industries as anti-cancer, anti-hyperglycemic and anti-inflammatory supplements.

Keywords: essential oils; *Melissa officinalis*; oregano; lavender; Chios Mastic (Mastiha); glucocorticoid receptor; anti-inflammatory actions; apoptosis; anti-hyperglycemic actions

Citation: Georgantopoulos, A.; Vougioukas, A.; Kalousi, F.D.; Tsialtas, I.; Psarra, A.-M.G. Comparative Studies on the Anti-Inflammatory and Apoptotic Activities of Four Greek Essential Oils: Involvement in the Regulation of NF-κB and Steroid Receptor Signaling. *Life* **2023**, *13*, 1534. https://doi.org/10.3390/life13071534

Academic Editors: Salvador González and Azahara Rodríguez-Luna

Received: 28 May 2023
Revised: 30 June 2023
Accepted: 4 July 2023
Published: 10 July 2023

Copyright: © 2023 by the authors. Licensee MDPI, Basel, Switzerland. This article is an open access article distributed under the terms and conditions of the Creative Commons Attribution (CC BY) license (https://creativecommons.org/licenses/by/4.0/).

1. Introduction

Essential oils are extremely complex mixtures of volatile compounds particularly abundant in aromatic plants, which are mainly secondary metabolites, such as terpenoids and terpenes, which are biogenerated by the mevalonate pathway [1]. Aromatic monoterpenes and sesquiterpenes are the most abundant ones [2,3]. Essential oils are isolated from various anatomical parts of these plants and are produced via extraction, compression or/and distillation methods. They have been used for many years in different industrial sectors, such as cosmetics and personal care products, and are also widely used as food flavoring additives. The anti-microbial and anti-fungal effects of certain essential oils are well-known. Nowadays, an increasing number of studies have uncovered essential oils' anti-inflammatory, anti-oxidant and/or anti-cancer activities, which are attributed to their chemical composition of individual compounds or compounds in the mixture [4,5].

Essential oils' compounds, such as terpenes and triterpenes, are steroid-like compounds that could possibly interfere with steroid receptor signaling. Glucocorticoid and estrogen receptors are steroid receptors with remarkable anti-inflammatory activities, mainly via their interaction with the inflammatory factor NF-κB and the suppression of its activity [6–11]. In this study, four essential oils from three Greek aromatic plants, namely Oregano, *Melissa officinalis* and Lavender cultured in Thessaly, a Greek agricultural mainland region, and Chios Mastic essential oil from the resin of the endemic *Pistacia. Lentiscus L. var. Chia*, cultured exclusively in the southern part of the Greek island of Chios, were evaluated for their anti-inflammatory actions with respect to their interference with steroid receptor signaling. Particularly, the investigation of the potential dissociative activities of these essential oils on the steroid receptors' actions that lead to the suppression of TNF-α-induced NF-κB activation (and thus anti-inflammatory activities), with no or limited effects on the steroid receptors' transactivation (and thus possible gluconeogenic activities, as regards the glucocorticoid receptor) raised our interest [10,12]. Moreover, based on steroid receptors' crucial role in the regulation of energy metabolism and apoptosis, acting as transcription factors and direct or indirect regulators of energy metabolism- and apoptosis-related gene expression [13–16], the possible interference of the essential oils with the steroid hormones, especially the glucocorticoids signaling pathway, is an interesting issue to be explored.

Chios Mastic oil is produced through the distillation of Chios Mastic gum, the air-dried resinous substance from *Pistacia Lentiscus L. var. Chia* (mastic tree). The mastic tree is native to the Mediterranean region and is principally cultivated on the southern part of the Greek island of Chios. Compounds such as a-pinene and β-myrcene are predominantly found in Mastic oil [17]. Studies have shown that Chios Mastic oil may exhibit anti-oxidant, anti-lipidic and anti-cancer activities [18–21]. In addition, accumulating evidence has proven the anti-inflammatory actions of essential oil from the mastic tree, via inflammatory factors, including Interleukin-6 (IL-6) and Tumor Necrosis factor-alpha (TNF-α) reduction [22].

Similarly, *Melissa officinalis* essential oil has been shown to exhibit strong anti-bacterial and anti-fungal activities. In particular, it has been shown to inhibit the growth of several species of $Gram^+$ and $Gram^-$ bacteria, as well as fungi [23]. Moreover, beyond its anti-microbial actions, many other biological actions and therapeutic properties of *Melissa officinalis* essential oil have been uncovered, such as its anti-cancer activity, highlighting its potential pharmaceutical applications [24,25]. Additionally, possible anti-hyperglycemic and neuroprotective actions of *Melissa officinalis* essential oil have been proposed, attributed to its effect on the reduction in PEPCK and glucose 6-phosphatase (G6Pase) [26] and hypoxia-inducible factor 1 (HIF-1) [27], respectively.

As regards lavender essential oil, it has been used for many decades in the field of aromatherapy due to its positive effect on increasing the duration of sleep, its muscle relaxant properties as well as its anxiolytic and sedative effects [28,29]. In addition, recent studies have shown lavender essential oil's positive effects on wound healing [30]. Moreover, there are both in vitro and in vivo studies demonstrating the pro-apoptotic and anti-proliferative effects of lavender essential oil [31].

Oregano oil, in most cases, is prepared by drying the leaves and stems of the plants. Then, through steam distillation, a concentrated essential oil is produced. The chemical composition and content of oregano oil depend on a variety of factors, such as cultivation and growth conditions [32]. Oregano oil is proposed to exhibit anti-microbial, anti-inflammatory and anti-cancer activities, which are likely attributed to its composition of terpenoids [33–35]. In the same context, an increasing number of studies indicate its dose-dependent anti-proliferative properties [34] and its effect on the induction of apoptosis [36,37]. More interestingly, the anti-inflammatory effects of oregano essential oil have been suggested, in line with its chemical composition, enriched in carvacrol and thymol, which are involved in the reduction in reactive oxygen species (ROS) and nitric oxide levels, crucial mediators of inflammation [36].

Thus, many studies in the literature uncovered the plethora of the biological activities of essential oils from oregano, *Melissa officinalis*, lavender, and Chios Mastic. Nevertheless, none or a limited number of them have focused on the interference of these essential oils with steroid receptor signaling and, more precisely, on the characterization of the biochemical mechanisms and the possible biological impact of this action.

In this comparative study, the pro-apoptotic, anti-inflammatory and possible anti-hyperglycemic actions of four Greek essential oils, namely Chios Mastic Oil, *Melissa Officinalis* essential oil, oregano essential oil and lavender essential oil were evaluated. Emphasis was given to the investigation of the essential oils' potential involvement in the regulation of the inflammatory factor NF-κB and the glucocorticoid and estrogen receptors (ERs) signaling pathway. To this purpose, comparative studies on the effect of the four essential oils on cell viability, interference with Estrogen receptor alpha (ERα), Estrogen receptor beta (ERβ), Glucocorticoid receptor and NF-κB signaling in the human embryonic kidney HEK-293 cell line were performed, applying MTT, luciferase assays, and Western blot analysis.

2. Materials and Methods

2.1. Chemicals

Dulbecco's Modified Eagle Medium (DMEM), Trypsin, and Fetal Bovine Serum (FBS) were obtained from Thermo Fisher Scientific (Thermo Fisher Scientific GmbH, Basel, Switzerland). Molecular protein weight marker was purchased from Proteintech (Rosemont, IL, USA, North America). TNF-α was purchased from PeproTech EC, Ltd. (London, UK). Cocktail protease inhibitors were purchased from Roche (Mannheim, Germany). Reporter lysis 5× buffer and luciferin were purchased from Promega Coorporation (Madison, WI, USA). DEX and Estradiol (E2) were obtained from Sigma-Aldrich (St. Louis, MO, USA). The Greek Oregano, Lavender, and *Melissa officinalis* essential oils were provided by Tharros Company Aromatic Plants Products, Larissa Greece, and the essential oil from Chios Mastic tree was provided by the Chios Mastic Growers Association and Mastiha Shop. Essential oils used in the study were 100% natural products without further impurities, as indicated by the chemical analysis, provided by the companies (Supplementary Data S1). For the biochemical assessment, essential oils were diluted in dimethyl sulfoxide (DMSO) (1 V/1 V) and were subsequently added to culture media in the indicated dilutions.

2.2. Cell Culture

HEK-293 cell line was obtained from the American type culture collection (ATCC). HEK-293 non-cancerous cells were used in this study due to their high efficiency in transfection experiments, their considerable endogenous GR and NF-κB levels, and the extent of their use in receptor signaling studies and drug testing. Cells were cultured in 4.5 g/L glucose DMEM with phenol red, 10% *v*/*v* FBS, 2 mM L-glutamine and 100 units/mL penicillin–streptomycin at a temperature of 37 °C and 5% CO_2. For hormone depletion, 48 h before the treatment, cells were cultured in 4.5 g/L glucose DMEM without phenol red, 10% *v*/*v* charcoal–dextran stripped FBS, 2 mM L-glutamine and 100 units/mL penicillin–streptomycin.

2.3. Antibodies

GR monoclonal antibody and polyclonal antibodies against PEPCK and the p65 subunit of NF-κB were purchased from Santa Cruz Biotechnology (Inc., Europe, Heidelberg, Germany). Procaspase-3 polyclonal antibodies specific and monoclonal antibodies against procaspase-9 were obtained from Cell Signaling Technology (Leiden, The Netherlands). Monoclonal antibody against β-actin was obtained from Sigma Aldrich (Sigma Aldrich, St. Louis, MO, USA).

2.4. Cell Viability Assay

MTT assay was applied, as previously described [38], to investigate the effects of essential oils on HEK-293 cell viability. Briefly, 8×10^3 HEK-293 cells were plated on 96-well plates and cultured in DMEM 4.5 g/L glucose, 10% *v*/*v* FBS, 2 mM L-glutamine and 100 units/mL penicillin/streptomycin. After 24 h, cells were treated for 48 h with essential oils from oregano and *Melissa officinalis*, at concentrations of 23 μg/mL, 47 μg/mL, and 94 μg/mL, (dilution 1/40,000, 1/20,000, and 1/10,000 *v*/*v*, respectively). Essential oil of Chios Mastic was tested at concentrations of 21 μg/mL, 42 μg/mL, and 85 μg/mL (dilution 1/40,000, 1/20,000 and 1/10,000 *v*/*v*, respectively). At the same time, lavender essential oil was evaluated at concentrations of 23 μg/mL, 47 μg/mL, 94 μg/mL, and 187 μg/mL (dilution 1/40,000, 1/20,000 1/10,000, 1/5000 *v*/*v*, respectively). Control, untreated cells were incubated with DMSO at a final dilution of 1/1000 for 24 h or 48 h. Then, the medium was replaced with fresh medium containing MTT reagent at a final concentration of 0.5 mg/mL. Upon 3 h incubation and the removal of the medium, the produced formazan crystals were dissolved in 100% isopropanol, and the absorbance was measured at 570 nm and 690 nm, as a reference, in a multimode plate reader (EnSpire, PerkinElmer, Beaconsfield, UK). The intensity of the colored product (optical density, OD) is directly proportional to the number of viable cells present in the culture. Viability = Mean OD sample/Mean OD control × 100. Cell viability was expressed as the viability of the cells treated with various concentrations of the respective essential oil compared to the cell viability of the vehicle-treated (control) cells. The viability of control cells was considered 100%.

2.5. Regulation of ERs, GR and NF-κB Transcriptional Activity via Essential Oils

For the evaluation of the ERs, GR and NF-κB transcriptional activities, the co-transfection of HEK-293 cells with the respective steroid receptors- or NF-κB- luciferase reporter gene constructs (currying the respective transcription factor response element at the promoter of the luciferase gene) and a β-galactosidase reporter gene construct, was performed via the application of the calcium phosphate transfection method, as previously described [39]. For assessment of ERα or ERβ transcriptional activation, HEK293 cells that expressed low levels of ERs were also co-transfected with a pEGFC2ERα or pEGFPC2ERβ construct for ERα or ERβ expression, respectively. Briefly, 3×10^4 cells were grown on 24-well plates and co-transfected with either an estrogen receptor luciferase reporter gene (ER-Luc) construct and a pEGFPC2ERα or pEGFPC2ERβ construct for ERα or ERβ activity assessment, or an MMTV-GREs -Luc (Glucocorticoid Response Elements luciferase) construct for GR activity assessment, or an NF-κB-REs-Luc (NF-κB Response Elements luciferase) construct for NF-κB activity measurements and a β-galactosidase reporter construct, applying for the normalization of the results. Then, the cells were treated with the indicated amounts of essential oils in the presence or absence of (a) 10^{-7} M DEX for the GR transactivation measurements, (b) 10^{-9} M E2 for the ERs transactivation measurements, and (c) 20 ng/mL TNF-α for the evaluation of the NF-κB transcriptional activation. After 6 h incubation, cells were washed in PBSx1 and then lysed in reporter lysis buffer (Promega). The assessment of the activity of the expressed luciferase and β-galactosidase activity was followed. The light emission was measured using a chemiluminometer (LB 9508, www.berthhold.com, accessed on 5 August 2014), and the relative luciferase activity was expressed as normalized luciferase activity against β-galactosidase activity. The relative luciferase activity in control vehicle-treated cells was set as 1. Folds reduction or the induction of the luciferase activity by EOs was expressed compared to controls.

2.6. Western Blot Analysis

For Western blot analysis, 2×10^5 HEK-293 cells were grown on 6-well plates for 48 h in DMEM 4.5 g/L without phenol red, supplemented with 10% *v*/*v* dextran–charcoal stripped FBS. Then, cells were treated with 23 μg/mL and 47 μg/mL of the essential oils of oregano, *Melissa officinalis* and lavender, or with 21 μg/mL and 42 μg/mL of the Chios Mastic essential oil, in the presence or absence of 10 nM DEX, for 48 h. Then, cells were washed with PBSx1 and subsequently lysed in a lysis buffer containing 20 mM Tris pH: 7.5, 250 nM NaCl, 0.5% *v/v* Triton-X, 3 mM EDTA, supplemented with cocktail protease inhibitors. Protein determination was achieved by applying a Bradford assay, and then, cell extracts were electrophoresed on discontinuous SDS-PAGE and Western blotted, as previously described [40]. Enhanced chemiluminescence was used for the detection of protein bands. The expressed protein levels of β-actin were evaluated for the normalization of the expressed protein levels of GR, PEPCK, p65, procaspase-9 and procaspase-3. Quantification of band intensity was carried out by applying ImageJ (v1.52 p) analysis (NIH, Bethesda, MD, USA). Relative protein levels were expressed as band intensity normalized against the respective band's intensity of β-actin. Relative protein levels in control vehicle-treated cells were set as 1.

2.7. Statistical Analysis

All results are expressed as ± SD. Data were analyzed through *t*-test or two-way Analysis of Variance (two-way ANOVA), followed by Tukey's post hoc test, using StatPlus LE 7.3.0 Software (AnalystSoft, Brandon, FL, USA). Significant differences were considered at a two-tailed p value < 0.05.

3. Results

3.1. Essential Oils Effects on HEK-293 Cell Viability

To evaluate the possible effect of the essential oils on HEK-293 cells viability, an MTT assay was applied. HEK-293 cells were incubated with essential oils at a concentration range of 23 μg/mL to 187 μg/mL for 48 h, as indicated in Figure 1. As shown in Figure 1A,B, no statistically significant reduction in cell viability was observed in the Chios Mastic essential oil and the oregano essential oil at a concentration range of 21 to 85 μg/mL (Figure 1A) and 23 to 94 μg/mL (Figure 1B), respectively. Lavender essential oil (Figure 1D) showed moderate reduction in cell viability by up to approximately 10–20%, at a concentration range from 47 μg/mL to 187 μg/mL (Figure 1D), whereas the highest cytotoxicity was observed in the *Melissa officinalis* essential oil (Figure 1C), which exhibited 35% and 50% reduction in cell viability at concentration of 47 μg/mL and 94 μg/mL, respectively, compared to the control vehicle (1/1000 *v/v* DMSO) treated cells.

3.2. Chios Mastic Essential Oil Caused Moderate Reduction in the E2-Induced Transcriptional Activation of ERα

Taking into account the crucial role of steroid hormones, especially that of the glucocorticoids and estrogens, in the regulation of immune responses in conjunction with the emerging anti-inflammatory activities of essential oils, especially that of the mastic essential oil [41], we studied the possible interference of the mastic essential oil with the regulation of the ERs transcriptional activation. Thus, ERα transcriptional activation was evaluated in the presence or absence of the Mastic essential oil at a concentration range of 21 to 85 μg/mL and/or E2 10^{-9} M, applying ERE-dependent luciferase reporter gene assay, as described in the experimental section. As shown in Figure 2, Chios Mastic essential did not induce ERα transcriptional activation. A negligible suppression, by up to 20%, of the E2-induced ERα transcriptional activation was observed in Mastic essential oil at the highest concentration examined (85 μg/mL). Similarly, no statistically significant effect on the ERβ transcriptional activation was observed in the Mastic essential oil. Likewise, no effect on the E2-induced ERα transcriptional activation was observed in the essential oils from lavender, oregano and *Melissa officinalis* (Supplementary Figure S1). Thus, the involvement of the essential oils' effect on ERs signaling was not further evaluated.

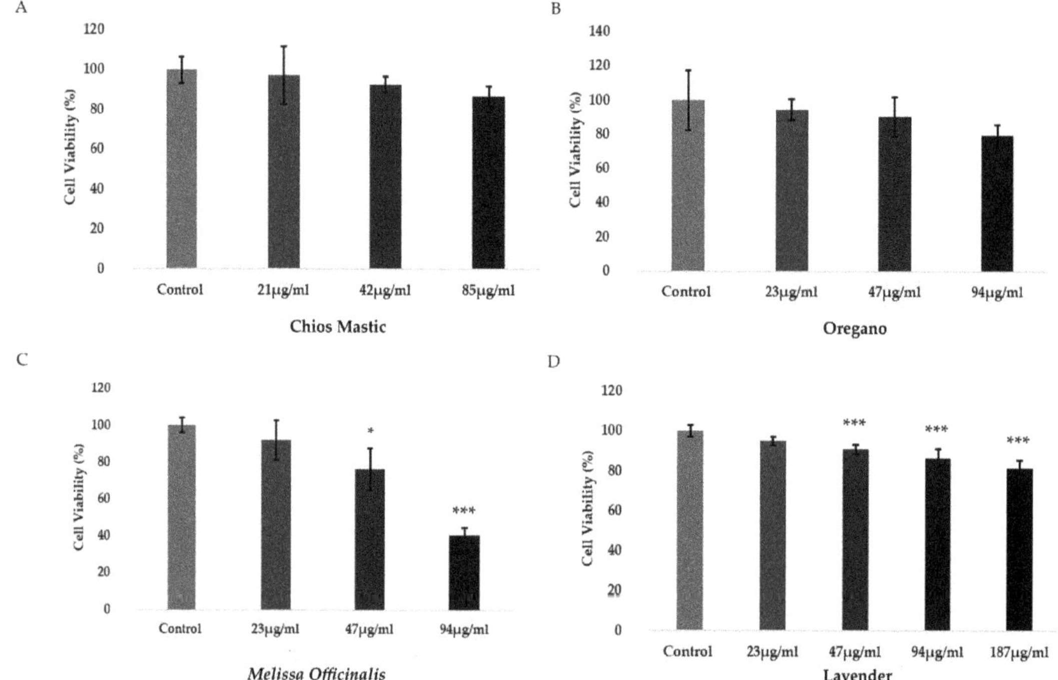

Figure 1. Evaluation of the cytotoxic effects of the essential oils from (**A**) Chios Mastic, (**B**) Oregano, (**C**) *Melissa officinalis* and (**D**) Lavender on HEK-293 cells. Cytotoxicity was assessed by applying MTT assay to HEK-293 cells subjected to essential oils treatment for 48 h. Viability of control vehicle-treated cells was considered 100%. Relative cell viability (Cell viability, %) is expressed as the viability of the cells treated with various concentrations of the respective essential oil compared to the viability of the control cells. Data were analyzed by *t*-test and are expressed as mean ± SD (n = 5), * $p < 0.05$, *** $p < 0.001$ compared to the respective control.

3.3. Chios Mastic, Oregano, *Melissa officinalis* and Lavender Essential Oils Regulate the DEX-Induced Transcriptional Activation of GR

Since we did not observe a significant effect of essential oils on ER activity, the possible effect of the essential oils on the transcriptional regulation of GR was assessed by applying a GRE-dependent luciferase reporter gene and β-galactosidase reporter gene assay to the HEK-293 cells. Thus, HEK-293 cells grown in hormone-depleted medium for 48 h were subsequently co-transfected with the respective constructs, as described in the experimental section. Then, cells were subjected to treatment with the essential oils at no cytotoxic concentrations, as indicated by the cytotoxicity assessment upon 24 h treatment (Supplementary Figure S2), in the presence or absence of 1 μM DEX for 6 h (Figure 3). As shown in Figure 3, DEX induced 3.5–5 fold increases in GR transcriptional activity, as was expected, while neither the induction nor suppression of GR transcriptional activation was observed in terms of the essential oils. Interestingly, oregano (Figure 3B) and *Melissa officinalis* (Figure 3C) essential oils suppressed the DEX-induced GR transcriptional activation. Specifically, the oregano essential oil caused approximately 15% and 25% suppression of the DEX-induced GR transcriptional activation at the concentrations of 47 μg/mL and 94 μg/mL, respectively. Essential oil from *Melissa officinalis* exhibited the highest suppressive effect, causing a 60% reduction in the DEX-induced GR transcriptional activation at a concentration of 94 μg/mL. On the contrary, a synergistic effect on the DEX-induced GR transcriptional activation was observed in the essential oil from lavender, leading to a statistically significant increase in DEX-induced GR transcriptional activation, by 25%, at a concentration of 184 μg/mL (Figure 3D). A similar effect (increase by 30%) was also observed in the Chios Mastic essential oil at the low concentration of 21 μg/mL (Figure 3A).

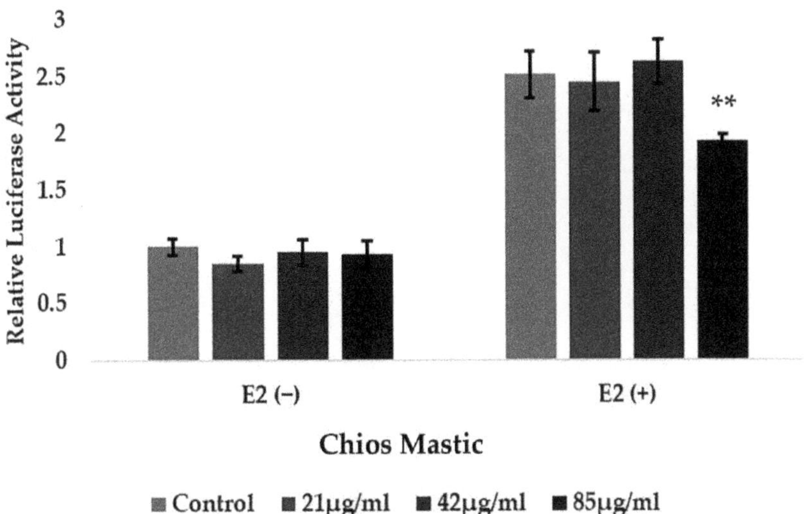

Figure 2. Chios Mastic essential oil caused moderate reduction in the E2-induced ERα transcriptional activation. Cells grown in hormone-free medium for 48 h were transiently co-transfected with an ERE-dependent Luciferase reporter gene construct, a PEGFPC2ERα and a β-galactosidase reporter construct. Subsequently, cells were treated with the indicated amounts of the Chios Mastic essential oil, in the presence or absence of E2, for 6 h. Then, cells were harvested and lysed. Assessment of the luciferase and β-galactosidase activity was followed in cell extracts. Relative luciferase activity was expressed as normalized against the β-galactosidase activity. Relative luciferase activity in control cells was set as 1. Data were analyzed by two-way ANOVA and are expressed as mean ± SD, (n = 6), ** $p < 0.01$.

3.4. *Effect of Essential Oils from Chios Mastic, Oregano, Melissa officinalis and Lavender on GR and PEPCK Protein Levels*

To further analyze the potential effects of the essential oils on GR signaling and to investigate the possible dissociative activity of the essential oils on GR transactivation (and thus GR gluconeogenic activities) and transrepression (suppression of the TNF-α-induced NF-κB activation, and thus GR anti-inflammatory activities), comparative studies of essentials oils effect on the regulation of GR and PEPCK protein levels was performed, applying Western blot analysis. Interestingly, as it is shown in Figure 4, essential oils induced a decrease in GR protein levels. More specifically, Chios Mastic and oregano essential oils caused a 30% reduction in GR protein levels at a concentration of approximately 45 μg/mL. A reduction in GR protein levels by 20% was also observed in the Chios Mastic oil at the lower concentration of 21 μg/mL (Figure 4A). Similarly, essential oil from *Melissa officinalis* caused a 40% reduction in GR protein levels at the concentration range from 23 μg/mL to 47 μg/mL. Reduction in GR protein levels by the oregano essential oil and the *Melissa officinalis* essential oil may be associated with the respective essential oils' observed suppression of the DEX-induced GR transcriptional activation. Moreover, a 20% reduction in GR protein levels was also induced by the lavender essential oil at a concentration of 47 μg/mL (Figure 4B). A further decrease, by 30%, in GR protein levels was observed in the presence of DEX.

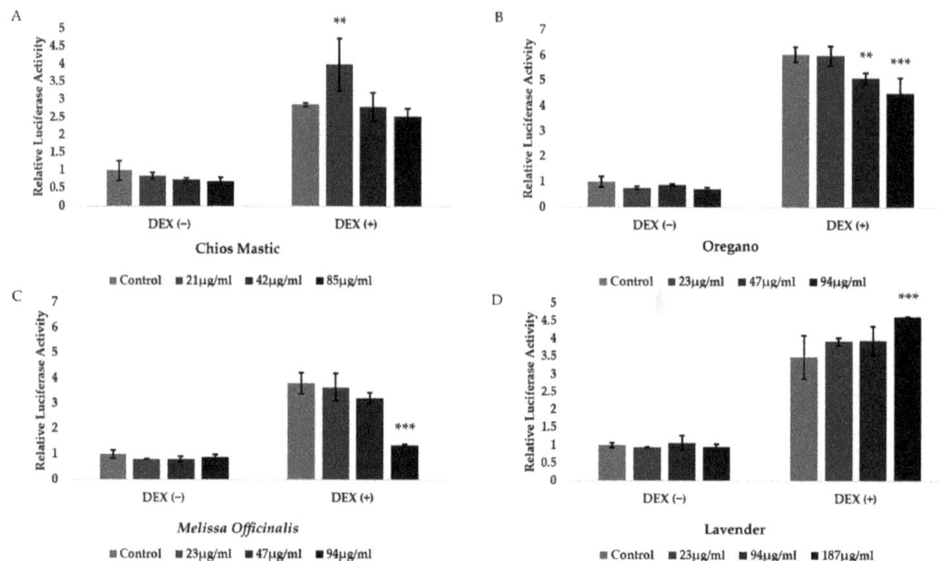

Figure 3. Regulation of the DEX induced GR transcriptional activation by the essential oils from (**A**) Chios Mastic, (**B**) Oregano, (**C**) *Melissa officinalis*, and (**D**) Lavender in HEK-293 cells. Cells grown in hormone-free medium were transiently co-transfected with a GRE-Luc reporter gene construct and a β-galactosidase reporter construct. Subsequently, cells were treated with the indicated amounts of the essential oils, in the presence or absence of 1 µM DEX, for 6 h. Then, cells were harvested and lysed. Assessment of the luciferase and β-galactosidase activity in cell extracts was followed. Relative luciferase activity was expressed as normalized against β-galactosidase activity. Relative luciferase activity in control cells was set as 1. Data were analyzed by two-way ANOVA and are expressed as mean ± SD, (n = 6), ** $p < 0.01$; *** $p < 0.001$.

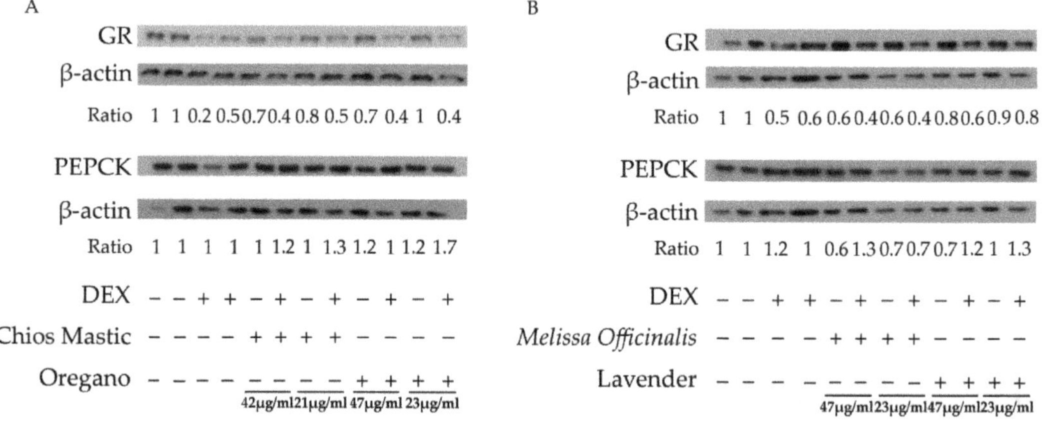

Figure 4. Regulation of GR and PEPCK protein levels by the essential oils from (**A**) Chios Mastic and Oregano, (**B**) *Melissa officinalis* and Lavender. Cells grown in hormone-free medium were incubated with the essential oils in the absence or presence of DEX 10^{-8} M, for 48 h in a hormone-free medium. Data were expressed as the ratios of bands intensity of GR and PEPCK normalized against the respective band's intensity of β-actin. Relative band intensity of control cells was set as 1. The uncropped blots are shown in Figure S4.

In the same context, the potential role of essential oils in the regulation of PEPCK, which constitutes a GR target and a key gluconeogenic enzyme [42,43], was assessed. In accordance with the essential oils-induced reduction in GR protein levels, *Melissa officinalis* and lavender essential oils caused a reduction in PEPCK protein levels. Specifically, *Melissa officinalis* essential oil caused 30% and 40% reductions in PEPCK protein levels at the concentrations of 23 μg/mL and 47 μg/mL, respectively (Figure 4B). Lavender essential oil caused a 30% reduction in PEPCK protein levels at the concentration of 47 μg/mL (Figure 4B). Co-administration with DEX reversed the suppressive effect of *Melissa officinalis* essential oil on the PEPCK protein level, causing a 30% increase compared to the control, (70% increase compared to the *Melissa officinalis*-induced reduction) at a concentration of 47 μg/mL. A similar effect was observed during the co-administration of lavender essential oil with DEX. Thus, an increase in PEPCK protein levels was observed to be approximately 20–30% at a concentration range of 23–47 μg/mL of lavender essential oil (Figure 4B) when administered with DEX compared to controls (60% increase compared to the lavender-induced reduction). A similar effect on the induction of PEPCK protein expression was observed upon co-administration of Chios Mastic essential oil with DEX (Figure 4A), supporting the observed strengthening effect of the lavender (Figure 4B) and Chios Mastic (Figure 4A) essential oils on the DEX-induced GR transcriptional activation. At both concentrations examined, essential oil from Chios Mastic did not cause any effect on PEPCK protein levels, whereas essential oil from oregano caused a 20% increase compared to controls (Figure 4A).

3.5. Anti-Inflammatory Activities of Essential Oils from Chios Mastic, Oregano and Melissa officinalis via Suppression of NF-κB Transcriptional Activation

To evaluate the possible anti-inflammatory activities of the essential oils from Chios Mastic, oregano and *Melissa officinalis*, an NF-κB-associated luciferase/β-galactosidase reporter gene assay was applied. Thus, the effect of the essential oils on the TNF-α-induced NF-κB transcriptional activation was assessed. Results from the study revealed the suppressive effect of the essential oils on the TNF-α-induced NF-κB transcriptional activation, uncovering their potential anti-inflammatory activities. Specifically, essential oil from Chios Mastic caused approximately 40% and 60% statistically significant inhibition of the TNF-α-induced NF-κB transcriptional activation at the concentrations of 42 μg/mL and 85 μg/mL, respectively (Figure 5A). Similarly, statistically significant suppression of the TNF-α-induced NF-κB transcriptional activation, by approximately 50% and 70% (Figure 5B), by oregano essential oil at concentrations of 47 μg/mL and 94 μg/mL, respectively, was observed. Essential oil from *Melissa officinalis* also exhibited suppression of the TNF-α-induced NF-κB transcriptional activation, although to a lower extent. Thus, essential oil from *Melissa officinalis* caused approximately 15% and 30% reduction in the NF-κB transcriptional activation at concentrations of 47 μg/mL and 94 μg/mL, respectively (Figure 5C). No anti-inflammatory activity was observed in the Lavender essential oil at a concentration range of 23 to 94 μg/mL (Supplementary Figure S3).

To evaluate whether the EOs-induced suppression of the NF-κB transcriptional activity was associated with the regulation of the protein levels of the p65 subunit of NF-κB, Western blot analysis of p65 was performed in protein extracts from HEK-293 cells treated with essential oils at a concentration range indicated in Figure 6, in the absence or presence of DEX for 48 h. As shown in Figure 6, a reduction in p65 protein levels was observed in the essential oils from Mastiha resin, oregano and lavender by 20–40% at a concentration range of approximately 22 to 45 μg/mL. Essential oil from *Melissa officinalis* showed an increase in p65 protein levels, which is in accordance with its just moderate activity on the suppression of the NF-κB activity.

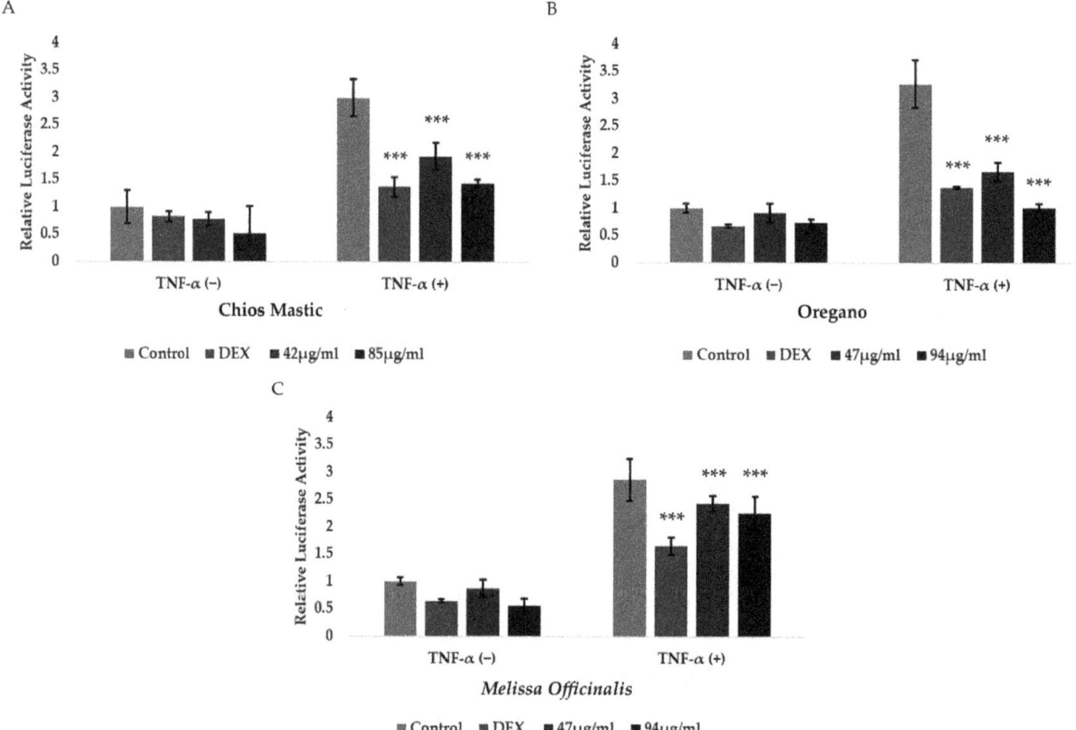

Figure 5. Anti-inflammatory activities of the essential oils from the Chios Mastic, Oregano and *Melissa officinalis*. Suppression of the TNF-α induced NF-κB transcriptional activation by the essential oils from the (**A**) Chios Mastic, (**B**) Oregano and (**C**) *Melissa officinalis*. HEK-293 cells were co-transfected with the NF-κB-luciferase reporter and the β-galactosidase reporter constructs and subsequently were treated with the indicated concentrations of the essential oils, or with 1 μM DEX, for 6 h, in the presence or absence of 20 ng/mL TNF-α, at hormone-free medium. Assessment of the luciferase and the β-galactosidase activity was followed. Results were expressed as relative luciferase activity normalized against β-galactosidase activity. Relative luciferase activity in control cells was set as 1. Data were analyzed by two-way ANOVA and are expressed as mean ± SD, (n = 6), *** $p < 0.001$.

3.6. Regulation of the Mitochondrial-Dependent Apoptosis by the Essential Oils

Furthermore, to assess the possible pro-apoptotic activities of the essential oils, Western blot analysis of procaspase-3 and procaspase-9 protein levels was performed in extracts from HEK-293 cells treated with essential oils, at the indicated concentrations (Figure 7), and/or 10 nM DEX for 48 h, in hormone-free medium. As shown in Figure 7B, 47 μg/mL of essential oils from lavender and *Melissa officinalis* showed a 30% reduction in procaspase-9 protein levels. This effect was also accompanied by a reduction in procaspase-3 protein levels. More specifically, 47 μg/mL of essential oil from lavender caused a 20% reduction in procaspase-3 protein compared to the control. *Melissa officinalis* essential oil also exhibited 20% to 50% reductions in procaspase-3 protein levels. Essential oils from Chios Mastic and oregano showed no remarkable effects on the procaspase-9 protein levels. In contrast, 42 and 47 μg/mL of the essential oils of Chios Mastic and Oregano caused 20% and 30% reduction in caspase-3 levels, respectively (Figure 7A).

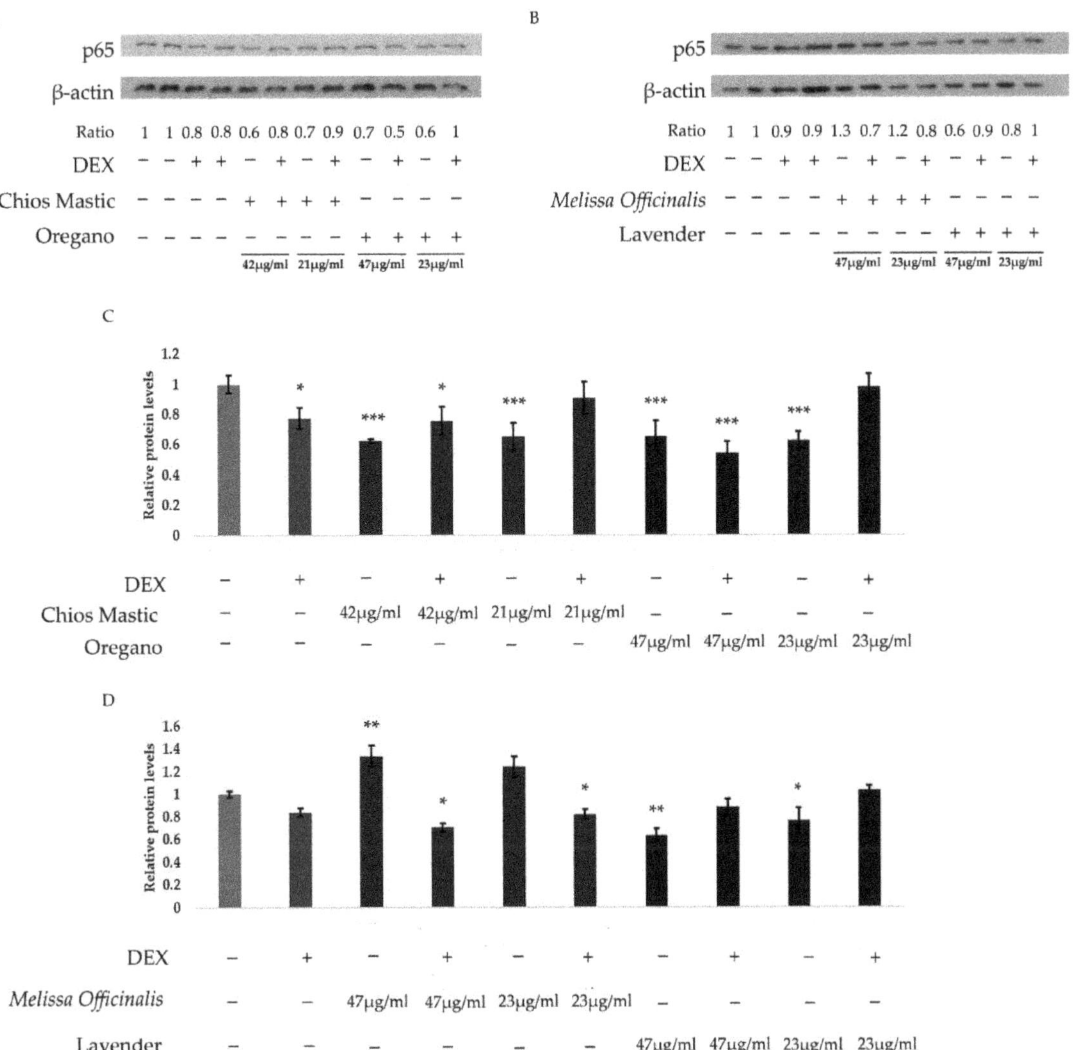

Figure 6. Regulation of the p65 protein levels by essential oils. Representative images from Western blot analysis of the p65 subunit of NF-κB in HEK-293 cells treated for 48 h in a hormone-free medium with the essential oils from (**A**) Chios Mastic, at concentrations of 21 μg/mL and 42 μg/mL, and with the essential oil from oregano, and (**B**) Melissa officinalis and lavender, at concentrations of 23 μg/mL and 47 μg/mL. Results were expressed as the ratios of the p65 bands intensity normalized against the respective band intensity of the β-actin. Relative band intensity of control cells was set as 1. (**C**,**D**) Quantification of the results in A and B, respectively. Data are expressed as means of the ratios ± SD, ($n = 3$), * $p < 0.05$; ** $p < 0.01$; *** $p < 0.001$, compared to control vehicle-treated cells. The uncropped blots are shown in Figure S4.

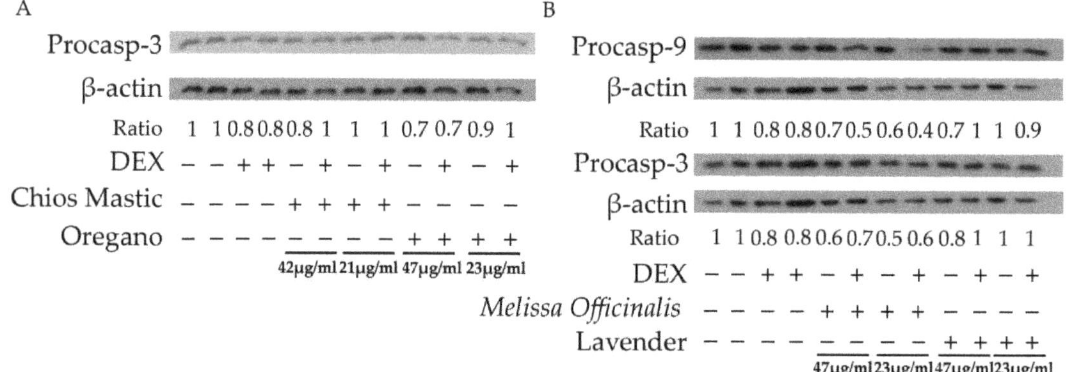

Figure 7. Evaluation of pro-apoptotic actions of the essential oils in HEK-293 cells. Western blot analysis of procaspase-9 and procaspase-3 in extracts from HEK-293 cells treated with essential oils ((**A**), Chios Mastic and Oregano; (**B**), *Melissa Officinalis* and Lavender), at the indicated concentrations, in the absence or presence of DEX for 48 h. Data are expressed as ratios of the levels of the apoptosis-associated molecules against the respective levels of β-actin. Relative band intensity of control vehicle-treated cells was set as 1. The uncropped blots are shown in Figure S4.

4. Discussion

In recent years, scientific studies that demonstrate the therapeutic benefits of chemical compounds of plant origin and particularly of essential oils from plants have been constantly growing [1,32,44–47]. In this study, four Greek essential oils were assessed for their biological actions in the HEK-293 cell line. Specifically, Chios Mastic, *Melissa officinalis*, oregano and lavender essential oils have been investigated with respect to their anti-proliferative and anti-inflammatory actions and their possible interference with estrogen and glucocorticoid receptors signaling. GRs and ERs are crucial regulators of cell physiology, affecting many cellular functions, including growth and development, cellular metabolism, and apoptosis [14,48,49]. In addition, those receptors are well-known for their anti-inflammatory actions via interference with NF-κB signaling and the control of the expression of many inflammation-related genes. Moreover, glucose homeostasis and lipid and protein metabolism are highly affected by glucocorticoids and estrogens [50,51]. Considering the high use of glucocorticoids for pharmaceutical purposes, due to their strong anti-inflammatory actions, which are accompanied by increased glucose synthesis, the possible interference of essential oils with the steroid hormone receptors signaling may uncover novel plant-derived steroids-like compounds as lead molecules for the development of selective steroid receptor regulators with increased desired anti-inflammatory activities but with reduced adverse side effects, such as increased glucose synthesis, myopathy, osteoporosis and hypothalamus–pituitary–adrenal axis dysregulation [10,52].

In this frame, comparative studies on the anti-inflammatory activities and interference with the glucocorticoid signaling of essential oils from Chios Mastic, *Melissa officinalis*, oregano and lavender were performed.

Applying luciferase assay and Western blot analysis, the effect of essential oils on the TNF-α-induced NF-κB activation and the regulation of the protein levels of the p65 subunit of NF-κB were evaluated. Our results showed a suppressive effect on the TNF-α-induced NF-κB activation by the essential oils from Chios mastic, oregano and *Melissa officinalis*. Specifically, essential oils from Chios mastic and oregano were the most active ones, whereas essential oil from *Melissa officinalis* exhibited lower activity. The anti-inflammatory activity of the essential oils could be attributed to essential oils compounds, such as α-thujone, β-thujone, camphor, caryophyllene and terpenoids that have been reported to exert suppression of NF-κB activity [53,54]. Interestingly, essential oils from oregano and Chios mastic, which were the most active ones, are also the most enriched in thujene and camphene, respectively (Supplementary Data S1, Essential oils chemical composition). Moreover, the reduction in NF-κB activity by the Chios mastic and oregano essential oils was accompanied by a reduction in the p65 subunit of NF-κB. Thus, the suppressive effect of the essential oils from Chios mastic and oregano on the NF-κB activity could be attributed to essential oils' suppressive effect both on NF-κB transcriptional activation and protein levels. In the same frame, *Melissa officinalis*, which showed limited anti-inflammatory action, caused an increase in p65 protein levels.

As regards the interference of essential oils with glucocorticoid signaling, and thus the regulation of NF-κB signaling and glucose synthesis, our results showed that essential oils did not induce GR transactivation. For the first time, our study revealed that essential oils from Chios Mastic and Lavender caused a moderate increase in DEX-induced GR transcriptional activation. However, essential oils from oregano and *Melissa officinalis* suppressed the DEX-induced GR transcriptional activation, in accordance with previous observation [55,56]. The enhancement of DEX-induced GR transcriptional activation by Chios mastic oil may lead to the glucocorticoid-induced regulation of inflammatory and pro-inflammatory molecules expression, which is also responsible for the anti-inflammatory actions of GR [57]. Moreover, for the first time, we showed a reduction in GR protein levels by the essential oils. Considering that GR is involved in the regulation of NF-κB activity, reduction in GR protein levels may also affect essential oils' anti-inflammatory actions. As regards the interference of essential oils with glucocorticoids and thus glucose synthesis, our results revealed the regulation of GR and its target PEPCK protein levels by the essential oils. Particularly, a reduction in GR protein levels was observed in all the essential oils examined, indicating that this action may constitute a common action of essential oils. Chios mastic oil's suppressive effect on GR protein levels may compensate for its promoting effect on GR transcriptional activation, resulting in no effect of Chios mastic essential oil on PEPCK protein levels, and thus no induction of glucose synthesis. A decrease in GR protein levels by essential oils from oregano, *Melissa officinalis*, and lavender was followed by a decrease in PEPCK protein levels, highlighting essential oils' potential anti-gluconeogenic actions. An antagonistic effect of the synthetic glucocorticoid dexamethasone on this action was uncovered, as regards *Melissa officinalis* and lavender essential oil. Thus, the reversal of the essential oils-induced decrease in PEPCK was observed when administered with DEX, corroborating the possible interference of the essential oils with glucocorticoids signaling. A reduction in GR protein levels by the oregano and *Melissa officinalis* essential oils may be responsible for the suppressive effect of the oils on the DEX-induced GR transcriptional activation. The interference of essential oils with glucocorticoids signaling is also in accordance with data from the literature demonstrating that essential oils' compounds such as α-Pinene, limonene, α-thujene, myrcene, sabinene, and para-cymene, are responsible for the calming effect of essentials oils via the interference and suppression of the HPA axis [53,58,59]. In this context, lavender essential oil is shown to cause a reduction in the stress hormone cortisol [60–62]. Thus, the effect of essential oils on GR signaling could be exerted both via the regulation of glucocorticoid levels and GR transcriptional activation and/or protein levels. A similar antagonistic action of essential oils compounds on steroid signaling has been observed using germacrene analogs, which have been proposed to exert anti-androgenic activities [63]. Lavender essential oil is also proposed to be involved in prepubertal gynecomastia via its anti-androgenic and estrogenic activities [64,65]. The interference of lavender essential oil with estrogen signaling is proposed to be beneficial for perimenopausal women by inducing an increase in estrogen levels and relieving perimenopausal symptoms [66]. Docking analysis verified the ability of thymol and carvacrol, compounds of essential oils, to bind to estrogen receptors [67]. Nevertheless, the estrogenic activity of lavender was not confirmed in a rat model [68] and in hormone-dependent (MCF-7) and -independent (MDA-MB-231) cell lines [69]. In the same frame, in this study, no interference of Chios Mastic, *Melissa officinalis*, oregano and lavender essential oils with estrogen signaling, either by ERα or ERβ, was found.

Essential oils are also well-known for their anti-proliferative and apoptotic activities. In this study, the MTT assay revealed cytotoxic activities of essential oils from *Melissa officinalis* and lavender, whereas essential oils from Chios Mastic and oregano showed no cytotoxicity at a concentration range from approximately 20 μg/mL to 85 μg/mL. Similarly, caspase-3 activation was not observed in the Chios Mastic and oregano essential oil. However, in accordance with the evaluation of the essential oils' cytotoxic effects, caspase-3 activation was observed in HEK-293 cells treated with *Melissa officinalis* and lavender at a concentration range from 47 to 187 μg/mL, indicating apoptosis activation. Most importantly, caspase-3 activation was accompanied by caspase-9 activation, revealing mitochondrial-dependent apoptosis activation. The lavender essential oil-induced caspase-9 activation supports previously reported observations applying annexin staining [31]. Chios Mastic and oregano essential oils also induced caspase-9 activation. Glucocorticoids are well known to induce mitochondrial-dependent apoptosis in a tissue-specific manner, including epithelial cells of the digestive system [16]. Thus, the apoptotic activities of the essential oils from lavender, Chios Mastic and oregano might be exerted, among others, via interference with glucocorticoid signaling and could have application in glucocorticoid-based cancer treatment. Anti-proliferative activities of essential oils from oregano are also reported in the literature [34,70,71] and may be associated with their effect on the induction of apoptosis via the intrinsic and/or extrinsic pathway. Moreover, there is supporting evidence for essential oils' compounds like carvacrol, limonene, citral, thymol and terpenoid analogues, such as Terpinen-4-olto are involved in the apoptotic mechanism in many cell types. Characteristic examples are murine mesothelioma (AE17), melanoma cells (B16-F10), and fibroblasts (L929), colon cancer (LS174T) cells, breast cancer (MCF-7) cell line, human metastatic breast cancer (MDA-MB 231) cell line and human promyelocytic leukemia (HL-60) cells [70,72].

To conclude, in this comparative study, the anti-inflammatory and apoptotic activities of the Greek Oregano, *Melissa officinalis*, Lavender and Chios Mastic essential oils were evaluated in relation to steroid hormones signaling interference. Estrogenic activity of the essential oils was not detected, whereas interference with glucocorticoid signaling was observed, affecting both glucocorticoid receptor activity and protein levels. This action also came with the regulation of the GR target gene expression, uncovering potential anti-gluconeogenic activities of the essential oils via possible interference with glucocorticoid signaling. The suppression of TNF-α-induced NF-κB activation that was not accompanied by the activation of GR transcriptional activation may possibly indicate essential oils' potential applications to the treatment of immune system disorders, minimizing the adverse side effects of glucocorticoids. Moreover, the anti-inflammatory activities of essential oils are revealed to be exerted both by suppression of the NF-κB activity and NF-κB protein levels. Similar to glucocorticoids, the apoptotic activities of the essential oils are exerted, at least in part, via activation of the mitochondrial-dependent apoptosis. Interestingly, a comparative evaluation of the biological actions of EOs revealed that Chios Mastic (Mastiha) and oregano EOs exhibited considerable anti-inflammatory activities. The former showed a reduction both in NF-κB activity and protein levels. Mastic oil also caused a reduction in GR protein levels that may compensate for its boosting effect on the DEX-induced GR transcriptional activation, ending up in no induction of the gluconeogenic PEPCK protein levels that constitute a GR target. Oregano, *Melissa officinalis* and lavender EOs suppressed the transcriptional activation of the GR. Furthermore, the most active, *Melissa officinalis* EO, showed a reduction both in GR and PEPCK protein levels. Thus, the anti-inflammatory and anti-gluconeogenic activities of the EOs were uncovered, possibly via the regulation of GR signaling. Moreover, the cytotoxic actions of *Melissa officinalis* and lavender essential oils via the induction of mitochondrial-dependent apoptosis were revealed. Our results highlight these essentials oils' anti-inflammatory, anti-gluconeogenic and apoptotic actions in relation to their implication on the regulation of steroid hormones actions, uncovering their use in steroid therapy with many potential applications in pharmaceutical and health industries as anti-cancer, anti-hyperglycemic and anti-inflammatory supplements.

Supplementary Materials: The following supporting information can be downloaded at: https://www.mdpi.com/article/10.3390/life13071534/s1, Supplementary Data S1: Chemical composition of essential oils; Supplementary Figure S1: Effect of essential oils on ERs transcriptional activation; Figure S2: Cytotoxic effects of essentials oils upon 24 h incubation; Figure S3: Effect of lavender essential oil on NF-κB transcriptional activation. Figure S4: Representative images used in western blot analysis.

Author Contributions: Conceptualization, A.-M.G.P.; methodology, A.-M.G.P.; validation, A.G., A.V., F.D.K. and I.T.; formal analysis, A.G. and A.V.; investigation, A.G. and A.V.; data curation, A.G. and A.-M.G.P.; writing, original draft preparation, A.G. and A.-M.G.P.; writing, review and editing, A.G. and A.-M.G.P.; visualization, A.G.; supervision, A.-M.G.P.; project administration A.-M.G.P.; funding acquisition, A.-M.G.P. All authors have read and agreed to the published version of the manuscript.

Funding: The research was supported by the Hellenic Foundation for Research and Innovation (HFRI) under the HFRI PhD Fellowship grant (Fellowship Number: 466 to F.D.K., Fellowship Number: 1434 to I.T.) and the project "Synthetic Biology: From omics technologies to genomic engineering" (OMIC-ENGINE) (MIS 5002636) which is implemented under the action "reinforcement of the Research and Innovation Infrastructure" funded by the Operational Programme "Competitiveness, Enterpreneurship and Innovation" (NSRF 2014-2020) and co-financed by Greece and the European Union (European Regional Development Fund) (Fellowship to A.G). This research was also funded by two Postgraduate Programs, "Application of Molecular Biology-Genetics-Diagnostic Biomarkers" and "Biotechnology-Quality assessment in Nutrition and the Environment", Department of Biochemistry and Biotechnology, University of Thessaly (to A.-M.G.P.).

Data Availability Statement: All data, tables, and figures are original. Details on data analysis are available from the corresponding author upon reasonable request.

Acknowledgments: Authors would like to thank the "Chios Gum Mastic Growers Association" and the "mastihashop" for their kind donation of Chios Mastic oil and Tharros Company Aromatic Plants Products, Larissa Greece, for Oregano, *Melissa officinalis* and Lavender essential oils donation.

Conflicts of Interest: The authors declare no conflict of interest. The funders had no role in the design of the study; in the collection, analyses, or interpretation of data; in the writing of the manuscript; or in the decision to publish the results.

Abbreviations

ATCC	American type culture collection
DEX	Dexamethasone
DMEM	Dulbecco's Modified Eagle Medium
DMSO	Dimethyl Sulfoxide
E2	Estradiol
EOs	Essential Oils
ER	Estrogen receptor
ERE	Estrogen Response Elements
ERα	Estrogen receptor alpha
ERβ	Estrogen receptor beta
FBS	Fetal Bovine Serum
G6Pase	Glucose 6-phosphatase
GR	Glucocorticoids Receptor
GREs -Luc	Glucocorticoid Response Elements-luciferase
HEK-293	Human Embryonic Kidney-293
HIF-1α	Hypoxia inducible factor 1 alpha
HPA	Hypothalamus-Pituitary-Adrenal
IL-6	Interleukin-6
NF-κB	Nuclear Factor Kappa B
NF-κB-REs-Luc	NF-κB Response Elements-luciferase
OD	Optical Density
PEPCK	Phoshoenolpyruvate carboxykinase
ROS	Reactive Oxygen Species
TNF-α	Tumor Necrosis factor alpha

References

1. Dhifi, W.; Bellili, S.; Jazi, S.; Bahloul, N.; Mnif, W. Essential Oils'Chemical Characterization and Investigation of Some Biological Activities: A Critical Review. *Medicines* **2016**, *3*, 25. [CrossRef] [PubMed]
2. Masyita, A.; Sari, R.M.; Astuti, A.D.; Yasir, B.; Rumata, N.R.; Emran, T.B.; Nainu, F.; Simal-Gandara, J. Terpenes and terpenoids as main bioactive compounds of essential oils, their roles in human health and potential application as natural food preservatives. *Food Chem. X* **2022**, *13*, 100217. [CrossRef] [PubMed]
3. Loza-Tavera, H. Monoterpenes in essential oils. Biosynthesis and properties. *Adv. Exp. Med. Biol.* **1999**, *464*, 49–62.
4. Elshafie, H.S.; Camele, I. An Overview of the Biological Effects of Some Mediterranean Essential Oils on Human Health. *BioMed Res. Int.* **2017**, *2017*, 9268468. [CrossRef] [PubMed]
5. Benny, A.; Thomas, J. Essential Oils as Treatment Strategy for Alzheimer's Disease: Current and Future Perspectives. *Planta Med.* **2019**, *85*, 239–248. [PubMed]
6. Barnes, P.J. Anti-inflammatory Actions of Glucocorticoids: Molecular Mechanisms. *Clin. Sci.* **1998**, *94*, 557–572. [CrossRef]
7. Auphan, N.; DiDonato, J.A.; Rosette, C.; Helmberg, A.; Karin, M. Immunosuppression by glucocorticoids: Inhibition of NF-kappa B activity through induction of I kappa B synthesis. *Science* **1995**, *270*, 286–290. [CrossRef]
8. Straub, R.H. The complex role of estrogens in inflammation. *Endocr. Rev.* **2007**, *28*, 521–574. [CrossRef]
9. Stice, J.P.; Knowlton, A.A. Estrogen, NFkappaB, and the heat shock response. *Mol. Med.* **2008**, *14*, 517–527. [CrossRef]
10. Sundahl, N.; Bridelance, J.; Libert, C.; De Bosscher, K.; Beck, I.M. Selective glucocorticoid receptor modulation: New directions with non-steroidal scaffolds. *Pharmacol. Ther.* **2015**, *152*, 28–41. [CrossRef]
11. Cain, D.W.; Cidlowski, J.A. Specificity and sensitivity of glucocorticoid signaling in health and disease. *Best Pract. Res. Clin. Endocrinol. Metab.* **2015**, *29*, 545–556. [CrossRef] [PubMed]
12. Georgatza, D.; Gorgogietas, V.A.; Kylindri, P.; Charalambous, M.C.; Papadopoulou, K.K.; Hayes, J.M.; Psarra, A.-M.G. The triterpene echinocystic acid and its 3-O-glucoside derivative are revealed as potent and selective glucocorticoid receptor agonists. *Int. J. Biochem. Cell Biol.* **2016**, *79*, 277–287. [CrossRef] [PubMed]
13. McColl, A.; Michlewska, S.; Dransfield, I.; Rossi, A.G. Effects of Glucocorticoids on Apoptosis and Clearance of Apoptotic Cells. *Sci. World J.* **2007**, *7*, 1165–1181. [CrossRef] [PubMed]
14. Psarra, A.-M.G.; Sekeris, C.E. Nuclear receptors and other nuclear transcription factors in mitochondria: Regulatory molecules in a new environment. *Biochim. Biophys. Acta (BBA)—Mol. Cell Res.* **2008**, *1783*, 1–11. [CrossRef]
15. da Silva, J.S.; Montagnoli, T.L.; Rocha, B.S.; Tacco, M.L.C.A.; Marinho, S.C.P.; Zapata-Sudo, G. Estrogen Receptors: Therapeutic Perspectives for the Treatment of Cardiac Dysfunction after Myocardial Infarction. *Int. J. Mol. Sci.* **2021**, *22*, 525. [CrossRef]
16. Gruver-Yates, A.L.; Cidlowski, J.A. Tissue-Specific Actions of Glucocorticoids on Apoptosis: A Double-Edged Sword. *Cells* **2013**, *2*, 202–223. [CrossRef]
17. Paraschos, S.; Magiatis, P.; Gikas, E.; Smyrnioudis, I.; Skaltsounis, A.-L. Quality profile determination of Chios mastic gum essential oil and detection of adulteration in mastic oil products with the application of chiral and non-chiral GC–MS analysis. *Fitoterapia* **2016**, *114*, 12–17. [CrossRef]
18. Vallianou, I.; Peroulis, N.; Pantazis, P.; Hadzopoulou-Cladaras, M. Camphene, a plant-derived monoterpene, reduces plasma cholesterol and triglycerides in hyperlipidemic rats independently of HMG-CoA reductase activity. *PLoS ONE* **2011**, *6*, e20516. [CrossRef]
19. Spyridopoulou, K.; Tiptiri-Kourpeti, A.; Lampri, E.; Fitsiou, E.; Vasileiadis, S.; Vamvakias, M.; Bardouki, H.; Goussia, A.; Malamou-Mitsi, V.; Panayiotidis, M.I.; et al. Dietary mastic oil extracted from Pistacia lentiscus var. chia suppresses tumor growth in experimental colon cancer models. *Sci. Rep.* **2017**, *7*, 3782. [CrossRef]
20. Xanthis, V.; Fitsiou, E.; Voulgaridou, G.P.; Bogadakis, A.; Chlichlia, K.; Galanis, A.; Pappa, A. Antioxidant and Cytoprotective Potential of the Essential Oil Pistacia lentiscus var. chia and Its Major Components Myrcene and alpha-Pinene. *Antioxidants* **2021**, *10*, 127. [CrossRef]
21. Georgiadis, I.; Karatzas, T.; Korou, L.-M.; Katsilambros, N.; Perrea, D. Beneficial Health Effects of Chios Gum Mastic and Peroxisome Proliferator-Activated Receptors: Indications of Common Mechanisms. *J. Med. Food* **2015**, *18*, 1–10. [CrossRef] [PubMed]
22. Magkouta, S.; Stathopoulos, G.; Psallidas, I.; Papapetropoulos, A.; Kolisis, F.N.; Roussos, C.; Loutrari, H. Protective effects of mastic oil from Pistacia lentiscus variation chia against experimental growth of lewis lung carcinoma. *Nutr. Cancer* **2009**, *61*, 640–648. [CrossRef]
23. Mimica-Dukic, N.; Bozin, B.; Sokovic, M.; Simin, N. Antimicrobial and antioxidant activities of Melissa officinalis L. (Lamiaceae) essential oil. *J. Agric. Food Chem.* **2004**, *52*, 2485–2489. [CrossRef]
24. de Sousa, A.C.; Alviano, D.S.; Blank, A.F.; Alves, P.B.; Alviano, C.S.; Gattass, C.R. Melissa officinalis L. essential oil: Antitumoral and antioxidant activities. *J. Pharm. Pharmacol.* **2004**, *56*, 677–681. [CrossRef] [PubMed]
25. de Carvalho, N.C.; Corrêa-Angeloni, M.J.; Leffa, D.D.; Moreira, J.; Nicolau, V.; de Aguiar Amaral, P.; Rossatto, A.E.; de Andrade, V.M. Evaluation of the genotoxic and antigenotoxic potential of Melissa officinalis in mice. *Genet. Mol. Biol.* **2011**, *34*, 290–297. [CrossRef]
26. Chung, M.J.; Cho, S.-Y.; Bhuiyan, M.J.H.; Kim, K.H.; Lee, S.-J. Anti-diabetic effects of lemon balm (*Melissa officinalis*) essential oil on glucose- and lipid-regulating enzymes in type 2 diabetic mice. *Br. J. Nutr.* **2010**, *104*, 180–188. [CrossRef]

27. Ayat, M.; Tameh, A.A.; Ghahremani, M.H.; Akbari, M.; Mehr, S.E.; Khanavi, M.; Hassanzadeh, G. Neuroprotective properties of Melissa officinalis after hypoxic-ischemic injury both in vitro and in vivo. *DARU J. Pharm. Sci.* **2012**, *20*, 42.
28. Prusinowska, R.; Śmigielski, K.; Stobiecka, A.; Kunicka-Styczyńska, A. Hydrolates from lavender (*Lavandula angustifolia*)—Their chemical composition as well as aromatic, antimicrobial and antioxidant properties. *Nat. Prod. Res.* **2016**, *30*, 386–393. [CrossRef]
29. Donelli, D.; Antonelli, M.; Bellinazzi, C.; Gensini, G.F.; Firenzuoli, F. Effects of lavender on anxiety: A systematic review and meta-analysis. *Phytomedicine* **2019**, *65*, 153099. [CrossRef]
30. Samuelson, R.; Lobl, M.; Higgins, S.; Clarey, D.; Wysong, A. The Effects of Lavender Essential Oil on Wound Healing: A Review of the Current Evidence. *J. Altern. Complement. Med.* **2020**, *26*, 680–690. [CrossRef] [PubMed]
31. Zhao, Y.; Chen, R.; Wang, Y.; Qing, C.; Wang, W.; Yang, Y. In Vitro and In Vivo Efficacy Studies of *Lavender angustifolia* Essential Oil and Its Active Constituents on the Proliferation of Human Prostate Cancer. *Integr. Cancer Ther.* **2017**, *16*, 215–226. [CrossRef] [PubMed]
32. De Falco, E.; Mancini, E.; Roscigno, G.; Mignola, E.; Taglialatela-Scafati, O.; Senatore, F. Chemical Composition and Biological Activity of Essential Oils of *Origanum vulgare* L. subsp. vulgare L. under Different Growth Conditions. *Molecules* **2013**, *18*, 14948–14960. [CrossRef] [PubMed]
33. Kountouri, A.M.; Gioxari, A.; Karvela, E.; Kaliora, A.C.; Karvelas, M.; Karathanos, V.T. Chemopreventive properties of raisins originating from Greece in colon cancer cells. *Food Funct.* **2013**, *4*, 366–372. [CrossRef]
34. Spyridopoulou, K.; Fitsiou, E.; Bouloukosta, E.; Tiptiri-Kourpeti, A.; Vamvakias, M.; Oreopoulou, A.; Papavassilopoulou, E.; Pappa, A.; Chlichlia, K. Extraction, Chemical Composition, and Anticancer Potential of *Origanum onites* L. Essential Oil. *Molecules* **2019**, *24*, 2612. [CrossRef] [PubMed]
35. Rodriguez-Garcia, I.; Silva-Espinoza, B.; Ortega-Ramirez, L.; Leyva, J.; Siddiqui, M.W.; Valenzuela, M.R.C.; Gonzalez-Aguilar, G.; Zavala, J.F.A. Oregano Essential Oil as an Antimicrobial and Antioxidant Additive in Food Products. *Crit. Rev. Food Sci. Nutr.* **2016**, *56*, 1717–1727. [CrossRef] [PubMed]
36. Leyva-López, N.; Gutiérrez-Grijalva, E.P.; Vazquez-Olivo, G.; Heredia, J.B. Essential Oils of Oregano: Biological Activity beyond Their Antimicrobial Properties. *Molecules* **2017**, *22*, 989. [CrossRef]
37. Misharina, T.A.; Burlakova, E.B.; Fatkullina, L.D.; Alinkina, E.S.; Vorob'eva, A.K.; Medvedeva, I.B.; Erokhin, V.N.; Semenov, V.A.; Nagler, L.G.; Kozachenko, A.I. Effect of oregano essential oil on the engraftment and development of Lewis carcinoma in F1 DBA C57 black hybrid mice. *Prikl. Biokhim. Mikrobiol.* **2013**, *49*, 432–436. [CrossRef]
38. Mosmann, T. Rapid colorimetric assay for cellular growth and survival: Application to proliferation and cytotoxicity assays. *J. Immunol. Methods* **1983**, *65*, 55–63. [CrossRef]
39. Kalousi, F.D.; Pollastro, F.; Christodoulou, E.C.; Karra, A.G.; Tsialtas, I.; Georgantopoulos, A.; Salamone, S.; Psarra, A.G. Apoptotic, Anti-Inflammatory Activities and Interference with the Glucocorticoid Receptor Signaling of Fractions from Pistacia lentiscus L. var. chia Leaves. *Plants* **2022**, *11*, 934. [CrossRef] [PubMed]
40. Tsialtas, I.; Georgantopoulos, A.; Karipidou, M.E.; Kalousi, F.D.; Karra, A.G.; Leonidas, D.D.; Psarra, A.-M.G. Anti-Apoptotic and Antioxidant Activities of the Mitochondrial Estrogen Receptor Beta in N2A Neuroblastoma Cells. *Int. J. Mol. Sci.* **2021**, *22*, 7620. [CrossRef]
41. Miyamoto, T.; Okimoto, T.; Kuwano, M. Chemical Composition of the Essential Oil of Mastic Gum and their Antibacterial Activity Against Drug-Resistant Helicobacter pylori. *Nat. Prod. Bioprospect.* **2014**, *4*, 227–231. [CrossRef]
42. Cassuto, H.; Kochan, K.; Chakravarty, K.; Cohen, H.; Blum, B.; Olswang, Y.; Hakimi, P.; Xu, C.; Massillon, D.; Hanson, R.W.; et al. Glucocorticoids Regulate Transcription of the Gene for Phosphoenolpyruvate Carboxykinase in the Liver via an Extended Glucocorticoid Regulatory Unit. *J. Biol. Chem.* **2005**, *280*, 33873–33884. [CrossRef] [PubMed]
43. Oray, M.; Abu Samra, K.; Ebrahimiadib, N.; Meese, H.; Foster, C.S. Long-term side effects of glucocorticoids. *Expert Opin. Drug Saf.* **2016**, *15*, 457–465. [CrossRef]
44. Miguel, M.G. Antioxidant and Anti-Inflammatory Activities of Essential Oils: A Short Review. *Molecules* **2010**, *15*, 9252–9287. [CrossRef] [PubMed]
45. Tariq, S.; Wani, S.; Rasool, W.; Shafi, K.; Bhat, M.A.; Prabhakar, A.; Shalla, A.H.; Rather, M.A. A comprehensive review of the antibacterial, antifungal and antiviral potential of essential oils and their chemical constituents against drug-resistant microbial pathogens. *Microb. Pathog.* **2019**, *134*, 103580. [CrossRef]
46. Sharifi-Rad, J.; Sureda, A.; Tenore, G.C.; Daglia, M.; Sharifi-Rad, M.; Valussi, M.; Tundis, R.; Sharifi-Rad, M.; Loizzo, M.R.; Ademiluyi, A.O.; et al. Biological Activities of Essential Oils: From Plant Chemoecology to Traditional Healing Systems. *Molecules* **2017**, *22*, 70. [CrossRef]
47. Sousa, V.I.; Parente, J.F.; Marques, J.F.; Forte, M.A.; Tavares, C.J. Microencapsulation of Essential Oils: A Review. *Polymers* **2022**, *14*, 1730. [CrossRef]
48. Chen, P.; Li, B.; Ou-Yang, L. Role of estrogen receptors in health and disease. *Front. Endocrinol.* **2022**, *13*, 839005. [CrossRef] [PubMed]
49. Evans, R.M. The Nuclear Receptor Superfamily: A Rosetta Stone for Physiology. *Mol. Endocrinol.* **2005**, *19*, 1429–1438. [CrossRef] [PubMed]
50. Jensen, E.V. Steroid hormone receptors. *Curr. Top Pathol.* **1991**, *83*, 365–431. [PubMed]

51. Bai, C.; Schmidt, A.; Freedman, L.P.; Hasson, S.A.; Fogel, A.I.; Wang, C.; MacArthur, R.; Guha, R.; Heman-Ackah, S.; Martin, S.; et al. Steroid Hormone Receptors and Drug Discovery: Therapeutic Opportunities and Assay Designs. *ASSAY Drug Dev. Technol.* **2003**, *1*, 843–852. [CrossRef] [PubMed]
52. Quattrocelli, M.; Zelikovich, A.S.; Salamone, I.M.; Fischer, J.A.; McNally, E.M. Mechanisms and Clinical Applications of Glucocorticoid Steroids in Muscular Dystrophy. *J. Neuromuscul. Dis.* **2021**, *8*, 39–52. [CrossRef] [PubMed]
53. Valderrama, L.D.R.L. Effects of essential oils on central nervous system: Focus on mental health. *Phytother. Res.* **2021**, *35*, 657–679. [CrossRef] [PubMed]
54. Ramalho, T.R.; Oliveira, M.T.; Lima, A.L.; Bezerra-Santos, C.R.; Piuvezam, M.R. Gamma-Terpinene Modulates Acute Inflammatory Response in Mice. *Planta Med.* **2015**, *81*, 1248–1254. [CrossRef] [PubMed]
55. Bartoňková, I.; Dvořák, Z. Assessment of endocrine disruption potential of essential oils of culinary herbs and spices involving glucocorticoid, androgen and vitamin D receptors. *Food Funct.* **2018**, *9*, 2136–2144. [CrossRef] [PubMed]
56. Johnson, S.A.; Rodriguez, D.; Allred, K. A Systematic Review of Essential Oils and the Endocannabinoid System: A Connection Worthy of Further Exploration. *Evid.-Based Complement. Altern. Med.* **2020**, *2020*, 8035301. [CrossRef]
57. Baschant, U.; Culemann, S.; Tuckermann, J. Molecular determinants of glucocorticoid actions in inflammatory joint diseases. *Mol. Cell. Endocrinol.* **2013**, *380*, 108–118. [CrossRef]
58. Morrone, L.A.; Rombolà, L.; Pelle, C.; Corasaniti, M.T.; Zappettini, S.; Paudice, P.; Bonanno, G.; Bagetta, G. The essential oil of bergamot enhances the levels of amino acid neurotransmitters in the hippocampus of rat: Implication of monoterpene hydrocarbons. *Pharmacol. Res.* **2007**, *55*, 255–262. [CrossRef]
59. Costa, C.A.R.d.A.; Kohn, D.O.; de Lima, V.M.; Gargano, A.C.; Flório, J.C.; Costa, M. The GABAergic system contributes to the anxiolytic-like effect of essential oil from Cymbopogon citratus (lemongrass). *J. Ethnopharmacol.* **2011**, *137*, 828–836. [CrossRef]
60. Atsumi, T.; Tonosaki, K. Smelling lavender and rosemary increases free radical scavenging activity and decreases cortisol level in saliva. *Psychiatry Res.* **2007**, *150*, 89–96. [CrossRef]
61. Toda, M.; Morimoto, K. Effect of lavender aroma on salivary endocrinological stress markers. *Arch. Oral Biol.* **2008**, *53*, 964–968. [CrossRef] [PubMed]
62. Qadeer, S.; Emad, S.; Perveen, T.; Yousuf, S.; Sheikh, S.; Sarfaraz, Y.; Sadaf, S.; Haider, S. Role of ibuprofen and lavender oil to alter the stress induced psychological disorders: A comparative study. *Pak. J. Pharm. Sci.* **2018**, *31*, 1603–1608.
63. Srivflai, J.; Khorana, N.; Waranuch, N.; Wisuitiprot, W.; Suphrom, N.; Suksamrarn, A.; Ingkaninan, K. Germacrene Analogs are Anti-androgenic on Androgen-dependent Cells. *Nat. Prod. Commun.* **2016**, *11*, 1225–1228. [PubMed]
64. Luderer, U.; Dv, H.; N, L.; Ks, K.; Ca, B. Faculty Opinions recommendation of Prepubertal gynecomastia linked to lavender and tea tree oils. *N. Engl. J. Med.* **2007**, *356*, 479–485.
65. Ramsey, J.T.; Li, Y.; Arao, Y.; Naidu, A.; A Coons, L.; Diaz, A.; Korach, K.S. Lavender Products Associated With Premature Thelarche and Prepubertal Gynecomastia: Case Reports and Endocrine-Disrupting Chemical Activities. *J. Clin. Endocrinol. Metab.* **2019**, *104*, 5393–5405. [CrossRef]
66. Shinohara, K.; Doi, H.; Kumagai, C.; Sawano, E.; Tarumi, W. Effects of essential oil exposure on salivary estrogen concentration in perimenopausal women. *Neuro Endocrinol. Lett.* **2017**, *37*, 567–572.
67. Zhang, X.; Peng, Y.; Wu, C. Chicken embryonic toxicity and potential in vitro estrogenic and mutagenic activity of carvacrol and thymol in low dose/concentration. *Food Chem. Toxicol.* **2021**, *150*, 112038. [CrossRef] [PubMed]
68. Politano, V.T.; McGinty, D.; Lewis, E.M.; Hoberman, A.M.; Christian, M.S.; Diener, R.M.; Api, A.M. Uterotrophic Assay of Percutaneous Lavender Oil in Immature Female Rats. *Int. J. Toxicol.* **2013**, *32*, 123–129. [CrossRef]
69. Simões, B.M.; Kohler, B.; Clarke, R.; Stringer, J.; Frazer, L.N.; Young, K.; Rautemaa-Richardson, R.; Zucchini, G.; Armstrong, A.; Howell, S.J. Estrogenicity of essential oils is not required to relieve symptoms of urogenital atrophy in breast cancer survivors. *Ther. Adv. Med. Oncol.* **2018**, *10*, 1758835918766189. [CrossRef]
70. Yin, Q.-H.; Yan, F.-X.; Zu, X.-Y.; Wu, Y.-H.; Wu, X.-P.; Liao, M.-C.; Deng, S.-W.; Yin, L.-L.; Zhuang, Y.-Z. Anti-proliferative and pro-apoptotic effect of carvacrol on human hepatocellular carcinoma cell line HepG-2. *Cytotechnology* **2012**, *64*, 43–51. [CrossRef]
71. Balusamy, S.R.; Perumalsamy, H.; Huq, A.; Balasubramanian, B. Anti-proliferative activity of Origanum vulgare inhibited lipogenesis and induced mitochondrial mediated apoptosis in human stomach cancer cell lines. *Biomed. Pharmacother.* **2018**, *108*, 1835–1844. [CrossRef] [PubMed]
72. Gautam, N.; Mantha, A.K.; Mittal, S. Essential Oils and Their Constituents as Anticancer Agents: A Mechanistic View. *BioMed Res. Int.* **2014**, *2014*, 154106. [CrossRef] [PubMed]

Disclaimer/Publisher's Note: The statements, opinions and data contained in all publications are solely those of the individual author(s) and contributor(s) and not of MDPI and/or the editor(s). MDPI and/or the editor(s) disclaim responsibility for any injury to people or property resulting from any ideas, methods, instructions or products referred to in the content.

life

MDPI

Review

Clinical Applications of *Polypodium leucotomos* (Fernblock®): An Update

Azahara Rodríguez-Luna [1,2,*], Alicia Zamarrón [3], Ángeles Juarranz [3] and Salvador González [4]

1 Department of Basic Health Sciences, Faculty of Health Sciences, Universidad Rey Juan Carlos (URJC), 28933 Alcorcón, Spain
2 Department of Pharmacology, Faculty of Pharmacy, University of Seville, 41012 Seville, Spain
3 Department of Biology, Faculty of Sciences, Autónoma University of Madrid (UAM), 28049 Madrid, Spain; aliszm@gmail.com (A.Z.); angeles.juarranz@uam.es (Á.J.)
4 Department of Medicine and Medical Specialties, Alcalá de Henares University, 28805 Madrid, Spain; salvagonrod@gmail.com
* Correspondence: azahararodriguezluna@gmail.com

Abstract: Exposure to sun radiation leads to higher risk of sunburn, pigmentation, immunosuppression, photoaging and skin cancer. In addition to ultraviolet radiation (UVR), recent research indicates that infrared radiation (IR) and visible light (VIS) can play an important role in the pathogenesis of some of these processes. Detrimental effects associated with sun exposure are well known, but new studies have shown that DNA damage continues to occur long after exposure to solar radiation has ended. Regarding photoprotection strategies, natural substances are emerging for topical and oral photoprotection. In this sense, Fernblock®, a standardized aqueous extract of the fern *Polypodium Leucotomos* (PLE), has been widely administered both topically and orally with a strong safety profile. Thus, this extract has been used extensively in clinical practice, including as a complement to photodynamic therapy (PDT) for treating actinic keratoses (AKs) and field cancerization. It has also been used to treat skin diseases such as photodermatoses, photoaggravated inflammatory conditions and pigmentary disorders. This review examines the most recent developments in the clinical application of Fernblock® and assesses how newly investigated action mechanisms may influence its clinical use.

Keywords: *Polypodium leucotomos*; Fernblock®; photoprotection; photodermatoses; photoaging; hyperpigmentation

Citation: Rodríguez-Luna, A.; Zamarrón, A.; Juarranz, Á.; González, S. Clinical Applications of *Polypodium leucotomos* (Fernblock®): An Update. *Life* **2023**, *13*, 1513. https://doi.org/10.3390/life13071513

Academic Editor: Stefania Lamponi

Received: 20 June 2023
Revised: 30 June 2023
Accepted: 3 July 2023
Published: 5 July 2023

Copyright: © 2023 by the authors. Licensee MDPI, Basel, Switzerland. This article is an open access article distributed under the terms and conditions of the Creative Commons Attribution (CC BY) license (https://creativecommons.org/licenses/by/4.0/).

1. Introduction

Photoprotection is the first-line prevention strategy to avoid the development of skin cancer and premature aging. Ultraviolet radiation (UVR) promotes skin cancer by inducing DNA damage, triggering inflammatory processes and causing immunosuppression and plays part in premature aging through alterations in extracellular matrix network and remodeling. Most of these detrimental effects are mainly mediated by generation of reactive oxygen species (ROS) and the consequent oxidative stress. Adoption of specific behaviors (such as wearing protective clothes, hats and sunglasses, and avoiding excessive sun exposure) and the use of topical sunscreens are the most common measures to counteract the harmful effects of UVR. Although traditional sunscreens are a critical component in all photoprotective regimens, they have limitations (inadequate application and need for frequent reapplication, short half-life, lack of photostability and insufficient protection against all wavelengths, among others) and have also increasingly been questioned for their safety and their impact on the environment [1,2]. In this sense, photoprotection can be provided not only by topical sunscreens but also by oral administration of substances (such as polyphenols, carotenoids and other antioxidants) that are being identified as systemic photoprotection agents in humans [2]. A well-known photoprotective agent is the standardized aqueous extract from the leaves of the fern *Polypodium leucotomos* (PLE or

Fernblock® (trademark name)). *Polypodium leucotomos* (PL) is a fern of the Polypodiaceae family, genus Phlebodium, native to Central and South America, where it has had a historical role in traditional medicine, especially for the treatment of skin diseases. A standardized aqueous extract from the leaves of the fern PL (PLE), rich in polyphenols and specifically in phenolic acids, has been developed to exploit the photoprotective properties of the plant and to provide a steady phenolic content. This extract was introduced as Fernblock® in Europe in the year 2000, both in topical and oral forms, and is currently available in more than 26 countries, including the U.S., as a dietary supplement, since 2006.

Phenolic compounds identified in PLE are 4-hydroxybenzoic acid, 3,4-dihydroxybenzoic acid (protocatechuic acid), 4-hydroxy-3-methoxybenzoic acid (vanillic acid), 3,4-dihydroxycinnamic acid (caffeic acid), 4-hydroxycinnamic acid (p-coumaric),3-methoxy-4-hydroxycinnamic acid (ferulic acid), 4-hydroxycinnamoylquinic acid and five chlorogenic acid isomers [3]. Of these, ferulic and caffeic acids are the most potent antioxidants. However, it is important to note that González et al. demonstrated in 2018 that there were significant differences between different PL extracts which could be attributed to the specific plant part used, the method of extraction and the plant's origin and growth conditions. Generally, extracts from the leaves are more potent and yield more meaningful outcomes. Nonetheless, this research also suggested that other moieties, whether antioxidant or not, may have a critical role in the function of these extracts as dietary supplements with antiaging and antioxidant properties [4,5]. As mentioned earlier, this extract is rich in polyphenols and specifically in phenolic acids and has been developed to exploit the photoprotective properties of the PL fern and to provide a steady phenolic content. Its solid mechanism of action, its success in clinical trials, and the increased social interest in natural substances, such as polyphenols, have placed PLE as an interesting photoprotective, antioxidant and anti-inflammatory option (Figure 1). In this regard, numerous studies have been carried out to prove the role of the aqueous PLE in photoprotection, which have been summarized by many authors [6,7].

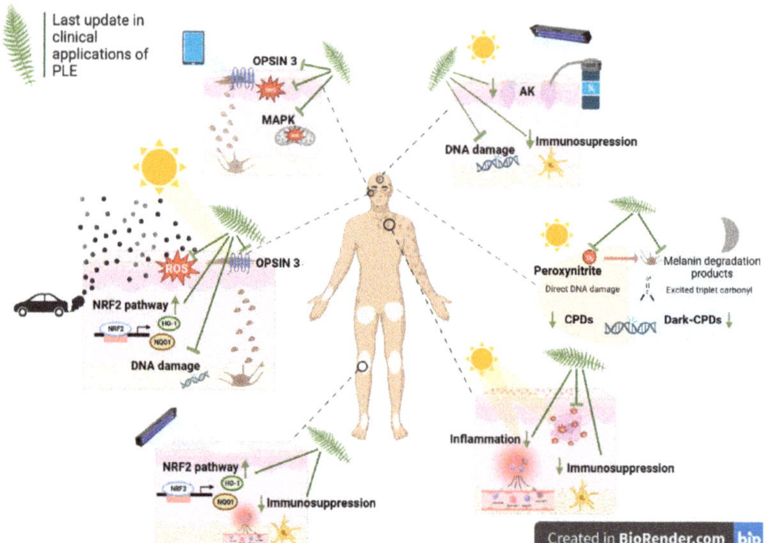

Figure 1. The graphical summary highlights the latest relevant studies concerning Fernblock®, represented by the image of the fern, and examines their implications for its clinical application. The image includes abbreviations such as MAPK (mitogen-activated protein kinase), AK (actinic keratosis), NRF2 (nuclear factor erythroid-2-related factor 2), and CPDs (cyclobutane pyrimidine dimers), which correspond to specific molecular pathways or biomarkers related to Fernblock® and its effects. Created with BioRender.com.

The aim of this review is to establish the current state of the art regarding the uses of PLE and provide an interpretive synthesis that describes how recent advances may influence its clinical applications. In certain cases, referencing older studies considered fundamental in the study area can also provide valuable insights and improve our understanding of newly obtained findings. Additionally, we aim to determine the potential future directions of PLE research, with special emphasis on the role of Fernblock® as an adjuvant in photoprotection. Finally, we will explore the newly researched action mechanisms that may impact its clinical management.

2. Materials and Methods

This state-of-the-art narrative review was developed in accordance with the guidelines published by Barry et al. (2022) [8]. The first step in our search process was to build an initial pool of articles, for which we searched Pubmed, Scopus and Google Scholar databases, though we prioritized Pubmed Central to be the National Library of Medicine of reference. The main criterion to select articles was the inclusion of PLE, using the following keywords: *Polypodium leucotomos*, Fernblock®, photoprotection, photodermatosis, skin pigmentation and skin cancer. In all searches, keyword combinations included either the term "*Polypodium leucotomos*" or the term "Fernblock®". We limited our search to articles published between 2019 and 2023 and written in English. From the 224 collected articles, 53 were excluded due to various reasons, such as duplication, erratum, being outside the scope, or unavailability of full-text copies. Among the 171 selected articles that met the specified criteria, 34 were eliminated due to their brief mention of PLE without providing significant relevance. The remaining articles, totaling 120, were included in the tables, while 17 articles are cited throughout the text as they contained more comprehensive information regarding PLE.

3. Results and Discussion

3.1. PLE Photoprotective Activity

Numerous compounds have been demonstrated to have a protective effect against various harmful effects of UV radiation, including photocarcinogenesis, sunburn, photoaging, and UVB-mediated phototoxicity. These effects result from the modulation of different pathways as demonstrated in different in vitro and in vivo models [9]. Briefly, we can affirm that PLE's activity has been widely studied, and it has been attributed to several action mechanisms. PLE improves the skin's endogenous antioxidant system by reducing lipid peroxides and neutralizing ROS and free radicals, especially superoxide anions and hydroxyl radicals generated after exposure to UV and VIS radiation, unlike traditional antioxidants such as vitamin C, E, or carotenoids, which are mainly effective against singlet oxygen. Moreover, Fernblock® increases the activity of the nuclear factor erythroid-2-related factor 2 (NRF2) transcription factor and its associated antioxidant targets, which is linked to its capacity to decrease inflammation, melanin production, and overall cell damage [10].

In the context of UVR-induced inflammation, the basis of its anti-inflammatory properties is its ability to inhibit the expression of the tumor necrosis factor α (TNF-α), iNOS, redox-sensitive transcriptional factors, activator protein 1 (AP-1) and nuclear factor kB (NF-kB) [11]. PLE also decreases the expression of COX-2 and PGE2 [12]. Over the years, many indicators have been examined which provide essential information for verifying the photoprotective effect that has been demonstrated through clinical and preclinical studies. However, the effect of PLE on AP-1 and NF-kB expression after exposure to solar simulated radiation (SSR) cannot be explained only by the antioxidant action of PLE since treatment with a bona fide antioxidant does not decrease AP-1 and NF-kB expression in human keratinocytes subjected to SSR [11].

3.2. Clinical Applications

The successful results of Fernblock® in clinical trials, together with its action mechanisms and growing interest in natural substances like polyphenols, have positioned PLE as a promising option for photoprotection, antioxidant and anti-inflammatory treatment and as an adjuvant therapy for various pathologies [2,6]. In the upcoming sections, we will examine the research progress supporting its application in these conditions, and we will suggest a future scenario based on evidence.

3.2.1. Oncodermatology

In the field of oncology, a recent review by Calzari et al. (2023) elucidates the ways in which PLE functions and assesses its applications in oncodermatology, with reference to both in vitro and in vivo research [13]. However, this review is not the only one, as the trend in recent years has been to publish literature reviews demonstrating the photoprotective effects of PLE, thereby confirming its therapeutic potential against various types of cancerous growths. Alongside these reviews, there have been seven experimental studies, four of which were clinical and three preclinical, which can be found summarized in Table 1.

Accordingly, a wide range of earlier studies are recognized as crucial references that have significantly advanced our understanding of the potential of Fernblock® in the field of cancer prevention in both mice and humans. PLE inhibited UVR-mediated DNA damage and mutagenesis through a double mechanism that consisted of prevention of cyclobutane pyrimidine dimer (CPDs) accumulation and reduction of 8-OH-dG and H2Ax, thus preventing oxidative damage [4]. Also, PLE decreased UVA-dependent mitochondrial DNA damage by reducing common deletions (CD) [14]. In vitro and in vivo studies suggest that PLE may have a role in the treatment of UV-induced skin inflammation and cancer, probably due to its antioxidant and p53-activating properties [15]. It is important to note that the extracellular matrix (ECM) provides structural integrity to the tissue and is remodeled during skin aging/photoaging and cancer. In vitro experiments showed that PLE directly inhibited the enzymatic activity and expression of MMPs in melanoma cells and fibroblasts and stimulated the expression of tissue inhibitors of metalloproteinases (TIMPs) in melanoma cells, reducing melanoma cell growth and ECM remodeling [16,17]. Also, clinical studies shown that PLE reduces epidermal cell proliferation and the number of cyclin D1- and PCNA-positive epidermal cells caused by UVR exposure [18,19]. In relation to the process of cancerization, previous research has shown that taking oral PLE supplements following PDT can improve AK clearance and reduce recurrence as compared to PDT alone [20].

However, three recent studies have been conducted to determine the effectiveness of PLE in the context of cancerization not only by oral administration but also by topical administration. The first of them investigated the effectiveness of a new medical device (NMD) in treating the field cancerization in 30 patients with Aks after PDT. The NMD contained a complex of DNA-repair enzymes, UV-filters and Fernblock®, while the control group received a standard sunscreen (SS). The study utilized clinical, dermoscopic, reflectance confocal microscopy (RCM) and histological evaluations to assess the outcomes and found that after six and twelve months of treatment, the SS group showed a significant increase in the number of AKs compared to the NMD group. The NMD group also showed a significant reduction in the extension and grade of atypia compared to the SS group. Histopathological evaluation showed an improvement in keratinocyte atypia grade in all groups after six months of PDT, but p53 expression was significantly lower in the NMD group at twelve months compared to the SS group. Overall, the NMD was well-tolerated with no serious adverse events reported [21]. Another recent prospective clinical study evaluated the effectiveness of the same formulation in individuals with AK who underwent cryotherapy. The evaluation involved measuring changes in the AKASI score (Actinic Keratosis Area and Severity Index) and utilized non-invasive line-field confocal-optical coherence tomography (LC-OCT) analysis. The findings revealed that the use of the sunscreen containing DNA-repairing enzymes and PLE significantly reduced the AKASI score after 3 and 12 months treatment compared to the control group. Consequently, the study concluded that the PLE-based sunscreen considerably improved AKASI score among individuals receiving cryotherapy treatment [22]. Finally, a recent prospective, multicenter, randomized controlled trial was conducted to compare not only the effectiveness of a sunscreen with Fernblock® vs. one without but also the impact of oral photoprotection for managing AKs in elderly individuals with severe actinic damage. The group that received both topical sunscreen with Fernblock® and Fernblock® oral supplementation showed the most significant improvements in AK and field cancerization parameters compared to control group (which used a standard topical sunscreen). These results suggest that combining oral and topical photoprotection leads to superior clinical and anatomical outcomes [23]. In summary, these studies provide evidence to suggest that both oral and topical PLE could be utilized as an adjuvant treatment option for field cancerization. However, it is necessary to conduct further research in order to validate its effectiveness when compared to established and widely accepted medications considered to be the gold standard.

Regarding melanoma, Aguilera et al. (2013) also investigated the protective role of oral administration of PLE in patients at risk of malignant melanoma (MM) and in the interaction between MC1R polymorphisms and the cyclin-dependent kinase (CDK) inhibitor 2A gene (CDKN2A) status with MED 25–50%. Among patients with familial MM, those individuals with mutations in CDKN2A and/or MC1R had greater differences regarding the response to treatment with PLE [24]. According to these results, the authors indicated that patients with higher UVR sensitivity (lower basal MED) would benefit the most from oral PLE treatment. These results are intriguing, and thus studies with long-term PLE administration in patients with a high risk of developing MM would be important to expand and confirm these data. No recent clinical studies have been reported in this field that confirm the effectiveness of PLE in preventing melanoma. However, there are new pre-clinical studies which help uncover new mechanisms of action in relation to this matter (Table 1). Within the scope of the latest in vitro studies, three principal works have been incorporated to the scientific approach. The first one explores the potential of a dietary supplement containing sulforaphane (SFN) and Fernblock® extract, in terms of its antioxidant, antineoplastic and antiaging properties. The study analyzed the impact of SFN/FB combination on MMPs, ROS production, and IL-1β secretion in human normal keratinocytes. The combination of these actives was found to be more effective than each on its own in inhibiting melanoma cell migration in vitro, MMP-1, -2, -3 and -9 production, inflammasome activation and IL-1β secretion. Moreover, when used in normal keratinocytes with a pro-inflammatory stimulus like TNF-α, SFN/FB was more efficient in inhibiting MMP-1 and -3 production and IL-1β secretion than SFN or FB alone. Based on these results, the authors suggested that SFN/FB-based supplements could be used as potential preventive measures against skin aging and as adjuvants in the treatment of advanced melanoma [25].

The second work represents an important step forward with respect to understanding the mechanisms involved in DNA damage, and in particular the formation of dark CPDs. Initially discovered by Premi el al (2015), this recent work performed by Portillo-Esnaola et al. (2021) confirms that UVA radiation triggers DNA damage in melanocytes even hours after sun exposure has ended due to increased production of nitrogen reactive species (NO•, O2− and ONOO−), which is linked with the increased formation of CPDs and dark-CPDs. UVA-induced significant dark-CPD formation was observed as soon as 3 h after exposure and the highest peak of dark-CPD formation was obtained 24 h after exposure. However, pre-treatment with Fernblock® (0.3–0.75 mg/mL) was found to reduce the production of these reactive species and the formation of dark-CPDs due to its antioxidant and scavenging properties. We now understand that PLE not only prevents sunlight-induced DNA damage but also offers protection against it even after exposure to solar irradiation. This suggests that Fernblock® could be a promising candidate to complement traditional sunscreens in providing long-lasting skin protection against dark-CPD formation formed after irradiation [26,27].

The third study, performed by Gallego-Rentero et al. (2022) is related to DNA damage induced by photopollution. The interaction of UVA radiation with environmental pollutants, specifically those of a polycyclic aromatic hydrocarbon (PAH) nature such as benzo[a]pyrene (BaP), produces what is known as photopollution. BaP acts as a photosensitizer and upon absorption of UVA radiation it causes increased cell damage in vitro and tumorigenicity in mice even at non-toxic concentrations. Thus, the study evaluated the protective effect of Fernblock® against the combination of pollution and UVA radiation in human keratinocyte and mouse melanocyte cell lines. This preclinical study demonstrated the efficacy of Fernblock® in preventing changes in cellular structure, viability, oxidative stress, and DNA damage. These findings provide strong evidence that Fernblock® induces the priming of cells, rapidly promoting the activation of repair mechanisms and efficient elimination of oxidized derivatives that appear in the nuclear DNA as a result of sequential exposure to BaP and UVA light [28].

In order to enhance photoprotection, it is crucial to explore innovative methods that move beyond conventional measurements of minimal erythema dose (MED). The first evaluation focuses on demonstrating the clinical impact of assessing the immunomodulatory and preventive effects of DNA damage through in vivo studies. Thus, Schalka and Donato (2019) clinically evaluated the efficacy of an SPF 90 sunscreen with PLE in protecting against sun-induced skin damage vs. the same formulation without PLE. The presence of PLE provided additional protection, further reducing erythema, pigmentation, DNA damage, collagen breakdown and immunosuppression vs. placebo [29]. One of the most significant findings from this study is the marked reduction of p53 in skin areas protected with SPF 90 sunscreen containing Fernblock® indicating reduced DNA alteration. These findings were completed by Aguilera et al. (2021), who conducted an in vitro study that analyzed the impact of Fernblock® as a part of topical sunscreen in protecting the skin from photoimmunosuppression and other detrimental biological effects caused by exposure to UV radiation. In addition to the biological activity demonstrated in previous studies, the UV absorption properties of PLE provide an additional booster effect to topical sunscreens, increasing SPF and UVAPF and enhancing protection against not only erythema and permanent pigment darkening reaction but also against immunosuppression [30].

Regarding xeroderma pigmentosum (XP), data has shown that PLE reduces UVR-induced COX-2 levels, at least in part through activation of p53, and decreased epidermal cell proliferation induced by UVR in a mouse model [15]. A case report on XP treatment demonstrated the efficacy of a topical film-forming medical device containing a DNA-repair enzyme, photolyase (Repairsomes®) and very high protection UV filters in preventing the growth of skin cancer lesions in patients with XP [31].

Finally, it is worth highlighting the recent findings of Lacerda et al. (2023) in the field of oral cancer prevention. Their study demonstrated that PLE has the ability to suppress oral cancer cell growth in vitro in SCC-9, SCC-15 and SCC-25 cell lines and prevent tumor development in vivo in mice with induced oral carcinogenesis. A decrease in the expression of Ki67 and PCNA proliferating markers as well as in N-cad (Cdh2), Vim and Twist markers related to migration was observed in tongue tissues. Therefore, PLE may have a beneficial effect on immune and inflammatory responses related to oral tumors and could be a promising natural therapeutic approach for preventing and treating oral cancer due its immunomodulatory activity [32].

Table 1. This table provides a summary of the scientific articles published in the last five years in the field of oncodermatology, specifically highlighting the references to, and conclusions about, PLE treatment.

Oncodermatology			
Design	**Pathology/Focus**	**Summary/Outcome**	**Study Reference**
Review	General oncodermatology	This review reports the mechanisms through which *Polypodium leucotomos* acts to evaluate its uses in oncodermatology with references to in vitro and in vivo studies.	[13]
Review and book chapter	Continuing medical education about skin cancer and sunscreen use	These reviews provide evidence-based recommendations for the use of sunscreen as a preventive strategy against skin cancer while also considering potential risks and environmental impacts associated with the use of some chemical sunscreen filters. PLE is included as a reference oral sunscreen technology for prevention of photodamage.	[33–37]
Reviews and book chapter	Botanical interventions for photoprotection and skin cancer	These works review the main actives derived from plants with scientific evidence as treatment in photoprotection and offer an overview of cancer and phytotherapy. Specifically, they review the existing literature on the properties of PLE and its potential therapeutic effects in preventing skin damage. These reviews include studies conducted in vitro, in vivo and clinical trials.	[38–42]
Review	Preventive interventions for keratinocyte carcinoma	This manuscript examines the potential of pharmaceuticals, plant-derived phytochemicals and vitamins for preventing keratinocyte carcinoma. One such reference photoprotectant is PLE, which has been shown to inhibit the development of tumors and acute UV-induced damage in humans.	[43]
Clinical study	Field cancerization	This clinical study suggests that a new medical device treatment containing Fernblock® (NMD) is a useful treatment method for improving the precancerous field and preventing the development of new AKs.	[21]
Clinical study	Oral cancer	The findings indicate that PLE has the ability to suppress oral cancer cell growth in vitro and prevent tumor development in vivo. Thus, PLE could be a promising natural therapeutic approach for preventing and treating oral cancer.	[32]
Preclinical study	Skin cancer markers	This in vitro study suggests that FB could be a promising candidate to complement traditional sunscreens in providing long-lasting skin protection against dark-CPDs formation after irradiation.	[26]
Preclinical study	Melanoma	This in vitro research suggests that supplements containing sulforaphane/FB could be used to prevent skin aging and help treat advanced melanoma.	[25]
Preclinical study	Skin cancer induced by photopollution	This preclinical study demonstrates the efficacy of PLE in preventing changes in cellular structure, viability, oxidative stress and activation of the melanogenic signaling pathway caused by exposure to both BaP and UVA light.	[28]

Table 1. *Cont.*

Oncodermatology			
Design	**Pathology/Focus**	**Summary/Outcome**	**Study Reference**
Actinic keratosis			
Review	Actinic keratosis	This review article examines in vitro experiments and clinical trials that utilize evidence-based therapeutic methods before or after photodynamic therapy (PDT). Specifically, the effectiveness of topical treatments and oral supplementation, such as diclofenac, imiquimod and PLE, among others, as well as mechanical-physical treatments, are evaluated.	[44]
Review	Actinic keratosis	In this article, the authors offer expert opinions and practical insights into the treatment of actinic keratosis and field cancerization using monotherapy or a combination of therapies among which PLE is cited. The primary objective is to achieve improved, quicker and more tolerable clinical outcomes.	[45]
Review	Actinic keratosis	This review discusses various physical ablative techniques and drug preparations available for treatment. It emphasizes the need for careful evaluation of efficacy, toxicity and tolerability data, as well as practical considerations such as treatment protocols and patient preferences, to achieve maximal adherence and prevent treatment failure. It includes PLE as a chemopreventive treatment tool against the development of AK.	[46]
Xeroderma pigmentosum			
Review	Xeroderma pigmentosum	The purpose of this review is to present the symptoms, diagnosis, and treatment of XP. It also includes oral PLE as a treatment adjuvant due to its chemoprotective, antioxidative, anti-inflammatory and immunomodulatory properties. All these effects have the potential to lessen the phototoxic effects of UVR and thus reduce UVR-induced skin damage and cancer.	[47]

3.2.2. Photodermatoses and Photoaggravated Skin Diseases

Photosensitivity occurs when there is an abnormal reaction between a specific component of the sun's electromagnetic spectrum such as UVR (UVA, UVB) or VIS, and chromophores in the skin. UVA is the most common type of sunlight that leads to photosensitivity, while exposure to VIS may trigger a condition called porphyria. The causes of photodermatosis can be varied: some types are caused by autoimmune reactions, while others are triggered by drugs or connective tissue disease. Additionally, certain types of photodermatosis can be caused by abnormal inherited biochemical pathways [48]. It is widely recognized that exposure to UV radiation can cause changes to both the skin and overall immune system. Regulatory T (Treg) cells play a crucial role in maintaining immune homeostasis by suppressing immune responses to both self- and non-self antigens [49].

Regarding photo-immunosuppression, a multitude of older studies represent essential references that have greatly enhanced our comprehension of the potential of PLE in treating photodermatoses. PLE is endowed with immunomodulatory properties acting as a photoimmunoprotective agent through different mechanisms. PLE prevents UCA isomerization from trans to cis isomer which is a triggering event of skin immunosuppression [50]. PLE also prevents epidermal Langerhans cells (eLC) depletion caused by

UV irradiation in vivo [51]. Multiple molecular mechanisms may underlie the increase in survival of dendritic cells, including inhibition of UCA isomerization as mentioned above, inhibition of iNOS expression [11] and improvement of endogenous systemic antioxidant systems [52]. Consequently, PLE is able to reduce the infiltration of neutrophils and macrophages [12] and decrease inflammatory molecules in humans [19] and mice [51], thus inhibiting mast cells and leukocyte extravasation in the irradiated area. The immunomodulation of these markers has a significant clinical impact because reducing the infiltration of neutrophils and macrophages can directly affect the inflammatory response of the skin to UV or visible light radiation. This exaggerated response is observed in immunologically mediated photodermatoses (previously referred to as idiopathic), and controlling the release of these markers can directly impact these skin disorders, as they are aggravated by greater inflammatory responses.

Concerning photodermatoses and photoaggravated diseases, there is a need to increase clinical research on various conditions such as lupus erythematosus, polymorphous light eruption (PMLE), rosacea, solar urticaria, different forms of dermatitis, and psoriasis, among others. Given the known action mechanisms of PLE and its demonstrated beneficial effects in the prevention of photodermatoses, expanding the existing clinical evidence on the impact of PLE in these pathologies would be worthwhile, especially considering their high prevalence [53]. Another important area to continue research is photodermatosis induced by chemical agents, as it is one of the primary issues faced by society. There are over 300 drugs classified as phototoxic, which undergo a photochemical reaction when the skin is exposed to radiation and become chemically excited. This leads to a reaction with other molecules in the skin environment, such as free radicals, proteins and enzymes, causing phototoxicity and cell damage. This can result in photodermatosis and may also lead to DNA damage and skin cancer. Moreover, these drugs can generate inflammatory reactions known as photoallergy. Phototoxic drugs are widely used and are taken by a significant proportion of the population. Among the phototoxic drugs are antihypertensive drugs, antidiabetic drugs, NSAIDs, antibacterial drugs and others [54]. On light of all this, this review highlights the potential clinical effects of PLE on various types of photodermatosis, providing a summary table with the most recent studies conducted on the topic (Table 2).

Polymorphous Light Eruption and Actinic Prurigo

Polymorphous light eruption (PMLE) is a frequently occurring skin condition caused by an immune response triggered by exposure to UVR from sunlight, which can lead to a range of alterations due to the breakdown in the body's ability to suppress the immune response. It has been found that the combination of Fernblock®, nicotinamide, vitamin D and zinc can reduce the intensity of pruritus and the severity of flare-ups in 87% of PMLE patients [55].

Although less common than other photodermatoses, actinic prurigo is also characterized as a condition where exposure to sunlight triggers an immune response. As in the case of PMLE, the treatment of actinic prurigo with PLE leads to a significant improvement in this disease: duration and severity of skin eruptions are reduced [56]. Recent research suggests that Th1 cells and TNF-α are significant contributors to the development of actinic prurigo. As a result, PLE may have a vital role in reducing actinic prurigo by regulating the immune response [57].

Solar Urticaria and Photosensitive Lupus

Solar urticaria (SU) is a rare chronic acquired photosensitivity disorder that causes recurrent episodes of urticaria rash on skin areas exposed to sunlight. Although usually a benign condition, it may be extremely disabling and limiting for patients. Its pathogenesis has not yet been adequately understood and it can be difficult to diagnose. In 2019, Photiou et al. performed a retrospective review of 83 patients identified as having SU. Of the total number of patients who underwent the monochromator test, SU was confirmed in 58%, and most of them reacted to VIS and UVA or UVA alone. Among the treatment strategies for SU, antihistamines and sun avoidance are the most prescribed. In difficult-to-treat SU patients who do not respond to these strategies, other options such as the monoclonal anti-immunoglobulin E antibody omalizumab have been shown to be effective. Also, PLE is suggested as a safe treatment option in SU [58].

The findings of this study are consistent with prior research conducted by Caccialanza et al., reflected in two clinical studies in 2007 and 2011. These studies assessed the effectiveness of orally administered PLE in the treatment of SU. The first study involved two patients diagnosed with SU: the intake of PLE at a dose of 480 mg/day significantly reduced the skin reaction to sunlight and improved the related symptoms such as prurito [59,60]. In the second study, the four SU patients enrolled were also treated with 480 mg/day of oral PLE and the same benefits were observed, without side effects or tolerance problems [59,60]. All this makes PLE an effective and safe treatment for photoprotection in this idiophatic photodermatosis [59,60].

Photosensitive lupus is another form of photo-exacerbated dermatosis. It consists of a multifactorial inflammatory and autoimmune disease with a variety of clinical manifestations of differing severity. In 2022, Malara et al. conducted a review of the effects of the anti-inflammatory drug thalidomide in patients with discoid lupus erythematosus (DLE) who were refractory to different medications. In two of the patients included in this study who present painful erythematous lesions, 50 mg daily of thalidomide was administered along with PLE capsules and sunscreen with remarkable clinical results. Patients who received PLE experienced a longer-lasting clearance of symptoms. As incorporating this natural active helped to reduce the side effects associated with thalidomide treatment, PLE could serve as a safety measure to lower thalidomide dosage [61].

Additionally, Segars et al. hypothesized the use of PLE as an immunomodulatory agent to adjunct the treatment of subacute cutaneous lupus erythematosus (SCLE), another type of photosensitive lupus [62]. Previous data from a case report by Breithaupt et al. (2012) presented new evidence of a beneficial clinical effect of PLE in a patient with moderately controlled subacute cutaneous lupus erythematosus on hydroxychloroquine. This patient achieved near total remission with the addition of PLE to her treatment regimen, suggesting it might have future applications in photosensitizing dermatoses, including other forms of cutaneous lupus erythematosus [63].

The clinical findings of these studies align with the mechanisms discussed in a recent review by McCarty et al., indicating that PLE, with its high content of phenolic compounds and antioxidant activity, has the potential to suppress superoxide anions, lipoperoxides, and hydroxyl radicals. Additionally, PLE shows promise as an anti-inflammatory agent and an immune modulator with therapeutic applications [9].

Table 2. This table provides a summary of the scientific articles published in the last five years in the field of photodermatosis, specifically including references to, and conclusions about, PLE treatment.

Photodermatoses and Photoaggravated Skin Diseases			
Design	**Pathology/Focus**	**Summary/Outcome**	**Study Reference**
Review	UVB phototoxicity	The objective of the text is to explore and discuss the potential of various nutraceuticals in preventing or mitigating the effects of phototoxicity caused by UVB radiation. The text provides the mechanisms by which these nutraceuticals, including spirulina, soy isoflavones and PLE, among others, may offer protection against UVB-induced sunburn, photoaging and NMSC.	[9]
Review	Photodermatoses	This chapter outlines various topical and systemic agents that can trigger phototoxic and photoallergic reactions. In terms of treatment, the chapter mentions that PLE can be used as a systemic antioxidant in conjunction with PUVA to manage cases of PMLE.	[64]

Table 2. *Cont.*

Photodermatoses and Photoaggravated Skin Diseases			
Design	**Pathology/Focus**	**Summary/Outcome**	**Study Reference**
Review	Idiopathic photodermatoses	This chapter provides a clinical approach to managing idiopathic photodermatoses, including conditions such as PMLE, actinic prurigo and idiopathic solar urticaria, among others. Preventing and managing these conditions involves implementing photoprotective measures and increasing the skin's tolerance to sunlight through the use of narrow-band UVB therapy and other forms of phototherapy or photochemotherapy when necessary. In addition, topical or systemic antioxidants like PLE may be helpful in certain cases.	[65]
Review	Diet and Photodermatoses	Prior studies have explored the connection between diet and several skin conditions, including rosacea, hidradenitis suppurativa, herpes labialis and vitiligo. The authors consolidate the findings from existing literature to create clear and concise guidance regarding dietary supplements that could be beneficial or harmful. By doing so, they provide healthcare professionals with evidence-based recommendations to assist their patients, including PLE as a recommended supplement in the treatment of vitiligo.	[66]
Case report	Photodermatoses	The case study involves a 55-year-old man who experienced a severe and painful skin eruption with erythema and blisters in sun-exposed areas one month after starting vandetanib treatment. Despite treatment with steroids and avoiding sun exposure, the condition did not improve until the patient began taking oral supplements of PLE. This case highlights the potential of PLE as a safe and effective photoprotective agent for treating refractory phototoxic reactions.	[67]
Polymorphous light eruption and actinic prurigo			
Review	PMLE	The purpose of the review is to provide a better understanding of the molecular pathogenesis of PMLE by examining the immunological disturbances associated with the disease. The authors emphasize the potential of PLE as an immunomodulatory and antioxidant agent and suggest it could be used as a preventive therapeutic approach for PMLE treatment.	[53]
Review	PMLE	The goal of this article is to provide readers with the latest information on PMLE with regards to its epidemiology, clinical presentation, underlying pathophysiology, available treatments and prognosis. PLE is presented as a potential treatment for PMLE, and the review cites open-label studies showing that this supplement can reduce the severity, frequency and rapidity of onset of PMLE reactions.	[68]
Review	Actinic prurigo	The aim of this study is to provide a summary of current knowledge related to two types of photodermatoses—actinic prurigo (AP) and hydroa vacciniforme (HV), both of which typically develop during childhood. Among suggested treatment, botanical agents such as PLE may be beneficial in reducing photosensitivity in certain skin conditions like PMLE and solar urticaria. However, further studies are needed to suggest their usefulness in treating AP.	[57]

Table 2. *Cont.*

Photodermatoses and Photoaggravated Skin Diseases			
Design	**Pathology/Focus**	**Summary/Outcome**	**Study Reference**
Case report	Actinic prurigo	In this report, the authors describe the successful use of PLE in an 11-year-old girl with AP. PLE treatment led to a significant reduction in her symptoms and no negative side effects were observed. PLE has a wide-ranging impact on the immune system and acts as an antioxidant by promoting an anti-inflammatory environment.	[56]
Solar Urticaria and Photosensitive Lupus			
Restrospective analysis	Solar urticaria (SU)	The authors conducted a retrospective analysis in 83 patients with SU. Among the 60 patients who underwent monochromator testing, 35 were confirmed to have SU, with most reacting to VIS and UVA, or UVA alone. The mainstay of treatment for SU is antihistamines and sun avoidance. However, for patients who do not respond to these treatments, other options such as omalizumab may be of potential interest. Also, PLE is sugested as a treatment option for SU, without side effects.	[58]
Review	Lupus	This review analyses natural actives traditionally used to treat rheumatological conditions, including antimalarials, which could also be beneficially indicated for cutaneous lupus eryhtematosus (CLE). It also suggests their combination with PLE as a photoprotective supplement to control photosensitivity.	[69]
Review	Lupus	This text reviews the available evidence regarding local and systemic therapies for CLE and provides healthcare professionals with alternative treatment options for patients who were previously treated with quinacrine, which is currently unavailable in the USA. Among these options, PLE is proposed with a level of evidence of 5 in accordance to the levels adapted from the Oxford Centre for Evidence-Based Medicine.	[70]
Clinical cases	Lupus	This article examines the use of thalidomide in the treatment of discoid lupus erythematosus (DLE) and discusses four case studies that demonstrate its success. Two of the case studies included the addition of PLE. Patients who received PLE experienced a longer duration for complete clearance of symptoms; moreover, incorporating PLE helped to lower the thalidomide dosage and thus reduce its side effects.	[61]
Clinical study	PMLE	In this prospective study, a standardized extract of *P. leucotomos*, along with nicotinamide, vitamin D and zinc was orally administered to 15 patients suffering PMLE. These patients had not achieved symptom control through the use of only topical photoprotection. Administering a standardized extract of *P. leucotomos*, nicotinamide, vitamin D and zinc orally in conjunction with appropriate topical photoprotection offers a safe and effective alternative for preventing and minimizing the frequency and severity of outbreaks in individuals with PMLE.	[55]
Other photodermatoses: Chronic Actinic Dermatitis			

Table 2. *Cont.*

Photodermatoses and Photoaggravated Skin Diseases			
Design	**Pathology/Focus**	**Summary/Outcome**	**Study Reference**
Case report	Chronic actinic dermatitis	In this study, a case of a patient with chronic actinic dermatitis (CAD) who showed only partial improvement with dupilumab is described. Initial management included sun avoidance and photoprotective therapy, which included topical Fernblock®, among others. The CAD did not improve, and the treatment continued with topical corticosteroids, immunomodulators, and systemic immunosuppressive agents. The continued implementation of photoprotection measures such as oral supplements, including oral and topical PLE, is recommended due to their proven efficacy as adjuvants to the above-mentioned pharmacological treatments.	[71]
Other studies with PL extracts in photodermatoses: non- Fernblock® PL extracts			
Review	Rosacea	The purpose of this study is to explain the origin of rosacea, with a particular focus on the influence of UV radiation and exposome on the development of this skin condition. Additionally, this review highlights the importance of non-pharmacological approaches, with specific emphasis on photoprotection strategies in managing rosacea, using, for example, an extract of *P. leucotomos.*	[72]

3.2.3. Pigmentary Disorders

The color of skin is determined by several pigments, one of which is the melanin produced by melanocytes, whose primary function is to protect from UVR. Pigmentary skin disorders are frequently encountered in the practice of adult medicine and include both hypopigmentation and hyperpigmentation alterations. Although most of these disorders are rarely associated with health risks or systemic diseases, they can sometimes lead to severe of life-threatening pathologies such as melanoma. Moreover, they can lead to negative effects on quality of life and become a source of discomfort and emotional stress for patients. Despite their frequency, these disorders remain challenging to treat [73].

Among these skin disorders, vitiligo and melasma are two of the most frequent. Both affect the skin's appearance and can have notably adverse effects on an individual's health-related quality of life. Melasma is in fact the skin pigmentary disorder with the highest incidence [74].

Vitiligo

Vitiligo is a common depigmenting skin disorder characterized by the selective loss of melanocytes. It has recently been reported that the combination of NB-UVB and oral administration of PLE accelerates repigmentation and increases total repigmented area. Also, PLE can improve the efficacy of photo(chemo)therapy for vitiligo by reducing negative side effects and improving tolerance. PLE does this by preventing sunburn and phototoxic reactions, as shown in both in vitro and in vivo studies involving human and animal models [75]. A recent review suggests that innovative drug delivery methods have the potential to enhance the delivery of topical medications by improving their localization within the epidermis and increasing their effectiveness, providing an interesting possibility for future improvements in PLE's efficacy [76].

Moreover, a recent review discussed the role of NRF2-ARE (antioxidant response element) pathway in the pathogenesis of vitiligo, since this pathway is involved in cellular defense against oxidative stress. The review listed several agents known as NRF2 activators that included PLE and may represent a potential therapeutic strategy for this pigmentary disorder [77]. In this sense, recent studies have demonstrated the ability of PLE to modulate NRF2 pathway, which could potentially help explain the positive clinical effects of PLE treatment in patients with vitiligo [10] (Table 3). Although further research is required to validate this hypothesis, this rationale could prove valuable not only in the treatment of vitiligo but also in other photodermatoses [78].

Hyperpigmentation Disorders

Hyperpigmentation is a term used to describe a skin condition in which an excess of pigment production occurs, resulting in dark spots or areas of the skin that appear darker than the rest. Hyperpigmentation can be caused by exposure to environmental pollution, hormonal therapies, cosmetics, contraceptive pills, pregnancy, photosensitizing agents and genetic susceptibility [79]. Some common forms of hyperpigmentation include melasma, environmental lentigo and post-inflammatory hyperpigmentation [80]. In recent studies on PLE and its ability to prevent hyperpigmentation, primary emphasis has been on melasma and the various factors that contribute to increased pigmentation (Table 3). This section will delve into the discussion of these studies and the factors involved in promoting hyperpigmentation.

Pigmentation resulting from exposure to VIS is a significant concern and thus a prominent area of research. VIS has been demonstrated to have various biological impacts on the skin, including DNA damage caused by the generation of ROS, the activation of pro-inflammatory cytokines, exacerbation of photo-induced conditions, and the promotion of pigmentation in melano-competent individuals. Thus, Mohammad et al., (2019) conducted a clinical study to assess the effectiveness of oral PLE treatment in preventing the adverse effects induced by VIS. The study involved twenty-two participants with Fitzpatrick skin phototypes IV–VI. On day 0, the subjects were exposed to VIS radiation. Immediately, as well as 24 h and 7 days after radiation, clinical evaluations using the Investigator's Global Assessment (IGA) scoring system and spectroscopic assessments were conducted. The participants were then given a 28-day supply of PLE (480 mg daily). All subjects experienced immediate pigment darkening, persistent pigment darkening, and delayed tanning both before and after taking PLE, but instrumental assessments showed a statistically significant decrease in pigment darkening and delayed tanning in the PLE group. Results of this research indicate that PLE has an impact on VIS-induced effects. Therefore, PLE could be utilized as a supplementary approach to conventional photoprotection methods in order to safeguard against the detrimental effects of VIS [81]. These clinical findings can be explained by in vitro studies that have explored the potential mechanisms by which PLE prevents pigmentation caused by VIS. These studies have specifically examined the role of VIS, particularly blue light, in activating pathways associated with melanogenesis. A recent in vitro study conducted by Portillo et al. (2021) examined the effectiveness of Fernblock® in mitigating pigmentation induced by blue light emitted by digital devices. The study revealed that Fernblock® acts through multiple mechanisms, including the modulation of the p38 melanogenic signaling pathway, inhibition of photooxidation of melanin precursors and reduction of opsin 3 expression. These findings underscore the potential of Fernblock® as a protective agent against the detrimental effects of visible light, particularly blue light [82]. While multiple in vitro studies have also shown that blue light can activate mechanisms associated with melanogenesis [83,84], further research is necessary to gain a better understanding of how chronic exposure and the prevalence of electronic devices in modern life can potentially impact melanogenesis. The controversy surrounding this topic remains significant, highlighting the need for more studies to provide comprehensive information about the clinical implications of long-term exposure to blue light and its effects on melanin production, particularly in heavily pigmented individuals who are especially prone to skin hyperpigmentation [85–88].

Recent studies have also focused on investigating the role of pollution in melanin production. Observations have revealed that the prevalence of pigmentation is higher in animals residing in polluted areas, particularly in urban-industrial sites. As an increasing number of individuals are exposed to elevated levels of air pollution, there is a possibility that environmental pollutants can influence melanogenesis in human skin. Epidemiological studies have indicated that exposure to air pollutants associated with traffic, such as diesel exhaust particles, is correlated with an increased occurrence of clinical manifestations of hyperpigmentation [89]. Ahn et al. observed that RNA-sequencing data from melanocytes exposed to particulate matter (PM) revealed an increase in the expression of molecules associated with the unfolded protein response. Notably, the IRE1α signaling pathway consistently showed upregulation and was found to be involved in PM-induced melanogenesis [90]. Similarly, the already mentioned study conducted by Gallego-Rentero et al. (2022) in skin cancer prevention demonstrated that skin exposure to photopollution not only results in DNA damage and oxidative stress but also triggers the activation of the melanogenesis pathway. This study provided compelling evidence of significant upregulation of opsin-3 in cells treated with BaP and subsequently exposed to UVA. However, when cells were pre-treated with Fernblock® before irradiation, the expression of opsin-3 remained similar to basal level. This study revealed the capacity of Fernblock® to prevent the activation of melanogenesis through the modulation of opsin-3 expression, specifically in response to photopollution [29].

With respect to melasma, the considerable quantity of existing evidence of PLE's efficacy in preventing and reducing this disorder has meant that no new specific studies have been conducted over the past few years. The most recent studies have been narrative reviews that emphasize the use of PLE as an effective oral treatment for this pathology based on the evaluation of various measures such as the MASI (Melasma Area and Severity Index), MI (Melasma Index), melasma area and pigmentary intensity, among others. These reviews encompass a wide range of research in this area, including a notable reference study conducted by Goh et al. (2018). The study specifically validated the safety and effectiveness of oral PLE treatment as an adjuvant in the management of melasma. It was found to be particularly effective when used in combination with topical hydroquinone and sunscreen [91]. However, it is important to note that other recent studies have been conducted in this field using different PL extracts than Fernblock®. This is the case of the study conducted by Piquero-Casals et al. (2020) that examined the clinical outcomes of a combination of trichloroacetic acid, phytic acid and ascorbic acid peel as well as oral antioxidant supplements (including a non-detailed extract from PL) and topical treatments, for individuals with difficult-to-treat melasma. The findings indicate that this approach could be an effective solution for managing this patient group [92]. Research in the field of new active ingredients with antioxidant and antimelanogenic properties is ongoing on a daily basis, with the goal of discovering novel compounds for use in the cosmetics industry. As evidence of this, a recent article compared and assessed the protective potential of various extracts from Spanish ferns, using the standardized PLE as a reference. The research concluded that all fern extracts exhibit antioxidant activity and have the potential to inhibit hyperpigmentation through their anti-tyrosinase activity. Moreover, it concluded that hydrophilic extracts are more potent and effective than lipophilic extracts [93].

Recent studies are also beginning to investigating other fields, including the effects of agents like pollution, photopollution and blue light on pigmentation [86,89]. There is growing interest in exploring the potential of PLE in offering protection against these factors as they are closely associated with the development of pigmentary disorders and photoaging. As a result, the focus has expanded to explore novel areas that can contribute to a deeper understanding of pigmentation-related mechanisms and the potential benefits of PLE in this context.

Table 3. This table provides a summary of the scientific articles published in the last five years in the field of pigmentary disorders, specifically including references to, and conclusions about, PLE treatment.

Pigmentary Disorders			
Design	**Pathology/Focus**	**Summary/Outcome**	**Study Reference**
Reviews and book chapters	Pigmentary disorders	Pigmentary disorders (melasma, vitiligo, periocular hyperpigmentation, pigmented contact dermatitis and lichen planus pigmentosus) are over-represented in women in most societies. Their mechanisms and future therapies, including PLE, are reviewed.	[94,95]
Review	Pathways and ingredients involved in pigmentary disorders	The text provides an overview of the role oxidative stress plays in melanogenesis, particularly in response to skin exposure to UVR and VIS. It also discusses various pathways involved in pigmentary disorders. Additionally, the text offers guidance on effective approaches to modulate melanogenesis, including the use of vitamins, PLE, niacinamide and other options, such as lightening agents, that can aid in the better management of pigmentary disorders.	[80,96–100]
Review	Visible light and hyperpigmentation	These reviews focus on the role of VIS in hyperpigmentation disorders (melasma and PIH) and analyze the direct and indirect effects of blue light emitted by digital devices reported on in vitro and in vivo studies. Recent advances in understanding the protective role of PLE against UVA and VIS have led to it being cited as a reference agent in preclinical and clinical studies.	[95,101]
Review	Management of hyperpigmentation: dermatological procedures	This review focuses on the medical and dermoaesthetic procedures for hyperpigmentation and on challenges (resistance, recurrence, adverse effects) in the management of pigmentary disorders such as melasma and PIH. PLE is included as a reference compound due to studies that suggest its beneficial activity in treating dyschromias.	[102–104]
Review	Management of hyperpigmentation: topical	The review presents alternative ways to manage hyperpigmentation, including PLE, glutathione and thiamidol. It also provides a table summarizing the scientific evidence supporting their effectiveness.	[105]
Review	Management of hyperpigmentation: oral	The texts provide a review of literature on oral treatments for hyperpigmentation, with a specific focus on examining the clinical evidence that supports the use of several oral treatments, including PLE and others.	[106,107]
Review	Management of PIH	These reviews describe the first-line treatments for epidermal PIH (based on topical or oral skin lightening agents) and the available adjunctive therapies for patients refractory to first-line treatment or for dermal PIH (peels, laser, etc.). They also analyze the use of sunscreens for the treatment of melasma and PIH. PLE is included as a systemic skin-lightening agent, among others. It is noted that PLE is the only ingredient listed that does not have any reported adverse effects.	[108–110]

Table 3. *Cont.*

Pigmentary Disorders			
Design	**Pathology/Focus**	**Summary/Outcome**	**Study Reference**
Clinical study	Skin pigmentation induced by VIS + UVA light	This study evaluates the role of topical antioxidants in protecting against VIS + UVA-induced effects in skin phototypes I-VI. Topical antioxidants inhibit erythema in phototypes I-III and reduce pigmentation in phototypes IV-VI. PLE is used as a reference compound to compare its antioxidant properties with those of other compounds.	[111]
Melasma			
Reviews and book chapters	Management of melasma: focus on topical and oral treatments	These articles discuss various techniques for treating melasma and evaluate the effectiveness and safety of ingredients like hydroquinone and tranexamic acid, as well as delivery systems that improve the depigmentation activity of certain agents. Additionally, all these reviews place PLE among the most effective oral agents recommended for melasma.	[79,112–119] Sistematic review and metanalysis: [120–122] Book chapters: [123,124]
Review	Management of melasma: focus on dermatological procedures	The main objective of these reviews was to analyze the available evidence on the efficacy and safety of microneedling alone or in combination with topical agents in reducing pigmentation and improving the quality of life for adult patients with melasma. Oral PLE treatment was included as a therapeutic option for melasma, suggesting that combined therapies tend to produce better results compared to monotherapy.	[125–127]
Review	Melasma	The review aims to provide a comprehensive understanding of the role of oxidative stress in melasma and the potential therapeutic benefits of various antioxidants for individuals with this condition. Here, PLE is considered a principal antioxidant for treating melasma, and a summary of clinical studies is included in these documents.	[117,128]
Review	Melasma pathways	The manuscript provides a review of the processes and pathways responsible for skin pigmentation, specifically the changes in melanogenesis that lead to melasma and resulting hyperpigmentation. The paper also discusses current treatments and therapies, including those administered topically, orally and through phototherapy, with a particular focus on the effects of cosmetics. PLE is included as a plant-based oral treatment.	[129]
Clinical study	Pigmentation disorders	This study aims to assess the effectiveness of PLE in preventing VIS-induced effects in human skin. PLE treatment induces a significant decrease in persistent pigment darkening and delayed tanning and reduces expression of several damage markers. The study provides scientific evidence to position PLE as a treatment for pigmentation disorders.	[81]
Vitiligo			

Table 3. *Cont.*

Pigmentary Disorders			
Design	**Pathology/Focus**	**Summary/Outcome**	**Study Reference**
Review	Vitiligo treatments	These reviews discuss the pathogenesis of vitiligo and focus on treatment options for the disease including standard drug treatments, phototherapy (NB-UVB and PUVA) and the effectiveness of antioxidant therapies. Moreover, in the majority of cases, antioxidant therapies on their own are not capable of producing significant clinical improvements, except perhaps in mild cases, and they must be used alongside standard drug treatments in order to achieve noticeable outcomes, such as PLE in concomitance with NB-UVB or PUVA.	[76,130–138] Focused on PUVA: [139–141]
Review	Safety in vitiligo treatments	These reviews concentrate on the potential harm related to the use of medicinal plants, including PLE, and offer a summary of adverse drug reactions (ADRs) that have been reported in national and global individual case safety report databases.	[142,143]
Review	The role of NRF2-ARE in vitiligo	This paper examines the role of NRF2 in vitiligo and reviews several agents known as NRF2 activators, including PLE. It suggests that PLE's efficacy in the treatment of vitiligo is the results of its activation of the NRF2 pathway.	[77]
Clinical study	PLE in vitiligo	This study involved 44 patients with generalized vitiligo who received either combined treatment of NB-UVB phototherapy and oral PLE or NB-UVB phototherapy and placebo. The results showed that oral PLE combined with NB-UVB improved repigmentation and increased response rate as compared to NB-UVB alone.	[75]
Other studies with PL extracts in melasma: non- Fernblock® PL extracts			
Case report	Melasma	The aim of this research was to examine the practical outcomes of a treatment plan that combined multiple interventions: one conducted at home and the other performed in a clinical setting. The protocol consisted of using a peel containing trichloroacetic acid, phytic acid and ascorbic acid, in combination with oral antioxidant supplements containing an extract from PL. Additionally, topical products were provided to individuals with persistent melasma. The findings of the study suggest that this treatment protocol could represent an effective approach for managing melasma.	[92]
Chemical assay with different PL extracts	Photoaging, hyperpigmentation	This study evaluates the antioxidant capacity of different hydrophilic and lipophilic fern extracts. All ferns present antioxidant activity and potential to inhibit hyperpigmentation (antityrosinase activity). This report concluded that hydrophilic extracts are more potent and effective.	[93]

3.2.4. Extrinsic Aging

Over the years, researchers have studied how factors such as sun exposure, air pollution, hormonal changes, diet and psychological factors can affect the skin by causing hyperpigmentation, worsening photodermatosis, promoting wrinkles or leading to the development of cancerous lesions [144]. While pollution can exacerbate certain skin conditions like atopy or eczema, a correlation has been demonstrated between pollution and the premature appearance of aging signs [145]. In recent years, there has been a growing interest in evaluating the potential preventive effects of PLE against pollution and solar-radiation-induced photoaging, resulting in an increase in the number of studies in this area (Table 4). In an in vitro study, human keratinocytes were treated with PLE and subsequently exposed to UVB radiation or fine particulate pollutants (PM2.5). The results showed that PLE promoted the NRF2 pathway and its downstream targets, counteracting oxidative stress and thus inducing an increase in cell viability. Additionally, PLE was found to decrease IL6, IL8 and melanin production induced by UVB exposure. These findings suggest that PLE may protect against photooxidative stress and other environmental pollutants [10]. Also, chronic exposure to UVR from sun is well-known to contribute to photoaging, inducing alterations which include clinical and histological changes and increased inflammation and oxidative stress. With respect to photoaging and activation of the melanogenic signaling pathway by photopollution, Fernblock® was found to prevent morphological changes in mitochondria when exposed to a combination of BaP and UVA. It also inhibits the overproduction of ROS generated by the exposure to photopollution in both melanocyte and keratinocyte cell lines. These findings suggest that Fernblock® has the potential to mitigate oxidative stress-induced damage and mitochondrial injury resulting from the combined effect of these harmful agents [28].

Although not as well-researched as UVR, VIS (and more concretely its blue component) and IR radiation have also been shown to cause skin alterations that lead to photoaging. In this last 5 years, research in this area has been directed towards investigating PLE's action mechanisms against VIS, with a particular focus on its protective effects against blue light (Table 4). Not only recently, but beginning in 2018, Zamarrón et al. performed an in vitro study to assess the protective potential of PLE against harmful effects induced by VIS and infrared A (IRA) radiation. PLE mitigates VIS/IRA-induced alterations in different ECM components (MMP-1, CTSK, fibrillins 1 and 2 and elastin) and prevents damage induced by both radiations on cell morphology and viability. All this places PLE as a potential preventive strategy against damage caused by exposure to VIS and IR light [146]. Related to these findings, in 2019 a pilot research was conducted to evaluate the potential effect of oral Fernblock® on photodamage induced by IR and VIS in human volunteers. After oral administration of Fernblock®, the participants were exposed to the combination of IR/VIS, and the expression of MMP1 from skin biopsies was analyzed. The results showed that Fernblock® significantly reduced the overexpression of MMP1 induced by IR/VIS, indicating its potential antiphotoaging effect [147]. In another study previously mentioned in the pigmentation disorders section, 22 participants were exposed to VIS and then observed for 7 days to establish a baseline response. They then received a 28-day supply of PLE (480 mg daily) after the month was up, and VIS was radiated on the other side of the participants' backs. Immunohistochemistry results indicated that PLE also attenuates VIS-induced damages such as inflammation (reduction of COX-2), alteration on structural integrity (MMP-1,2,9, which are responsible for the initial degradation of collagen) and alterations in pigmentation (MART-1) [81]. Results of these two studies reflect PLE's ability to mitigate photoaging-related ECM degradation associated with VIS/IR exposure.

Concerning blue light from the VIS spectrum, excessive exposure blue light from digital devices could also play a role in premature aging. In this sense, the results from the already-mentioned in vitro work performed by Portillo et al. (2021) demonstrated that a dose of blue light simulating our usual exposure to digital devices causes damage to cell morphology and viability, even inducing alterations to mitochondrial membrane potential. The study demonstrated that preventive treatment with PLE reversed these effects, mitigating oxidative stress and promoting reversal to mitochondrial membrane potential basal status. [89].

Another study carried out with a different extract of *P. leucotomos* evaluated the photoprotective properties of an oral food supplement containing: vitamins A, C, D3,

E, selenium, lycopene, lutein, green tea, *P. leucotomos* and grape extracts. Oral intake of this supplement increases MED and ferric reducing antioxidant power (FRAP). In general, it improves antioxidant status of skin and exerts photoprotective effects against photoaging [148].

In conclusion, research on the role of PLE in extrinsic aging has experienced the most significant growth compared to its other clinical applications. In addition, numerous studies have been dedicated to investigating mechanisms by which PLE exerts its protective effects against novel agents involved in aging such as pollution or VIS/blue/IR radiation (Table 4).

Table 4. This table provides a summary of the scientific articles published in the last five years in the field of extrinsic aging, specifically including references to, and conclusions about, PLE treatment.

Extrinsic Aging			
Design	**Pathology/Focus**	**Summary/Outcome**	**Study Reference**
Review	Skin aging	The present study reviews the literature on the underlying causes and pathophysiological processes of skin aging, healthy skin aging, and basic protective antiaging approaches. PLE is regarded as a reference antioxidant due to its phenolic content and it can be used orally or topically to counteract skin aging due to its capacity to reduce the harmful effects of UVR and its photoprotective, antioxidant, anti-inflammatory and antiaging properties.	[149]
Review	Photoaging	These reviews evaluate the ability of sunscreens to protect against photoaging and analyze the ideal characteristics of sunscreen, taking into account the impact of VIS and IR on skin aging (apart from UVR). This document includes PLE as a reference compound for oral photoprotection and its role in the prevention of photoaging.	[7,150,151]
Review	Photoaging pathway	This article discusses the mechanisms of photoaging, specifically in human dermal fibroblasts. PLE is included for its role in preventing photoaging caused by VIS and IR radiation by decreasing MMP-1 and Cat K levels and preventing changes in the expression of fibrillin 1, fibrillin 2, and elastin.	[152]
Review	Ingredients and photoaging	These reviews provide a summary of the pathways involved in skin aging and explore various therapeutic approaches that utilize natural actives. The reviews cite PLE as one of the main natural actives, extensively researched and with robust scientific evidence supporting its use and its role as a reference product.	[153–163]
Clinical study	Sunburn/photoaging/skin cancer	This study compares the efficacy of two identical sunscreens, one containing PLE and the other not. The presence of PLE in the formulation provided a significantly greater reduction in skin damage triggered by solar radiation (reduction of erythema, pigmentation, DNA damage, collagen breakdown and immunosuppression).	[29]
Clinical study	Photoaging induced by IR and VIS	This pilot study evaluates the effect of oral PLE against the IR and VIS-induced photodamage. PLE attenuates IR/VIS-induced MMP1 overexpression. This study reinforces the anti-photoaging potential of PLE.	[147]

Table 4. *Cont.*

Extrinsic Aging			
Design	**Pathology/Focus**	**Summary/Outcome**	**Study Reference**
Clinical study	Photoaging induced by VIS	Twenty-two participants were exposed to VIS before taking PLE and then observed for 7 days to establish a baseline response. After 28 days of taking PLE, VIS was administered to the opposite side of the participant's back. Instrumental assessments showed a statistically significant decrease in persistent pigment darkening and delayed tanning in patients after PLE administration.	[81]
In vitro study	Photoaging induced by blue light	This study assesses the capacity of PLE to reduce pigmentation induced by blue light from digital devices. PLE prevents cell death, alteration of mitochondrial morphology and phosphorylation of p38 triggered by blue light. PLE also prevents melanin photodegradation through regulation of opsin-3 in melanocytes.	[82]
In vitro study	Pollution and aging	This study evaluates the potential of PLE to protect against xenotoxic stress related to exposure to fine particulate pollutants. PLE can reduce pollution-induced stress through modulation of NRF2 pathway.	[10]
Other studies with PL extracts in photoaging: non- Fernblock® PL extracts			
Clinical study	Photoaging	This study evaluates the photoprotective properties of an oral food supplement containing: vitamins A, C, D3, E, selenium, lycopene, lutein, green tea, *P. leucotomos* and grape extracts. Oral intake of this supplement increases MED and FRAP. In general, it improves antioxidant status of skin and exerts photoprotective effects.	[148]

4. Conclusions and Future Perspectives

Strong scientific evidence supports the photoprotective properties of Fernblock® in the prevention, attenuation and even reversal of phototoxic effects caused by solar radiation in the skin. In recent years, the pivotal role of the NRF2 in skin health has garnered significant attention. Its involvement in processes related to photoprotection, cancer prevention, and mitigation of photoaging has led to the exploration of its pharmacological modulation as a novel research approach [164]. Of particular interest is thus the recent findings regarding Fernblock®'s ability to modulate NRF2 [10]. A recent study by Tabolacci et al. (2023) demonstrated that UVA radiation upsets the redox balance, resulting in a notable decrease in the concentration of GSH. This decrease in GSH is believed to be closely associated with the modulation of the NRF2 pathway [165]. Based on these findings, the suggested crucial role of Fernblock® in modulating the endogenous antioxidant systems of the skin (including CAT, GSH and GSSR) [166] may be directly linked to the modulation of NRF2. This connection could potentially explain Fernblock®'s efficacy in protecting against damage induced by solar radiation and other harmful agents. Also, the NRF2 signaling pathway exerts a negative regulatory effect on pro-inflammatory cytokines, chemokine releasing factors, MMPs and other inflammatory mediators such as COX-2 and iNOS. These mediators, either directly or indirectly, impact the relevant NF-kB and MAPK pathways as well as other networks involved in inflammation [167]. Fernblock®'s NRF2-modulatory mechanism can also be observed through its ability to inhibit transcriptional activation of NF-kB [11]. This underlines the extract's role in modulating the interaction between these signaling mechanisms. These findings not only suggest the prevention of inflammation and oxidative stress, but there is also evidence that directly associates the modulation of DNA damage response by NRF2 through MAPK signaling [168]. In addition to discovering the role of NRF2 in DNA repair associated with the MAPK mechanism, another recent study revealed that the antioxidant and DNA protective effects are achieved through the modulation of the PI3K-AKT-NRF2 pathway. As a result of this modulation, a reduction in UVB-induced CPD formation was observed, leading to an enhancement in the activity and efficiency of the nucleotide excision repair (NER) pathway involved in DNA repair in irradiated skin cells [169]. In this regard, strong data have demonstrated that there are numerous mechanisms that may be involved in the regulation of NER pathway, with NRF2 being one of the key players [170]. Considering all these evidences, we can hypothesize that the effect of Fernblock® on NRF2 modulation may be closely linked to its ability to repair DNA, potentially establishing a direct relationship with the reduction of CPDs not only at the time of irradiation but also after exposure (dark-CPDs) [28]. Perhaps both mechanisms could provide an explanation for Fernblock®'s modulator role over markers such as H2AX or p53, facilitating promoting DNA repair. Thus, the observed functions can be directly linked to the abundant polyphenols content found in the Fernblock® extract. This family of chemical compounds has substantial evidence supporting its ability to mitigate the impact of UV radiation on various aspects, including DNA repair, reduction of cellular antioxidant levels, modulation of antioxidant signal transduction pathways, control of immunological response and protection of the extracellular matrix [171]. Understanding the rationale behind these action mechanisms may also help elucidate the effect of Fernblock® in conditions such as XP or AK, where deficient NER mechanisms exist.

Additionally, there is a growing interest in investigating the potential of NRF2 modulation in mitigating skin aging, both in relation to chronological aging and as a response to photodamage. The research on Fernblock® thus appears particularly relevant as it increases cellular antioxidant defenses, promotes DNA damage repair, reduces inflammation and stimulates skin repair [172]. NRF2 has been also associated with cutaneous pigmentation disorders that arise from redox imbalances, particularly in vitiligo as well as chronological hair greying. The dysregulation of NRF2 signaling has been implicated in these conditions, suggesting that oxidative stress and altered redox processes contribute to the loss of pigmentation in the skin and hair and promotes collagen degeneration [173]. Understanding the role of NRF2 in these disorders can provide insights into potential therapeutic approaches aimed at restoring normal pigmentation and collagen production in the management of these conditions [174].

An increasing amount of evidence suggests that the NRF2 signaling pathway is also dysregulated in many types of cancer, leading to abnormal expression of NRF2-dependent genes. Additionally, inflammation plays a crucial role in diseases associated with oxidative stress, particularly in cancer [167]. Therefore, we strongly believe that further investigation into the role of Fernblock® in the NRF2 pathway would be highly compelling, as well as assessing its impact on the development of AKs and its potential progression to skin cancer.

The primary prevention strategy for these skin alterations typically involves the use of specific UVA and UVB filters for photoprotection. It is obvious, however, that Fernblock® can improve photoprotection: not only does it counteract oxidative stress caused by solar radiation and aging, but it has also shown efficacy in reducing inflammation, melanogenesis and overall cellular damage in cultured keratinocytes exposed to pollutant particles in experimental models [28]. Furthermore, immunoprotection emerges as an additional marker of protection. Consequently, incorporating compounds that can modulate the immune response becomes crucial in preventing abnormal reactions that patients with dermatoses may experience upon sunlight exposure. Conducting further clinical research on Fernblock® in photodermatoses thus becomes essential to provide further evidence supporting the potential immunomodulatory effects of this standardized extract.

In summary, there are several potential areas for future exploration regarding the role of Fernblock®. One area of interest is further investigating the relationship between Fernblock® and NRF2 and its impact on other mechanisms such as DNA repair and autophagy. Understanding how Fernblock® influences these processes can provide valuable insights into its broader effects on cellular health and skin protection. Additionally, exploring the potential effect of Fernblock® in new pathologies, such as photodermatoses or pigmentary disorders, could uncover novel therapeutic applications for this ingredient. Continued research in these areas will contribute to a deeper understanding of Fernblock®'s mechanisms and potential benefits in various skin-related conditions.

Author Contributions: Conceptualization, A.R.-L. and S.G.; methodology, A.R.-L. and A.Z.; writing—original draft preparation, A.R.-L. and A.Z.; writing—review and editing, A.R.-L., A.Z., Á.J. and S.G.; supervision, S.G.; project administration, S.G. All authors have read and agreed to the published version of the manuscript.

Funding: This research received no external funding.

Conflicts of Interest: S.G. has a consultant role for Cantabria Labs. The remaining authors declare no conflicts of interest.

References

1. Yeager, D.G.; Lim, H.W. What's New in Photoprotection: A Review of New Concepts and Controversies. *Dermatol. Clin.* **2019**, *37*, 149–157. [CrossRef]
2. Parrado, C.; Philips, N.; Gilaberte, Y.; Juarranz, A.; González, S. Oral Photoprotection: Effective Agents and Potential Candidates. *Front. Med.* **2018**, *5*, 188. [CrossRef]
3. García, F.; Pivel, J.P.; Guerrero, A.; Brieva, A.; Martínez-Alcázar, M.P.; Caamaño-Somoza, M.; González, S. Phenolic Components and Antioxidant Activity of Fernblock, an Aqueous Extract of the Aerial Parts of the Fern Polypodium Leucotomos. *Methods Find. Exp. Clin. Pharmacol.* **2006**, *28*, 157–160. [CrossRef] [PubMed]
4. González, S.; Lucena, S.R.; Delgado, P.; Juarranz, A. Comparison of Several Hydrophilic Extracts of Polypodium Leucotomos Reveals Different Antioxidant Moieties and Photoprotective Effects in Vitro. *J. Med. Plants Res.* **2018**, *13*, 336–345. [CrossRef]
5. Del Rosso, J.Q. Polypodium Leucotomos Extract (PLE): New Study Gives Evidence-Based Insight into Ain't Nothing Like the Real Thing. *J. Clin. Aesthet. Dermatol.* **2019**, *12*, 45. [PubMed]
6. Parrado, C.; Nicolas, J.; Juarranz, A.; Gonzalez, S. The Role of the Aqueous Extract Polypodium Leucotomos in Photoprotection. *Photochem. Photobiol. Sci.* **2020**, *19*, 831–843. [CrossRef]
7. Pourang, A.; Dourra, M.; Ezekwe, N.; Kohli, I.; Hamzavi, I.; Lim, H.W. The Potential Effect of Polypodium Leucotomos Extract on Ultraviolet- and Visible Light-Induced Photoaging. *Photochem. Photobiol. Sci.* **2021**, *20*, 1229–1238. [CrossRef]
8. Barry, E.S.; Merkebu, J.; Varpio, L. State-of-the-Art Literature Review Methodology: A Six-Step Approach for Knowledge Synthesis. *Perspect. Med. Educ.* **2022**, *11*, 281–288. [CrossRef]
9. McCarty, M.F.; Benzvi, C.; Vojdani, A.; Lerner, A. Nutraceutical Strategies for Alleviation of UVB Phototoxicity. *Exp. Dermatol.* **2023**, *6*, 722–730. [CrossRef]
10. Delgado-Wicke, P.; Rodríguez-Luna, A.; Ikeyama, Y.; Honma, Y.; Kume, T.; Gutierrez, M.; Lorrio, S.; Juarranz, Á.; González, S. Fernblock® Upregulates NRF2 Antioxidant Pathway and Protects Keratinocytes from PM2.5-Induced Xenotoxic Stress. *Oxid. Med. Cell. Longev.* **2020**, *2020*, 2908108. [CrossRef]
11. Jańczyk, A.; Garcia-Lopez, M.A.; Fernandez-Peñas, P.; Alonso-Lebrero, J.L.; Benedicto, I.; López-Cabrera, M.; Gonzalez, S. A Polypodium Leucotomos Extract Inhibits Solar-Simulated Radiation-Induced TNF-α and INOS Expression, Transcriptional Activation and Apoptosis. *Exp. Dermatol.* **2007**, *16*, 823–829. [CrossRef] [PubMed]
12. Zattra, E.; Coleman, C.; Arad, S.; Helms, E.; Levine, D.; Bord, E.; Guillaume, A.; El-Hajahmad, M.; Zwart, E.; Van Steeg, H.; et al. Polypodium Leucotomos Extract Decreases UV-Induced Cox-2 Expression and Inflammation, Enhances DNA Repair, and Decreases Mutagenesis in Hairless Mice. *Am. J. Pathol.* **2009**, *175*, 1952–1961. [CrossRef] [PubMed]
13. Calzari, P.; Vaienti, S.; Nazzaro, G. Uses of Polypodium Leucotomos Extract in Oncodermatology. *J. Clin. Med.* **2023**, *12*, 673. [CrossRef] [PubMed]
14. Villa, A.; Viera, M.H.; Amini, S.; Huo, R.; Perez, O.; Ruiz, P.; Amador, A.; Elgart, G.; Berman, B. Decrease of Ultraviolet A Light-Induced "Common Deletion" in Healthy Volunteers after Oral Polypodium Leucotomos Extract Supplement in a Randomized Clinical Trial. *J. Am. Acad. Dermatol.* **2010**, *62*, 511–513. [CrossRef] [PubMed]
15. Rodríguez-Yanes, E.; Juarranz, Á.; Cuevas, J.; Gonzalez, S.; Mallol, J. Polypodium Leucotomos Decreases UV-Induced Epidermal Cell Proliferation and Enhances P53 Expression and Plasma Antioxidant Capacity in Hairless Mice. *Exp. Dermatol.* **2012**, *21*, 638–640. [CrossRef] [PubMed]
16. Philips, N.; Smith, J.; Keller, T.; Gonzalez, S. Predominant Effects of Polypodium Leucotomos on Membrane Integrity, Lipid Peroxidation, and Expression of Elastin and Matrixmetalloproteinase-1 in Ultraviolet Radiation Exposed Fibroblasts, and Keratinocytes. *J. Dermatol. Sci.* **2003**, *32*, 1–9. [CrossRef]

17. Philips, N.; Conte, J.; Chen, Y.J.; Natrajan, P.; Taw, M.; Keller, T.; Givant, J.; Tuason, M.; Dulaj, L.; Leonardi, D.; et al. Beneficial Regulation of Matrixmetalloproteinases and Their Inhibitors, Fibrillar Collagens and Transforming Growth Factor-β by Polypodium Leucotomos, Directly or in Dermal Fibroblasts, Ultraviolet Radiated Fibroblasts, and Melanoma Cells. *Arch. Dermatol. Res.* **2009**, *301*, 487–495. [CrossRef]
18. Kohli, I.; Shafi, R.; Isedeh, P.; Griffith, J.L.; Al-Jamal, M.S.; Silpa-archa, N.; Jackson, B.; Athar, M.; Kollias, N.; Elmets, C.A.; et al. The Impact of Oral Polypodium Leucotomos Extract on Ultraviolet B Response: A Human Clinical Study. *J. Am. Acad. Dermatol.* **2017**, *77*, 33–41.e1. [CrossRef]
19. Middelkamp-Hup, M.A.; Pathak, M.A.; Parrado, C.; Goukassian, D.; Rius-Díaz, F.; Mihm, M.C.; Fitzpatrick, T.B.; González, S. Oral Polypodium Leucotomos Extract Decreases Ultraviolet-Induced Damage of Human Skin. *J. Am. Acad. Dermatol.* **2004**, *51*, 910–918. [CrossRef]
20. Auriemma, M.; Di Nicola, M.; Gonzalez, S.; Piaserico, S.; Capo, A.; Amerio, P. Polypodium Leucotomos Supplementation in the Treatment of Scalp Actinic Keratosis. *Dermatol. Surg.* **2015**, *41*, 898–902. [CrossRef]
21. De Unamuno Bustos, B.; Aguilera, N.C.; Azorín García, I.; Andrino, A.C.; Ros, M.L.; Rodrigo, R.; Vitale, M.; González, S.; Botella Estrada, R. Long-Term Efficacy of a New Medical Device Containing Fernblock ® and DNA Repair Enzyme Complex in the Treatment and Prevention of Cancerization Field in Patients with Actinic Keratosis. *J. Clin. Exp. Dermatol. Res.* **2019**, *10*, 499. [CrossRef]
22. Lamberti, A.; Cartocci, A.; Donelli, C.; Cortonesi, G.; Trovato, E.; Milani, M.; Rubegni, P.; Cinotti, E. *Prevention Strategies in Patients Affected by Actinic Keratosis of the Head: A 12-Month, Prospective, Assessor-Blinded, Controlled Study with Lesion-Directed Treatment Associated with Medicalized Photoprotection*; Longdom Publishing SL: Barcelona, Spain, 2022; Volume 13, p. 5.
23. Pellacani, G.; Peris, K.; Ciardo, S.; Pezzini, C.; Tambone, S.; Farnetani, F.; Longo, C.; Chello, C.; Gonzalez, S. The Combination of Oral and Topical Photoprotection with a Standardized Polypodium Leucotomos Extract Is Beneficial against Actinic Keratosis. *Photodermatol. Photoimmunol. Photomed.* **2023**, 1–8. [CrossRef]
24. Aguilera, P.; Carrera, C.; Puig-Butille, J.A.; Badenas, C.; Lecha, M.; González, S.; Malvehy, J.; Puig, S. Benefits of Oral Polypodium Leucotomos Extract in MM High-Risk Patients. *J. Eur. Acad. Dermatol. Venereol.* **2013**, *27*, 1095–1100. [CrossRef]
25. Serini, S.; Guarino, R.; Vasconcelos, R.O.; Celleno, L.; Calviello, G. The Combination of Sulforaphane and Fernblock® XP Improves Individual Beneficial Effects in Normal and Neoplastic Human Skin Cell Lines. *Nutrients* **2020**, *12*, 1608. [CrossRef] [PubMed]
26. Portillo-Esnaola, M.; Rodríguez-Luna, A.; Nicolás-Morala, J.; Gallego-Rentero, M.; Villalba, M.; Juarranz, Á.; González, S. Formation of Cyclobutane Pyrimidine Dimers after UVA Exposure (Dark-CPDs) Is Inhibited by an Hydrophilic Extract of Polypodium Leucotomos. *Antioxidants* **2021**, *10*, 1961. [CrossRef] [PubMed]
27. Premi, S.; Wallisch, S.; Mano, C.M.; Weiner, A.B.; Wakamatsu, K.; Bechara, E.J.H.; Halaban, R.; Brash, D.E. Chemiexcitation of Melanin Derivatives Induces DNA Photoproducts Long after UV Exposure. *Science* **2015**, *347*, 842–847. [CrossRef] [PubMed]
28. Gallego-Rentero, M.; Nicolás-Morala, J.; Alonso-Juarranz, M.; Carrasco, E.; Portillo-Esnaola, M.; Rodríguez-Luna, A.; González, S. Protective Effect of the Hydrophilic Extract of Polypodium Leucotomos, Fernblock®, against the Synergistic Action of UVA Radiation and Benzo[a]Pyrene Pollutant. *Antioxidants* **2022**, *11*, 2185. [CrossRef] [PubMed]
29. Schalka, S.; Donato, L.C. Evaluation of Effectiveness of a Sunscreen Containing Polypodium Leucatomos Extract in Reducing the Sun Damage to the Skin. *Surg. Cosmet. Dermatol.* **2019**, *11*, 310–318. [CrossRef]
30. Aguilera, J.; Vicente-Manzanares, M.; de Gálvez, M.V.; Herrera-Ceballos, E.; Rodríguez-Luna, A.; González, S. Booster Effect of a Natural Extract of Polypodium Leucotomos (Fernblock®) That Improves the UV Barrier Function and Immune Protection Capability of Sunscreen Formulations. *Front. Med.* **2021**, *8*, 684665. [CrossRef]
31. González-Morán, A.; Piquero-Casals, J. Use of a Topical Film-Forming Medical Device Containing Repairsomes® in a Patient with Xeroderma Pigmentosum to Avoid Progression to Skin Cancerization. *Clin. Cosmet. Investig. Dermatol.* **2020**, *13*, 677–681. [CrossRef]
32. Lacerda, P.A.; Oenning, L.C.; Bellato, G.C.; Lopes-Santos, L.; de Antunes, N.J.; Mariz, B.A.L.A.; Teixeira, G.; Vasconcelos, R.; Simões, G.F.; de Souza, I.A.; et al. Polypodium Leucotomos Targets Multiple Aspects of Oral Carcinogenesis and It Is a Potential Antitumor Phytotherapy against Tongue Cancer Growth. *Front. Pharmacol.* **2023**, *13*, 1098374. [CrossRef]
33. González, S.; De Gálvez, M.V.; De Troya, M.; Rodríguez-Luna, A.; Calzavara-Pinton, P. Personalized Medical Photoprotection: Determining Optimal Measures for Susceptible Patient Groups. *Open Dermatol. J.* **2023**, *17*, 1–7. [CrossRef]
34. González, S.; Aguilera, J.; Berman, B.; Calzavara-Pinton, P.; Gilaberte, Y.; Goh, C.L.; Lim, H.W.; Schalka, S.; Stengel, F.; Wolf, P.; et al. Expert Recommendations on the Evaluation of Sunscreen Efficacy and the Beneficial Role of Non-Filtering Ingredients. *Front. Med.* **2022**, *9*, 790207. [CrossRef]
35. Perez, M.; Abisaad, J.A.; Rojas, K.D.; Marchetti, M.A.; Jaimes, N. Skin Cancer: Primary, Secondary, and Tertiary Prevention. Part I. *J. Am. Acad. Dermatol.* **2022**, *87*, 255–268. [CrossRef] [PubMed]
36. Passeron, T.; Lim, H.W.; Goh, C.L.; Kang, H.Y.; Ly, F.; Morita, A.; Ocampo Candiani, J.; Puig, S.; Schalka, S.; Wei, L.; et al. Photoprotection According to Skin Phototype and Dermatoses: Practical Recommendations from an Expert Panel. *J. Eur. Acad. Dermatol. Venereol.* **2021**, *35*, 1460–1469. [CrossRef] [PubMed]
37. Sander, M.; Sander, M.; Burbidge, T.; Beecker, J. The Efficacy and Safety of Sunscreen Use for the Prevention of Skin Cancer. *CMAJ* **2020**, *192*, E1802–E1808. [CrossRef]

38. Philips, N.; Richardson, R.; Siomyk, H.; Bynum, D.; Gonzalez, S. "Skin Cancer, Polyphenols, and Oxidative Stress" or Counteraction of Oxidative Stress, Inflammation, Signal Transduction Pathways, and Extracellular Matrix Remodeling That Mediate Skin Carcinogenesis by Polyphenols. In *Cancer*; Academic Press: Cambridge, MA, USA, 2021; pp. 439–450. [CrossRef]
39. Araújo Lacerda, P.; Marinho Ottoni Costa, L.; Cuoghi Bellato, G.; Ayaka Yamashita, M.; Lopes-Santos, L.; Augusto, T.M.; Karla Cervigne, N. Perspectives on Cancer and Phytotherapy: An Overview Focusing on Polypodium Leucotomos Therapeutic Properties. *J. Cancer Prev. Curr. Res.* **2021**, *12*, 9–18. [CrossRef]
40. Subhadarshani, S.; Athar, M.; Elmets, C.A. Photocarcinogenesis. *Curr. Dermatol. Rep.* **2020**, *9*, 189–199. [CrossRef]
41. Bhatia, B.K.; Lim, H.W.; Hamzavi, I.H. *Comprehensive Dermatologic Drug Therapy*; Wolverton, E.S., Jashin, J., Wu, J.J., Eds.; Elsevier: Amsterdam, The Netherlands, 2020; Volume 23, pp. 263–270.
42. Zimmerman, C. Herbs for Low-Risk Skin Cancers and Precancers. *Altern. Complement. Ther.* **2019**, *25*, 163–166. [CrossRef]
43. Pihl, C.; Togsverd-Bo, K.; Andersen, F.; Haedersdal, M.; Bjerring, P.; Lerche, C.M. Keratinocyte Carcinoma and Photoprevention: The Protective Actions of Repurposed Pharmaceuticals, Phytochemicals and Vitamins. *Cancers* **2021**, *13*, 3684. [CrossRef]
44. Piaserico, S.; Mazzetto, R.; Sartor, E.; Bortoletti, C. Combination-Based Strategies for the Treatment of Actinic Keratoses with Photodynamic Therapy: An Evidence-Based Review. *Pharmaceutics* **2022**, *14*, 1726. [CrossRef] [PubMed]
45. Piquero-Casals, J.; Morgado-Carrasco, D.; Gilaberte, Y.; Del Rio, R.; Macaya-Pascual, A.; Granger, C.; López-Estebaranz, J.L. Management Pearls on the Treatment of Actinic Keratoses and Field Cancerization. *Dermatol. Ther.* **2020**, *10*, 903–915. [CrossRef] [PubMed]
46. Calzavara-Pinton, P.; Calzavara-Pinton, I.; Rovati, C.; Rossi, M. Topical Pharmacotherapy for Actinic Keratoses in Older Adults. *Drugs Aging* **2022**, *39*, 143–152. [CrossRef]
47. Leung, A.K.C.; Barankin, B.; Lam, J.M.; Leong, K.F.; Hon, K.L. Xeroderma Pigmentosum: An Updated Review. *Drugs Context* **2022**, *11*, 1–17. [CrossRef] [PubMed]
48. Oakley, A.M.; Badri, T.; Harris, B.W. *Photosensitivity*; StatPearls Publishing: Treasure Island, FL, USA, 2022.
49. Ali, N.; Rosenblum, M.D. Regulatory T Cells in Skin. *Immunology* **2017**, *152*, 372. [CrossRef]
50. Capote, R.; Alonso-Lebrero, J.L.; García, F.; Brieva, A.; Pivel, J.P.; González, S. Polypodium Leucotomos Extract Inhibits Trans-Urocanic Acid Photoisomerization and Photodecomposition. *J. Photochem. Photobiol. B* **2006**, *82*, 173–179. [CrossRef]
51. Mulero, M.; Rodríguez-Yanes, E.; Nogués, M.R.; Giralt, M.; Romeu, M.; González, S.; Mallol, J. Polypodium Leucotomos Extract Inhibits Glutathione Oxidation and Prevents Langerhans Cell Depletion Induced by UVB/UVA Radiation in a Hairless Rat Model. *Exp. Dermatol.* **2008**, *17*, 653–658. [CrossRef]
52. Rodríguez-Yanes, E.; Cuevas, J.; González, S.; Mallol, J. Oral Administration of Polypodium Leucotomos Delays Skin Tumor Development and Increases Epidermal P53 Expression and the Anti-Oxidant Status of UV-Irradiated Hairless Mice. *Exp. Dermatol.* **2014**, *23*, 526–528. [CrossRef]
53. Kadurina, M.; Kazandjieva, J.; Bocheva, G. Immunopathogenesis and Management of Polymorphic Light Eruption. *Dermatol. Ther.* **2021**, *34*, e15167. [CrossRef]
54. George, E.A.; Baranwal, N.; Kang, J.H.; Qureshi, A.A.; Drucker, A.M.; Cho, E. Photosensitizing Medications and Skin Cancer: A Comprehensive Review. *Cancers* **2021**, *13*, 2344. [CrossRef]
55. Valladares Narganes, L.M. Actividad Fotoprotectora Del Extracto Estandarizado de Polypodium Leucotomos, Nicotinamida, Vitamina D y Cinc En La Prevención y La Reducción de Brotes En Los Pacientes Con Erupción Polimorfa Lumínica. *Piel Form. Contin. En Dermatol.* **2022**, *37*, 7–12, ISSN 0213-9251. [CrossRef]
56. Stump, M.; Dhinsa, H.; Powers, J.; Stone, M. Attenuation of Actinic Prurigo Eruptions with Polypodium Leucotomos Supplementation. *Pediatr. Dermatol.* **2022**, *39*, 145–146. [CrossRef] [PubMed]
57. Adler, B.L.; DeLeo, V.A. Actinic Prurigo and Hydroa Vacciniforme. *Curr. Dermatol. Rep.* **2019**, *8*, 63–72. [CrossRef]
58. Photiou, L.; Foley, P.; Ross, G. Solar Urticaria—An Australian Case Series of 83 Patients. *Australas. J. Dermatol.* **2019**, *60*, 110–117. [CrossRef]
59. Caccialanza, M.; Percivalle, S.; Piccinno, R.; Brambilla, R. Photoprotective Activity of Oral Polypodium Leucotomos Extract in 25 Patients with Idiopathic Photodermatoses. *Photodermatol. Photoimmunol. Photomed.* **2007**, *23*, 46–47. [CrossRef]
60. Caccialanza, M.; Recalcati, S.; Piccinno, R. Oral Polypodium Leucotomos Extract Photoprotective Activity in 57 Patients with Idiopathic Photodermatoses. *G. Ital. Dermatol. Venereol.* **2011**, *146*, 85–87.
61. Malara, G.; Verduci, C.; Altomonte, M.; Cuzzola, M.; Trifiro, C.; Politi, C.; Tripepi, G. Thalidomide and Discoid Lupus Erythematosus: Case Series and Review of Literature. *Drugs Context* **2022**, *11*, 2021-9-8. [CrossRef]
62. Segars, K.; McCarver, V.; Miller, R.A. Dermatologic Applications of Polypodium Leucotomos: A Literature Review. *J. Clin. Aesthet. Dermatol.* **2021**, *14*, 50.
63. Breithaupt, A.D.; Jacob, S.E. Subacute Cutaneous Lupus Erythematosus: A Case Report of Polypodium Leucotomos as an Adjuvant Therapy—PubMed. *Cutis* **2012**, *89*, 183–184.
64. High, W.A.; Lori, D. Dermatology Secrets E-Book. Prok-Google Libros. Available online: https://books.google.es/books?hl=es&lr=&id=CJ8FEAAAQBAJ&oi=fnd&pg=PA144&dq=polypodium+leucotomos+photoprotection&ots=k7113VhiFc&sig=1gSDkgnMjeprMY1jmV2b0IB4iWY#v=onepage&q=polypodium%20leucotomos%20photoprotection&f=false (accessed on 4 May 2023).
65. Hölzle, E.; Dawe, R. The Idiopathic Photodermatoses and Skin Testing. *Harper's Textb. Pediatr. Dermatol.* **2019**, 943–956. [CrossRef]

66. Jamgochian, M.; Alamgir, M.; Rao, B. Diet in Dermatology: Review of Diet's Influence on the Conditions of Rosacea, Hidradenitis Suppurativa, Herpes Labialis, and Vitiligo. *Am. J. Lifestyle Med.* **2021**, *17*, 152–160. [CrossRef]
67. Korman, A.M.; Reynolds, K.A.; Nabhan, F.; Konda, B.; Shah, M.H.; Kaffenberger, B.H. Vandetanib-Induced Phototoxic Drug Eruption Treated with Polypodium Leucotomos Extract: A Case Report and Review of the Literature. *J. Clin. Aesthet. Dermatol.* **2019**, *12*, 35. [PubMed]
68. Artz, C.E.; Farmer, C.M.; Lim, H.W. Polymorphous Light Eruption: A Review. *Curr. Dermatol. Rep.* **2019**, *8*, 110–116. [CrossRef]
69. Lubov, J.E.; Jamison, A.S.; Baltich Nelson, B.; Amudzi, A.A.; Haas, K.N.; Richmond, J.M. Medicinal Plant Extracts and Natural Compounds for the Treatment of Cutaneous Lupus Erythematosus: A Systematic Review. *Front. Pharmacol.* **2022**, *13*, 188. [CrossRef]
70. Yan, D.; Borucki, R.; Sontheimer, R.D.; Werth, V.P. Candidate Drug Replacements for Quinacrine in Cutaneous Lupus Erythematosus. *Lupus Sci. Med.* **2020**, *7*, e000430. [CrossRef] [PubMed]
71. Verma, L.; Pratt, M. A Case Report of Therapeutically Challenging Chronic Actinic Dermatitis. *SAGE Open Med. Case Rep.* **2019**, *7*, 2050313X1984523. [CrossRef]
72. Morgado-Carrasco, D.; Granger, C.; Trullas, C.; Piquero-Casals, J. Impact of Ultraviolet Radiation and Exposome on Rosacea: Key Role of Photoprotection in Optimizing Treatment. *J. Cosmet. Dermatol.* **2021**, *20*, 3415–3421. [CrossRef]
73. Shahbazi, A.; Zargar, S.J.; Aghdami, N.; Habibi, M. The Story of Melanocyte: A Long Way from Bench to Bedside. *Cell Tissue Bank.* **2023**. [CrossRef]
74. Zhu, Y.; Zeng, X.; Ying, J.; Cai, Y.; Qiu, Y.; Xiang, W. Evaluating the Quality of Life among Melasma Patients Using the MELASQoL Scale: A Systematic Review and Meta-Analysis. *PLoS ONE* **2022**, *17*, e0262833. [CrossRef]
75. Pacifico, A.; Damiani, G.; Iacovelli, P.; Conic, R.R.Z.; Gonzalez, S.; Morrone, A. NB-UVB plus Oral Polypodium Leucotomos Extract Display Higher Efficacy than NB-UVB Alone in Patients with Vitiligo. *Dermatol. Ther.* **2021**, *34*, e14776. [CrossRef]
76. Qadir, A.; Ullah, S.N.M.N.; Jahan, S.; Ali, A.; Khan, N. Drug Delivery of Natural Products through Nano-Carriers for Effective Vitiligo Therapy: A Compendia Review. *J. Cosmet. Dermatol.* **2022**, *21*, 5386–5404. [CrossRef] [PubMed]
77. Lin, X.; Meng, X.; Song, Z.; Lin, J. Nuclear Factor Erythroid 2-Related Factor 2 (Nrf2) as a Potential Therapeutic Target for Vitiligo. *Arch. Biochem. Biophys.* **2020**, *696*, 108670. [CrossRef] [PubMed]
78. Kahremany, S.; Hofmann, L.; Gruzman, A.; Dinkova-Kostova, A.T.; Cohen, G. NRF2 in Dermatological Disorders: Pharmacological Activation for Protection against Cutaneous Photodamage and Photodermatosis. *Free Radic. Biol. Med.* **2022**, *188*, 262–276. [CrossRef]
79. Hatem, S.; El Hoffy, N.M.; Elezaby, R.S.; Nasr, M.; Kamel, A.O.; Elkheshen, S.A. Background and Different Treatment Modalities for Melasma: Conventional and Nanotechnology-Based Approaches. *J. Drug Deliv. Sci. Technol.* **2020**, *60*, 101984. [CrossRef]
80. Nautiyal, A.; Wairkar, S. Management of Hyperpigmentation: Current Treatments and Emerging Therapies. *Pigment. Cell Melanoma Res.* **2021**, *34*, 1000–1014. [CrossRef]
81. Mohammad, T.F.; Kohli, I.; Nicholson, C.L.; Do, G.T.; Chaowattanapanit, S.; Nahhas, A.F.; Braunberger, T.L.; Lim, H.W.; Mda, I.H.H. Oral Polypodium Leucotomos Extract and Its Impact on Visible Light-Induced Pigmentation in Human. *Subjects. J. Drugs Dermatol.* **2019**, *18*, 1198–1203.
82. Portillo, M.; Mataix, M.; Alonso-Juarranz, M.; Lorrio, S.; Villalba, M.; Rodríguez-Luna, A.; González, S. The Aqueous Extract of Polypodium Leucotomos (Fernblock®) Regulates Opsin 3 and Prevents Photooxidation of Melanin Precursors on Skin Cells Exposed to Blue Light Emitted from Digital Devices. *Antioxidants* **2021**, *10*, 400. [CrossRef]
83. Regazzetti, C.; Sormani, L.; Debayle, D.; Bernerd, F.; Tulic, M.K.; De Donatis, G.M.; Chignon-Sicard, B.; Rocchi, S.; Passeron, T. Melanocytes Sense Blue Light and Regulate Pigmentation through Opsin-3. *J. Investig. Dermatol.* **2018**, *138*, 171–178. [CrossRef]
84. Zhou, S.; Yamada, R.; Sakamoto, K. Low Energy Multiple Blue Light-Emitting Diode Light Irradiation Promotes Melanin Synthesis and Induces DNA Damage in B16F10 Melanoma Cells. *PLoS ONE* **2023**, *18*, e0281062. [CrossRef]
85. Ceresnie, M.S.; Patel, J.; Lim, H.W.; Kohli, I. The Cutaneous Effects of Blue Light from Electronic Devices: A Systematic Review with Health Hazard Identification. *Photochem. Photobiol. Sci.* **2023**, *22*, 457–464. [CrossRef]
86. de Gálvez, E.N.; Aguilera, J.; Solis, A.; de Gálvez, M.V.; de Andrés, J.R.; Herrera-Ceballos, E.; Gago-Calderon, A. The Potential Role of UV and Blue Light from the Sun, Artificial Lighting, and Electronic Devices in Melanogenesis and Oxidative Stress. *J. Photochem. Photobiol. B* **2022**, *228*, 112405. [CrossRef] [PubMed]
87. Ramser, A.; Casey, A. Blue Light and Skin Health. *J. Drugs Dermatol.* **2022**, *21*, 962–966. [CrossRef]
88. Suitthimeathegorn, O.; Yang, C.; Ma, Y.; Liu, W. Direct and Indirect Effects of Blue Light Exposure on Skin: A Review of Published Literature. *Skin. Pharmacol. Physiol.* **2022**, *35*, 305–318. [CrossRef]
89. Grether-Beck, S.; Felsner, I.; Brenden, H.; Marini, A.; Jaenicke, T.; Aue, N.; Welss, T.; Uthe, I.; Krutmann, J. Air Pollution-Induced Tanning of Human Skin. *Br. J. Dermatol.* **2021**, *185*, 1026–1034. [CrossRef]
90. Ahn, Y.; Lee, E.J.; Luo, E.; Choi, J.; Kim, J.Y.; Kim, S.; Kim, S.H.; Bae, Y.J.; Park, S.; Lee, J.; et al. Particulate Matter Promotes Melanin Production through Endoplasmic Reticulum Stress—Mediated IRE1α Signaling. *J. Investig. Dermatol.* **2022**, *142*, 1425–1434.e6. [CrossRef] [PubMed]
91. Goh, C.-L.; Chuan, S.Y.; Tien, S.; Thng, G.; Vitale, M.A.; Delgado-Rubin, A. Double-Blind, Placebo-Controlled Trial to Evaluate the Effectiveness of Polypodium Leucotomos Extract in the Treatment of Melasma in Asian Skin: A Pilot Study-PubMed. *J. Clin. Aesthet. Dermatol.* **2018**, *11*, 14–19.

92. Piquero-Casals, J.; Granger, C.; Piquero-Casals, V.; Garre, A.; Mir-Bonafé, J.F. A Treatment Combination of Peels, Oral Antioxidants, and Topical Therapy for Refractory Melasma: A Report of 4 Cases. *Clin. Cosmet. Investig. Dermatol.* **2020**, *13*, 209–213. [CrossRef] [PubMed]
93. Farràs, A.; Cásedas, G.; Les, F.; Terrado, E.M.; Mitjans, M.; López, V. Evaluation of Anti-Tyrosinase and Antioxidant Properties of Four Fern Species for Potential Cosmetic Applications. *Forests* **2019**, *10*, 179. [CrossRef]
94. Sinha, S.; Sarkar, R.; Upadhyaya, A. Pigmentary Disorders in Women. In *Skin Diseases in Females*; Springer: Berlin/Heidelberg, Germany, 2022; pp. 181–223. [CrossRef]
95. Bhatia, B.K.; Huggins, R.H.; Tisack, A. Pigmentary Disorders. In *Practical Guide to Dermatology*; Springer: Berlin/Heidelberg, Germany, 2020; pp. 213–222. [CrossRef]
96. Xing, X.; Dan, Y.; Xu, Z.; Xiang, L. Implications of Oxidative Stress in the Pathogenesis and Treatment of Hyperpigmentation Disorders. *Oxid. Med. Cell Longev.* **2022**, *2022*, 7881717. [CrossRef]
97. Kasraee, B. Skin Depigmenting Agents: Where Do We Stand? In *Pigmentation Disorders Etiology and Recent. Advances in Treatments*; Intechopen: London, UK, 2022. [CrossRef]
98. Shah, S.; Shah, R.M.; Patel, S.; Patel, S.; Doshi, S.; Lio, P. Integrative Approaches to Hyperpigmentation Therapy. *J. Intergrative Dermatol.* **2022**.
99. Nahhas, A.F.; Abdel-Malek, Z.A.; Kohli, I.; Braunberger, T.L.; Lim, H.W.; Hamzavi, I.H. The Potential Role of Antioxidants in Mitigating Skin Hyperpigmentation Resulting from Ultraviolet and Visible Light-Induced Oxidative Stress. *Photodermatol. Photoimmunol. Photomed.* **2019**, *35*, 420–428. [CrossRef]
100. Mohiuddin, A.K. Skin Lightening & Management of Hyperpigmentation. *Pharm. Sci. Anal. Res. J.* **2019**, *2019*, 180020.
101. Cohen, L.; Brodsky, M.A.; Zubair, R.; Kohli, I.; Hamzavi, I.H.; Sadeghpour, M. Cutaneous Interaction with Visible Light: What Do We Know. *J. Am. Acad. Dermatol.* **2020**, *20*, 30551-X. [CrossRef] [PubMed]
102. Ko, D.; Wang, R.F.; Ozog, D.; Lim, H.W.; Mohammad, T.F. Disorders of Hyperpigmentation. Part II. Review of Management and Treatment Options for Hyperpigmentation. *J. Am. Acad. Dermatol.* **2023**, *88*, 291–320. [CrossRef] [PubMed]
103. Liu, W.; Krutmann, J.; Tian, Y.; Granger, C.; Piquero-Casals, J.; Trullàs, C.; Passeron, T.; Lim, H.W.; Lai, W. Commentary: Facial Aesthetic Dermatological Procedures and Photoprotection in Chinese Populations. *Dermatol. Ther.* **2023**, *13*, 13–27. [CrossRef]
104. Sowash, M.; Alster, T. Review of Laser Treatments for Post-Inflammatory Hyperpigmentation in Skin of Color. *Am. J. Clin. Dermatol.* **2023**, *24*, 381–396. [CrossRef]
105. Charoo, N.A. Hyperpigmentation: Looking beyond Hydroquinone. *J. Cosmet. Dermatol.* **2022**, *21*, 4133–4145. [CrossRef]
106. Shimshak, S.J.E.; Tolaymat, L.M.; Haga, C.B.; Dawson, N.L.; Gillis, M.S.; Yin, M.; Kirsch, B.; Cooper, M.; Sluzevich, J.C. A Review of Oral Therapies for the Treatment of Skin Hyperpigmentation. *J. Cultan. Med. Surg.* **2021**, *26*, 169–175. [CrossRef]
107. Wang, B.; An, X.; Qu, L.; Wang, F. Review on Oral Plant Extracts in Skin Whitening. *Food Sci. Technol.* **2022**, *42*, e83922. [CrossRef]
108. Garg, S.; Tuknayat, A.; Garg, S.; Tuknayat, A. Tips for Managing Post-Inflammatory Hyperpigmentation of Acne. *Cosmoderma* **2021**, *1*, 28. [CrossRef]
109. Fatima, S.; Braunberger, T.; Mohammad, T.; Kohli, I.; Hamzavi, I. The Role of Sunscreen in Melasma and Postinflammatory Hyperpigmentation. *Indian. J. Dermatol.* **2020**, *65*, 5. [CrossRef]
110. Sarkar, R.; Das, A. Postinflammatory Hyperpigmentation: What We Should Know. *Pigment. Int.* **2019**, *6*, 57.
111. Lyons, A.B.; Zubair, R.; Kohli, I.; Nahhas, A.F.; Braunberger, T.L.; Mokhtari, M.; Ruvolo, E.; Lim, H.W.; Hamzavi, I.H. Mitigating Visible Light and Long Wavelength UVA1-Induced Effects with Topical Antioxidants. *Photochem. Photobiol.* **2022**, *98*, 455–460. [CrossRef] [PubMed]
112. Tan, S.; Aw, D. A Review of Oral Treatments for Melasm. *Hong Kong J. Dermatol. Venereol.* **2021**, *29*, 62–70.
113. Yuniandari, A.A.A.A.N.; Wijayanti, D. Evaluating The Efficacy And Safety Of Tranexamic Acid To Hydroquinone And Triple Combination Cream In The Treatment Of Melasma. *J. Health Sains* **2022**, *3*, 1643–1657.
114. Jung Thapa, R.; Chandra Karki, B.; Chandra Yadav, S. Management of Melasma: Emerging Facts. *J. Tumor Sci. Res.* **2022**, *1*, 1–6. [CrossRef]
115. McKesey, J.; Tovar-Garza, A.; Pandya, A.G. Melasma Treatment: An Evidence-Based Review. *Am. J. Clin. Dermatol.* **2020**, *21*, 173–225. [CrossRef]
116. Grimes, P.E.; Ijaz, S.; Nashawati, R.; Kwak, D. New Oral and Topical Approaches for the Treatment of Melasma. *Int. J. Womens Dermatol.* **2019**, *5*, 30–36. [CrossRef]
117. Babbush, K.M.; Babbush, R.A.; Khachemoune, A. Treatment of Melasma: A Review of Less Commonly Used Antioxidants. *Int. J. Dermatol.* **2021**, *60*, 166–173. [CrossRef]
118. Cassiano, D.P.; Espósito, A.C.C.; da Silva, C.N.; Lima, P.B.; Dias, J.A.F.; Hassun, K.; Miot, L.D.B.; Miot, H.A.; Bagatin, E. Update on Melasma—Part II: Treatment. *Dermatol. Ther.* **2022**, *12*, 1989–2012. [CrossRef]
119. Katoch, S.; Sarkar, R. Approach to a Case of Melasma. *Pigment. Int.* **2021**, *8*, 195.
120. Liu, Y.; Wu, S.; Wu, H.; Liang, X.; Guo, D.; Zhuo, F. Comparison of the Efficacy of Melasma Treatments: A Network Meta-Analysis of Randomized Controlled Trials. *Front. Med.* **2021**, *8*, 1587. [CrossRef] [PubMed]
121. Neagu, N.; Conforti, C.; Agozzino, M.; Marangi, G.F.; Morariu, S.H.; Pellacani, G.; Persichetti, P.; Piccolo, D.; Segreto, F.; Zalaudek, I.; et al. Melasma Treatment: A Systematic Review. *J. Dermatol. Treat.* **2022**, *33*, 1816–1837. [CrossRef] [PubMed]
122. Ayu, S.; Suryantari, A.; Putu, N.; Sweta, T.B.; Veronica, E.; Ngurah, G.; Rai, B.; Hartawan, M.; Luh, N.; Ratih, P.; et al. Systematic Review of Melasma Treatments: Advantages and Disadvantages. *Bali Dermatol. Venereol.* **2020**, *3*, 37–51. [CrossRef]

123. Da Cunha, M.G.; da Silva Urzedo, A.P. Melasma: A Review about Pathophysiology and Treatment. In *Pigmentation Disorders Etiology and Recent. Advances in Treatments*; Intechopen: London, UK, 2022. [CrossRef]
124. Datuin-De Leon, M.S.L.; Handog, E.B. Oral Agents in the Treatment of Melasma. In *Melasma: A Monograph*; Sarkar, R., Ed.; Jaypee Brothers Medical Pub Location: New Delhi, India, 2020; ISBN 978-93-89188-81-3.
125. Brasil dos Santos, J.; Nagem Lopes, L.P.; de Lima, G.G.; Teixeira da Silva, R.; da Silva e Souza Lorca, B.; Miranda Pinheiro, G.; Faria de Freitas, Z.M. Microneedling with Cutaneous Delivery of Topical Agents for the Treatment of Melasma: A Systematic Review. *J. Cosmet. Dermatol.* **2022**, *21*, 5680–5695. [CrossRef]
126. Kamal, K.; Heitmiller, K.; Christman, M. Lasers, Lights, and Compounds for Melasma in Aesthetics. *Clin. Dermatol.* **2022**, *40*, 249–255. [CrossRef] [PubMed]
127. Mahajan, V.K.; Patil, A.; Blicharz, L.; Kassir, M.; Konnikov, N.; Gold, M.H.; Goldman, M.P.; Galadari, H.; Goldust, M. Medical Therapies for Melasma. *J. Cosmet. Dermatol.* **2022**, *21*, 3707–3728. [CrossRef]
128. Babbush, K.M.; Khachemoune, A.; Babbush Bs, K.M.; Babbush, R.A.; Facms, F. The Therapeutic Use of Antioxidants for Melasma. *J. Drugs Dermatol.* **2020**, *19*, 788–792. [CrossRef]
129. Maddaleno, A.S.; Camargo, J.; Mitjans, M.; Vinardell, M.P. Melanogenesis and Melasma Treatment. *Cosmetics* **2021**, *8*, 82. [CrossRef]
130. Guarneri, F.; Bertino, L.; Pioggia, G.; Casciaro, M.; Gangemi, S. Therapies with Antioxidant Potential in Psoriasis, Vitiligo, and Lichen Planus. *Antioxidants* **2021**, *10*, 1087. [CrossRef]
131. Bouceiro Mendes, R.; Alpalhão, M.; Filipe, P. UVB Phototherapy in the Treatment of Vitiligo: State of the Art and Clinical Perspectives. *Photodermatol. Photoimmunol. Photomed.* **2022**, *38*, 215–223. [CrossRef]
132. Wang, E.; Rodrigues, M.; Dermatology, C. An Update and Review of Narrowband Ultraviolet B Phototherapy for Vitiligo. *Dermatol. Rev.* **2022**, *3*, 326–335. [CrossRef]
133. Sun, M.C.; Xu, X.L.; Lou, X.F.; Du, Y.Z. Recent Progress and Future Directions: The Nano-Drug Delivery System for the Treatment of Vitiligo. *Int. J. Nanomed.* **2020**, *15*, 3267–3279. [CrossRef]
134. Leung, A.K.C.; Lam, J.M.; Leong, K.F.; Hon, K.L. Vitiligo: An Updated Narrative Review. *Curr. Pediatr. Rev.* **2020**, *17*, 76–91. [CrossRef] [PubMed]
135. Ali, S.A.; Parveen, N.; Ali, A.S. Promoting Melanocyte Regeneration Using Different Plants and Their Constituents. In *Cancer Therapy*; Bentham Science: Sharjah, United Arab Emirates, 2019; Volume 3, pp. 247–276. [CrossRef]
136. Taïeb, A.; Picardo, M. Management Overview. In *Vitiligo*; Springer: Cham, Switzerland, 2019; pp. 345–351.
137. Picardo, M.; Taïeb, A. *Vitiligo*; Picardo, M., Taïeb, A., Eds.; Springer International Publishing: Cham, Switzerland, 2019; ISBN 978-3-319-62958-2.
138. Böhm, M.; Schunter, J.A.; Fritz, K.; Salavastru, C.; Dargatz, S.; Augustin, M.; Tanew, A. S1 Guideline: Diagnosis and Therapy of Vitiligo. *J. Dtsch. Dermatol. Ges.* **2022**, *20*, 365–378. [CrossRef] [PubMed]
139. Maghfour, J.; Hamzavi, I.H.; Mohammad, T.F. An Updated Review on Systemic and Targeted Therapies for Vitiligo. *Dermatol. Rev.* **2022**, *3*, 313–325. [CrossRef]
140. Dutta, R.; Kumar, T.; Ingole, N. Diet and Vitiligo: The Story So Far. *Cureus* **2022**, *14*, e28516. [CrossRef]
141. Arora, P. Nutraceuticals in Vitiligo: Not Just "Designer" Foods. *Pigment. Int.* **2022**, *9*, 147. [CrossRef]
142. Lotti, T.; Agarwal, K.; Podder, I.; Satolli, F.; Kassir, M.; Schwartz, R.A.; Wollina, U.; Grabbe, S.; Navarini, A.A.; Mueller, S.M.; et al. Safety of the Current Drug Treatments for Vitiligo. *Expert Opin. Drug Saf.* **2020**, *19*, 499–511. [CrossRef]
143. Hussain, I. The Safety of Medicinal Plants Used in the Treatment of Vitiligo and Hypermelanosis: A Systematic Review of Use and Reports of Harm. *Clin. Cosmet. Investig. Dermatol.* **2021**, *14*, 261–284. [CrossRef] [PubMed]
144. Passeron, T.; Krutmann, J.; Andersen, M.L.; Katta, R.; Zouboulis, C.C. Clinical and Biological Impact of the Exposome on the Skin. *J. Eur. Acad. Dermatol. Venereol.* **2020**, *34*, 4–25. [CrossRef] [PubMed]
145. Marrot, L. Pollution and Sun Exposure: A Deleterious Synergy. Mechanisms and Opportunities for Skin Protection. *Curr. Med. Chem.* **2018**, *25*, 5469–5486. [CrossRef] [PubMed]
146. Zamarrón, A.; Lorrio, S.; González, S.; Juarranz, Á. Fernblock Prevents Dermal Cell Damage Induced by Visible and Infrared A Radiation. *Int. J. Mol. Sci.* **2018**, *19*, 2250. [CrossRef] [PubMed]
147. Truchuelo, M.T.; Jimenez, N.; Dias, I.J.; Gallego-Rentero, M.; Alonso-Juarranz, M.; Gonzalez, S. A Pilot Study to Assess the Effects of an Oral Photo Protector of Botanical Origin against Visible and Infrared Radiations in Human Volunteers. *J. Dermatol. Dermatol. Dis.* **2019**, *6*, 1–3.
148. Granger, C.; Aladren, S.; Delgado, J.; Garre, A.; Trullas, C.; Gilaberte, Y. Prospective Evaluation of the Efficacy of a Food Supplement in Increasing Photoprotection and Improving Selective Markers Related to Skin Photo-Ageing. *Dermatol. Ther.* **2020**, *10*, 163–178. [CrossRef]
149. Bay, E.Y.; Topal, I.O. Aging Skin and Anti-Aging Strategies. *Explor. Res. Hypothesis Med.* **2022**. [CrossRef]
150. Guan, L.L.; Lim, H.W.; Mohammad, T.F. Sunscreens and Photoaging: A Review of Current Literature. *Am. J. Clin. Dermatol.* **2021**, *22*, 819. [CrossRef] [PubMed]
151. Furukawa, J.Y.; Martinez, R.M.; Morocho-Jácome, A.L.; Castillo-Gómez, T.S.; Pereda-Contreras, V.J.; Rosado, C.; Velasco, M.V.R.; Baby, A.R. Skin Impacts from Exposure to Ultraviolet, Visible, Infrared, and Artificial Lights—A Review. *J. Cosmet. Laser Ther.* **2021**, *23*, 1–7. [CrossRef]
152. Lee, L.Y.; Liu, S.X. Pathogenesis of Photoaging in Human Dermal Fibroblasts. *Int. J. Dermatol. Venereol.* **2021**, *3*, 37–42. [CrossRef]

153. Tanveer, M.A.; Rashid, H.; Tasduq, S.A. Molecular Basis of Skin Photoaging and Therapeutic Interventions by Plant-Derived Natural Product Ingredients: A Comprehensive Review. *Heliyon* **2023**, *9*, e13580. [CrossRef]
154. Bharadvaja, N.; Gautam, S.; Singh, H. Natural Polyphenols: A Promising Bioactive Compounds for Skin Care and Cosmetics. *Mol. Biol. Rep.* **2023**, *50*, 1817–1828. [CrossRef] [PubMed]
155. Rincón-Valencia, S.; Mejía-Giraldo, J.C.; Puertas-Mejía, M.Á. Algae Metabolites as an Alternative in Prevention and Treatment of Skin Problems Associated with Solar Radiation and Conventional Photo-Protection. *Braz. J. Pharm. Sci.* **2023**, *58*, e201046. [CrossRef]
156. Torres-Contreras, A.M.; Garcia-Baeza, A.; Vidal-Limon, H.R.; Balderas-Renteria, I.; Ramírez-Cabrera, M.A.; Ramirez-Estrada, K. Plant Secondary Metabolites against Skin Photodamage: Mexican Plants, a Potential Source of UV-Radiation Protectant Molecules. *Plants* **2022**, *11*, 220. [CrossRef] [PubMed]
157. González, S.; Parrado, C.; Juarranz, Á. Introduction to Special Issue "Recent Advances in Skin Anti-Aging Agents". *Plast. Aesthet. Res.* **2022**, *9*, 32. [CrossRef]
158. Cândido, T.M.; Ariede, M.B.; Lima, F.V.; de Souza Guedes, L.; Velasco, M.V.R.; Baby, A.R.; Rosado, C. Dietary Supplements and the Skin: Focus on Photoprotection and Antioxidant Activity—A Review. *Nutrients* **2022**, *14*, 1248. [CrossRef]
159. Li, L.; Chong, L.; Huang, T.; Ma, Y.; Li, Y.; Ding, H. Natural Products and Extracts from Plants as Natural UV Filters for Sunscreens: A Review. *Animal Model. Exp. Med.* **2022**, *6*, 183–195. [CrossRef]
160. Song, C.; Lorz, L.R.; Lee, J.; Cho, J.Y. In Vitro Photoprotective, Anti-Inflammatory, Moisturizing, and Antimelanogenic Effects of a Methanolic Extract of Chrysophyllum Lucentifolium Cronquist. *Plants* **2021**, *11*, 94. [CrossRef]
161. Thompson, K.G.; Kim, N. Dietary Supplements in Dermatology: A Review of the Evidence for Zinc, Biotin, Vitamin D, Nicotinamide, and Polypodium. *J. Am. Acad. Dermatol.* **2021**, *84*, 1042–1050. [CrossRef]
162. Torres, A.E.; Luk, K.M.; Lim, H.W. Botanicals for Photoprotection. *Plast. Aesthet. Res.* **2020**, *7*, 57. [CrossRef]
163. Calzavara-Pinton, P.; Calzavara-Pinton, I.; Arisi, M.; Rossi, M.T.; Scapagnini, G.; Davinelli, S.; Venturini, M. Cutaneous Photoprotective Activity of a Short-Term Ingestion of High-Flavanol Cocoa: A Nutritional Intervention Study. *Photochem. Photobiol.* **2019**, *95*, 1029–1034. [CrossRef] [PubMed]
164. Tsuruta, Y.; Katsuyama, Y.; Okano, Y.; Ozawa, T.; Yoshimoto, S.; Ando, H.; Masaki, H.; Ichihashi, M. Possible Involvement of Dermal Fibroblasts in Modulating Nrf2 Signaling in Epidermal Keratinocytes. *Biol. Pharm. Bull.* **2023**, *46*, 725–729. [CrossRef] [PubMed]
165. Tabolacci, E.; Tringali, G.; Nobile, V.; Duca, S.; Pizzoferrato, M.; Bottoni, P.; Maria Elisabetta, C. Rutin Protects Fibroblasts from UVA Radiation through Stimulation of Nrf2 Pathway. *Antioxidants* **2023**, *12*, 820. [CrossRef] [PubMed]
166. Gonzalez, S.; Gilaberte, Y.; Philips, N.; Juarranz, A. Fernblock, a Nutriceutical with Photoprotective Properties and Potential Preventive Agent for Skin Photoaging and Photoinduced Skin Cancers. *Int. J. Mol. Sci.* **2011**, *12*, 8466–8475. [CrossRef] [PubMed]
167. Ahmed, S.M.U.; Luo, L.; Namani, A.; Wang, X.J.; Tang, X. Nrf2 Signaling Pathway: Pivotal Roles in Inflammation. *Biochim. Biophys. Acta Mol. Basis Dis.* **2017**, *1863*, 585–597. [CrossRef]
168. Jeayeng, S.; Wongkajornsilp, A.; Slominski, A.T.; Jirawatnotai, S.; Sampattavanich, S.; Panich, U. Nrf2 in Keratinocytes Modulates UVB-Induced DNA Damage and Apoptosis in Melanocytes through MAPK Signaling. *Free Radic. Biol. Med.* **2017**, *108*, 918. [CrossRef]
169. Tanveer, M.A.; Rashid, H.; Nazir, L.A.; Archoo, S.; Shahid, N.H.; Ragni, G.; Umar, S.A.; Tasduq, S.A. Trigonelline, a Plant Derived Alkaloid Prevents Ultraviolet-B-Induced Oxidative DNA Damage in Primary Human Dermal Fibroblasts and BALB/c Mice via Modulation of Phosphoinositide 3-Kinase-Akt-Nrf2 Signalling Axis. *Exp. Gerontol.* **2023**, *171*, 112028. [CrossRef]
170. Kobaisi, F.; Fayyad, N.; Rezvani, H.R.; Fayyad-Kazan, M.; Sulpice, E.; Badran, B.; Fayyad-Kazan, H.; Gidrol, X.; Rachidi, W. Signaling Pathways, Chemical and Biological Modulators of Nucleotide Excision Repair: The Faithful Shield against UV Genotoxicity. *Oxid. Med. Cell. Longev.* **2019**, *2019*, 4654206. [CrossRef] [PubMed]
171. Bosch, R.; Philips, N.; Suárez-Pérez, J.A.; Juarranz, A.; Devmurari, A.; Chalensouk-Khaosaat, J.; González, S. Mechanisms of Photoaging and Cutaneous Photocarcinogenesis, and Photoprotective Strategies with Phytochemicals. *Antioxidants* **2015**, *4*, 248–268. [CrossRef]
172. Ma, L.; Chen, Y.; Gong, Q.; Cheng, Z.; Ran, C.; Liu, K.; Shi, C. Cold Atmospheric Plasma Alleviates Radiation-Induced Skin Injury by Suppressing Inflammation and Promoting Repair. *Free Radic. Biol. Med.* **2023**, *204*, 184–194. [CrossRef]
173. Rojo De La Vega, M.; Krajisnik, A.; Zhang, D.D.; Wondrak, G.T. Nutrients Targeting NRF2 for Improved Skin Barrier Function and Photoprotection: Focus on the Achiote-Derived Apocarotenoid Bixin. *Nutrients* **2017**, *9*, 1371. [CrossRef]
174. Yokoi, K.; Yasumizu, Y.; Ohkura, N.; Shinzawa, K.; Okuzaki, D.; Shimoda, N.; Ando, H.; Yamada, N.; Fujimoto, M.; Tanemura, A. Increased Anti-Oxidative Action Compensates for Collagen Tissue Degeneration in Vitiligo Dermis. *Pigment. Cell Melanoma Res.* **2023**, 1–10. [CrossRef] [PubMed]

Disclaimer/Publisher's Note: The statements, opinions and data contained in all publications are solely those of the individual author(s) and contributor(s) and not of MDPI and/or the editor(s). MDPI and/or the editor(s) disclaim responsibility for any injury to people or property resulting from any ideas, methods, instructions or products referred to in the content.

Review

Triterpenes as Potential Drug Candidates for Rheumatoid Arthritis Treatment

Célia Faustino , Lídia Pinheiro * and Noélia Duarte *

iMed.ULisboa, Research Institute for Medicines, Faculdade de Farmácia, Universidade de Lisboa, Avenida Prof. Gama Pinto, 1649-003 Lisbon, Portugal; cfaustino@ff.ulisboa.pt
* Correspondence: lpinheiro@ff.ulisboa.pt (L.P.); mduarte@ff.ulisboa.pt (N.D.)

Abstract: Rheumatoid arthritis (RA) is a chronic autoimmune inflammatory disease characterized by joint inflammation, swelling and pain. Although RA mainly affects the joints, the disease can also have systemic implications. The presence of autoantibodies, such as anti-cyclic citrullinated peptide antibodies and rheumatoid factors, is a hallmark of the disease. RA is a significant cause of disability worldwide associated with advancing age, genetic predisposition, infectious agents, obesity and smoking, among other risk factors. Currently, RA treatment depends on anti-inflammatory and disease-modifying anti-rheumatic drugs intended to reduce joint inflammation and chronic pain, preventing or slowing down joint damage and disease progression. However, these drugs are associated with severe side effects upon long-term use, including immunosuppression and development of opportunistic infections. Natural products, namely triterpenes with anti-inflammatory properties, have shown relevant anti-arthritic activity in several animal models of RA without undesirable side effects. Therefore, this review covers the recent studies (2017–2022) on triterpenes as safe and promising drug candidates for the treatment of RA. These bioactive compounds were able to produce a reduction in several RA activity indices and immunological markers. Celastrol, betulinic acid, nimbolide and some ginsenosides stand out as the most relevant drug candidates for RA treatment.

Keywords: rheumatoid arthritis; inflammation; triterpenes; celastrol; betulinic acid; ginsenosides; saponins

Citation: Faustino, C.; Pinheiro, L.; Duarte, N. Triterpenes as Potential Drug Candidates for Rheumatoid Arthritis Treatment. *Life* **2023**, *13*, 1514. https://doi.org/10.3390/life13071514

Academic Editors: Azahara Rodríguez-Luna and Salvador González

Received: 25 May 2023
Revised: 30 June 2023
Accepted: 4 July 2023
Published: 5 July 2023

Copyright: © 2023 by the authors. Licensee MDPI, Basel, Switzerland. This article is an open access article distributed under the terms and conditions of the Creative Commons Attribution (CC BY) license (https://creativecommons.org/licenses/by/4.0/).

1. Introduction

Nature has always been the foundation for the discovery of folk medical treatments and many drugs used in modern medicine. Currently, the use of natural products and natural supplements is progressively increasing, and their scientific validation is a priority to guarantee the safe use of these products. In addition, natural products derived from plants, marine organisms and microorganisms, as well as their synthetic derivatives designed based on their distinctive pharmacophores, play a pivotal role in the process of drug discovery and development. This contribution is reflected by the significant number of drug molecules recently introduced to the market, as extensively emphasized in various reviews [1–4]. Notably, a considerable 41% of small-molecule anti-cancer drugs approved between 1981 and 2019 possess structures derived from natural products (e.g., paclitaxel, vincristine and etoposide). The impact of natural products extends beyond cancer therapeutics and encompasses other therapeutic areas such as cardiovascular diseases (e.g., statins, digoxin and warfarin), multiple sclerosis (e.g., fingolimod), protozoal infections (e.g., quinine and artemisinin) and a plethora of other infectious diseases [1,5].

Earth's biodiversity is still far from being fully explored in terms of discovering new bioactive compounds. The structural diversity of secondary metabolites is a rich biogenetic supply for the discovery of novel drugs when compared to synthetic molecules, offering hit and lead compounds for rational drug design [6]. Among the diverse families of natural products (e.g., terpenoids, steroids, phenolic compounds and alkaloids), triterpenes are an important group of phytochemicals, possessing a wide array of biological effects, which have been extensively documented in the scientific literature [7–11]. Among these effects, the anti-microbial [12,13], anti-tumor [14–16], anti-diabetic [10], anti-cholesterol [17], anti-inflammatory [18] and immunomodulatory [19] activities have gathered considerable attention within the pharmaceutical area [20,21]. Particularly, numerous scientific studies have highlighted the potent anti-inflammatory properties of triterpenes, making them potentially relevant in the treatment of inflammatory conditions such as arthritis and related diseases [22].

Arthritis is an acute or chronic joint disease usually associated with joint stiffness, pain, inflammation, swelling and decreased range of motion [23,24]. There are more than 100 different types of arthritis, the most common being non-inflammatory degenerative arthritis known as osteoarthritis [23,24]. Rheumatoid arthritis (RA) is the most common autoimmune inflammatory type of arthritis. Inflammatory arthritis can also be caused by other factors, such as crystal deposition-induced inflammation (e.g., gout, pseudogout) or infections (e.g., septic arthritis). Inflammatory arthritis has also been associated with other autoimmune connective tissue diseases (e.g., systemic lupus erythematosus) and extra-articular comorbidities [23,24].

RA is an important cause of disability and its prevalence varies globally (Table 1), with higher rates in industrialized countries, which could possibly be explained by the higher exposures to environmental factors. Nevertheless, other risk factors are also considered in the development of RA, such as advancing age, female sex, smoking and stress, among others (Table 2). RA most commonly affects the joints, but it is also considered a systemic disease because it can also affect other organs, such as the cardiovascular or respiratory system (Table 2) [23]. This chronic autoimmune condition represents a substantial health, social and economic burden, resulting in chronic pain and disability, impacting work performance and interfering with daily tasks, decreasing the patient's quality of life and contributing to anxiety and depression [23].

Table 1. Global prevalence, incidence and years lived with disability (YLDs) attributable to RA for men, women and both genders in 2019 with percentage change (numbers in parentheses) between 2010 and 2019. Data from Global Burden of Disease Collaborative Network, 2020 [25].

Gender	Prevalence Cases (Millions)	Incidence Cases (Millions)	YLDs Counts (Millions)
Male	5.39 (23.9%)	0.330 (20.1%)	0.716 (23.3%)
Female	13.2 (21.7%)	0.744 (17.3%)	1.72 (21.2%)
Overall	18.6 (22.3%)	1.07 (18.1%)	2.43 (21.8%)

Table 2. Summary of symptoms, risk factors and common comorbidities of RA.

Commonly affected joints	Hands, wrists, knees and feet, typically in symmetrical pattern.
Symptoms	Pain, tenderness, early morning stiffness lasting 30 min or longer and swelling involving multiple (peripheral) joints bilaterally, low-grade fever, fatigue and weight loss.
Main risk factors	Advancing age, female sex, positive family history/genetics, overweight/obesity, smoking, particulate matter exposure, infectious agents, microbiome dysbiosis, stress and pro-inflammatory diet (rich in fried foods, processed foods, refined carbohydrates, sodas and red meat).
Common comorbidities	Cardiovascular disease, lymphoma, interstitial lung disease, pulmonary fibrosis, vasculitis, metabolic syndrome, type 2 diabetes, atherosclerosis, osteoporosis, anemia, dry keratoconjunctivitis and depression.

Currently, RA treatment is based on anti-inflammatory drugs and disease-modifying anti-rheumatic drugs, aiming at reducing joint inflammation and pain, protecting joints and other tissues from permanent damage and slowing the progression of RA. The sustained use of these drugs is associated with severe side effects such as stomach upset, heartburn, internal bleeding, osteoporosis, adrenal suppression or development of opportunistic infections. Furthermore, some drugs are very expensive and non-effective in a percentage of RA patients [26,27]. Therefore, the discovery of new drugs with fewer side effects is essential and should embrace several approaches, including the study of natural products and/or their synthetic derivatives. In recent years, several reviews have reported the anti-RA effects of natural compounds and herbal drugs [28–32]. However, the information is scattered amongst the diverse compound families and plant sources. As far as we know, a comprehensive review gathering the most recent studies on triterpenoids with RA-related effects is still missing. This review covers and discusses the latest results on triterpenes, natural products with anti-inflammatory properties, which have been shown to be effective against RA both *in vitro* and *in vivo* in several animal models. For a better comprehension several aspects of RA will firstly be addressed, including the etiology, pathogenesis, current treatment and a summary of the different animal models used in the *in vivo* studies.

2. Materials and Methods

The literature search was carried out during January 2023 using PubMed, Web of Science and ScienceDirect, and an appropriate combination of keywords and truncations adapted for each database was used (for example, combinations of triterpenes with arthritis, rheumatoid arthritis, inflammation and treatment). Only peer-reviewed research articles in the English language and published in a six-year timespan (2017–2022) were considered. The studies were individually screened by the authors based on quality, accuracy and relevance to the aim of the review. Mendeley reference manager software (2020) was used to manage the references and eliminate duplicates.

3. Rheumatoid Arthritis

RA is a systemic autoimmune and chronic inflammatory disease that primarily affects the joints, causing inflammation and swelling of the synovium with subsequent destruction of articular structures, pain and disability [24,33]. Typically, RA symmetrically affects small peripheral joints (hands, wrists and feet) but may progress to involve proximal joints if not treated [24,33]. The acute-phase response to inflammation is signaled by raised serum levels of C-reactive protein and increased erythrocyte sedimentation rate, which are relevant disease assessment biomarkers. Systemic inflammation associated with RA is responsible for extra-articular comorbidities, including cardiovascular disease, lymphoma, interstitial lung disease, pulmonary fibrosis, vasculitis, metabolic syndrome, type 2 diabetes, atherosclerosis, osteoporosis, anemia, dry keratoconjunctivitis and depression, resulting in increased morbidity and mortality in RA patients [24,33,34].

The presence of autoantibodies against post-translational modified proteins, namely anti-citrullinated protein antibodies (ACPAs), usually measured as anti-cyclic citrullinated peptide antibodies, is a hallmark of the disease, along with less specific autoantibodies that bind the Fc region of immunoglobulin G (IgG), known as rheumatoid factors (RFs), of various isotypes (e.g., IgM, IgG and IgA) [33,35,36]. These antibodies can be found in 50–70% of RA patients [33,37] and are currently used as biomarkers for diagnostic purposes. Based on the presence or absence of these antibodies in serum, RA can be subdivided in seropositive or seronegative, respectively [33]. Furthermore, RF is a predictive factor for occurrence of rheumatoid nodules, which are the most common extra-articular feature of RA [34]. The presence of this autoantibody has been detected in approximately 90% of RA patients with nodular disease [34]. Autoantibodies can already be detected decades before disease onset [37] and seropositivity is associated with a more aggressive RA phenotype and increased mortality [36,37].

3.1. Etiology

RA prevalence increases with population aging, peaking in the 60–64 and 65–69 age groups for women and men, respectively, according to 2019 data [38]. Women are 2–3 times more likely to develop RA than men (Table 1). Sex hormones may play a role in disease development since susceptibility to RA increases in post-menopausal women while breastfeeding has been associated with a decreased risk of developing RA [39].

Although the etiology of RA is still unknown, disease onset and progression are likely the result of an interplay between (epi)genetic and environmental factors and the presence or absence of autoantibodies. The heritability of RA is around 50% for seropositive RA and about 20% for seronegative RA [40]. Genetic predisposition for developing RA has been mainly associated with human leukocyte antigen (HLA) class II genotypes, namely HLA-DRB1 alleles of the major histocompatibility complex (MHC), which share a conserved amino acid sequence in their peptide-binding groove, known as the "shared epitope" [41,42]. Shared epitope-positive HLA-DRB1 alleles are associated with ACPA production and an increased risk of developing severe seropositive RA [36,41,42]. Several non-HLA-related genetic associations in RA have also been detected, such as polymorphisms in PTPN22, a shared autoimmunity gene also associated with systemic lupus erythematosus, type 1 diabetes mellitus, juvenile idiopathic arthritis and vasculitides involved in the regulation of both T cells and B cells, which is linked to an increased risk of severe seropositive RA, especially in Caucasians and Africans [43]. Similarly, single-nucleotide polymorphisms in the TNFAIP3 gene locus are related to both inflammatory and autoimmune diseases and have been associated with RA susceptibility [41]. TNFAIP3 encodes the (de)ubiquitinating enzyme A20 that inhibits tumor necrosis factor (TNF)-induced activation of nuclear factor kappa-B (NF-κB), and TNFAIP3 gene-deficient mice develop spontaneous arthritis [41]. Epigenetic factors are also relevant contributors to the disease pathogenesis, for instance, the unique DNA methylome pattern of RA fibroblast-like synoviocytes (FLSs) is different from that of osteoarthritic FLSs, and this persistent differential methylation contributes to the aggressive proliferative phenotype of RA FLSs [44].

Smoking, fine particulate matter exposure and periodontal disease are known environmental risk factors for developing RA [24,33,45]. Lung exposure to smoke, silica dust and other particulate air pollutants can induce the expression of calcium-dependent peptidyl-arginine deiminases (PADs), which convert arginine to citrulline, thus increasing protein citrullination and triggering ACPA production in genetically susceptible individuals [42,45,46]. Similarly, aberrant citrullination of endogenous peptides by Porphyromonas gingivalis PAD, a major causative agent of periodontitis, may be involved in breach of tolerance to citrullinated proteins in RA [47,48]. Other infectious agents, such as mycobacteria or Epstein–Barr virus, can trigger RA via molecular mimicry [33,42]. Recently, gut microbiome dysbiosis has been implicated in early RA [49], corroborating data from animal models of arthritis [50]. Furthermore, alterations in common oral, gastrointestinal and pulmonary microbial populations have been associated with ACPA status [51].

3.2. Pathogenesis

Both adaptative and innate immune systems are involved in the pathogenesis of RA. A pre-RA phase comprises early generation of ACPAs that bind citrullinated residues on many self-proteins, including collagen type II (CII), vimentin, α-enolase, fibronectin, fibrinogen and histones [35,36,42,47].

Mucosal surfaces, especially the lung, are potential trigger sites [35,36,47], consistent with mucosal microbiota disturbance and smoking as environmental risk factors for developing RA [36,51]. Therefore, a systemic break in tolerance occurs prior to onset of joint pathophysiology. Expansion of T cell-mediated autoimmunity through epitope spreading to additional self-antigens present in joints can then lead to onset of synovitis while formation of immune complexes between ACPAs and citrulline-containing antigens that further bind RF leads to abundant complement activation, thus potentiating the inflammatory response [33,35,36,42].

The primary manifestation of RA is autoimmune-mediated synovitis characterized by large-scale infiltration of leukocytes into the synovium, including autoreactive T cells (especially T helper (Th) cells Th1 and Th17) and B cells, macrophages, mast cells and neutrophils (the latter largely resident in the synovial fluid), accompanied by substantial release of inflammatory mediators, including cytokines, chemokines, eicosanoids, growth factors, vasoactive amines, matrix metalloproteinases (MMPs) and reactive oxygen species (ROS) [35,36,52]. Pro-inflammatory cytokines, particularly interleukin (IL)-6, induce the synthesis of acute-phase proteins (including C-reactive protein) involved in the acute-phase response. IL-6 as well as TNF-α, IL-17, IL-1 and transforming growth factor beta (TGF-β) can also induce osteoclastogenesis by enhancing the expression of receptor activator of nuclear factor kappa-B ligand (RANKL) in osteoblasts, FLSs, activated T cells and mature B cells. Binding of RANKL to its receptor, RANK, on monocytes and macrophages triggers differentiation to bone-resorbing osteoclasts, leading to bone erosion observed in RA [35,36,52].

In the inflamed RA synovium, activated FLSs adopt an apoptosis-resistant and aggressive proliferative phenotype leading to pannus formation with production of pro-inflammatory cytokines (e.g., TNF-α, IL-6 and IL-1β) and chemokines (e.g., IL-8, CCL2, CCL5 and CXCL10), extracellular matrix-degrading enzymes and pro-angiogenic factors resulting in chondrocyte apoptosis, cartilage matrix degradation and activation of endothelial cells [35,36,52]. Vascular endothelial growth factor (VEGF)-mediated angiogenesis and increase in vascular permeability promote further infiltration of leukocytes into the hypoxic synovium milieu, leading to synovial hyperplasia, joint swelling and systemic chronic pain [36]. Moreover, the invasive RA FLSs can migrate and infiltrate distant joints, resulting in symmetrical joint damage typical in RA [52].

Immune cells including $CD4^+$ T, $CD8^+$ T, NK and B cells are also involved in the complex pathogenesis of RA. Among them, $CD4^+$ T cells stand out in relieving the pathological process of the disease. $CD4^+CD25^+$ regulatory T (T_{reg}) cells, which have immunosuppressive functions, are part of the $CD4^+$ T cell subset [53]. Expression of the specific nuclear transcription factor Foxp3 in $CD4^+CD25^+$ T_{reg} cells is a pivotal element for preserving inhibitory activity [53].

3.3. Treatment

Nowadays, RA can be effectively managed with different medication modalities. In addition, the adoption of a healthy lifestyle, including regular exercise, no smoking, reduced stress and an anti-inflammatory diet, such as the Mediterranean diet, rich in fruits, vegetables, whole grains, nuts, fish and olive oil, can also help in the treatment of disease [45]. Early diagnosis and treatment are essential to achieve remission or low disease activity. Initial treatment involves the use of disease-modifying anti-rheumatic drugs (DMARDs) able to delay or even halt disease progression, preventing radiographic progression and improving function and quality of life [26,33]. These are often used in combination with non-steroidal anti-inflammatory drugs (NSAIDs) or low-dose glucocorticoids (e.g., prednisone, prednisolone, dexamethasone, betamethasone and triamcinolone) to reduce pain and inflammation while the disease remains active. Glucocorticoids, although providing rapid symptomatic relief and useful in episodes of high disease activity ("flares"), are associated with serious long-term adverse events, including adrenal suppression [26,33]. DMARDs are immunosuppressive and immunomodulatory agents classified as either synthetic or biologic (Table 3). The former includes conventional synthetic DMARDs, like methotrexate (MTX), and targeted synthetic DMARDs, which are Janus kinase (JAK) inhibitors, for oral administration [26]. The Janus kinase inhibitors (JAKis) are orally available tsDMARDs that antagonize the activation of the intracellular cytoplasmatic enzymes JAKs, which control various biological functions, such as triggering the inflammatory cascade in immune cells. As a new type of DMARD, the JAKi targets a specific and critical pathway regarding the pattern of RA development and progress [54].

Table 3. Major classes of disease-modifying anti-rheumatic drugs (DMARDs) currently in the market.

Synthetic DMARDs		Biologic DMARDs			
Conventional DMARDs	Targeted synthetic DMARDs (JAK inhibitors)	TNF inhibitors	IL-6R inhibitors	T cell co-stimulation inhibitors	B cell-depleting agents
Methotrexate leflunomide sulfasalazine hydroxy-chloroquine	Tofacitinib, baricitinib, filgotinib, upadacitinib, peficitinib	Etanercept, infliximab, adalimumab and biosimilars golimumab, certolizumab, pegol	Tocilizumab, sarilumab	Abatacept	Rituximab and biosimilars

IL-6R, interleukin-6 receptor; JAK, Janus kinase; TNF, tumor necrosis factor.

MTX is the most often used DMARD due to its efficacy to achieve remission or slow disease activity, and MTX plus a glucocorticoid is recommended as first-line RA therapy [26,33]. Insufficient response to this treatment within 3–6 months requires addition of a targeted synthetic DMARD or a biologic one [26]. Although JAKis have the maximum therapeutic effect when administered concomitantly with MTX, in patients where csDMARDs cannot be used as co-medication or in cases of poor prognostic condition, JAKis have shown a marked efficacy when used as monotherapy [55]. The recent trend is to start JAKis combined with MTX, followed by MTX reduction/discontinuation after achieving a sufficient therapeutic effect [55]. Following current therapeutic guidelines in the 2020 updated European League Against Rheumatism (EULAR) and the 2015 American College of Rheumatology guidelines, the combination of bDMARDs and tsDMARDs with conventional synthetic DMARDs (csDMARDs) is the most effective therapeutic approach for RA [55].

Biologic DMARDs are highly specialized genetically engineered proteins for parenteral administration that target specific soluble inflammatory mediators, immune cells or signaling pathways involved in RA pathogenesis [26,33,47]. These biological response modifiers include TNF inhibitors, IL-6 receptor (IL-6R) inhibitors, T cell co-stimulation inhibitors (abatacept, binds to CD80/CD86 on antigen-presenting cells, modulating T cell activation) and B cell-depleting agents (rituximab, anti-CD20 monoclonal antibody), being an effective second-line treatment for pathogenesis [26,33,47]. IL-1 inhibitors, such as the IL-1 receptor antagonist (IL-1Ra) anakinra, have also been licensed for RA treatment. However, lower efficacy compared with other biologic DMARDs and a dose schedule requiring daily subcutaneous injections do not recommend its use [26,33,47].

DMARDs are associated with several adverse events, including malignancies, major adverse cardiovascular events, venous thromboembolism and increased risk of serious infections (more frequent with biologics), including tuberculosis reactivation [26,33]. Safety aspects, patient clinical history and cost of therapy must be considered in DMARD selection, though the introduction of biosimilar DMARDs contributed to a reduction in the price of biologics [26]. DMARDs may be tapered (by reducing the dose or increasing the interval between doses) during sustained remission but should not be stopped [26,33].

4. Animal Models of Rheumatoid Arthritis

Animal models of RA are valuable resources for studying the disease pathogenesis and testing novel anti-RA drug candidates. Both spontaneous and induced experimental models have been used in RA research. Spontaneous RA can be modeled using genetically modified mice, such as human TNF transgenic mice, IL-1Ra knockout mice, double transgenic K/BxN (showing cross-reactive autoantibodies against glucose-6-phosphate isomerase) and SKG transgenic mice [56,57]. The latter develop T cell-mediated chronic and progressive autoimmune polyarthritis, spontaneously and upon stimulation with intraperitoneal zymosan injection [56,57].

Antibodies, antigens and adjuvants are usually used to induce RA in animal models [57]. The first established animal model of RA was adjuvant arthritis (AA) induced in rats by a single subcutaneous injection of complete Freund's adjuvant (CFA), consisting of a suspension of heat-killed *Mycobacterium tuberculosis* in mineral oil injected into the rat's hindfoot or tail root [56,57]. CFA induces polyarthritis 10–45 days after immunization due to T cell response to the mycobacterial heat shock protein Hsp65. Additionally, some adjuvants without immunogenic properties can also induce arthritis in susceptible animal strains, including incomplete Freund's adjuvant (IFA), which lacks mycobacteria [56,57].

In the antigen-induced arthritis (AIA) model, an antigen, such as ovalbumin or bovine serum albumin, is intra-articularly injected into the knee joint of animals (mice, rats or rabbits) after previous sensitization by subcutaneous injection of the protein emulsified in CFA [56,57]. Boosting of the immune response is achieved by concomitant intraperitoneal administration of heat-inactivated *Bordetella pertussis*. AIA is a T cell-dependent monoarthritis model and T cell-mediated flares can be induced by local or systemic rechallenge with low-dose antigen [57]. Modified antigens, e.g., methylated proteins, are used to induce chronic arthritis [56,57].

Collagen-induced arthritis (CIA) is the gold standard *in vivo* model of RA, mainly characterized by breach of tolerance and production of autoantibodies against self-collagen, resembling human RA [56–58]. Typically, susceptible mice strains are immunized with bovine, murine or chicken CII emulsified in CFA and injected intradermally into the mouse's tail [58]. Rats are generally susceptible to adjuvant-induced arthritis, after being immunized with an emulsion of CII in IFA subcutaneously injected at the base of the tail [58]. The development of CIA is associated with both B cell and T cell responses with production of anti-CII antibodies and collagen-specific T cells [56]. A booster immunization with an emulsion of CII in IFA is frequently applied following primary immunization (on the 14th or 21st day for mice and the 7th day for rats) to ensure high CIA incidence [58]. Clinical signs of polyarthritis appear 21–28 days (mice) or 2–3 weeks (rats) after the first immunization, depending on the strain [58]. This model has also been expanded to non-human primates [57].

On the other hand, in the collagen antibody-induced arthritis (CAIA) model, a simple mouse model of RA, arthritis is induced by tail vein administration of a cocktail of anti-CII monoclonal antibodies, usually followed by intraperitoneal injection of lipopolysaccharide (LPS) to enhance the incidence and severity of the disease [57]. The CAIA model has several advantages over the classic CIA model, such as rapid disease onset (24–48 h after LPS injection), synchronicity and the capacity to use genetically modified mice, including gene knockout and transgenic mice [57].

Other experimentally induced inflammatory models of RA include streptococcal cell wall-, proteoglycan- and zymosan-induced arthritis. A single intraperitoneal injection of streptococcal cell wall peptidoglycan–polysaccharide polymers induces a cycle of exacerbation and remission of inflammatory arthritis in the peripheral joints of rodents [56,57]. Mice immunized with intraperitoneal injection of human proteoglycans isolated from cartilage of RA patients submitted to joint replacement surgery develop autoantibodies and inflammatory polyarthritis [57]. Intra-articular injection of zymosan, a polysaccharide from the cell wall of *Saccharomyces cerevisiae*, induces chronic proliferative inflammatory monoarthritis following complement activation via the alternative pathway [56,57]. Although none of the developed experimental models can perfectly reproduce the pathophysiology of human RA, they are useful tools for identification of new targets and development of novel therapies, as exemplified by cytokine inhibitors [56].

5. Triterpenes and Some Biosynthetic Considerations

Triterpenes are a large and structurally diverse group of natural compounds, widely distributed through the plant kingdom [59]. They can be classified as primary metabolites, e.g., phytosterols that are structural constituents of the cell membranes and ubiquitous in all plant organisms, and secondary metabolites that are generally restricted to some plant families and genera [21,60]. According to the isoprene biogenetic rule, triterpenes derive from an all-*trans* squalene C30 precursor [21]. Squalene is derived from two farnesyl diphosphate units (C15) by a tail-to-tail coupling catalyzed by squalene synthase. Cyclization of squalene proceeds in the vast majority of cases, by its oxidation to squalene 2,3-epoxide catalyzed by squalene epoxidase. The polycyclic structure adopted from squalene depends on the conformation in which the squalene chain can be folded on the oxidosqualene cyclase enzyme surface, into chair or boat conformations, or with a part remaining unfolded. The formation of the polycyclic triterpenic scaffold can be rationalized by a sequence of cyclizations, usually initiated by acid-catalyzed ring opening of the squalene epoxide and through a series of carbocation intermediates in a stepwise sequence, giving rise to more than 200 distinct triterpene skeletons [21,61,62]. A deeper explanation of triterpene biosynthesis is beyond the scope of this work and further details, including genes and enzymes regulating the biosynthetic pathways, can be found in several excellent reviews [20,59–62].

Most triterpenes have tetracyclic (C6-C6-C6-C5; e.g., dammarane, cucurbitane, lanostane and cycloartane types), and pentacyclic (C6-C6-C6-C6-C5 or C6-C6-C6-C6-C6; e.g., oleanane, ursane, lupane, friedelane, hopane and taraxastane types) scaffolds (Figure 1), but acyclic, monocyclic, bicyclic and hexacyclic structures have also been isolated [21]. Triterpenes may have a variety of oxygenated functional groups and unsaturations, giving rise to a high number of structurally diverse compounds. They can also be found in either free or glycosidic form (saponins), where one or more sugar residues are covalently linked to the triterpenic nucleus. Saponins are amphiphilic compounds due to the lipophilic sapogenin and the hydrophilic sugar side chain(s), forming stable soap-like foams in solution [21]. Even though saponins are highly toxic when injected in the bloodstream, causing hemolysis of the red cells by increasing the permeability of the plasma membrane, they are relatively harmless when taken orally. The toxicity is minimized after ingestion by low absorption and by the acid-catalyzed hydrolysis that releases the aglycone and the molecules of sugar [21].

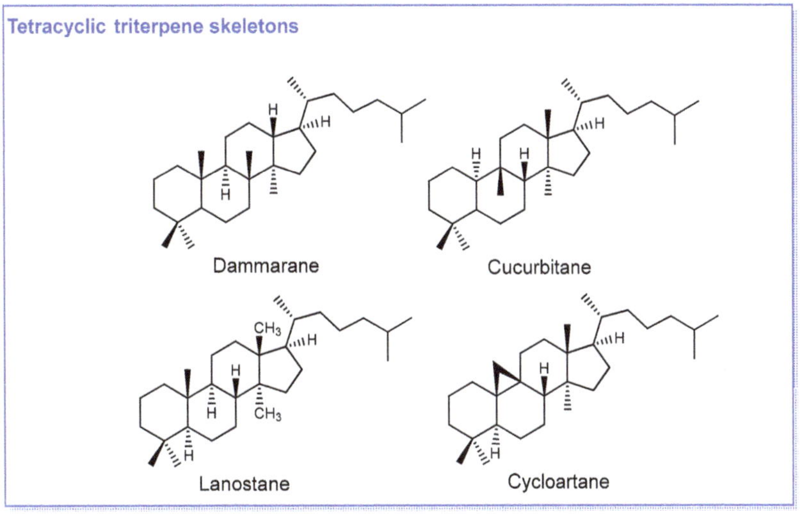

Figure 1. *Cont.*

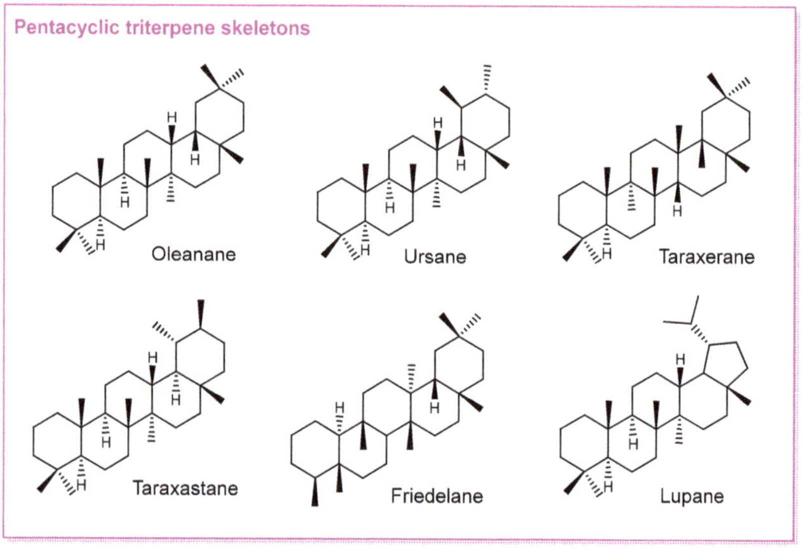

Pentacyclic triterpene skeletons

Oleanane Ursane Taraxerane

Taraxastane Friedelane Lupane

Figure 1. Structures of the main tetracyclic and pentacyclic triterpene skeletons.

6. Triterpenes with Rheumatoid Arthritis-Related Effects

Herein, 36 triterpenic compounds with RA-related *in vitro* and/or *in vivo* effects reported in the literature from 2017 to 2022 are presented (Figures 2–5 and Tables 4–6). The triterpenes are divided into three major classes: pentacyclic triterpenes (Figure 2 and Table 4), tetracyclic and rearranged triterpenes (Figure 3 and Table 5) and triterpenic saponins (Figures 4 and 5 and Table 6).

6.1. Pentacyclic Triterpenes

Celastrol (**1**), also known as tripterine, is a nor-triterpene quinone methide with the friedelane skeleton found in *Tripterygium wilfordii*, known as "Thunder God Vine", a vine commonly grown is southeast China and used in traditional Chinese medicine for the treatment of RA and other autoimmune and inflammatory diseases [63]. Recent studies suggest that NLRP3 inflammasome-induced inflammation is involved in the pathogenesis of RA [47]. Celastrol (**1**) treatment significantly reduced the secretion of IL-1β and IL-18 in the serum of CFA-induced rats and in supernatants of human mononuclear macrophages (THP-1 cells) due to inhibition of the NF-κB pathway and hindering of NLRP3 inflammasome activation [63]. **1** also suppressed ROS production induced by LPS and adenosine triphosphate (ATP) in THP-1 cells [63] and prevented NLRP3 inflammasome activation *in vitro* by inhibiting complex formation between NLRP3 and ASC adaptor protein [64], essential for recruitment of caspase-1 and maturation of IL-1β. **1** also inhibited TNF-α-induced proliferation of FLSs, enhanced autophagosome levels and expression of autophagy-related proteins (LC3, p62 and Beclin-1) and increased the LC3-II/LC3-I ratio [65]. Furthermore, the autophagy inhibitor 3-methyladenine significantly reversed effects of **1** on the expression of autophagy-related proteins [65]. In CIA mice, **1** attenuated disease severity via upregulation of autophagy through inhibition of the PI3K/Akt/mTOR axis [65]. Autophagy dysregulation has been implicated in several autoimmune diseases, including RA. Enhanced autophagy contributes to RA FLS hyperplasia and apoptosis resistance, production of citrullinated peptides, osteoclastogenesis and bone resorption, resulting in severe bone and cartilage damage [66].

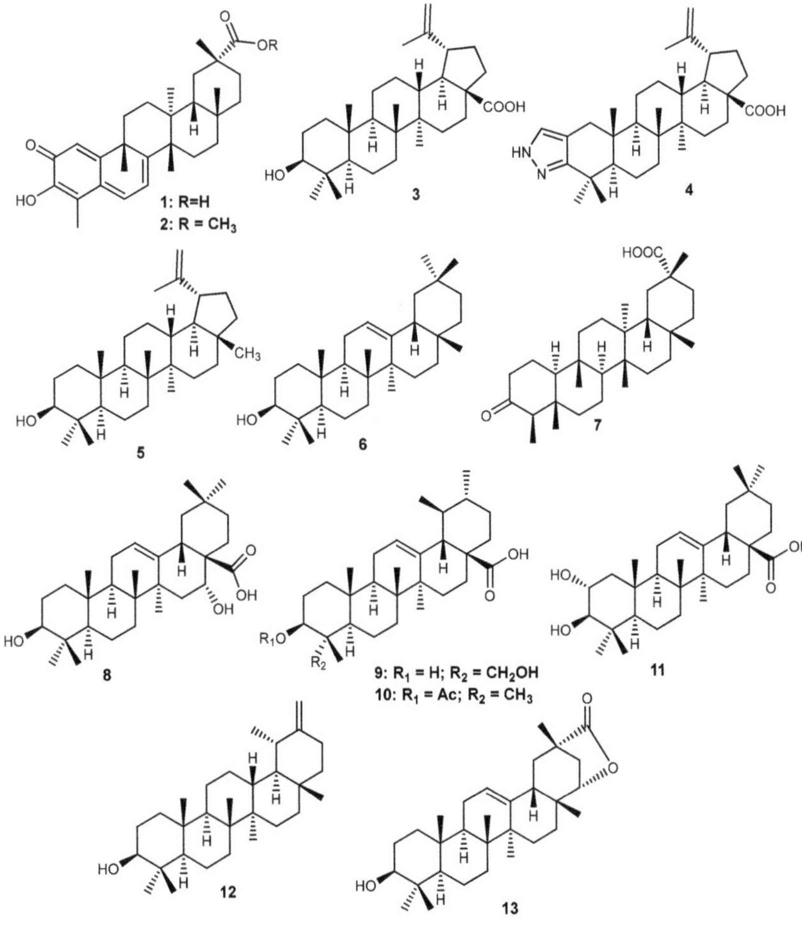

Figure 2. Structures of pentacyclic triterpenes (**1–13**) with activity on RA.

The co-administration of **1** and diclofenac has been routinely used in Chinese medicine for the treatment of RA. In order to shed light on the possible interaction potential of the two drugs, Wang et al. studied the *in vivo* effects of diclofenac on the pharmacokinetic profiles of **1** in rats [67]. When co-administered, several pharmacokinetic parameters significantly change, in particular, the C_{max} and the AUC_0 of **1** decreased from 66.93 ± 10.28 to 41.25 ± 8.06 μg/L and 765.84 ± 163.61 to 451.33 ± 110.88 (μg × h/L), respectively. On the other hand, T_{max} increased from 6.05 ± 1.12 to 7.82 ± 1.15 h, and oral clearance increased from 1.29 ± 0.15 to 2.27 ± 0.31 L/h/kg. Moreover, it was found that the efflux ratio of **1** across the Caco-2 cell model increased when co-administered with diclofenac. In this way, the authors concluded that diclofenac could decrease the exposure of **1** in rats. It was also suggested that this effect could be carried out by decreasing the intestinal absorption of celastrol (**1**) through induction of P-glycoprotein (P-gp) activity [67].

To evaluate the progression of the disease and the response of RA patients to treatment, several biomarkers have been used, such as RF and ACPAs, although they can also be found in other autoimmune diseases. In this way, Dudics et al. studied the microRNA profile of immune (lymphoid) cells of arthritic Lewis rats and celastrol (**1**)-treated arthritic rats, in order to evaluate its ability as a novel RA biomarker [68]. Using combined miRNA–microarray technology and bioinformatics-based analysis, it was found that eight specific miRNAs (miR-22, miR-27a, miR-96, miR-142, miR-223, miR-296, miR-298 and miR-451) and their target genes are crucially involved in functional pathways for RA pathogenesis. In particular, miR-22, miR-27a, miR-96, miR-142, miR-223, miR-296, miR-298 and miR-451 were modulated by celastrol (**1**) treatment. Through the quantitation of these miRNAs in serum samples of control, arthritic and celastrol (**1**)-treated rats, in the peak phase of adjuvant-induced arthritis, it was found that miR-142, miR-155, miR-212 and miR-223 levels were higher in arthritic *vs.* control rats, further validating their value as circulating biomarkers to assess arthritis progression and response to therapy [68].

Fang et al. aimed at studying the effect of **1** on activated RA FLSs obtained from synovial biopsies of human RA patients [69]. Several assays were carried out in order to assess proliferation, invasion and expression of pro-inflammatory cytokines and to screen for differentially expressed genes. The authors found that **1** significantly modulated the RA–FLS activation status by reducing the proliferation and invasion of the cells. Moreover, a change in the expression of several chemokine genes, including CCL2, CXCL10, CXCL12, CCR2 and CXCR4, was also observed. This finding could be useful for therapy since chemokines could be responsible for the arthritis pain by promoting leukocyte infiltration and synoviocyte proliferation and activation. In particular, the release of CCL2 and CXCL12 proteins from RA FLS cells was significantly downregulated by celastrol (**1**) treatment. Celastrol (**1**) treatment also diminished the activation and translocation of NF-κB p65, which is known to participate in the regulation of many cytokines, adhesion molecules, chemokines, receptors and adaptive enzymes in arthritis [69].

Inhibition of oxidative stress underlies the improvement observed in CIA rats treated with **1** (1 mg/kg) in a study carried out by Gao et al. [70]. **1** enhanced the superoxide dismutase activity and significantly inhibited the levels of malondialdehyde, superoxide anions and NADPH oxidase activity [70]. Reduction of arthritis scores and spleen and thymus indexes was also observed, as well as the suppression of serum levels of TNF-α, IL-1β, IL-6 and interferon gamma (IFN-γ), which could be attributed to the downregulation of inflammatory mediators [70].

The mechanistic complexity of **1**, due to its multiple targets, was analyzed by Song et al. by employing a network pharmacological approach. The authors identified probable molecular targets of the compound and the interaction pathways related to their roles, investigating the networks formed by those pathways [71]. Using a web-based bioinformatics application (ingenuity pathway analysis), pathways and networks were built grounded in the functions of the human genes appertaining to RA and the selected potential targets. The networks comprised cell movement, immune cell trafficking, hematological system development and function, inflammatory response, connective tissue disorders, organismal injury and abnormalities and cell-to-cell signaling and interactions. Results indicated that MMP-9, COX-2, c-Myc, TGF-β, c-JUN, JAK-1, JAK-3, IKK-β, SYK, MMP-3, JNK and MEK1 were the direct targets of **1** in RA. Being high-degree nodes in RA-associated networks probably affected by **1**, COX-2, IKK-β, JNK and MEK1 were selected for docking studies [71]. Results of the pathway analysis obtained by Song et al. suggested that **1** can regulate the functions of Th1 and Th2 cells, fibroblasts, macrophages and endothelial cells, which would explain its therapeutic effects against RA [71].

Pristimerin (**2**) is the celastrol methyl ester, a natural triterpene found in plants of the Celastraceae and Hippocrateaceae families. In TNF-α-stimulated human RA FLSs, treatment with **2** decreases cell viability and migration in a dose-dependent manner [72]. According to cell metabolomics analysis, the effects involved phospholipid and fatty acid biosynthesis, glutathione metabolism and amino acid metabolic pathways [72]. *In vivo*, compound **2** ameliorated arthritis symptoms and reduced serum levels of TNF-α and NO and synovial expressions of p-Akt and p-ERK in the CFA-induced arthritis rat model. Network pharmacology analysis showed that the effects were mediated through the MAPK/ERK1/2 and PI3K/Akt pathways and direct binding to TNF-α [72].

The effects of betulinic acid (**3**) on the proliferation, migration and inflammatory response of RA FLSs were studied by Wang and Zhao [73]. Compound **3** inhibited the proliferation, migration and invasion of RA FLSs in a dose- and time-dependent manner at non-cytotoxic concentrations (5–20 μM). It also decreased MMP expression and inhibited the production of TNF-α-induced inflammatory cytokines, namely of IL-6 and IL-8. The PI3K/Akt signaling pathway plays a significant role in regulating inflammation, proliferation and migration of RA FLSs and in signal activation of NF-κB, being highly expressed in the synovial tissues of RA patients. Betulinic acid (**3**) also avoided activation of the Akt/NF-κB pathway and can be considered a potential therapeutic agent for the treatment of RA [73].

Huimin et al. explored the protective effects of **3** on CFA-induced rats, observing the significant inhibition activity of the drug regarding the arthritis index, toe swelling, joint pathology and hemorheology [74]. Serum and synovial levels of Il-6, IL-1β and TNF-α were improved following treatment with **3**. Since Rho and Rho-associated protein kinase (ROCK) control the production of inflammatory cytokines, the anti-inflammatory mechanism of **3** was investigated through the Rho/ROCK/NF-κB activation by treating rats with fasudil (a ROCK inhibitor). Protein levels of RhoA, ROCK1 and ROCK2 were downregulated, leading to the blockage of phosphorylation of IKKα, IKKβ, IκB and NF-κB. The results provided information about the mechanism of compound **3** on RA, which may be related to the downregulation of ROCK/NF-κB signaling pathways [74].

Since RA FLSs display an aggressive phenotype, which is linked to cartilage and bone destruction, Li et al. examined the effects of **3** on the migration and invasion of RA FLSs (prepared from synovial tissue specimens of diagnosed RA patients), seeking a mechanistic understanding of the therapeutic potential of **3** [75]. Treatment with **3** restrained the migratory and invasion capacity of RA FLSs and decreased the formation of actin stress fibers and actin cytoskeleton score [75]. Considering the TNF-α-induced RA FLSs, treatment with **3** led to a significant decrease in the mRNA expression of IL-1β, IL-6, IL-8 and IL-17A, as well as to a decrease in phosphorylated IKK, IκBα and NF-κB and to a reduction of the NF-κB accumulation. These results suggest that the inhibition of NF-κB signaling pathways by **3** causes the inhibition of migration, invasion, actin cytoskeleton reorganization and interleukin expression of RA FLSs [75].

RA is highly associated with increased risk of cardiovascular disease, with RA patients being almost twice as likely to develop heart disease as compared with the general population [76]. Besides the traditional risk factors, chronic inflammation associated with RA appears to promote atherosclerosis, and in both diseases similar pathophysiologic processes are recognized, including increased expression of cellular adhesion molecules, pronounced infiltration by macrophages and Th1 cells, neovascularization and collagen degradation mediated by MMPs [77]. Statins are HMG CoA reductase inhibitors widely used for treatment of hyperlipidemia and prevention of cardiovascular disease, and their anti-inflammatory properties have been proved to be associated with several molecular mechanisms such as suppression of chemokine and pro-inflammatory cytokine synthesis, MMP inhibition, reduced MHC-II expression induced by IFN-γ and reduced expression of CD40 on macrophages and other smooth muscle cells [78,79]. The synergist effect of oral co-administration of **3** (2 mg/kg) and fluvastatin (5 mg/kg) was studied *in vivo* using a CIA rat model, and several physical, morphological and biochemical parameters were collected [80]. Combined treatment with **3** and fluvastatin showed a decrease in the severity of arthritic index values and inhibition of paw edema (89%) after 60 days when compared with the single administration of the drugs (80% and 74%, respectively) or the control group without treatment. A reduction of RF, C-reactive protein, total lipids and ACPAs, as well as an increased activity of catalase, superoxide dismutase and glutathione peroxidase enzymes, in the different tissues was also observed in the rats treated with a combination of both drugs. Moreover, it was also found that the expression of the anti-inflammatory cytokine IL-10 was increased in the co-treated group, while the expression of Toll-like receptor

(TLR) 2 and TLR4, IL-1β, TNF-α, IFN-γ, cell adhesion molecules and nuclear translocation of NF-κB in the aorta decreased, when compared to the single-treated groups [80].

Taking into consideration that betulinic acid (**3**) can be regarded as a lead compound for further development of potential anti-inflammatory agents, several derivatives having heterocyclic rings fused at C-2 and C-3 were synthesized and assayed as inhibitors of osteoclast differentiation and bone resorption [81]. The most potent compound, pyrazole derivative **4**, exhibited potent inhibitory activity on RANKL-induced osteoclast formation (IC_{50} = 0.09 μM), being 200-fold more active than the parent triterpene **3**. In a later work, Chen et al. studied the modulation activity of **4** on T cell differentiation and proliferation and potential anti-rheumatic effects in a CIA mouse model [82]. When compared to the control group that received no treatment, the severity of symptoms was significantly attenuated in treated mice, that showed a mean arthritis score of 2.63 on day 41 (control group: 6.88). Further radiological and histopathological analysis corroborated these findings, since considerably less articular damage was observed and arthritis cartilage destruction and inflammatory cell infiltration were highly decreased, possibly due to inhibition of Th1 and Th17 differentiation, enhanced IL-4, IL-10, IL-13 expression and increased $CD4^+$ $Foxp3^+$ cells [82].

Lupeol (**5**), a lupane triterpenoid with antioxidant and anti-inflammatory properties found in many edible fruits and vegetables, inhibited PI3K/Akt signaling in CIA rats [83]. Lupeol significantly reduced paw edema, reverted the high levels of biochemical markers (RF, C-reactive protein and ceruloplasmin) and pro-inflammatory mediators (TNF-α, IL-6 and PGE2) in the rat serum and enhanced apoptosis by downregulating Bcl-2 protein expression while upregulating Bax, caspase-3 and caspase-9 [83]. However, the overall effects were inferior to those of indomethacin, the NSAID used as positive control [83].

β-amyrin (**6**) and polpunonic acid (**7**) are found in the root bark of *Ziziphus abyssinica* (Hochst Ex A. Rich), a recognized medicinal plant widely distributed in the tropical regions of the world, showing antioxidant, anti-bacterial and anti-plasmodial activities, among others [84]. Henneh et al. were able to isolate them as pure chemical entities and determine their absolute configuration, examining possible therapeutic effects in RA in a CFA-induced arthritis rat model [84]. Compounds **6** and **7** (at equal doses) reversed the changes induced in the RA model (considering body weight, paw thickness, erythema and arthritic index). Histopathological examinations of rat hind paws showed a significant reduction of cartilage erosion and subchondral cyst and Weichselbaum's lacunae formation, with an effect dependent on the type of compound and the doses of administration. There was also evidence of bone remodeling and decreased bone cavitation after treatment with both compounds, most pronounced for **6** [84].

Echinocystic acid (**8**) isolated from the bark of *Albizia julibrissin* Durazz was able to ameliorate arthritic symptoms induced in transgenic SKG mice after a single intraperitoneal injection of zymosan. The treatment with **8** reduced inflammatory cell infiltration, pro-inflammatory cytokine levels, synovial hyperplasia and bone loss in mouse paw tissues [85]. These effects have been attributed to inhibition of both IL-6- and TGF-β-induced Th17 cell differentiation, namely by suppression of phosphorylation of STAT3. In TNF-α-activated human RA FLSs (MH7A cells), administration of **8** reduced both protein and mRNA expression of inflammatory cytokines (IL-6 and IL-1β) by downregulating MAPK and NF-κB signaling pathways [85].

Bone homeostasis depends on the balance between osteoclast-mediated bone resorption and osteoblast-mediated bone formation. Excessive osteoclast activity has been associated with RA, osteoarthritis, osteoporosis and other bone-related diseases [36,52,86]. 23-Hydroxyursolic acid (**9**) isolated from *Viburnum lutescens* was found to inhibit RANKL-induced osteoclastogenesis *in vitro* by decreasing the number of tartrate-resistant acid phosphatase (TRAP)-positive osteoclasts and F-actin ring formation [87]. Actin ring formation is a characteristic marker of bone resorption activity of mature osteoclasts. Compound **9** also inhibited RANKL-induced phosphorylation of ERK and JNK, IκBα degradation, c-Fos expression, activation of the nuclear factor NFATc1 and expression of its target genes [87]. Oral administration of **9** to mice conferred protection against LPS-induced osteoclast formation and bone loss [87].

The study conducted by Lee et al. compared the *in vitro* and *in vivo* effects of ursolic acid-3-acetate (**10**) and dexamethasone, using TNF-α-stimulated human FLSs and a murine model of RA [88]. The treated rats showed a decrease in clinical symptoms, including clinical arthritis score, disease incidence and paw thickness, which were confirmed by microPET imaging. A decrease in serum IgG1 and IgG2a levels was also observed. Characteristic RA histological and radiological changes, such as hyperplasia, pannus formation, cartilage destruction and bone erosion in the joint, were improved, with results comparable to the anti-inflammatory drug dexamethasone. On the other hand, the *in vitro* studies revealed a reduction of Th1/Th17 phenotype $CD4^+$ T lymphocyte expansion, pro-inflammatory cytokines (IL-1β, IL-6, IFN-γ and IL-17) and MMP-1/3 production in the knee joint tissue and RA synovial fibroblasts, through the downregulation of IKKα/β, IκBα and NF-κB [88].

Maslinic acid (**11**), a pentacyclic triterpenoid found in olive (*Olea europaea*) fruit, displays a vast number of therapeutic properties, including preventing and mitigating arthritis in animals and humans, particularly in relation to knee joint arthritis symptoms [89]. Using the CAIA mouse model of RA, Shimazu et al. clarified the molecular mechanisms implicated in the anti-arthritic properties of **11**. Arthritis symptoms were mitigated, and the gene expression of inflammatory cytokines in synovial membranes was inhibited downstream of NF-κB signaling, with **11** also inactivating the TLR signaling pathway. Treatment of CAIA mice with **11** (200 mg/kg) downregulated the expression of the mRNA encoding LTA4 hydrolase, which catalyzes the hydrolysis of LTA4 to LTB4, a chemotactic factor whose overproduction is involved in RA. **11** suppressed the production of LTB4 by acting through the glucocorticoid receptor, as expression levels of several genes controlled by this receptor were altered by **11** [89]. Upregulation of the mRNAs encoding MMP-2 and MMP-9 was observed, along with the upregulation of the expression levels of transcripts encoding tissue inhibitor of metalloproteinases (TIMP)-1, TIMP-2 and TIMP-4, where the proteinase/inhibitor imbalance can facilitate proteolysis in the cartilage of arthritis [89]. The anti-arthritis efficacy of compound **11** thus appears to be grounded in the suppression of synovial inflammation through the inactivation of TLRs, the downregulation of leukotrienes via the glucocorticoid receptor and the promotion of tissue formation with the repair of damaged cartilage [89].

Taraxasterol (**12**) is a taraxastane-type triterpenoid mostly isolated from Chinese medicinal *Taraxacum officinale*, exhibiting anti-inflammatory and antioxidant activities in several disorders [90]. Literature reports have been pointing to its ability to lower pro-inflammatory cytokines and mediators in LPS-induced RAW 264.7 cells *in vitro* and in the ovalbumin-induced asthma mouse model [90]. *In vitro* and *in vivo* studies of **12** in IL-1β-stimulated human RA FLSs and CIA mice, respectively, allowed Chen et al. to investigate the anti-inflammatory effects and subjacent mechanisms of **12** on RA [90]. Since the inflammatory responses in RA FLSs are mostly modulated by NF-κB and the NLRP3 inflammasome [90], the inhibition of NF-κB/NLRP3 pathways is therefore a potential therapeutic approach in RA management. In fact, **12** suppressed NF-κB activation in human RA FLSs, inhibiting the IL-1β-induced IκB degradation and nuclear translocation of p65 in the studied cell line. Results showed that **12** can modulate TGF-β-activated kinase 1 (TAK1) activation (which in turn regulates NF-κB activation), probably exerting its anti-inflammatory activity by modulating the TAK1/IκB/IKK pathway in human RA FLSs [90]. Compound **12** suppressed the expression of NLRP3 inflammasome (reported to be well associated with NF-κB signal transduction) and its modulators, such as TXNIP and ACS, both in human RA FLSs and CIA mice, thereby decreasing cleaved caspase-1 levels; thus, anti-inflammatory effects of **12** could be related to the inhibition of NLRP3 inflammasome signaling. Treatment of CIA mice with **12** mitigated joint destruction and other clinical RA manifestations, downregulated NF-κB and reduced the IL-1β-induced expressions of TNF-α, IL-6, IL-8, MMP-1 and MMP-3 [90].

Macrophage plasticity produces different functional phenotypes in reaction to specific stimuli. Macrophages can be polarized into the classical M1 or the alternative M2 phenotypes. Classically activated (M1) macrophages, induced by LPS or Th1 cytokines IFN-γ and granulocyte–macrophage colony-stimulating factor (GM-CSF), express MHC-II, inducible nitric oxide synthase (iNOS) and co-stimulation molecules like CD80 and CD86 for effective T cell antigen presentation and secrete pro-inflammatory cytokines (e.g., TNF-α, IL-1β, IL-6, IL-12 and IL-23) as well as NO and ROS which are essential for killing intracellular pathogens [36,91]. Alternatively, activated (M2) macrophages, stimulated mainly by Th2 cytokines IL-4 and IL-13 and by macrophage colony-stimulating factor (M-CSF), express mannose receptor CD206, IL-4 receptor, arginase 1 and peroxisome proliferator-activated receptor gamma (PPARγ) and produce anti-inflammatory cytokines (e.g., IL-10 and TGF-β) and trophic polyamines involved in tissue repair [36,91]. The M1/M2 polarization is imbalanced in RA, with higher expression of M1 macrophages in the synovial fluid of RA patients, which promotes osteoclastogenesis [86,91]. ACPAs in the RA synovial fluid can induce interferon regulatory factor 5 (IRF5), leading to increased polarization of peripheral blood monocytes into the M1-like phenotype and thus increasing the M1/M2 ratio [91]. Glucocorticoids and some DMARDs like MTX act by repolarizing M1-like macrophages of RA patients into the M2-like state [86,91]. Wilforlide A (**13**), a pentacyclic triterpenoid from *Tripterygium wilfordii* Hook F, delays the development of RA in CIA mice, inhibiting iNOS production (an M1 surface marker), pro-inflammatory M1 cytokines and chemokines in the mouse synovium [92]. Similarly, *in vitro* results showed that **13** hindered macrophage chemotaxis and M1 polarization in LPS/IFN-γ-stimulated THP-1 cells presumably through inactivation of the TLR4/NF-κB signaling pathway [92].

Table 4. Pentacyclic triterpenes with *in vitro*/*in vivo* RA-related effects.

Pentacyclic Triterpene	Cell Model/Animal Model/Dosage	Effects and Mode of Action	Ref.
Celastrol (**1**)	• TNF-α-stimulated FLSs; pre-treated with **1** (0, 25, 50 or 100 nM) for 2 h and stimulated with TNF-α (10 ng/mL) for 48 h • CIA in male DBA/1 SPF grade mice; intragastric administration of **1** (0, 0.5, 1 or 2 mg/kg/day), vehicle (0.5% CMC-Na) or MTX (2 mg/kg/day), on days 28 to 56 post-immunization	• *In vitro* inhibition of TNF-α-induced proliferation of FLSs • Decrease in p-mTOR, PI3K and p-AKT levels • Increase in autophagosome levels, LC3-II/LC-I ratio and Beclin-1 expression, *in vitro* and *in vivo* • *In vivo* inhibition of the production of pro-inflammatory cytokines TNF-α and IL-1β • Reduction of protein levels of PI3K, p-AKT, p-mTOR and p62 in joint tissue, thus ameliorating paw swelling and hind paw bone damage in CIA mice	[65]
	• LPS/ATP-stimulated human macrophages (THP-1 cells); incubation with PMA (100 nM) for 48 h and treated with **1** (0, 12.5, 25 or 50 nM) or dexamethasone (50 nM) for 1 h prior to incubation with LPS (1 μg/mL) for 24 h followed by ATP (5 mM) stimulation for 30 min • AA in male SD rats; injected with CFA in the left hind joint on day 1 and treated with **1** (0.5 or 1 mg/kg) or vehicle (0.9% saline), i.p., daily, from day 9 up to day 30	• Reduction of joint swelling, arthritis index score, inflammatory cell infiltration and synovial hyperplasia in CFA-induced rats • Decrease in levels of IL-1β and IL-18 in the rat serum and supernatants of THP-1 cells exposed to **1** • Inhibition of ROS production, blocking of NF-κB signaling and hindering the activation of the NLRP3 inflammasome	[63]
	• CIA male Wistar rats; intradermal injection twice at the base of tail with BTIIC emulsion with CFA (1 mg/mL); day 0 (200 μL) and day 7 (100 μL). Experiment I: CIA rats treated i.p. with **1** (1 mg/kg/day) or vehicle. Experiment II: CIA rats treated i.p. with **1** (1 mg/kg/day) and Ad-Nox4 (1 × 1010 TU/mL; tail vein) for 28 days	• Significant reduction of paw edema and arthritis scores. Improvement of the spleen and thymus indexes • Reduction of TNF-α, IL-1β, IL-6, IFN-γ levels in CIA rats • Increase in superoxide dismutase activity; reduction of malondialdehyde and superoxide anions levels and NADPH oxidase activity • Potential therapeutic effects on RA may be ascribed to downregulation of inflammatory cytokine levels and attenuation of oxidative stress	[70]
	• Caco-2 cell line; treated with increasing concentrations of **1** (1–10 μM for viability assays; 2 μM for P-gp efflux) • Male Sprague Dawley rats administered with **1** (1 mg/kg, control group) or both **1** (1 mg/kg) and diclofenac (10 mg/kg)	• Significant change in several pharmacokinetic parameters suggested a decreased intestinal absorption of **1**, through induction of P-gp	[67]
	• Male Lewis rats; i.p. administration of **1** (1 mg/kg) beginning at the onset of the disease and then daily for 3 days, followed by injection every other day until the day of euthanization. Control rats were injected with PBS-DMSO on the same days	• miRNAs (miR-22, miR-27a, miR-96, miR-142, miR-223, miR-296, miR-298 and miR-451) and their target genes in functional pathways important for RA pathogenesis • miR-22, miR-27a, miR-96, miR-142, miR-223 and miR-296 were modulated by **1** • Higher levels of serum miR-142, miR-155, miR-212 and miR-223 in arthritic *vs.* control rats	[68]

Table 4. *Cont.*

Pentacyclic Triterpene	Cell Model/Animal Model/Dosage	Effects and Mode of Action	Ref.
	• Human RA FLSs; treated with **1** (50 μg/mL) for 24 h; RA–FLS1 and RA–FLS2 cells treated with **1** (0.25–2 μM) for 24 h	• Impaired cell proliferation and cell cycle arrest and inhibition of RA FLS invasion • Reduction of secretion of IL-6, IL-8 and MCP-1 in a dose-dependent manner; no change in the secretion of IL-10 • Expression of some chemokines and chemokine receptors was altered significantly after treatment	[69]
Pristimerin (**2**)	• TNF-α-stimulated human RA FLSs (MH7A cells) at 20 ng/mL and treated with **2** (0, 0.5, 1 or 2 μM) for 24 h • AA male Wistar rats; intragastric administration of **2** (0.8 mg/kg/day), vehicle (0.3% CMC-Na) or MTX (0.6 mg/kg/day), for 28 days, starting the next day after CFA immunization	• Inhibition of viability and migration of TNF-α-stimulated MH7A cells (IC_{50} 1.403 μM) • Reduction of paw swelling, TNF-α and NO serum levels as well as p-Akt and p-ERK levels • Alteration of phospholipid and fatty acid biosynthesis, glutathione metabolism and amino acid metabolic pathways • Network pharmacology analysis and molecular docking studies showed that effects were mediated through the MAPK/ERK1/2, PI3K/Akt pathways and direct binding to TNF-α	[72]
Betulinic acid (**3**)	• RA FLSs; pre-treatment with **3** (5, 10, and 20 μM) for 1 h and then stimulated with TNF-α (10 ng/mL) for 24 h	• Inhibition of proliferation and migration of RA FLSs • Attenuation of TNF-α-enhanced MMP expression in RA FLSs • Inhibition of inflammatory response in RA FLSs exposed to TNF-α and prevention of the activation of Akt/NF-κB pathway	[73]
	• RA FLSs treated with DMSO or **3** (0, 2.5, 5, 10 μM) for 24 h. Stimulation with TNF-α (0 or 10 ng/mL) • CIA male DBA/1 mice; injected i.d. on day 0 with emulsion of BTIIC (100 mg) in CFA (1:1, *v*/*v*) and on day 21 with emulsion of BTIIC (100 mg) in IFA (1:1, *v*/*v*). CIA mice injected i.p. with **3** (20 mg/kg/day) or DMSO, for 21 days	• Suppression of the migratory capacity of RA FLSs • Downregulation of the mRNA expression of IL-1β, IL-6, IL-8 and IL-17A in TNF-α-induced RA FLSs • Decrease in TNF-α-induced activation of NF-κB signal pathway (phosphorylated NF-κB, IκBα and IKK) and the NF-κB nuclear accumulation • Inhibitory effect of NF-κB PDTC on the formation of actin stress fibers and actin cytoskeleton score of RA FLSs • Attenuation of synovitis, synovial hyperplasia and invasion into calcified cartilage and bone in CIA mice	[75]

Table 4. *Cont.*

Pentacyclic Triterpene	Cell Model/Animal Model/Dosage	Effects and Mode of Action	Ref.
	• CIA male rats twice immunized with BTIIC:CFA (1:1) injection into the right hind paw, back and tail (7 days, 2 weeks). On day 15, **3** (20 and 40 mg/kg/day, orally) or diclofenac sodium (5 mg/kg/day, orally) or ROCK inhibitor fasudil (5 mg/kg/day, i.p.) was administered for 4 weeks	• Inhibition of arthritis index, amelioration of joint pathology, diminished hind paw swelling, enhanced blood rheology and synovial cell apoptosis and re-establishment of cytokine negative regulation of ROCK/NF-κB signaling pathways • Decreased secretion of IL-6, IL-1β and TNF-α, inhibition of proliferation of synovial tissue, reduction of monocytes and lymphocytes • Decreased levels of RhoA, ROCK1, ROCK2, p-NF-κBp65 and p-IκBα levels. Mechanistically, **3** downregulated ROCK/NF-κB signaling pathways	[74]
	• CIA female albino rats; oral administration of **3** (2 mg/kg) and fluvastatin (5 mg/kg) from day 14 after arthritis induction until day 60	• Decrease in the severity of arthritic index values and inhibition of paw edema on combined treatment • Reduction of RF, C-reactive protein, total lipids and ACPAs; increased activity of catalase, superoxide dismutase and glutathione peroxidase enzymes and expression of the anti-inflammatory cytokine IL-10 • Decreased expression of TLR2 and TLR4, IL-1β, TNF-α, IFN-γ, cell adhesion molecules and nuclear translocation of NF-κB in aorta decreased, when compared to the single-treated groups	[80]
Betulinic acid derivative SH479 (**4**)	• CD4^{+} T cells and splenic lymphocytes of CIA mice treated with different concentrations • Male DBA/1J mice treated with 20 mg/kg of SH479 i.p. daily beginning from day 23 after arthritis induction	• *In vivo* inhibition of CD4^{+} T cell infiltration and cytokine production; inhibition of Th1 and Th17 differentiation as well as antigen-specific T cell proliferation • Decrease in arthritis scores as well as bone destruction and cartilage depletion in the CIA mouse model	[82]
Lupeol (**5**)	• CIA male SD rats; gastric administration of **5** (0 or 10 mg/kg) or indomethacin (3 mg/kg), from the 5th day to the 20th day after arthritic induction	• Reduction of paw edema • Inhibition of COX-2 and 5-LOX enzymes and reversion of the high serum levels of pro-inflammatory mediators (PGE2, TNF-α and IL-6), RF, C-reactive protein and ceruloplasmin • Downregulation of Bcl-2 protein expression and upregulation of Bax, caspase-3 and -9 through PI3K/Akt inhibition	[83]
β-amyrin (**6**) and Polpunonic acid (**7**)	• Sprague Dawley rats; intraplantar injection of CFA (100 μL) in AA rats and IFA (100 μL) in non-AA rats; rats treated with **6** (3, 10, 30 mg/kg, p.o.), **7** (3, 10, 30 mg/kg, p.o.) and dexamethasone (3 mg/kg, p.o.) or distilled water (10 mL/kg) once every day, for 14 days	• Reduction of the primary and secondary paw swelling and the arthritis score in the later stage of the adjuvant-induced arthritis (from 14.25 to 6.5) • Reversion of cartilage erosion and subchondral cyst and Weichselbaum's lacunae formation • Non-marked impact on general hematological and serum biochemical parameters due to treatment with **6**, **7** or dexamethasone	[84]

Table 4. *Cont.*

Pentacyclic Triterpene	Cell Model/Animal Model/Dosage	Effects and Mode of Action	Ref.
Echinocystic acid (**8**)	• TNF-α-stimulated human RA FLSs at 10 ng/mL for 24 h and treated with **8** (0, 5 or 10 μM) for additional 24 h • ZIA in female SKG/Jcl mice; oral administration of **8** (10 or 25 mg/kg) or vehicle (90% glyceryl trioctanoate and 10% DMSO) or MTX (10 mg/kg), i.p., daily, for 3 consecutive weeks, starting on the 21st day after single i.p. injection of zymosan A (2 mg/mice)	• Reduction of synovial hyperplasia, inflammatory cell infiltration and cartilage damage on ankle joints • Attenuated levels of pro-inflammatory cytokines (TNF-α, IL-6, IL-1β, IL-17A, IFN-γ and GM-CSF) and sustained reduction in joint swelling of arthritic hind paws, similar to MTX at the highest EA dose • Cellular reduction of both protein and mRNA expression of IL-6 and IL-1β by downregulating MAPK and NF-κB pathways • The effects were attributed to phosphorylation inhibition of STAT3 (but not JAK2) and subsequent suppression of IL-6- and TGF-β-induced Th17 cell differentiation	[85]
23-Hydroxyursolic acid (**9**)	• RAW264.7 cells and primary mouse BMDMs; incubation with **9** (0, 1, 3 or 10 μM) in the presence of RANKL (100 ng/mL) and M-CSF (30 ng/mL) for 4 days (RAW264.7) or 6 days (BMDMs) • LPS-stimulated ICR mice: oral administration of **9** (25 or 50 mg/kg) or vehicle (corn oil), 1 h before LPS (5 mg/kg, i.p.) injection and thereafter every other day for 8 days	• Inhibition of RANKL-induced osteoclastogenesis in RAW264.7 (IC_{50} = 1.9 ± 0.2 μM) and BMDMs (IC_{50} = 2.1 ± 0.3 μM) without affecting cell viability and protected mice against LPS-induced bone loss • Attenuation of osteoclast formation by inhibiting RANKL-mediated ERK and JNF phosphorylation, NF-κB signaling, c-Fos expression, NFATc1 activation and expression of osteoclast-specific marker genes (OSCAR, MMP-9, TRAP, DC-STAMP and CtsK), both *in vitro* and *in vivo*	[87]
Ursolic acid-3-acetate (**10**)	• Human RA FLSs treated with **10** up to 10 μM for cell viability assays; pre-treatment with **10** for 1 h and stimulated with TNF-α for 12 h	• Decrease in clinical arthritis symptoms, paw thickness, histological and radiological changes and serum IgG1 and IgG2a levels • Reduction of Th1/Th17 phenotype $CD4^+$ T lymphocyte expansion and inflammatory cytokine production • Decreased expression and production of inflammatory mediators, in the knee joint tissue and RA synovial fibroblasts, through the downregulation of IKKα/β, IκBα and NF-κB	[88]
Maslinic acid (**11**)	• CAIA male DBA/1J mice treated with **11** (200 mg/kg) by daily oral administration, from day 1 to day 11 • Mice injected i.p. with 1 mg of a CII monoclonal antibody on day 8 and 25 μg of LPS on day 11	• Lowering of arthritis score, paw thickness and front paw swelling on day 12 • Suppression of the gene expression of inflammatory cytokines downstream of NF-κB signaling and inactivation of the TLR signaling pathway • Downregulation of the expression levels of the genes encoding TNF-α, IL-1β, IL-5 and IL-12, and upregulation of IκBα transcript and protein expression • Decrease in the production of LTB4 and alteration of the gene expression of glucocorticoids	[89]

Table 4. *Cont.*

Pentacyclic Triterpene	Cell Model/Animal Model/Dosage	Effects and Mode of Action	Ref.
Taraxasterol (**12**)	• IL-1β-stimulated human RA FLSs pre-treated with **12** (0.3 to 30 μM), 1 h before the incubation with IL-1β (10 ng/mL) for 48 h	• Downregulation of IL-1β, increase in TNF-α, IL-6, IL-8, MMP-1 and MMP-3 levels in human RA FLSs and in joint tissues of CIA mice, in a dose-dependent manner • Inhibition of NF-κB activations and modulation of the TAK-1/IKK/IκB regulators in human RA FLSs and joint tissues of CIA mice, in a dose-dependent manner • NLRP3, TXNIP and ASC expressions were blocked and the maturation of caspase-1 was decreased, *in vitro* and *in vivo* • Reduction of clinical arthritis score and cartilage destruction in ankle joints of CIA mice • Potential therapeutic action of 12 by modulation of NF-κB/NLRP3 inflammasome pathways	[90]
Wilforlide A (**13**)	• LPS/IFN-γ-stimulated macrophages (THP-1 cells) treated with PMA (200 nM) for 3 days, then stimulated with LPS (1 μg/mL) and IFN-γ (100 ng/mL) and treated with **13** (0, 1, 5, 10, 20, 40, 80, 160 and 300 ng/mL) for 48 h	• Reduction of inflammatory infiltration, joint swelling and histological damage in the ankle joints of CIA mice • Inhibition of iNOS expression in activated macrophages of arthritic synovial joints, reduction of the high levels of pro-inflammatory cytokines (MCP1, GM-CSF and M-CSF) in joint synovium and enhanced expression of anti-inflammatory cytokines (IL-10 and TGF-β) in mouse serum • *In vitro* inhibition of M1 macrophage polarization by suppressing LPS/IFN-γ-induced TLR4 upregulation, IκBα degradation and NF-κB p65 activation	[92]

6.2. Tetracyclic and Rearranged Triterpenes

Antcin K (**14**) is a tetracyclic ergostane-type triterpenoid isolated from *Antrodia cinnamomea*, a mushroom endemic to Taiwan and used in folk medicine due to its antioxidant, anti-inflammatory and immunomodulatory activities [93]. Antcin K (**14**) decreased pro-inflammatory cytokine production in human RA FLSs by inhibiting the phosphorylation of focal adhesion kinase (FAK), PI3K, Akt and NF-κB. Moreover, **14** also ameliorated paw swelling, cartilage degeneration and bone erosion in the CIA mouse model [93].

Ganoderic acid A (**15**), a lanostane triterpenoid extracted from *Ganoderma lucidum* (an edible mushroom), has been traditionally used in East Asia to treat inflammatory, proliferative and immunological diseases without side effects, making it a potential therapeutic agent for RA [94,95]. Cao et al. evaluated the protective effects of **15** in CIA rats to explore its therapeutic role in RA [95]. A reduction in toe swelling and arthritis index was observed, as well as an improvement in joint pathological changes and hemorheology. Serum and synovium levels of IL-1β, IL-6 and TNF-α were markedly reduced in CIA rats, and oxidative stress was regulated. **15** substantially reduced p-STAT3 and suppressor of cytokine signaling 1 (SOCS1); these results indicate a downregulation of protein expression of p-JAK3 and p-STAT3, which may lead to the regulation of the JAK/STAT signaling pathway [95]. Furthermore, protein expression levels of p-NF-κB p65 and p-IκBα in joint synovial tissue of CIA rats were reduced by **15**. The therapeutic role may also be related to the regulation of the NF-κB signaling pathway [95].

Gedunin (**16**), a limonoid-type triterpenoid isolated from several genera of the Meliaceae family, such as the Indian neem tree (*Azadirachta indica*), antagonized ROS production and reduced pro-inflammatory cytokine levels and iNOS expression in LPS-stimulated macrophages (RAW264.7 cells), TNF-α-stimulated FLSs (MH7A cells) and IL-1β-stimulated primary RA FLSs [96]. Furthermore, **16** was able to reduce paw swelling, arthritis score and cytokine production in CIA mice [96]. The *in vitro* and *in vivo* anti-inflammatory and anti-arthritic effects of **16** were due to activation of the Nrf2 signaling through inhibition of Keap1, a key oxidative stress sensor protein, by inducing p62 expression and upregulation of anti-oxidative enzymes, including heme oxygenase (HO)-1 [96].

Other studies showed that 7-deacetyl-gedunin (**17**) isolated from the fruits of *Toona sinensis* (A. Juss.) Roem suppressed ROS production and inhibited proliferation of human RA FLSs isolated from the joint synovium cave of RA patients submitted to knee surgery [97]. Compound **17** also decreased pro-inflammatory cytokine release in human FLSs (MH7A cells) but with significant inhibition of cell viability [97]. Mechanistic studies revealed that **17** exerted anti-inflammatory effects by regulating antioxidative enzymes through Nrf2 activation by inhibiting Keap1 via inducing p62 expression and antioxidant response element (ARE)-driven gene transcription [97].

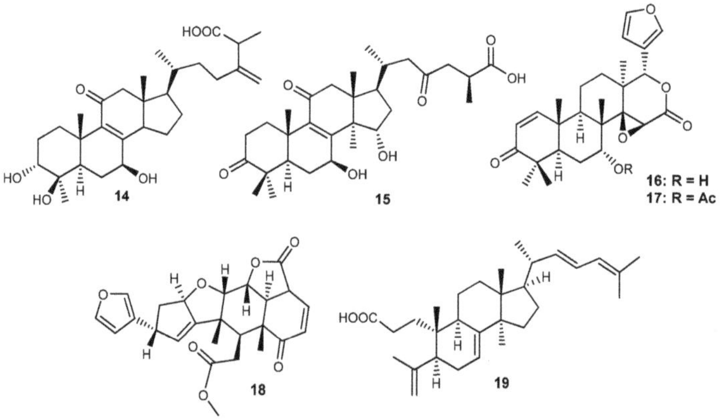

HOOC HO HO H OH 14 O OH OH O H OH 15 O O O O O H H OR 16: R = H 17: R = Ac

O H O H H O H H O H O O O 18 HOOC H H H 19

Figure 3. Structures of tetracyclic (**14–15**) and rearranged triterpenes (**16–19**) with activity on RA.

Nimbolide (**18**), a major limonoid from *Azadirachta indica*, dose-dependently reduced the expression of p38 MAPK and inhibited the phosphorylation of NF-κB in IL-β-stimulated rabbit FLSs (HIG-82 cells) [98]. In a rat model of inflammatory arthritis, **18** significantly reduced STAT3 phosphorylation, attenuating STAT3 signaling with simultaneous inhibition of Notch-1 transmembrane protein receptors and NF-κB activation, thus reducing oxidative stress and pro-inflammatory cytokine levels in synovial tissue of arthritic rats [98]. Furthermore, combination therapy with both **18** (3 mg/kg/day) and MTX (2 mg/kg/week) potentiated the anti-arthritic effects of MTX while reducing its hepatorenal toxicity in a rat model of RA, presumably through antioxidant and anti-inflammatory effects [98]. The efficiency of nimbolide (**18**) was also examined by Cui et al. against joint inflammation in CIA male albino rats [99]. Treatment with **18** (20 mg/kg) resulted in a substantial increase in body weight and a pronounced reduction in arthritic index score, thymus and spleen indices, hind paw volume and edema formation, comparable to diclofenac [99]. Serum levels of IL-1β, IL-6, IL-10 and TNF-α showed a marked reduction in arthritic rats, and the activities of antioxidant enzymes were significantly improved. Supplementation with **18** downregulated the protein expression of iNOS, NF-κB, p-IκBα, IKKα and COX-2, reinforcing the contribution of nimbolide to the therapeutic strategy against RA [99].

Heilaohuacid G (**19**) is a new 3,4-seco-lanostane type triterpenoid isolated from the roots of *Kadsura coccinea*, a medicinal plant distributed in South China and used in Tujia ethnomedicine to treat RA [100]. Biological activity screening tests revealed that **19** inhibited the proliferation of RA FLSs in a concentration-dependent manner, with IC_{50} values of 8.16 ± 0.47 μM [100]. Further studies showed that **19** induced RA FLS apoptosis and suppressed inflammatory responses in LPS-induced RA FLSs and macrophages (RAW264.7 cells) by inhibiting NF-κB signaling [100].

Table 5. Tetracyclic and rearranged triterpenes with *in vitro*/*in vivo* RA-related effects.

Tetracyclic and Rearranged Triterpenes	Cell Model/Animal Model/Dosage	Effects and Mode of Action	Ref.
Antcin K (**14**)	• Human RA FLSs (MH7A cells) treated with **14** (0, 0.3, 1, 3 or 10 μM) for 24 h • CIA C57BL/6J mice treated with **14** (0, 10 or 30 mg/kg), i.p., on alternated days for 4 weeks	• Inhibition of pro-inflammatory cytokines (TNF-α, IL-1β and IL-8) in human RA FLSs through downregulation of FAK, PI3K, Akt and NF-κB signaling pathways • Amelioration of paw swelling, cartilage damage and bone erosion in CIA mice and decreased serum levels of TNF-α, IL-1β, IL-6 and IL-8	[93]
Ganoderic acid A (**15**)	• Rats twice immunized with BTIIC:CFA (1:1) s.c. injection into the right hind paw, back and tail root (7 days, 2 weeks); on day 15, oral administration of **15** (20 and 40 mg/kg/day) or diclofenac sodium (5 mg/kg/day) or physiological saline, for 4 weeks	• Improvement of glossiness, food intake and body weight of rats • Reduction of swelling and limping of the hind feet, degree of toe swelling and joint inflammation • Decrease in TNF-α, IL-6 and IL-1β serum and synovium levels was observed. p-JAK3, p-STAT3, SOCS1, p-NF-κB p65 and p-IκBα protein expression levels were significantly reduced • The mechanism may lie in the downregulation of JAK/STAT and NF-κB signaling pathways	[95]
Gedunin (**16**)	• LPS-induced macrophages (RAW264.7 cells), TNF-α-stimulated FLSs (MH7A cells) and IL-1β-stimulated primary RA FLSs; cells pre-treated with **16** (0, 1, 5, 10, 25 or 50 μM) for 1 h and incubated with 100 ng/mL LPS (RAW264.7 cells), 10 ng/mL TNF-α (MH7A cells) or 2.5 ng/mL IL-1β (RA FLSs) for 24 h • CIA DBA/1 male mice; daily i.p. administration of **16** (2.5 or 5 mg/kg) or vehicle (saline, PEG400 and DMSO 6:3:1 *v*/*v*) or MTX (10 mg/kg), intragastrically, for 20 days	• Reduction of iNOS expression, inhibition of IL-1β, IL-6 and TNF-α secretion and antagonization of ROS production *in vitro* • Reduction of arthritis incidence, suppression of mRNA expression of IL-1β and amelioration of arthritis score, paw edema and bone erosion in CIA mice • Mechanistic *in vitro* studies showed that **16** downregulated Keap1 protein expression and upregulated that of Nrf2, HO-1, NQO1 and p62, in time- and dose-dependent manners	[96]
7-Deacetyl-gedunin (**17**)	• TNF-α- stimulated MH7A cells and IL-1β-stimulated human RA FLSs from the joints of RA patients; cells treated with **17** (0, 1, 2.5, 5, 10, 25, 50 75, 100 or 150 μM) for 24, 48 or 72 h after incubation with 10 ng/mL TNF-α (MH7A cells) or 2.5 ng/mL IL-1β (RA FLSs)	• Suppressed cell proliferation, inhibited ROS production and downregulated MMP-1, -3, -9 and -13 without cytotoxicity (IL-1β-treated cells) • Downregulation of IL-6 and IL-33 with inhibition of cell viability (TNF-α-treated cells) • Mechanistically, **17** increased the expression of anti-oxidative enzymes (HO-1 and NQO1) and p62, thus downregulating Keap1 and activating Nrf2	[97]

Table 5. *Cont.*

Tetracyclic and Rearranged Triterpenes	Cell Model/Animal Model/Dosage	Effects and Mode of Action	Ref.
Nimbolide (**18**)	• IL-1β stimulated rabbit FLSs (HIG-82 cells) pre-treated with **18** (0, 0.5 or 1 μM) for 24 h and stimulated with IL-1β (10 ng/mL) for next 6 h • AA Wistar rats injected with CFA (100 μL, i.a.) in the knee joint and treated with **18** (1 or 3 mg/kg) or vehicle (1% DMSO), i.p., daily, for 21 days	• Inhibition of the migration of FLSs *in vitro* (IC_{50} 3.29 ± 0.15 μM) and decreased expression levels of iNOS, COX-2, MMP-2 and p38. Suppressed nitroso-oxidative stress and reduced the levels of iNOS, COX-2, IL-6 and MMP-2, both *in vitro* and *in vivo* • Decrease in synovial hyperplasia, prevention of cartilage destruction, pain attenuation and amelioration of arthritis progression *in vivo* by abrogating the STAT3/NF-κB/Notch-1 signaling pathway in synovial tissue of arthritic rats	[98]
	• Rats injected with CFA (100 μL, i.d.) into the right hind footpad and treated with vehicle (DMSO), **18** (20 mg/kg/day) or diclofenac sodium (5 mg/kg/day), by oral gavage, for 25 days	• Significant body weight increase • Decrease in paw volume and arthritic index score and in activities of liver marker serum enzymes (SGOT, SGPT, ALP) • Reduction of serum levels of TNF-α, IL-6, IL-1β and IL-10. Decrease in MDA levels and enhancement in activities of antioxidant enzymes; outcomes were comparable to diclofenac sodium • **18** reduced the higher protein levels of COX-2, iNOS, NF-κB, p-IκBα and IKKα in CFA-induced RA rats	[99]
Heilaohuacid G (**19**)	• LPS-stimulated human RA FLSs and macrophages (RAW264.7 cells). Cells treated with **19** (0, 2.5, 5, 10 or 20 μM) for 24 h (RAW264.7) or 48 h (RA FLSs) and then incubated with LPS (100 ng/mL) for another 4 h	• Inhibition of RA FLS proliferation with IC_{50} value of 8.16 ± 0.47 μM • Induction of RA FLS apoptosis and inhibition of the secretion of pro-inflammatory cytokines • Reduced TNF-α and IL-6 in LPS-induced RA FLSs and RAW264.7 cells by suppressing NF-κB signaling	[100]

6.3. Triterpenic Saponins

Astragaloside (**20**), a saponin found in *Astragalus membranaceus*, suppressed excessive FLS proliferation in the AA rat model of RA through the inhibition of the expression of the long non-coding RNA (lncRNA) LOC100912373 and increased release of miR-17-5p, which binds to 3-phosphoinositide-dependent protein kinase 1 (PDK1) and prevents activation of the PDK1/Akt pathway [101]. Abnormal expression of non-coding RNAs, such as miRNAs and lncRNAs, has been implicated in the pathogenesis of autoimmune diseases, including RA [102]. lncRNAs are expressed by many immune system cells, including T and B lymphocytes, monocytes, macrophages and dendritic cells, and lncRNA dysregulation has been associated with autoimmunity onset [102]. Additionally, lncRNAs can act as molecular sponges, sequestering miRNAs and RNA-binding proteins, hindering interactions with their target RNAs [102]. The lncRNA LOC100912373 is a critical gene involved in RA pathogenesis since it can induce FLS proliferation by competing with miR-17-5p and thus promoting activation of the PDK1/Akt signaling pathway that contributes to RA development [103].

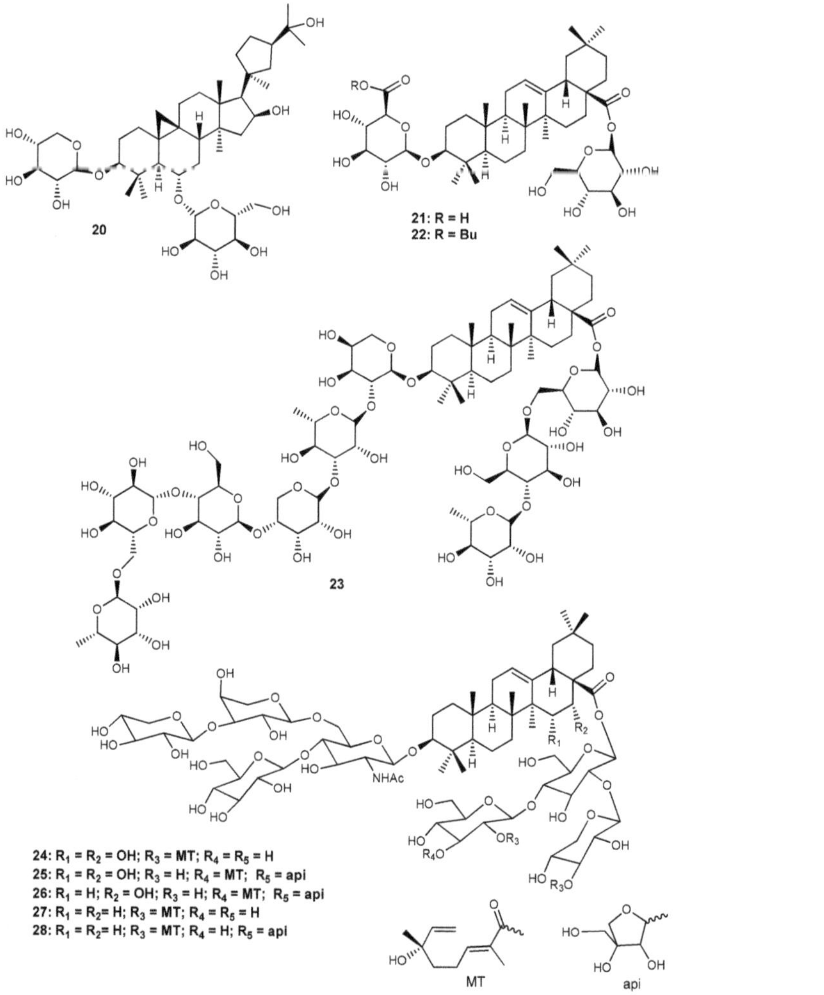

20

21: R = H
22: R = Bu

23

24: $R_1 = R_2 = OH$; $R_3 = MT$; $R_4 = R_5 = H$
25: $R_1 = R_2 = OH$; $R_3 = H$; $R_4 = MT$; $R_5 = api$
26: $R_1 = H$; $R_2 = OH$; $R_3 = H$; $R_4 = MT$; $R_5 = api$
27: $R_1 = R_2 = H$; $R_3 = MT$; $R_4 = R_5 = H$
28: $R_1 = R_2 = H$; $R_3 = MT$; $R_4 = H$; $R_5 = api$

MT

api

Figure 4. *Cont.*

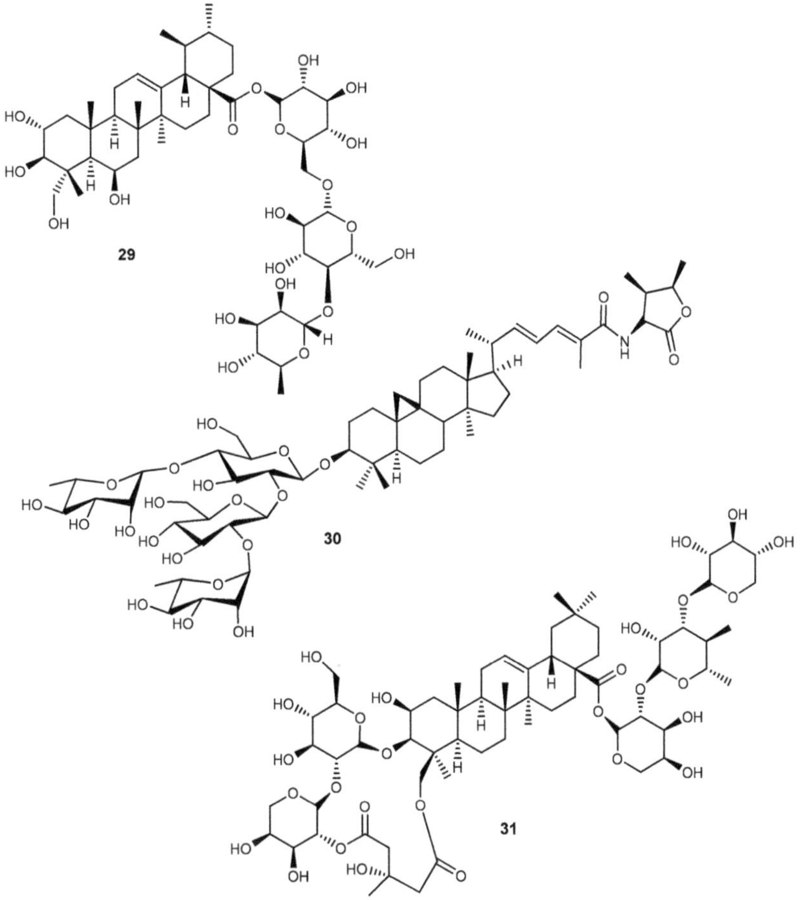

Figure 4. Structures of triterpenic saponins (**20–31**) with activity on RA.

Chikusetsusaponin IVa (**21**), an oleanane-type saponin from *Panax japonicus* C.A. Mey, alleviated RA symptoms in CIA mice [104]. Molecular docking and molecular dynamics simulations revealed that **21** can bind to RA core targets IFN-γ and IL-1β. The study results suggest that the *in vivo* anti-inflammatory and osteoprotective effects of **21** were due to inhibition of the JAK/STAT signaling pathway [104].

Immunopathology in RA is driven by a predominance of arthritogenic Th1 cells (secreting IFN-γ) and Th17 cells (secreting IL-17) over T_{reg} cells [105]. Cytokine IL-6 is critical for the differentiation of Th17 cells and the balance between pathogenic Th17 and protective T_{reg} [105]. Biologic DMARDs targeting the IL-6 receptor have been shown to improve signs and symptoms of RA. Chikusetsusaponin IVa butyl ester (**22**) is a triterpenoid saponin extracted from *Acanthopanas gracilistylus* and a small-molecule IL-6R inhibitor. IL-6R blockade by **22** inhibited Th17 cell differentiation, IL-17A secretion and STAT3 phosphorylation in mouse $CD4^+$ cells (under Th17 polarization conditions) *in vitro* and ameliorated RA symptoms in the CIA mouse model [106]. Thus, saponin **22** represents a promising agent for RA therapy.

Circular RNAs (circRNAs), which are endogenous non-coding RNAs forming stable covalently closed-loop structures, act as miRNA sponges and participate in the regulation of several cellular signaling pathways [102,107]. circRNAs are important epigenetic modulators of gene expression in inflammation and autoimmune regulation, closely associated with RA pathogenesis [102,107]. Clematichinenoside AR (**23**) is a triterpenoid saponin isolated from the roots of *Clematis chinensis* Osbeck. Saponin **23** inhibited proliferation and inflammatory response in FLSs from RA patients *in vitro* and ameliorated RA pathology in CIA mice by combining with frizzled class receptor 4 (FZD4) and blocking the circular pleiotrophin (circPTN)/miR-145-5p/FZD4 signal axis [108]. The authors demonstrated that circPTN promoted FZD4 expression through sponging miR-145-5p with subsequent activation of the Wnt/β-catenin pathway [108]. Confocal microscopy showed that **23** downregulated the expression of β-catenin and its nuclear entry in FLSs by binding FZD4, thus inhibiting the Wnt pathway [108]. Compound (**23**) was also the focus of Xiong et al., who explored its protective action against human TNF-α-induced inflammation and cytotoxicity based on the accumulated evidence about the correlation between RA therapeutic effects and the antagonist effects against TNF-α in RA mouse models [109]. **23** markedly inhibited IL-6 and IL-8 release from recombinant human (rh) TNF-α-stimulated MH7A cells. Cartilage and bone destruction were reversed, probably through downregulation of MMP-1 expression and downregulation of p38 and ERK MAPK signal activation by **23** in rhTNF-α-induced MH7A cells [109]. Treatment of TNF-α-sensitive murine fibroblast L929 cells with **23** reduced the proliferation inhibition ratio caused by rhTNF-α/actinomycin D (ActD) and antagonized rhTNF-α-induced cytotoxicity. Morphological changes in apoptosis (including chromatin condensation, nuclear fragmentation and cell shrinkage) stimulated by rhTNF-α/ActD in L929 cells were attenuated after pre-treatment with **23** [109]. The antagonistic effect of **23** upon cytotoxicity might be ascribed to the degeneration of ROS and the raising of mitochondrial membrane potential, together with the inhibition of prolonged JNK activation following pre-treatment.

Entadaosides **24**–**28**, oleanane-type triterpene saponins isolated from the stems of *Entada phaseolides* (L.) Merr, possess anti-inflammatory properties and are used in traditional Chinese medicine for the treatment of arthritis [110]. All entadaoside saponins **24**–**28** were able to prevent RA progression and ameliorate hyperalgesia, paw swelling and joint destruction in CIA rats by reducing pro-inflammatory cytokine levels, upregulating ubiquitin-editing enzyme A20 expression, inhibiting p38 and ERK1/2 in the periphery and phosphorylation of p38 in the spinal cord [110].

Madecassoside (**29**) is a pentacyclic triterpenoid saponin present in *Centella asiatica*, with previously reported anti-inflammatory and anti-arthritis potential, among other important biological activities. It was also found to induce apoptosis of keloid fibroblasts and keratinocytes and to inhibit LPS-induced TNF-α production, as well as the migration of keloid fibroblasts [111]. Yu et al. used IL-1β stimulation to induce the invasion of FLSs, aiming at exploring the anti-arthritis mechanism of saponin **29** [112]. It was found that oral administration of the triterpenoid exerted a significant therapeutic effect, reducing the articular and bone tissue damage and decreasing hyperemia in the synovial tissue. A dose-dependent *in vitro* inhibitory effect on FLS invasion mediated by IL-1β was also observed, as well as a decrease in MMP-13 activity and mRNA level expression, possibly by preventing NF-κB translocation and phosphorylation.

Qiao et al. compared the anti-arthritis effect of madecassoside (**29**) and its metabolite madecassic acid in pseudo-germ-free CIA rats, discussing the influence of gut microbiota and the mechanism of **29** to stimulate T_{reg} cells [113]. Previous studies revealed the potential of **29** to increase the number of T_{reg} cells in the small intestine, improving the release of IL-10 through the increase in Foxp3^{+} T lymphocytes in the intestinal lamina propria. However, neither **29** nor its metabolite was able to foment the differentiation of T_{reg} cells and the expression of IL-10 in CD4^{+}T cells of CIA rats [113]. In the comparison study, oral administration of **29** was shown to mitigate the arthritis symptoms and the histological alterations in CIA rats, unlike intestinal madecassic acid, suggesting its functionality in the parent form. The increased number of T_{reg} cells by oral administration of **29** was observed mainly in the ileum but without a significant effect concerning T_{reg} cell differentiation and Foxp3 and IL-10 expression *in vitro*. The anti-arthritis effect of compound **29** was strongly influenced by gut microbiota; the sequencing of the 16S rRNA gene indicated that

29 antagonized the richness and diversity of gut microbiota in CIA rats, enhancing the level of n-butyric acid (which increased the immunosuppressive function of T_{reg} cells *in vitro*). The co-administration of heptanoyl CoA (a competitive inhibitor of butyrate synthase) confirmed the contribution of madecassoside-induced butyrate to the anti-arthritis action, as it caused the downregulation of ileum T_{reg} cell number and expression of Foxp3 and IL-10 [113].

Mussaendoside O (**30**), a *N*-triterpene cycloartane saponin isolated from *Mussaenda pubescens*, inhibits RANKL-induced osteoclastogenesis *in vitro* in a concentration-dependent manner [114]. Moreover, 30 attenuates LPS-induced bone resorption and osteoclast formation in mice by repressing RANKL-induced activation of p38 MAPK and JNK, preventing c-Fos activation and subsequent expression of NFATc1. Saponin **30** also diminished RANKL-induced increase in mRNA expression of NFATc1 target genes, including OSCAR, TRAP, DC-STAMP and cathepsin K [114].

Tubeimoside I (**31**) is a triterpenoid saponin previously isolated from *Bolbostemma paniculatum* tubers and found in several Chinese medicine preparations, with anti-inflammatory, anti-tumor and anti-viral activities [115]. The effect of this compound on RA was studied *in vivo* using a CIA rat model and *in vitro* using cultured FLSs [116]. The treatment with **31** suppressed the synovial inflammation and bone destruction in CIA rats in a dose-dependent manner, decreasing erythema and swelling at the doses of 5 and 10 mg/kg. These results were further confirmed by histopathological assays. Moreover, when compared with the control group, an important decrease in pro-inflammatory cytokine production was observed in the joint tissues of tubeimoside I-treated rats, including IL-1β, IL-6, IL-8 and TNF-α, and downregulation of MMP-9 expression. *In vitro* studies also showed that the compound suppressed the proliferation and migration of FLS cells, which are the main causes of synovial hyperplasia, contributing to the cartilage destruction and exacerbating joint damage. The authors suggested that the observed effects may be due to the inhibition of TNF-α-induced activation of NF-κB and MAPKs (p38 and JNK) [116].

Ginsenosides are glycosylated dammarane-type triterpenoids unique to ginseng species. Ginseng is a drug derived from the roots of *Panax ginseng*, used in traditional medicine to treat several diseases, including anemia, diabetes, gastritis and insomnia. It is also used as a general restorative, promoting health and longevity [21]. Based on the location and number of glycoside residues, ginsenosides can be further subdivided into protopanaxadiols (e.g., Rb1, Rb2, Rg3, Rg5 and Rh2), with glycoside residues attached to C-3 and/or C-20 positions, or protopanaxatriols (e.g., Rg1, Rg2 and Rh1), with an additional glycoside residue at C-6 (Figure 4).

Zhang et al. compared several ginsenosides (CK, Rg1, Rg3, Rg5 and Rb1; **32–36**) from *Panax ginseng* Meyer for their therapeutic effect on RA [117]. Ginsenoside CK (**32**) is the major metabolite of natural diol ginsenosides in the intestinal tract [117]. Among the tested ginsenosides, **32** was the most effective, showing strong anti-inflammatory and immunomodulating properties [117]. Ginsenoside CK (**32**) significantly inhibited cell proliferation and enhanced apoptosis of LPS-activated RAW264.7 and TNF-α-stimulated human umbilical vein endothelial cells (HUVECs) [117]. *In vivo*, **32** ameliorated swelling and joint functional impairment in CIA mice [117]. Moreover, **32** was able to increase $CD8^+$ T cells to downregulate the immune response and decrease the number of activated $CD4^+$ T cells and M1 macrophages, thus inhibiting pro-inflammatory cytokine secretion [117]. In an attempt to explore the mechanism of macrophage polarization and phagocytosis by compound **32**, Wang et al. concluded that, through β-arrestin2 regulation in peritoneal macrophages, the compound inhibited TLR4 coupling with Gαi and stimulated TLR4-Gαs coupling [118]. Due to the significant decrease in colocalization of β-arrestin2 and Gαi, owing to **32** treatment, their combined interaction foments the regulation of immune inflammation and the polarization of macrophages to M1. Potential therapeutic properties of **32** for RA therapy seem to be related to the reduction of M1 polarization and secretion of inflammatory cytokines, while overexpressing M2 and IL-10 levels to alleviate inflammation and repair bone tissue in CIA mice. Compound **32** also appears to restore B cell function, in addition to alleviating clinical manifestations of RA (such as the polyarthritis index, spleen and joint pathological scores and spleen index) and the level of serum antibodies in CIA mice [119]. Although IgD B cell receptor (BCR) endocytosis was promoted, it should be noted that the expression level of IgD-BCR did not change. **32** facilitated the co-localization between β-arrestin1 and IgD and between adaptor protein 2 (AP2) and IgD. Mechanistically, the IgD-BCR internalization, in a β-arrestin-AP2-dependent manner, led to the inhibition of B cell activation, which may explain the improvement observed in CIA model mice [119].

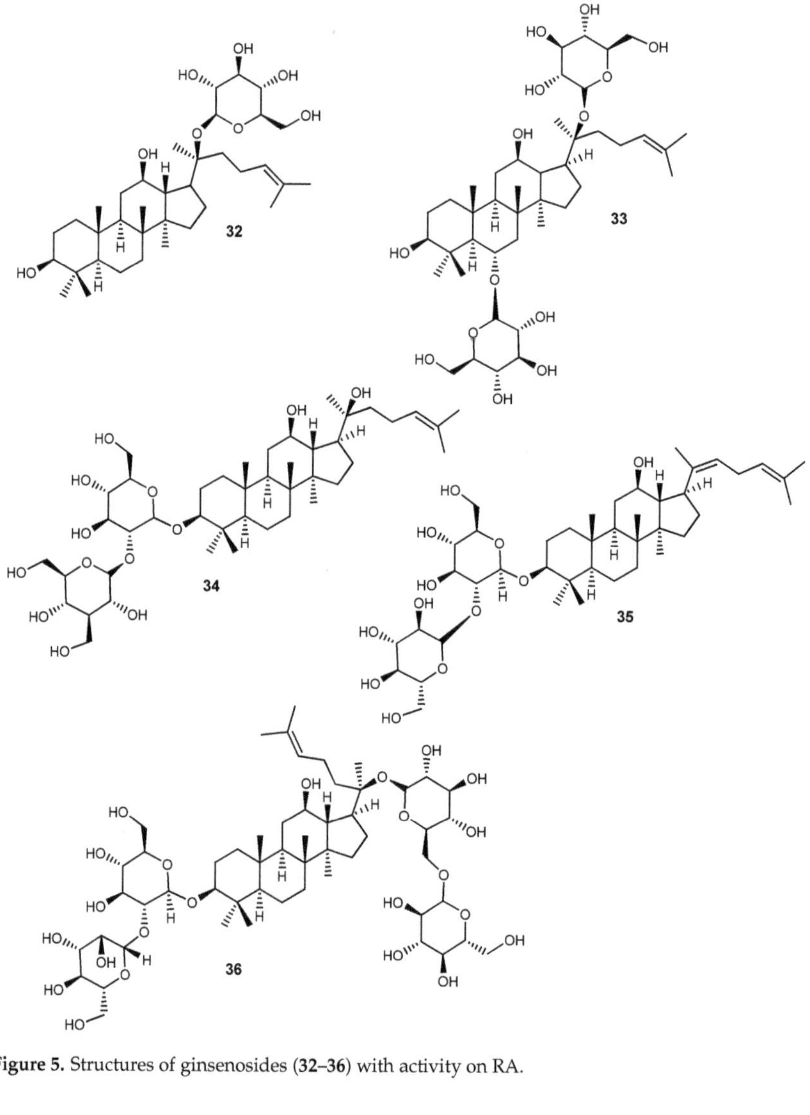

Figure 5. Structures of ginsenosides (**32**–**36**) with activity on RA.

Ginsenoside CK (**32**) is thus a potential candidate for RA therapy and is currently being tested as an anti-RA drug in China. Phase 1 clinical trials in healthy Chinese volunteers to evaluate the pharmacokinetics and safety of **32** showed that a single oral dose of a 200 mg tablet was well tolerated, reaching a maximum plasma concentration (C_{max}) of 796.8 ng/mL in 3.6 h (T_{max}) with a terminal half-life ($t_{1/2}$) of 27.7 h [120]. High-fat food was found to accelerate and increase absorption of **32** while plasma levels were slightly higher in women compared to men [120]. A double-blind, phase 2 study (NCT03755258) to evaluate the safety, efficacy and pharmacokinetics of ginsenoside CK (**32**) tablets in RA patients started in China in March 2017. RA patients (n = 128) were randomly assigned ginsenoside CK tablets (100, 200 or 300 mg) or placebo once daily, orally, for 12 weeks. However, the study was suspended after two years due to the high cost associated with manufacturing of **32**, essentially dependent on *Panax* plants, extraction and biotransformation of ginsenosides. Therefore, development of alternative production methods, such as microbial fermentation processes suitable for scale-up, is an attractive solution.

Many studies on the anti-inflammatory effect of ginsenoside Rg3 (**34**) have been described, emphasizing its ability to regulate NF-κB activity, causing the reduction of cytokine levels, to promote M2 macrophage polarization and to inhibit the inflammation process in the liver through the activation of the PI3K/AKT signaling pathway [53]. Considering the lack of mechanistic robustness regarding the effect of ginsenoside Rg3 in RA, Zhang et al. evaluated the anti-inflammatory effect of the compound **34** through a set of clinical features, pathological alterations and cytokine levels observed in RA mice. $CD4^+CD25^+Foxp3^+$ T_{reg} cell percentage was analyzed and a metabolomic analysis (GC-MS/MS) was performed, aiming to provide information on immunosuppressive activity and related mechanisms [53]. Treatment with **34** (25 mg/kg) led to a decrease in IL-6 and TNF-α levels and an increase in TGF-β and IL-10 levels, mirroring its anti-inflammatory potential. **34** regulated the pathways of oxidative phosphorylation and maintained peripheral immune tolerance in RA mice, enhancing the function of $CD4^+CD25^+Foxp3^+$ T_{reg} cells [53].

Table 6. Triterpenic saponins with in vitro/in vivo RA-related effects.

Triterpenic Saponins	Cell Model/Animal Model/Dosage	Effects and Mode of Action	Ref.
Astragaloside (**20**)	• AA rat FLSs incubated with **20** (0, 7.8, 15.6, 31.25, 62.5, 125, 250 or 500 mg/L) for 24, 48 or 72 h at 37 °C • AA in male SD SPF grade rats; rats were immunized with single CFA injection into the left hind foot and studied for 20 days	• Inhibition of FLS proliferation, reduction of lncRNA LOC100912373 expression, increased miR-17-5p expression and decreased PDK1 and p-AKT levels • Reversion of the effects of LOC100912373 overexpression on FLS proliferation and cell cycle progression by regulating the expression of LOC100912373 and the miR-17-5p/PDK1 axis	[101]
Chikusetsusaponin IVa (**21**)	• CIA in DBA/1J mice; treatment with **21** (50 or 100 mg/kg), dexamethasone (0.2 mg/kg) or saline (negative control) with additional treatments between days 28 and 40	• Reduction of arthritis index, joint synovial inflammation, paw edema and bone loss in CIA mice • Decrease in both rat serum levels and mRNA expression of inflammatory cytokines (IL-1β, IL-6, IFN-γ and TNF-α) and inhibition of protein expression levels of JAK1, JAK2, STAT3 and p-STAT3 in the rat synovial tissue • Molecular docking and molecular dynamics simulations revealed that **21** binds to IFN-γ and IL-1β	[104]
Chikusetsusaponin IVa butyl ester (**22**)	• Naïve $CD4^+$ cells from C57BL/6 mouse spleens, incubated with **22** (0, 2.5, 5.0, 7.5 or 10 μM) for 24 h before stimulation under Th17 polarizing conditions for 3 days • CIA DBA/1J mice treated with **22** (20 or 40 mg/kg) or vehicle (5% Solutol® HS), 6 days weekly for 8 weeks from first immunization	• Decrease in arthritis scores, inflammation scores of the ankle joints, hind paw swelling and ankle joint bone erosion in CIA mice • Reduction of Th17 cells and increased T_{reg} cells, reversing the abnormal Th17/T_{reg} ratio in the spleens of CIA mice • *In vitro* inhibition of Th17 cell differentiation, IL-17A secretion and STAT3 phosphorylation and decrease in mRNA levels of IRF4 and RORγT in splenic $CD4^+$ cells under Th17 polarization conditions	[106]
Clematichinenoside AR (**23**)	• FLSs from RA patients and CIA mice incubated with **23** (0.187 mg/L) for 36 h • CIA DBA/1 SPF grade mice administered **23** (0, 0.18, 0.37, 0.75 or 1.5 mg/kg) or MTX (0.75 mg/kg), by oral gavage, on the 28th day after first immunization	• Inhibition of the arthritis score of CIA mice, reduction of paw swelling and restored mouse body weight • Suppression of FLS proliferation, secretion inhibition of IL-1β, IL-6 and IL-8 and reduction of the expression of β-catenin, fibronectin and MMP-3 *in vitro* • Inhibition of the Wnt/β-catenin pathway by binding to FZD4 and blocking the circPTN/miR-145-5p/FZD4 signal axis	[108]

Table 6. *Cont.*

Triterpenic Saponins	Cell Model/Animal Model/Dosage	Effects and Mode of Action	Ref.
	• Human RA FLSs (MH7A cells) incubated with **23** (3, 10 or 30 μM) for 1 h, followed by exposure to rhTNF-α (10 ng/mL) for 24 h • TNF-α-sensitive mouse fibroblast (L929) cells pre-incubated for 1 h with **23** (1 10 or 100 μM), followed by rhTNF-α (5 ng/mL) stimulation in the presence of ActD (0.5 μg/mL) for 24 h	• Significant reduction of IL-6 secretion and IL-8 production in a concentration-dependent mode, in MH7A cells stimulated by recombinant human TNF-α • Decrease in rhTNF-α-induced MMP-1. Suppression of phosphorylated levels of p38 and ERK1/2 produced by rhTNF-α. Abolition of rhTNF-α-induced L929 cell cytotoxicity • Attenuation of L929 cells' morphological induced modifications (increase in cell density and decrease in apoptotic morphology levels) • Mechanistically, anti-destructive effects of **23** caused by rhTNF-α may be through the downregulation of MMP-1 expression, and the protective effects of murine L929 cells may lie in the suppression of JNK continuous phosphorylation	[109]
Entadaosides (**24**–**28**)	• CIA Wistar rats treated by oral gavage with each entadaoside (25, 50 or 100 mg/kg/day), celecoxib (18 mg/kg/day) or saline (negative control) for 3 weeks	• Reduction of mRNA levels and production of pro-inflammatory cytokines (TNF-α, IL-17) in synovial tissues and hind paw joint • Upregulation of A20 and inhibition of ERK1/2 activation in hind paw joints as well as p38, both in the periphery and spinal cord	[110]
Madecassoside (**29**)	• CIA-induced Wistar rats twice injected i.d. at the base of tail with emulsion CII in CFA (1 mg/mL), on day 0 (200 μL) and day 7 (100 μL). From day 14 to day 30, oral administration of **29** (30 mg/kg) or madecassic acid (15 mg/kg) or vehicle (CMC-Na). Co-administration of heptanoyl CoA (0.3 mg/kg) with MAD, through insertion of a Teflon canula into the anus (8 cm), from day 14 to day 34.	• Decrease in the maximum paw swelling and arthritis index score; improvement of body weight loss and histological changes (joints' synovial hyperplasia, inflammatory cell infiltration and cartilage and bone destruction) • Reversion of changes in gut microbiota, rise in acetic acid and butyric acid levels • Selective promotion of the production of T_{reg} cells in the parent form (**29**), although *in vitro* the effects on T_{reg} cell differentiation and the expression of Foxp3 and IL-10 were not so significant • Increase in the expression of T_{reg} cells and promotion of the expression of Foxp3 and IL-10 in rat ileum (rather than duodenum and jejunum), fomented by sodium butyrate (in a concentration-dependent mode)	[113]
	• AIA rat model treated by oral gavage with **29** (25 mg/kg) and dexamethasone (positive control, 0.5 mg/kg) for 13 days	• Inhibition of migration and invasion of FLSs induced by IL-1β; suppression of IL-1β-triggered FLS invasion through suppression of MMP-13 activity and transcription via inhibiting the MMP-13 promoter-binding activity of NF-κB and downregulating the translocation and phosphorylation of NF-κB	[112]

Table 6. *Cont.*

Triterpenic Saponins	Cell Model/Animal Model/Dosage	Effects and Mode of Action	Ref.
Mussaendoside O (**30**)	• Mouse BMDMs and RAW264.7 cells incubated with **30** (0, 0.3, 1 or 3 μM) in the presence of RANKL (100 ng/mL) and M-CSF (30 ng/mL) for 4 days (RAW264.7) or 7 days (BMDMs) • LPS-stimulated ICR mice treated with **30** (10 or 20 mg/kg) or vehicle (corn oil), orally, 1 h before the first injection of LPS (5 mg/kg, i.p.) and thereafter every other day for 8 days	• Inhibition of RANKL-induced osteoclast differentiation in BMDMs (IC_{50} 0.75 ± 0.15 μM) and RAW264.7 (IC_{50} 0.75 ± 0.15 μM), without decreasing cell viability • **30** failed to inhibit LPS-induced production of pro-inflammatory mediators (NO, iNOS, COX-2 and TNF-α) in RAW264.7 cells • Inhibition of RANKL-induced osteoclastogenesis *in vitro* was attributed to the impairing of c-Fos and subsequent NFATc1 expression • At 20 mg/kg, **30** significantly protected mice against LPS-induced bone loss presumably by suppressing c-Fos expression through inhibition of JNK and p38 MAPK pathways	[114]
Tubeimoside I (**31**)	• CIA Wistar rats; i.p. administration of **31** (1, 5 or 10 mg/kg/day)	• Synovial inflammation and bone destruction were suppressed in CIA rats in a dose-dependent manner • Decrease in pro-inflammatory cytokine production	[116]
Ginsenosides CK, Rg1, Rg3, Rg5 and Rb1 (**32–36**)	• LPS-activated RAW264.7 cells and TNF-α-stimulated HUVECs; cells treated with 100 ng/mL LPS (RAW264.7) or 10 ng/mL TNF-α (HUVECs) for 24 h followed by incubation with each ginsenoside (1.5625, 3.125, 6.25, 12.5, 25, 50, 100 or 200 μg/mL), MTX (positive control) or DMSO (negative control) • CIA male DBA/1 mice treated with 15 mg/kg ginsenosides (**32–36**) or vehicle (0.5% Tween-80), i.v., once every 2 days, 15 times, after onset of joint swelling	• All ginsenosides **32–36** showed good therapeutic effect on acute arthritis. Among the tested ginsenosides, **32** was the most effective • *In vitro*, **32** inhibited cell proliferation and enhanced apoptosis • *In vivo*, **32** reduced swelling and joint functional impairment in CIA mice • **32** increased $CD8^+$ T cells to downregulate the immune response and decreased the number of activated $CD4^+$ T cells and pro-inflammatory M1 macrophages, inhibiting the secretion of TNF-α and IL-6	[117]
Ginsenoside CK (**32**)	• CIA DBA/1 mice; CIA induced by two i.d. injections in the tail root with 100 μL emulsion of CII (1 mg/mL) and Calmette's vaccine (2 mg/mL), on days 0 and 21. On day 28, mice treated with intragastric administration of **32** (112 mg/kg/day) or MTX (2 mg/kg/day), for 24 days.	• **32** restored mouse body weight and alleviated symptoms of arthritis • Spleen index was attenuated, and proliferation of splenic and thymic lymphocytes was inhibited • Secretion of IL-1β, IL-17 and TNF-α was decreased. IL-10 level was promoted in serum and macrophage culture supernatants. M1 and M2 macrophages were diminished and augmented, respectively • Inhibition of the expression of Gαi, TLR4 and NF-κB, increasing Gαs level. The performance of **32** was similar to MTX. But unlike MTX, **32** inhibited the expression of β-arrestin2. Through β-arrestin2 regulation in macrophages, **32** inhibited TLR4–GαI coupling and promoted TLR4–Gαs coupling	[121]

Table 6. *Cont.*

Triterpenic Saponins	Cell Model/Animal Model/Dosage	Effects and Mode of Action	Ref.
	• CIA male DBA/1 mice; CIA induced by two i.d. injections in the tail root with 100 μL emulsion of CFA and CCII (at equal volumes), on days 0 and 21. Mice treated with **32** (28, 56 or 112 mg/kg) or MTX (2 mg/kg) or vehicle (CMC-Na), from day 28 to day 51	• Improvement of the polyarthritis index, swollen joint count, spleen and joint pathological scores and spleen index • Abnormal B cell spreading was inhibited. Production of serum antibodies (IgG1, IgG2a, anti-CII) was prevented, and the pathogenesis of CIA was improved. These outcomes were more pronounced with **32** (112 mg/kg) and in a similar trend to MTX • Homeostasis of B cell subsets (regulatory B cells, plasma cells, memory B cells, mature B and FO B cells) was restored in CIA mice • **32** promoted co-localizations between IgD and β-arrestin1 and between IgD and AP2. Although **32** did not alter IgD-BCR expression, it seemed to foment IgD-BCR internalization in a β-arrestin1-AP2-dependent manner	[119]
Ginsenoside Rg3 (**34**)	• Mice immunized with a single s.c. injection of CFA (100 μL) into the right hind footpad. On day 7, mice treated intragastrically with saline (100 μL) or **34** (25 mg/kg/day) for 16 days	• **34** reduced the swelling rates of RA mice, decreased the degree of cartilage destruction and vasodilation, diminished protein expression of TNF-α and IL-6 and raised the protein expression of IL-10 and TGF-β in the ankle joint • Enhancement of oxidative phosphorylation and reinforcement of the TCA cycle and the respiration of ETC. $CD4^+CD25^+Foxp3^+$ T_{reg} cell percentage was increased; lipids played a crucial role in the proliferation and differentiation of these cells	[53]

7. Conclusions and Future Perspectives

Morbidity and mortality associated with RA justify the continuous interest in the quest for new compounds with better efficacy and a mechanistic rationale, together with a deeper understanding of the anti-arthritis effects of novel isolated compounds or those already reported as therapeutic compounds in RA. Being a systemic autoimmune and chronic inflammatory disease, RA displays a significant increase in macrophages, chemokines, inflammation cytokines, B cells, $CD4^+$ T cells and autoantibodies. Current diagnostic biomarkers for RA consist of ACPAs and less specific RFs, along with synovial inflammation, cartilage and bone destruction and systemic disorders. Under inflammatory conditions, FLSs are implicated in the production of pro-inflammatory cytokines and chemokines, extracellular matrix-degrading enzymes and pro-angiogenic factors. These pro-inflammatory cytokines and chemokines include IL-1, IL-2, IL-3, IL-4, IL-6, IL-8, IL-17, IL-18, IFN-α and IFN-β, TNF-α, TGF-β, GM-CSF and macrophage inflammatory protein (MIP)-3α. The synovial inflammation results from the activation of NF-κB, production of PGE2 and upregulation of COX-2 expression, all of which are promoted by pro-inflammatory cytokines and chemokines. TLR signaling and the NLRP3 inflammasome also appear to have potential roles in the pathogenesis of RA. Current pharmacological approaches to managing RA involve DMARDs alone or in combination with NSAIDs or low-dose glucocorticoids. However, probable and considerable toxicity related to DMARDs affects their ability to treat the disease, fomenting the need to find new therapeutic options.

The use of medicinal plants to treat RA has a long-established tradition of efficacy. Yet, few types of natural medicines for RA treatment are available, most of which are involved in pre-clinical research. Celastrol, the principal active constituent of *Tripterygium wilfordii*, has been demonstrating great therapeutic potential in the treatment of RA, although its mechanism of action is far from being established. Nevertheless, its toxicity is a troublesome issue, affecting the gastrointestinal tract, liver and reproductive system. Increasing the number of bioactive compounds, with high quality and low toxicity, and identifying the mechanism(s) of action of plant components are of extreme importance, while also exploring new potential medicinal herbs. Natural compounds may be valuable alternative choices for drugs in RA treatment, either as adjuvants to conventional drugs or as therapeutic agents.

Triterpenes are a large and structurally diverse group of natural compounds with documented anti-inflammatory and immunomodulatory activities. In this review, thirty-six triterpenoids were divided into pentacyclic triterpenes, tetracyclic and rearranged triterpenes and triterpenic saponins and examined regarding their potential effects on RA, as they target and revert a large number of signaling pathways and cytokines, both *in vitro* and *in vivo* in several animal models of RA.

Anti-RA effects of the described triterpenoids mainly rely on their known anti-oxidative and anti-inflammatory properties. Triterpenoids can reduce oxidative stress by enhancing superoxide dismutase activity, inhibiting NADPH oxidase activity and decreasing malondialdehyde and superoxide anions levels [70,80]. The reported upregulation of anti-oxidative enzymes by triterpenoids has been attributed to activation of the Nrf2/ARE signaling pathway [96,97]. Triterpenoids are also capable of inhibiting COX-2 and 5-LOX enzymes [83], thus inhibiting the biosynthesis of prostaglandins and leukotrienes, respectively, which are important mediators of the inflammation process.

Among the different modes of action that have been described for anti-arthritic triterpenoids (Figure 6), inhibition of NF-κB signaling is the major one [63,74,75,80,85,87,88,90,92, 95,98–100,112,116]. Deregulated NF-κB activation is characteristic of chronic inflammatory diseases, such as RA [122]. The transcription factor NF-κB is known to play a pivotal role in the regulation of both innate and adaptive immune responses and it is a key mediator of the inflammatory process. NF-κB can induce the expression of several pro-inflammatory genes, including those encoding pro-inflammatory cytokines and chemokines, and is involved in the activation and differentiation of innate immune cells and inflammatory T cells [122]. Furthermore, triterpenoid inhibition of RANKL-induced osteoclastogenesis strongly contributes to the prevention of bone damage and disease progression in animal models of RA [87,114].

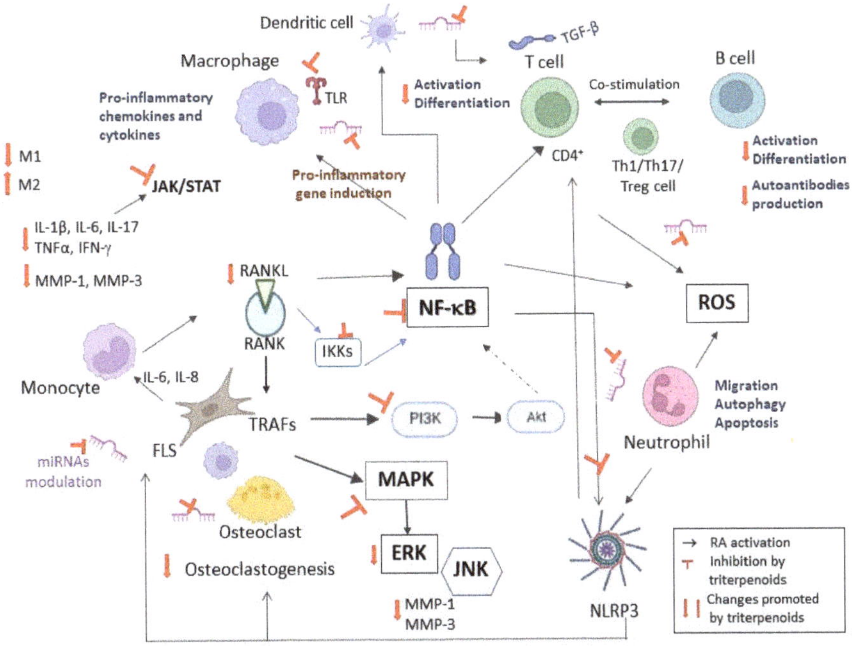

Figure 6. Main modes of action of anti-RA triterpenoids. Created with BioRender.com (accessed on 14 June 2023).

Other modes of action of anti-arthritic triterpenoids include inhibition of PI3K/Akt [53,65,72,83,93] and MAPK/ERK [72,85,109,110,116] signaling pathways, hindering of NLRP3 inflammasome activation [63,90] and modulation of miRNAs and their target genes involved in functional pathways relevant for RA pathogenesis [68]. Triterpenoid inactivation of TLR signaling [80,89,92,118], which hinders macrophage chemotaxis and M1 polarization, is another mechanism responsible for the anti-arthritic effects of this class of compounds. Suppression of both protein and mRNA expression of pro-inflammatory cytokines, such as IL-6 and IL-1β, through inhibition of the JAK/STAT signaling pathway has also been reported [85,95,98,99,104,106], with subsequent suppression of IL-6 and TGF-β-induced Th17 differentiation. The production of TNF-α-induced pro-inflammatory cytokines (IL-6 and IL-8) has also been inhibited by direct binding of the triterpenoid to TNF-α [72].

Additionally, these bioactive triterpenoids were able, in general, to produce a reduction in several RA activity indices, including paw edema, arthritis scores, body weight and hematological, biochemical and immunological markers. In the considered timespan of this review, pentacyclic triterpenes from *Tripterygium wilfordii*, such as celastrol, and betulinic acid, stand out as the most studied compounds with a deep investigation of their molecular mechanism. Nimbolide, a limonoid triterpene, has also been considered a potential therapeutic strategy against RA, and its contribution has been well addressed. Several ginsenoside compounds have been described as being effective in the treatment of RA, with ginsenoside CK appearing to have stronger anti-inflammation and immunomodulatory properties among them.

This review highlights the significant progress in the research concerning triterpenoids as potential agents in the management of RA. In the future, continued contributions from basic research, more comprehensive and in-depth research and well-controlled clinical trials are required. Knowledge gaps in triterpene mechanisms of action need to be addressed in future research. Cell and serum metabolomics profiling of the effects of some of the above triterpenoids has already been successfully established, paving the way for analytical profiling approaches such as metabolomics, proteomics or transcriptomics to provide mechanistic clarifications. Since gut microbiota plays a crucial role in health and disease, and some triterpenoids were shown to be affected by gut microbiome composition, this field could be further explored. Despite their numerous and/or potential pharmacological properties in the treatment of RA, triterpenoids show low bioavailability and toxicity. Toxicity evaluations have been lacking, which is expected to be handled in the future. On the other hand, investing more in the development of targeted drug delivery systems containing triterpenoids could overcome these significant drawbacks.

Author Contributions: Conceptualization, writing, review and editing, N.D., C.F. and L.P. All authors have read and agreed to the published version of the manuscript.

Funding: This research received no external funding.

Institutional Review Board Statement: Not applicable.

Informed Consent Statement: Not applicable.

Data Availability Statement: Data sharing not applicable.

Conflicts of Interest: The authors declare no conflict of interest.

Abbreviations

AA, adjuvant arthritis; ACPA, anti-citrullinated protein antibody; AIA, antigen-induced arthritis; Akt, protein kinase B; CAIA, collagen antibody-induced arthritis; CFA, complete Freund's adjuvant; CIA, collagen-induced arthritis; COX-2, cyclooxygenase-2; CII, collagen type II; DMARD, disease-modifying anti-rheumatic drug; ERK, extracellular signal-regulated kinase; FLS, fibroblast-like synoviocyte; HLA, human leukocyte antigen; IFN-γ, interferon gamma; IκB, inhibitor of nuclear factor κB; IKK, IκB kinase; IL, interleukin; iNOS, inducible nitric oxide synthase; JAK, Janus kinase; JNK, c-Jun N-terminal kinase; LPS, lipopolysaccharide; MAPK, mitogen-activated protein kinase; MHC, major histocompatibility complex; MMP, matrix metalloproteinase; mTOR, mammalian target of rapamycin; MTX, methotrexate; NF-κB, nuclear factor kappa-B; NSAID, non-steroidal anti-inflammatory drug; PI3K, phosphoinositide 3-kinase; RA, rheumatoid arthritis; RANKL, receptor activator of nuclear factor kappa-B ligand; RF, rheumatoid factor; ROS, reactive oxygen species; STAT, signal transducer and activator of transcription; TGF-β, transforming growth factor beta; TLR, Toll-like receptor; TNF, tumor necrosis factor.

References

1. Newman, D.J.; Cragg, G.M. Natural Products as Sources of New Drugs over the Nearly Four Decades from 01/1981 to 09/2019. *J. Nat. Prod.* **2020**, *83*, 770–803. [CrossRef] [PubMed]
2. Cragg, G.M.; Newman, D.J. Natural products: A continuing source of novel drug leads. *Biochim. Biophys. Acta Gen. Subj.* **2013**, *1830*, 3670–3695. [CrossRef] [PubMed]
3. Newman, D.J.; Cragg, G.M. Natural Products as Sources of New Drugs from 1981 to 2014. *J. Nat. Prod.* **2016**, *79*, 629–661. [CrossRef] [PubMed]
4. Newman, D.J. Natural products and drug discovery. *Natl. Sci. Rev.* **2022**, *9*, nwac206. [CrossRef]
5. Atanasov, A.G.; Zotchev, S.B.; Dirsch, V.M.; Orhan, I.E.; Banach, M.; Rollinger, J.M.; Barreca, D.; Weckwerth, W.; Bauer, R.; Bayer, E.A.; et al. Natural products in drug discovery: Advances and opportunities. *Nat. Rev. Drug Discov.* **2021**, *20*, 200–216. [CrossRef]
6. Howes, M.J.R.; Quave, C.L.; Collemare, J.; Tatsis, E.C.; Twilley, D.; Lulekal, E.; Farlow, A.; Li, L.; Cazar, M.E.; Leaman, D.J.; et al. Molecules from nature: Reconciling biodiversity conservation and global healthcare imperatives for sustainable use of medicinal plants and fungi. *Plants People Planet* **2020**, *2*, 463–481. [CrossRef]
7. Joshi, R.K. Bioactive Usual and Unusual Triterpenoids Derived from Natural Sources Used in Traditional Medicine. *Chem. Biodivers.* **2023**, *20*, e202200853. [CrossRef]

8. Podolak, I.; Janeczko, Z. Pharmacological Activity of Natural Non-glycosylated Triterpenes. *Mini Rev. Org. Chem.* **2014**, *11*, 280–291. [CrossRef]
9. Bildziukevich, U.; Wimmerová, M.; Wimmer, Z. Saponins of Selected Triterpenoids as Potential Therapeutic Agents: A Review. *Pharmaceuticals* **2023**, *16*, 386. [CrossRef]
10. Nazaruk, J.; Borzym-Kluczyk, M. The role of triterpenes in the management of diabetes mellitus and its complications. *Phytochem. Rev.* **2015**, *14*, 675–690. [CrossRef]
11. Catteau, L.; Zhu, L.; Van Bambeke, F.; Quetin-Leclercq, J. Natural and Hemi-Synthetic Pentacyclic Triterpenes as Antimicrobials and Resistance Modifying Agents against *Staphylococcus aureus*: A Review. *Phytochem. Rev.* **2018**, *17*, 1129–1163. [CrossRef]
12. Sycz, Z.; Tichaczek-Goska, D.; Wojnicz, D. Anti-Planktonic and Anti-Biofilm Properties of Pentacyclic Triterpenes—Asiatic Acid and Ursolic Acid as Promising Antibacterial Future Pharmaceuticals. *Biomolecules* **2022**, *12*, 98. [CrossRef] [PubMed]
13. Darshani, P.; Sen Sarma, S.; Srivastava, A.K.; Baishya, R.; Kumar, D. *Anti-Viral Triterpenes: A Review*; Springer: Cham, The Netherlands, 2022; Volume 21, ISBN 0123456789.
14. Ghante, M.H.; Jamkhande, P.G. Role of pentacyclic triterpenoids in chemoprevention and anticancer treatment: An overview on targets and underling mechanisms. *J. Pharmacopunct.* **2019**, *22*, 55–67. [CrossRef] [PubMed]
15. Bishayee, A.; Ahmed, S.; Brankov, N.; Perloff, M. Triterpenoids as potential agents for the chemoprevention and therapy of breast cancer. *Front. Biosci.* **2011**, *16*, 980–996. [CrossRef] [PubMed]
16. Ren, Y.; Kinghorn, A.D. Natural Product Triterpenoids and Their Semi-synthetic Derivatives with Potential Anticancer Activity. *Planta Med.* **2019**, *85*, 802–814. [CrossRef]
17. Mannino, G.; Iovino, P.; Lauria, A.; Genova, T.; Asteggiano, A.; Notarbartolo, M.; Porcu, A.; Serio, G.; Chinigò, G.; Occhipinti, A.; et al. Bioactive triterpenes of protium heptaphyllum gum resin extract display cholesterol-lowering potential. *Int. J. Mol. Sci.* **2021**, *22*, 2664. [CrossRef]
18. Jeong, G.-S.; Bae, J.-S. Anti-Inflammatory Effects of Triterpenoids; Naturally Occurring and Synthetic Agents. *Mini Rev. Org. Chem.* **2014**, *11*, 316–329. [CrossRef]
19. Renda, G.; Gökkaya, İ.; Şöhretoğlu, D. Immunomodulatory properties of triterpenes. *Phytochem. Rev.* **2022**, *21*, 537–563. [CrossRef]
20. Li, Y.; Wang, J.; Li, L.; Song, W.; Li, M.; Hua, X.; Wang, Y.; Yuan, J.; Xue, Z. Natural products of pentacyclic triterpenoids: From discovery to heterologous biosynthesis. *Nat. Prod. Rep.* **2023**. [CrossRef]
21. Dewick, P.M. *Medicinal Natural Products: A Biossynthetic Approach*, 3rd ed.; John Wiley & Sons: West Sussex, UK, 2009.
22. Miranda, R.D.S.; de Jesus, B.D.S.M.; da Silva Luiz, S.R.; Viana, C.B.; Adao Malafaia, C.R.; Figueiredo, F.D.S.; Carvalho, T.D.S.C.; Silva, M.L.; Londero, V.S.; da Costa-Silva, T.A.; et al. Antiinflammatory activity of natural triterpenes—An overview from 2006 to 2021. *Phyther. Res.* **2022**, *36*, 1459–1506. [CrossRef]
23. Arthritis Foundation Arthritis by the Numbers. Book of Trusted Facts & Figures 2020. Available online: https://www.arthritis.org/getmedia/73a9f02d-7f91-4084-91c3-0ed0b11c5814/abtn-2020-final.pdf (accessed on 15 February 2023).
24. Senthelal, S.; Li, J.; Ardeshirzadeh, S.; Thomas, M. *Arthritis*; StatPearls Publishing: Treasure Island, FL, USA, 2022.
25. Institute for Health Metrics and Evaluation Global Burden of Disease Study 2019 (GBD 2019) Disease and Injury Burden 1990–2019. Available online: https://www.healthdata.org/gbd/2019 (accessed on 12 March 2023).
26. Aletaha, D.; Smolen, J.S. Diagnosis and Management of Rheumatoid Arthritis: A Review. *JAMA J. Am. Med. Assoc.* **2018**, *320*, 1360–1372. [CrossRef]
27. Dudics, S.; Langan, D.; Meka, R.R.; Venkatesha, S.H.; Berman, B.M.; Che, C.T.; Moudgil, K.D. Natural products for the treatment of autoimmune arthritis: Their mechanisms of action, targeted delivery, and interplay with the host microbiome. *Int. J. Mol. Sci.* **2018**, *19*, 2508. [CrossRef]
28. Gandhi, G.R.; Jothi, G.; Mohana, T.; Vasconcelos, A.B.S.; Montalvão, M.M.; Hariharan, G.; Sridharan, G.; Kumar, P.M.; Gurgel, R.Q.; Li, H.B.; et al. Anti-inflammatory natural products as potential therapeutic agents of rheumatoid arthritis: A systematic review. *Phytomedicine* **2021**, *93*, 153766. [CrossRef]
29. Sharma, A.; Goel, A. Pathogenesis of rheumatoid arthritis and its treatment with anti-inflammatory natural products. *Mol. Biol. Rep.* **2023**, *50*, 4687–4706. [CrossRef] [PubMed]
30. Sharma, D.; Chaubey, P.; Suvarna, V. Role of natural products in alleviation of rheumatoid arthritis—A review. *J. Food Biochem.* **2021**, *45*, e13673. [CrossRef] [PubMed]
31. Moudgil, K.D.; Venkatesha, S.H. The Anti-Inflammatory and Immunomodulatory Activities of Natural Products to Control Autoimmune Inflammation. *Int. J. Mol. Sci.* **2023**, *24*, 95. [CrossRef] [PubMed]
32. Liu, X.; Wang, Z.; Qian, H.; Tao, W.; Zhang, Y.; Hu, C.; Mao, W.; Guo, Q. Natural medicines of targeted rheumatoid arthritis and its action mechanism. *Front. Immunol.* **2022**, *13*, 1–19. [CrossRef] [PubMed]
33. Smolen, J.S.; Aletaha, D.; McInnes, I.B. Rheumatoid arthritis. *Lancet* **2016**, *388*, 2023–2038. [CrossRef] [PubMed]
34. Conforti, A.; Di Cola, I.; Pavlych, V.; Ruscitti, P.; Berardicurti, O.; Ursini, F.; Giacomelli, R.; Cipriani, P. Beyond the joints, the extra-articular manifestations in rheumatoid arthritis. *Autoimmun. Rev.* **2021**, *20*, 102735. [CrossRef]
35. Alivernini, S.; Firestein, G.S.; McInnes, I.B. The pathogenesis of rheumatoid arthritis. *Immunity* **2022**, *55*, 2255–2270. [CrossRef]
36. Firestein, G.S.; McInnes, I.B. Immunopathogenesis of Rheumatoid Arthritis. *Immunity* **2017**, *46*, 183–196. [CrossRef] [PubMed]
37. Nell-Duxneuner, V.; Machold, K.; Stamm, T.; Eberl, G.; Heinzl, H.; Hoefler, E.; Smolen, J.; Steiner, G. Autoantibody profiling in patients with very early rheumatoid arthritis: A follow-up study. *Ann. Rheum. Dis.* **2010**, *69*, 169–174. [CrossRef] [PubMed]

38. Abbafati, C.; Abbas, K.M.; Abbasi-Kangevari, M.; Abd-Allah, F.; Abdelalim, A.; Abdollahi, M.; Abdollahpour, I.; Abegaz, K.H.; Abolhassani, H.; Aboyans, V.; et al. Global burden of 369 diseases and injuries in 204 countries and territories, 1990–2019: A systematic analysis for the Global Burden of Disease Study 2019. *Lancet* **2020**, *396*, 1204–1222. [CrossRef]
39. Raine, C.; Giles, I. What is the impact of sex hormones on the pathogenesis of rheumatoid arthritis? *Front. Med.* **2022**, *9*, 909879. [CrossRef] [PubMed]
40. Frisell, T.; Holmqvist, M.; Källberg, H.; Klareskog, L.; Alfredsson, L.; Askling, J. Familial risks and heritability of rheumatoid arthritis: Role of rheumatoid factor/anti-citrullinated protein antibody status, number and type of affected relatives, sex, and age. *Arthritis Rheum.* **2013**, *65*, 2773–2782. [CrossRef] [PubMed]
41. Padyukov, L. Genetics of rheumatoid arthritis. *Semin. Immunopathol.* **2022**, *44*, 47–62. [CrossRef]
42. Scherer, H.U.; Häupl, T.; Burmester, G.R. The etiology of rheumatoid arthritis. *J. Autoimmun.* **2020**, *110*, 102400. [CrossRef]
43. Abbasifard, M.; Imani, D.; Bagheri-Hosseinabadi, Z. PTPN22 gene polymorphism and susceptibility to rheumatoid arthritis (RA): Updated systematic review and meta-analysis. *J. Gene Med.* **2020**, *22*, e3204. [CrossRef]
44. Karami, J.; Aslani, S.; Tahmasebi, M.N.; Mousavi, M.J.; Sharafat Vaziri, A.; Jamshidi, A.; Farhadi, E.; Mahmoudi, M. Epigenetics in rheumatoid arthritis; fibroblast-like synoviocytes as an emerging paradigm in the pathogenesis of the disease. *Immunol. Cell Biol.* **2020**, *98*, 171–186. [CrossRef]
45. Schäfer, C.; Keyßer, G. Lifestyle Factors and Their Influence on Rheumatoid Arthritis: A Narrative Review. *J. Clin. Med.* **2022**, *11*, 7179. [CrossRef]
46. Klareskog, L.; Malmström, V.; Lundberg, K.; Padyukov, L.; Alfredsson, L. Smoking, citrullination and genetic variability in the immunopathogenesis of rheumatoid arthritis. *Semin. Immunol.* **2011**, *23*, 92–98. [CrossRef] [PubMed]
47. Guo, C.; Fu, R.; Wang, S.; Huang, Y.; Li, X.; Zhou, M.; Zhao, J.; Yang, N. NLRP3 inflammasome activation contributes to the pathogenesis of rheumatoid arthritis. *Clin. Exp. Immunol.* **2018**, *194*, 231–243. [CrossRef] [PubMed]
48. Montgomery, A.B.; Kopec, J.; Shrestha, L.; Thezenas, M.L.; Burgess-Brown, N.A.; Fischer, R.; Yue, W.W.; Venables, P.J. Crystal structure of Porphyromonas gingivalis peptidylarginine deiminase: Implications for autoimmunity in rheumatoid arthritis. *Ann. Rheum. Dis.* **2016**, *75*, 1255–1261. [CrossRef] [PubMed]
49. Zaiss, M.; Wu, J.; Mauro, D.; Schett, G.; Ciccia, F. The gut-joint axis in rheumatoid arthritis. *Nat. Rev. Rheumatol.* **2021**, *17*, 224–237. [CrossRef]
50. Liu, X.; Zeng, B.; Zhang, J.; Li, W.; Mou, F.; Wang, H.; Zou, Q.; Zhong, B.; Wu, L.; Wei, H.; et al. Role of the Gut Microbiome in Modulating Arthritis Progression in Mice. *Sci. Rep.* **2016**, *6*, 1–11. [CrossRef]
51. Arleevskaya, M.; Boulygina, E.; Brooks, W.H.; Renaudineau, Y. Microbiota analysis in rheumatoid arthritis: News and perspectives. *Clin. Microbiol. Infect. Dis.* **2019**, *4*, 1–4. [CrossRef]
52. Townsend, M.J. Molecular and cellular heterogeneity in the Rheumatoid Arthritis synovium: Clinical correlates of synovitis. *Best Pract. Res. Clin. Rheumatol.* **2014**, *28*, 539–549. [CrossRef]
53. Zhang, Y.; Wang, S.; Song, S.; Yang, X.; Jin, G. Ginsenoside Rg3 Alleviates Complete Freund's Adjuvant-Induced Rheumatoid Arthritis in Mice by Regulating CD4+CD25+Foxp3+Treg Cells. *J. Agric. Food Chem.* **2020**, *68*, 4893–4902. [CrossRef]
54. Angelini, J.; Talotta, R.; Roncato, R.; Fornasier, G.; Barbiero, G.; Cin, L.D.; Brancati, S.; Scaglione, F. JAK-inhibitors for the treatment of rheumatoid arthritis: A focus on the present and an outlook on the future. *Biomolecules* **2020**, *10*, 1002. [CrossRef]
55. Yamaoka, K.; Oku, K. JAK inhibitors in rheumatology. *Immunol. Med.* **2023**, 1–10. [CrossRef]
56. McNamee, K.; Williams, R.; Seed, M. Animal models of rheumatoid arthritis: How informative are they? *Eur. J. Pharmacol.* **2015**, *759*, 278–286. [CrossRef] [PubMed]
57. Zhao, T.; Xie, Z.; Xi, Y.; Liu, L.; Li, Z.; Qin, D. How to Model Rheumatoid Arthritis in Animals: From Rodents to Non-Human Primates. *Front. Immunol.* **2022**, *13*, 1–8. [CrossRef] [PubMed]
58. Brand, D.D.; Latham, K.A.; Rosloniec, E.F. Collagen-induced arthritis. *Nat. Protoc.* **2007**, *2*, 1269–1275. [CrossRef] [PubMed]
59. Xu, R.; Fazio, G.C.; Matsuda, S.P.T. On the origins of triterpenoid skeletal diversity. *Phytochemistry* **2004**, *65*, 261–291. [CrossRef]
60. Miettinen, K.; Iñigo, S.; Kreft, L.; Pollier, J.; De Bo, C.; Botzki, A.; Coppens, F.; Bak, S.; Goossens, A. The TriForC database: A comprehensive up-to-date resource of plant triterpene biosynthesis. *Nucleic Acids Res.* **2018**, *46*, D586–D594. [CrossRef]
61. Cárdenas, P.D.; Almeida, A.; Bak, S. Evolution of Structural Diversity of Triterpenoids. *Front. Plant Sci.* **2019**, *10*, 1523. [CrossRef]
62. Thimmappa, R.; Geisler, K.; Louveau, T.; O'Maille, P.; Osbourn, A. Triterpene biosynthesis in plants. *Annu. Rev. Plant Biol.* **2014**, *65*, 225–257. [CrossRef]
63. Jing, M.; Yang, J.; Zhang, L.; Liu, J.; Xu, S.; Wang, M.; Zhang, L.; Sun, Y.; Yan, W.; Hou, G.; et al. Celastrol inhibits rheumatoid arthritis through the ROS-NF-κB-NLRP3 inflammasome axis. *Int. Immunopharmacol.* **2021**, *98*, 107879. [CrossRef]
64. Yan, C.Y.; Ouyang, S.H.; Wang, X.; Wu, Y.P.; Sun, W.Y.; Duan, W.J.; Liang, L.; Luo, X.; Kurihara, H.; Li, Y.F.; et al. Celastrol ameliorates Propionibacterium acnes/LPS-induced liver damage and MSU-induced gouty arthritis via inhibiting K63 deubiquitination of NLRP3. *Phytomedicine* **2021**, *80*, 153398. [CrossRef]
65. Yang, J.; Liu, J.; Li, J.; Jing, M.; Zhang, L.; Sun, M.; Wang, Q.; Sun, H.; Hou, G.; Wang, C.; et al. Celastrol inhibits rheumatoid arthritis by inducing autophagy via inhibition of the PI3K/AKT/mTOR signaling pathway. *Int. Immunopharmacol.* **2022**, *112*, 109241. [CrossRef]
66. Ding, J.T.; Hong, F.F.; Yang, S.L. Roles of autophagy in rheumatoid arthritis. *Clin. Exp. Rheumatol.* **2022**, *40*, 2179–2187. [CrossRef] [PubMed]

67. Wang, Z.; Chen, D.; Wang, Z. Effects of diclofenac on the pharmacokinetics of celastrol in rats and its transport. *Pharm. Biol.* **2018**, *56*, 269–274. [CrossRef] [PubMed]
68. Dudics, S.; Venkatesha, S.H.; Moudgil, K.D. The micro-RNA expression profiles of autoimmune arthritis reveal novel biomarkers of the disease and therapeutic response. *Int. J. Mol. Sci.* **2018**, *19*, 2293. [CrossRef] [PubMed]
69. Fang, Z.; He, D.; Yu, B.; Liu, F.; Zuo, J.; Li, Y.; Lin, Q.; Zhou, X.; Wang, Q. High-throughput study of the effects of celastrol on activated fibroblast-like synoviocytes from patients with rheumatoid arthritis. *Genes* **2017**, *8*, 221. [CrossRef]
70. Gao, Q.; Qin, H.; Zhu, L.; Li, D.; Hao, X. Celastrol attenuates collagen-induced arthritis via inhibiting oxidative stress in rats. *Int. Immunopharmacol.* **2020**, *84*, 106527. [CrossRef] [PubMed]
71. Song, X.; Zhang, Y.; Dai, E.; Du, H.; Wang, L. Mechanism of action of celastrol against rheumatoid arthritis: A network pharmacology analysis. *Int. Immunopharmacol.* **2019**, *74*, 105725. [CrossRef]
72. Lv, M.; Liang, Q.; Luo, Z.; Han, B.; Ni, T.; Wang, Y.; Tao, L.; Lyu, W.; Xiang, J.; Liu, Y. UPLC-LTQ-Orbitrap-Based Cell Metabolomics and Network Pharmacology Analysis to Reveal the Potential Antiarthritic Effects of Pristimerin: In Vitro, In Silico and In Vivo Study. *Metabolites* **2022**, *12*, 839. [CrossRef]
73. Wang, J.; Zhao, Q. Betulinic acid inhibits cell proliferation, migration, and inflammatory response in rheumatoid arthritis fibroblast-like synoviocytes. *J. Cell. Biochem.* **2019**, *120*, 2151–2158. [CrossRef]
74. Huimin, D.; Hui, C.; Guowei, S.; Shouyun, X.; Junyang, P.; Juncheng, W. Protective effect of betulinic acid on Freund's complete adjuvant-induced arthritis in rats. *J. Biochem. Mol. Toxicol.* **2019**, *33*, e22373. [CrossRef]
75. Li, N.; Gong, Z.; Li, X.; Ma, Q.; Wu, M.; Liu, D.; Deng, L.; Pan, D.; Liu, Q.; Wei, Z.; et al. Betulinic acid inhibits the migration and invasion of fibroblast-like synoviocytes from patients with rheumatoid arthritis. *Int. Immunopharmacol.* **2019**, *67*, 186–193. [CrossRef]
76. Argnani, L.; Zanetti, A.; Carrara, G.; Silvagni, E.; Guerrini, G.; Zambon, A.; Scirè, C.A. Rheumatoid Arthritis and Cardiovascular Risk: Retrospective Matched-Cohort Analysis Based on the RECORD Study of the Italian Society for Rheumatology. *Front. Med.* **2021**, *8*, 745601. [CrossRef] [PubMed]
77. Skeoch, S.; Bruce, I.N. Atherosclerosis in rheumatoid arthritis: Is it all about inflammation? *Nat. Rev. Rheumatol.* **2015**, *11*, 390–400. [CrossRef]
78. Qasim, S.; Alamgeer; Kalsoom, S.; Shahzad, M.; Bukhari, I.A.; Vohra, F.; Afzal, S. Rosuvastatin Attenuates Rheumatoid Arthritis-Associated Manifestations via Modulation of the Pro-and Anti-inflammatory Cytokine Network: A Combination of in Vitro and in Vivo Studies. *ACS Omega* **2021**, *6*, 2074–2084. [CrossRef] [PubMed]
79. Koushki, K.; Shahbaz, S.K.; Mashayekhi, K.; Sadeghi, M.; Zayeri, Z.D.; Taba, M.Y.; Banach, M.; Al-Rasadi, K.; Johnston, T.P.; Sahebkar, A. Anti-inflammatory Action of Statins in Cardiovascular Disease: The Role of Inflammasome and Toll-Like Receptor Pathways. *Clin. Rev. Allergy Immunol.* **2021**, *60*, 175–199. [CrossRef] [PubMed]
80. Mathew, L.E.; Rajagopal, V.; Helen, A. Betulinic acid and fluvastatin exhibits synergistic effect on toll-like receptor-4 mediated anti-atherogenic mechanism in type II collagen induced arthritis. *Biomed. Pharmacother.* **2017**, *93*, 681–694. [CrossRef]
81. Xu, J.; Li, Z.; Luo, J.; Yang, F.; Liu, T.; Liu, M.; Qiu, W.W.; Tang, J. Synthesis and biological evaluation of heterocyclic ring-fused betulinic acid derivatives as novel inhibitors of osteoclast differentiation and bone resorption. *J. Med. Chem.* **2012**, *55*, 3122–3134. [CrossRef]
82. Chen, S.; Bai, Y.; Li, Z.; Jia, K.; Siwko, S.; Liu, M.; Jin, Y.; He, B.; Chen, H.; Liu, M.; et al. A betulinic acid derivative SH479 inhibits collagen-induced arthritis by modulating T cell differentiation and cytokine balance. *Biochem. Pharmacol.* **2017**, *126*, 69–78. [CrossRef]
83. Song, T.; Shi, R.; Vijayalakshmi, A.; Lei, B. Protective effect of lupeol on arthritis induced by type II collagen via the suppression of P13K/AKT signaling pathway in Sprague dawley rats. *Environ. Toxicol.* **2022**, *37*, 1814–1822. [CrossRef]
84. Henneh, I.T.; Huang, B.; Musayev, F.N.; Al Hashimi, R.; Safo, M.K.; Armah, F.A.; Ameyaw, E.O.; Adokoh, C.K.; Ekor, M.; Zhang, Y. Structural elucidation and in vivo anti-arthritic activity of β-amyrin and polpunonic acid isolated from the root bark of *Ziziphus abyssinica* HochstEx. A Rich (Rhamnaceae). *Bioorg. Chem.* **2020**, *98*, 103744. [CrossRef]
85. Cheng, Y.C.; Zhang, X.; Lin, S.C.; Li, S.; Chang, Y.K.; Chen, H.H.; Lin, C.C. Echinocystic Acid Ameliorates Arthritis in SKG Mice by Suppressing Th17 Cell Differentiation and Human Rheumatoid Arthritis Fibroblast-Like Synoviocytes Inflammation. *J. Agric. Food Chem.* **2022**, *70*, 16176–16187. [CrossRef]
86. Guo, Q.; Wang, Y.; Xu, D.; Nossent, J.; Pavlos, N.J.; Xu, J. Rheumatoid arthritis: Pathological mechanisms and modern pharmacologic therapies. *Bone Res.* **2018**, *6*, 15. [CrossRef]
87. Huong, L.T.; Gal, M.; Kim, O.; Tran, P.T.; Nhiem, N.X.; Van Kiem, P.; Van Minh, C.; Dang, N.H.; Lee, J.H. 23-Hydroxyursolic acid from Viburnum lutescens inhibits osteoclast differentiation in vitro and lipopolysaccharide-induced bone loss in vivo by suppressing c-Fos and NF-κB signalling. *Int. Immunopharmacol.* **2022**, *111*, 109038. [CrossRef]
88. Lee, J.Y.; Choi, J.K.; Jeong, N.H.; Yoo, J.; Ha, Y.S.; Lee, B.; Choi, H.; Park, P.H.; Shin, T.Y.; Kwon, T.K.; et al. Anti-inflammatory effects of ursolic acid-3-acetate on human synovial fibroblasts and a murine model of rheumatoid arthritis. *Int. Immunopharmacol.* **2017**, *49*, 118–125. [CrossRef]
89. Shimazu, K.; Fukumitsu, S.; Ishijima, T.; Toyoda, T.; Nakai, Y.; Abe, K.; Aida, K.; Okada, S.; Hino, A. The Anti-Arthritis Effect of Olive-Derived Maslinic Acid in Mice is Due to its Promotion of Tissue Formation and its Anti-Inflammatory Effects. *Mol. Nutr. Food Res.* **2019**, *63*, 1800543. [CrossRef] [PubMed]

90. Chen, J.; Wu, W.; Zhang, M.; Chen, C. Taraxasterol suppresses inflammation in IL-1β-induced rheumatoid arthritis fibroblast-like synoviocytes and rheumatoid arthritis progression in mice. *Int. Immunopharmacol.* **2019**, *70*, 274–283. [CrossRef] [PubMed]
91. Siouti, E.; Andreakos, E. The many facets of macrophages in rheumatoid arthritis. *Biochem. Pharmacol.* **2019**, *165*, 152–169. [CrossRef] [PubMed]
92. Cao, Y.; Liu, J.; Huang, C.; Tao, Y.; Wang, Y.; Chen, X.; Huang, D. Wilforlide A ameliorates the progression of rheumatoid arthritis by inhibiting M1 macrophage polarization. *J. Pharmacol. Sci.* **2022**, *148*, 116–124. [CrossRef] [PubMed]
93. Achudhan, D.; Liu, S.C.; Lin, Y.Y.; Huang, C.C.; Tsai, C.H.; Ko, C.Y.; Chiang, I.P.; Kuo, Y.H.; Tang, C.H. Antcin K Inhibits TNF-α, IL-1β and IL-8 Expression in Synovial Fibroblasts and Ameliorates Cartilage Degradation: Implications for the Treatment of Rheumatoid Arthritis. *Front. Immunol.* **2021**, *12*, 790925. [CrossRef]
94. Wang, T.; Lu, H. Ganoderic acid A inhibits ox-LDL-induced THP-1-derived macrophage inflammation and lipid deposition via Notch1/PPARγ/CD36 signaling. *Adv. Clin. Exp. Med.* **2021**, *30*, 1031–1041. [CrossRef]
95. Cao, T.; Tang, C.; Xue, L.; Cui, M.; Wang, D. Protective effect of Ganoderic acid A on adjuvant-induced arthritis. *Immunol. Lett.* **2020**, *226*, 1–6. [CrossRef]
96. Chen, J.Y.; Tian, X.Y.; Liu, W.J.; Wu, B.K.; Wu, Y.C.; Zhu, M.X.; Liu, J.W.; Zhou, X.; Zheng, Y.F.; Ma, X.Q.; et al. Importance of Gedunin in Antagonizing Rheumatoid Arthritis via Activating the Nrf2/ARE Signaling. *Oxid. Med. Cell. Longev.* **2022**, *2022*, 6277760. [CrossRef] [PubMed]
97. Chen, J.Y.; Zhu, G.Y.; Sun, Y.B.; Wu, Y.C.; Wu, B.K.; Zheng, W.T.; Ma, X.Q.; Zheng, Y.F. 7-deacetyl-gedunin suppresses proliferation of Human rheumatoid arthritis synovial fibroblast through activation of Nrf2/ARE signaling. *Int. Immunopharmacol.* **2022**, *107*, 108557. [CrossRef] [PubMed]
98. Anchi, P.; Swamy, V.; Godugu, C. Nimbolide exerts protective effects in complete Freund's adjuvant induced inflammatory arthritis via abrogation of STAT-3/NF-κB/Notch-1 signaling. *Life Sci.* **2021**, *266*, 118911. [CrossRef]
99. Cui, X.; Wang, R.; Bian, P.; Wu, Q.; Seshadri, V.D.D.; Liu, L. Evaluation of antiarthritic activity of nimbolide against Freund's adjuvant induced arthritis in rats. *Artif. Cells Nanomed. Biotechnol.* **2019**, *47*, 3391–3398. [CrossRef] [PubMed]
100. Yang, Y.; Jian, Y.Q.; Liu, Y.B.; Xie, Q.L.; Yu, H.H.; Wang, B.; Li, B.; Peng, C.Y.; Wang, W. Heilaohuacid G, a new triterpenoid from Kadsura coccinea inhibits proliferation, induces apoptosis, and ameliorates inflammation in RA-FLS and RAW 264.7 cells via suppressing NF-κB pathway. *Phyther. Res.* **2022**, *36*, 3900–3910. [CrossRef] [PubMed]
101. Jiang, H.; Fan, C.; Lu, Y.; Cui, X.; Liu, J. Astragaloside regulates lncRNA LOC100912373 and the miR-17-5p/PDK1 axis to inhibit the proliferation of fibroblast-like synoviocytes in rats with rheumatoid arthritis. *Int. J. Mol. Med.* **2021**, *48*, 1–10. [CrossRef]
102. Lodde, V.; Murgia, G.; Simula, E.R.; Steri, M.; Floris, M.; Idda, M.L. Long noncoding RNAs and circular RNAs in autoimmune diseases. *Biomolecules* **2020**, *10*, 1044. [CrossRef]
103. Fan, C.; Cui, X.; Chen, S.; Huang, S.; Jiang, H. LncRNA LOC100912373 modulates PDK1 expression by sponging miR-17-5p to promote the proliferation of fibroblast-like synoviocytes in rheumatoid arthritis. *Am. J. Transl. Res.* **2021**, *12*, 7709–7723.
104. Guo, X.; Ji, J.; Zhang, J.; Hou, X.; Fu, X.; Luo, Y.; Mei, Z.; Feng, Z. Anti-inflammatory and osteoprotective effects of Chikusetsusaponin IVa on rheumatoid arthritis via the JAK/STAT signaling pathway. *Phytomedicine* **2021**, *93*, 153801. [CrossRef]
105. Schinnerling, K.; Aguillón, J.C.; Catalán, D.; Soto, L. The role of interleukin-6 signalling and its therapeutic blockage in skewing the T cell balance in rheumatoid arthritis. *Clin. Exp. Immunol.* **2017**, *189*, 12–20. [CrossRef]
106. Yang, J.; Ling, Y.; Hua, D.; Zhao, C.; Sun, X.; Cai, X.; Chen, J.; Hu, C.; Cao, P. IL-6R blockade by chikusetsusaponin IVa butyl ester inhibits Th17 cell differentiation and ameliorates collagen-induced arthritis. *Cell. Mol. Immunol.* **2021**, *18*, 1584–1586. [CrossRef] [PubMed]
107. Li, Z.; Wang, J.; Lin, Y.; Fang, J.; Xie, K.; Guan, Z.; Ma, H.; Yuan, L. Newly discovered circRNAs in rheumatoid arthritis, with special emphasis on functional roles in inflammatory immunity. *Front. Pharmacol.* **2022**, *13*, 983744. [CrossRef] [PubMed]
108. Wang, X.; Zhou, D.; Zhou, W.; Liu, J.; Xue, Q.; Huang, Y.; Cheng, C.; Wang, Y.; Chang, J.; Wang, P.; et al. Clematichinenoside AR inhibits the pathology of rheumatoid arthritis by blocking the circPTN/miR-145-5p/FZD4 signal axis. *Int. Immunopharmacol.* **2022**, *113*, 109376. [CrossRef] [PubMed]
109. Xiong, Y.; Ma, Y.; Kodithuwakku, N.D.; Fang, W.; Liu, L.; Li, F.; Hu, Y.; Li, Y. Protective effects of Clematichinenoside AR against inflammation and cytotoxicity induced by human tumor necrosis factor-α. *Int. Immunopharmacol.* **2019**, *75*, 105563. [CrossRef]
110. Xiong, H.; Luo, M.; Ju, Y.; Zhao, Z.; Zhang, M.; Xu, R.; Ren, Y.; Yang, G.; Mei, Z. Triterpene saponins from Guo-gang-long attenuate collagen-induced arthritis via regulating A20 and inhibiting MAPK pathway. *J. Ethnopharmacol.* **2021**, *269*, 113707. [CrossRef]
111. Bandopadhyay, S.; Mandal, S.; Ghorai, M.; Jha, N.K.; Kumar, M.; Radha; Ghosh, A.; Proćków, J.; Pérez de la Lastra, J.M.; Dey, A. Therapeutic properties and pharmacological activities of asiaticoside and madecassoside: A review. *J. Cell. Mol. Med.* **2023**, *27*, 593–608. [CrossRef]
112. Yu, W.G.; Shen, Y.; Wu, J.Z.; Gao, Y.B.; Zhang, L.X. Madecassoside impedes invasion of rheumatoid fibroblast-like synoviocyte from adjuvant arthritis rats via inhibition of NF-κB-mediated matrix metalloproteinase-13 expression. *Chin. J. Nat. Med.* **2018**, *16*, 330–338. [CrossRef]
113. Qiao, S.; Lian, X.; Yue, M.; Zhang, Q.; Wei, Z.; Chen, L.; Xia, Y.; Dai, Y. Regulation of gut microbiota substantially contributes to the induction of intestinal Treg cells and consequent anti-arthritis effect of madecassoside. *Int. Immunopharmacol.* **2020**, *89*, 107047. [CrossRef]

114. Gal, M.; Kim, O.; Tran, P.T.; Huong, L.T.; Nhiem, N.X.; Van Kiem, P.; Dang, N.H.; Lee, J.H. Mussaendoside O, a N-triterpene cycloartane saponin, attenuates RANKL-induced osteoclastogenesis and inhibits lipopolysaccharide-induced bone loss. *Phytomedicine* **2022**, *105*, 154378. [CrossRef]
115. Wang, C.; Gao, M.; Gao, D.; Guo, Y.-H.; Gao, Z.; Gao, X.-J.; Wang, J.-Q. Tubeimoside-1: A review of its antitumor effects, pharmacokinetics, toxicity, and targeting preparations. *Front. Microbiol.* **2022**, *13*, 941270. [CrossRef]
116. Liu, Z.; Zhou, L.; Ma, X.; Sun, S.; Qiu, H.; Li, H.; Xu, J.; Liu, M. Inhibitory effects of tubeimoside I on synoviocytes and collagen-induced arthritis in rats. *J. Cell. Physiol.* **2018**, *233*, 8740–8753. [CrossRef] [PubMed]
117. Zhang, M.; Ren, H.; Li, K.; Xie, S.; Zhang, R.; Zhang, L.; Xia, J.; Chen, X.; Li, X.; Wang, J. Therapeutic effect of various ginsenosides on rheumatoid arthritis. *BMC Complement. Med. Ther.* **2021**, *21*, 1–12. [CrossRef] [PubMed]
118. Wang, R.; Zhang, M.; Hu, S.; Liu, K.; Tai, Y.; Tao, J.; Zhou, W.; Zhao, Z.; Wang, Q.; Wei, W. Ginsenoside metabolite compound-K regulates macrophage function through inhibition of β-arrestin2. *Biomed. Pharmacother.* **2019**, *115*, 108909. [CrossRef]
119. Zhang, M.; Hu, S.; Tao, J.; Zhou, W.; Wang, R.; Tai, Y.; Xiao, F.; Wang, Q.; Wei, W. Ginsenoside compound-K inhibits the activity of B cells through inducing IgD-B cell receptor endocytosis in mice with collagen-induced arthritis. *Inflammopharmacology* **2019**, *27*, 845–856. [CrossRef] [PubMed]
120. Chen, L.; Zhou, L.; Wang, Y.; Yang, G.; Huang, J.; Tan, Z.; Wang, Y.; Zhou, G.; Liao, J.; Ouyang, D. Food and sex-related impacts on the pharmacokinetics of a single-dose of ginsenoside compound K in healthy subjects. *Front. Pharmacol.* **2017**, *8*, 1–13. [CrossRef]
121. Wang, C.; Gao, Y.; Zhang, Z.; Chen, C.; Chi, Q.; Xu, K.; Yang, L. Ursolic acid protects chondrocytes, exhibits anti-inflammatory properties via regulation of the NF-κB/NLRP3 inflammasome pathway and ameliorates osteoarthritis. *Biomed. Pharmacother.* **2020**, *130*, 110568. [CrossRef]
122. Liu, T.; Zhang, L.; Joo, D.; Sun, S.C. NF-κB signaling in inflammation. *Signal Transduct. Target. Ther.* **2017**, *2*, e17023. [CrossRef]

Disclaimer/Publisher's Note: The statements, opinions and data contained in all publications are solely those of the individual author(s) and contributor(s) and not of MDPI and/or the editor(s). MDPI and/or the editor(s) disclaim responsibility for any injury to people or property resulting from any ideas, methods, instructions or products referred to in the content.

 life

Review

Plumbagin, a Natural Compound with Several Biological Effects and Anti-Inflammatory Properties

Giovannamaria Petrocelli [1], Pasquale Marrazzo [1,*], Laura Bonsi [1], Federica Facchin [1], Francesco Alviano [2,†] and Silvia Canaider [1,†]

1 Department of Medical and Surgical Sciences, University of Bologna, 40126 Bologna, BO, Italy; giovannam.petrocell2@unibo.it (G.P.); laura.bonsi@unibo.it (L.B.); federica.facchin2@unibo.it (F.F.); silvia.canaider@unibo.it (S.C.)

2 Department of Biomedical and Neuromotor Science, University of Bologna, 40126 Bologna, BO, Italy; francesco.alviano@unibo.it

* Correspondence: pasquale.marrazzo2@unibo.it

† These authors contributed equally to this work.

Abstract: Phytochemicals from various medicinal plants are well known for their antioxidant properties and anti-cancer effects. Many of these bioactive compounds or natural products have demonstrated effects against inflammation, while some showed a role that is only approximately described as anti-inflammatory. In particular, naphthoquinones are naturally-occurring compounds with different pharmacological activities and allow easy scaffold modification for drug design approaches. Among this class of compounds, Plumbagin, a plant-derived product, has shown interesting counteracting effects in many inflammation models. However, scientific knowledge about the beneficial effect of Plumbagin should be comprehensively reported before candidating this natural molecule into a future drug against specific human diseases. In this review, the most relevant mechanisms in which Plumbagin plays a role in the process of inflammation were summarized. Other relevant bioactive effects were reviewed to provide a complete and compact scenario of Plumbagin's potential therapeutic significance.

Keywords: plumbagin; plumbaginaceae; natural compound; inflammation; anti-inflammatory; antioxidant; antibacterial; antiparasitic; anti-senescence; stem cells

Citation: Petrocelli, G.; Marrazzo, P.; Bonsi, L.; Facchin, F.; Alviano, F.; Canaider, S. Plumbagin, a Natural Compound with Several Biological Effects and Anti-Inflammatory Properties. *Life* **2023**, *13*, 1303. https://doi.org/10.3390/life13061303

Academic Editors: Azahara Rodríguez-Luna and Salvador González

Received: 2 May 2023
Revised: 23 May 2023
Accepted: 29 May 2023
Published: 31 May 2023

Copyright: © 2023 by the authors. Licensee MDPI, Basel, Switzerland. This article is an open access article distributed under the terms and conditions of the Creative Commons Attribution (CC BY) license (https://creativecommons.org/licenses/by/4.0/).

1. Introduction

Over the years, a plethora of biocompounds derived from plants have been used for human healthcare owing to their antimicrobial, antioxidant, anticancer, and anti-inflammatory effects. Many biocompounds are usually plant secondary metabolites, whose biological features make them interesting for human supplementation and the development of new drugs. Among the plant secondary metabolites, naphthoquinones are the largest group [1]. Naphthoquinones are aromatic cyclic compounds that exert important pharmacological activities, such as anticancer [2] and antifertility [3], as well as broad antimicrobial [4], antibacterial [5], and anti-inflammatory effects [6]. The mechanism of action of naphthoquinones is mainly dependent on the redox state of the cells. In fact, these biocompounds inhibit the electron transport during oxidative phosphorylation and generate reactive oxygen radicals in aerobic conditions. At the molecular level, naphthoquinones act also as alkylating and intercalating agents in the DNA double helix [7].

Plumbagin (PB) is a naphthoquinone obtained from the roots of different medicinal plant families, such as *Plumbaginaceae*, *Droseraceae*, and *Ebenaceae*. *Plumbago zeylanica* L. has been considered the most known medicinal plant that contains PB [8,9]. Chemically, PB is structurally similar to vitamin K, while its half-life is rather short [10]. As for most plant-derived molecules, PB shows low solubility in water and, critically, limited oral bioavailability [11] as well as moderate toxicity [12]. Despite that, it still appears to be an intriguing candidate for the development of new therapeutic agents [13,14]. Several studies have shown the biological activities of this molecule, such as those antioxidant, antibacterial, antifungal, analgesic, anti-inflammatory, and anticancer [7,15,16]. In literature, the anticancer effect of PB has been deeply investigated [13,17], while there are no recent reviews focused on the anti-inflammatory activity of this compound.

Inflammation is a process underlying all chronic diseases, such as cancer, autoimmune disease, cardiovascular, and neurodegenerative ones. Over the years, several in vitro and in vivo studies have been carried out on inflammatory models to understand the mechanism underpinning the anti-inflammatory activity of PB. The studies showed how PB can be implicated in inflammation signaling by blocking Nuclear factor-kB (Nf-kB) and Mitogen-activated protein kinase (MAPK) pathways (throughout p38 and c-Jun N-terminal kinase (JNK) cassettes) [18–20]. This review aimed to summarize the current knowledge on the anti-inflammatory activity of PB, as well as on the cytoprotective and anti-senescent actions, with the goal of providing a comprehensive overview of all potential therapeutic applications of this biocompound, including in the regenerative medicine field.

2. The Cytoprotective Activity of PB

The cytoprotective activity of PB is strictly related to its antioxidant activity. In the literature, there are many studies that show PB cytoprotective activity, both in vitro and in vivo models. PB exerts its protective role against H_2O_2-induced damage in several cell models [21,22]. Pre-treatment with PB for 24 h in neuronal cells [22] or chondrocytes [19] exposed to H_2O_2 promoted cell viability to 100%. The increase in cell viability after PB pre-treatment was associated with an enhancement in antioxidant enzyme activities (superoxide dismutase—SOD, catalase—CAT, glutathione S-transferase—GST, glutathione peroxidase—GPx) and the expression of nuclear factor erythroid 2-related factor 2 (Nrf-2) [21–23]. Importantly, Nrf-2 plays a pivotal role in cellular resistance to oxidative agents. Those results have been replicated also in osteoblasts [24]. Indeed, PB pre-treatment increased cell viability (around 90%) and downregulated the expression of caspases 3, 8, and 9 in association with an increase in Nrf-2 expression [24]. The cytoprotective activity of PB occurred not only when it was administrated as pre-treatment, but also after the administration of H_2O_2, as demonstrated by Chu and co-workers, on rat nucleus pulposus cells [25]. Among the cytoprotective activities, PB exerts the antiapoptotic one. This is also related to a downregulation of the pro-apoptotic protein Bcl-2-Associated X-protein (Bax) and an upregulation of the anti-apoptotic protein B-cell lymphoma 2 (Bcl-2) [26]. The pathways implied in the anti-apoptotic effect of PB have also been validated in an ischemia mouse model [27]. It is well-known that apoptosis is strictly related to mitochondrial membrane potential, since a marked reduction of this potential leads to cell apoptosis. It has been shown that dexamethasone induced apoptosis in osteoblasts by reducing mitochondrial membrane potential and that this effect was restored by treatment with PB [24]. Similar results have been obtained on oxygen-glucose deprivation/reoxygenation (OGDR)-induced neuroinjury in human SH-SY5Y 2D [28] and a 3D culture model [29]. In vivo experiments, in addition to the neuroprotective [27] role of PB, showed hepatoprotective potential [30]. By using a murine model of acute liver injury, Wang and colleagues found enhanced liver regeneration after PB administration. This was related to an increase in hepatocyte proliferation and reduced apoptosis levels [30]. Autophagy as a protective mechanism induced by PB was reported by Pan and colleagues in rat hepatic tissue of bile duct ligation (BDL)-induced cholestatic liver injury [31].

Recently, researchers have shown interest in the potential cardioprotective role of PB. Cardiovascular disease is the world's leading cause of death [32]. Among the several causes, this disease is triggered by chemotherapeutic agents, such as doxorubicin. Doxorubicin-mediated cardiovascular toxicity is a restricting factor that limits the wide usage of this chemotherapy drug [33]. Recently, Li and co-workers showed an interesting protective role of PB against doxorubicin-stimulated cardiotoxicity in rats [34]. PB treatment reduced the circulant levels of cardiac damage markers (lactate—LDH, aspartate aminotransferase—AST, and creatine kinase—CK) in serum of PB-supplemented animals. Moreover, in rats, PB enhanced the expression of phosphatase and tensin homolog (pTEN) protein and inhibited the phosphoinositide-3-kinase/protein kinase B or Akt (PI3K/Akt) signaling pathway that was able to activate the apoptosis cascade [34]. The inhibition of apoptosis and ROS levels following PB treatment in cardiomyocytes was also investigated by Zhang and co-workers [35] by using H9c2 cardiomyocytes. In this study, the cytoprotective effect of PB was investigated in a model of oxidative stress induced by tertiary butyl hydrogen peroxide (THBP). Pre-treatment with PB increased cell viability, and decreased both ROS levels and apoptotic cell rate, according to the data of previous works [21,22,24]. The novelty of this work was the identification of a new mechanism of action whereby PB reduces apoptosis. In fact, the authors demonstrated an increase in the autophagic process in their in vitro model. In particular, the autophagic activity was stimulated by the inhibition of the p38 MAPK pathway, increasing microtubule-associated protein light chain 3 (LC3)II/LC3I ratio [35].

So far, the data obtained are encouraging. However, more in vivo studies must be carried out in order to ensure PB as a cytoprotective molecule in therapeutic applications.

3. The Anti-Inflammatory Activity of PB

As mentioned above, over the years the anti-inflammatory role of PB has gained interest due to a putative implication in the therapeutic field. Inflammation can be characterized by an imbalance of both reactive oxygen/nitric species (ROS/RNS) and pro-inflammatory cytokines. The equilibrium between pro-oxidant ROS or RNS and antioxidant agents (such as CAT, GPx, and SOD), cellular metabolism, and respiration determines the redox homeostasis of the cells. Redox status is also represented by the endogenous antioxidant glutathione GSH/GSSG ratio, which is central also for the immune system function [36]. Here, glutathione level plays a pivotal role in the activation, proliferation, and survival of cells. Checker and collaborators suggested that PB may induce a change in the redox status of murine lymphocytes [37]. In this cellular model, PB determined an increase in ROS levels and free thiols and altered the glutathionylation state of the proteins by binding the GSH. Moreover, after the PB treatment, the Nf-kB factor resulted in a glutathionylated form that prevented its nuclear translocation and the following activation of the survival cascade in the lymphocytes [37]. The same mechanism of action was described by Wang and colleagues in 2014 by using a RAW264.7 cellular model [19] and by Checker and co-worker in a mouse model of endotoxic shock [20]. In this model, the administration of endotoxins usually caused lethality in 80–90% of cases, while treatment with PB 2 h prior to lipopolysaccharides (LPS) administration increased the survival rate to approximately 90%. PB treatment reduced the circulating levels of pro-inflammatory cytokines such as tumor necrosis factor (TNF) α, interleukin (IL) 6, and IL1 β, as well as nitric oxide (NO). Moreover, the histopathological analysis showed an improvement in granulocyte and lymphocyte infiltration in the liver and lungs, respectively [20]. PB protection against sepsis and endotoxic shock was also validated by Zhang and colleagues [38]. The researchers investigated the protective effect during inflammation of PB on bone marrow-derived macrophages: PB inhibited the activation of pyruvate kinase M2 (PKM2) enzyme, necessary for the metabolic switch from oxidative phosphorylation to glycolysis, which usually occurs in the activated macrophages. In the sepsis mouse model, after PB treatment, a reduction of IL1 β levels, but also high mobility group box 1 (HMGB1) (late intermediate of inflammation) [38], was observed. All things considered, the anti-inflammatory action of PB seems to be related to its pro-oxidant role. By reducing free GSH levels, PB increased the production of ROS and Nf-kB oxidation, preventing its binding to DNA [20]. The nuclear translocation of Nf-kB is also related to the absence of NF-kappa-B inhibitor alpha (IkBa) protein. During the inflammatory response, IkBa is degraded. PB stabilized IkBa, preventing Nf-kB nuclear translocation [18,20]. Arruri and collaborators showed that PB treatment on a rat model of neuropathic pain led to a decrease of NO levels, as previously described [18,20], and of pro-inflammatory cytokines by inhibiting the Nf-kB pathway and activating Nrf-2 [39].

PB counteracts inflammation consequences not only by acting as a pro-oxidant effector. In fact, several works showed the antioxidant effect of this natural molecule in the context of inflammation models. Research on the anti-inflammatory effect of PB was performed on microglia [40], and it is known to stimulate an excessive amount of pro-inflammatory cytokines if activated for a long time, leading to the development of neurodegenerative disorders [41]. LPS-activated microglia showed an increased expression of inducible nitric oxide synthase (iNOS) [42] and of cytokines like IL1 α, granulocyte-colony stimulating factors (G-CSF), IL12, monocyte chemoattractant protein (MCP) 1, and MCP5 [40]: all these effects were reverted by PB treatment [40]. In human SH-SY5Y exposed to OGDR, PB exerted a protective role by inactivating the NLR family pyrin domain containing 3 (NLRP3)-induced inflammasome [28]. The antioxidant role of PB was linked to the upregulation of GSH and GPx enzyme levels [26] and to the inactivation of the Nf-kB/TNF α pathway [26,27]. Recently, Mahmoud and colleagues suggested an additional mechanism of PB as an anti-inflammatory modulator. According to their studies, the anti-inflammatory action of PB was associated with the inactivation of the interleukin-1 receptor-associated kinase (IRAK1), which, in physiological conditions, is implied in the production of IL8, IL6, IL1 β, and TNF α, [43] which are chemokines with a fundament role in inflammation onset.

3.1. PB in Osteoarthritis, Rheumatoid Arthritis, and Osteoporosis

Osteoarthritis is a chronic inflammatory disorder that is mainly found in aged people but is also linked to some disorders, such as obesity or a sedentary lifestyle. Prolonged oxidative stress in the chondrocytes leads to cartilage degeneration [44]. The identification of natural molecules that could be able to modulate oxidative stress and prevent the onset of this pathology is an intriguing field of research. Guo and co-workers investigated in vitro the antioxidant effect of PB on chondrocytes [21]. After H_2O_2 treatment, PB decreased ROS levels and lipid peroxidation and increased the enzymatic activity of GSH, SOD, GST, CAT, and GPx. This was related to the downregulation of Nf-kB, pro-inflammatory cytokines (TNF α, IL18, IL6), cyclooxygenase 2 (COX2), and iNOS, and the upregulation of Nrf2, an inducer of expression in many antioxidant enzymes [21,45]. However, there is still little in vivo evidence of the beneficial effect of PB for osteoarthritis. Therefore, further investigations are necessary to better understand the possible implications of PB in the treatment of this pathology.

Many recent studies highlighted the great potential of PB for the treatment of rheumatoid arthritis (RA), an autoimmune disease associated with chronic inflammation. Among the cytokines implied in the development of this disease, RANK-ligand (RANKL) and IL34 are the most relevant [46,47], playing a pivotal role in remodeling bone. Alteration of their expression can lead to bone loss and chronic inflammation. It was seen that RANKL exerted an osteoclastogenic action and was inhibited by osteoprotegerin (OPG) [48]. Patients with RA showed an increased expression of RANKL and a decreased expression of OPG, thus leading to bone loss [49,50]. RANKL secretion was also related to the expression of IL34. IL17 is a crucial factor regulating the expression of IL34. PB has already been shown to be an IL17 inhibitor [51]. These findings were confirmed by Cui and co-workers. Indeed, the suppression effect of PB on IL17 in synoviocytes derived from RA patients decreased the effects mediated by IL34 [52]. As a result, after PB treatment, synoviocytes showed a decreased expression of RANKL and an increased expression of OPG [52] These findings unraveled the potential use of PB for the treatment of this pathology; further investigations need to be conducted to better understand the underlying molecular mechanism of action of PB.

RA represents also a risk factor for osteoporosis. Osteoporosis usually occurs in women after menopause. However, it can be induced by prolonged treatment with glucocorticoids, since 30–50% of chronically treated patients develop this pathology [53]. In vitro studies on osteoblasts treated with dexamethasone provided downregulation models of osteogenic markers, such as osteocalcin (OCN), osteopontin (OPN), and Runt-related transcription factor 2 (Runx2) [24]. Pre-treatment with PB upregulated the expression of the osteogenic markers, suggesting a potential application of this molecule for the study of osteoporosis [24]. Bone remodeling is related to the Nf-kB pathway and the inflammatory level of the cell. Nf-kB promoted the osteoclastogenic process by inducing RANKL expression [54–56]. Shen and colleagues investigated the effect of PB on osteoclastogenic differentiation [57], demonstrating the involvement of Nf-kB during the process. PB inhibited the action of neutrophil inhibitory factor (NIF), a key factor in the processing of Nf-kB p100 subunit into p52 form, thus resulting in the indirect inhibition of osteoclastic differentiation [57]. A low dosage of PB (0.1–0.3 μM) on bone marrow mesenchymal stem cells increased Runx2 and alkaline phosphatase (ALP) expression but did not affect osterix (OSX) expression, which is an early osteogenic promoter of differentiation. Differently, PB 1 μM inhibited osteogenic differentiation. Hence, PB may promote late osteogenic differentiation and osteoblast maturation in a way that can be dosage-dependent. The authors also obtained interesting results in vivo by using an ovariectomized mouse model with bone impairment: PB-treated mice showed an increase in the thickness of trabeculae and in the number of osteoblasts [57]. Further studies need to be conducted in order to understand the optimal dosage of PB that could promote osteogenic differentiation in vivo and the possibility to use this molecule for therapeutical and regenerative medicine.

Sultanli and co-workers highlighted limitations regarding the usage of PB. According to the investigated cellular model, PB can alternatively promote or inhibit osteoclastic differentiation [58]. PB has been tested on macrophages derived from two different mouse models, C57BL/6 and BALB. In C57Bl/6-derived macrophages, PB promoted the osteoclastic differentiation, by upregulating acid phosphatase 5 (Acp5), Cathepsin-k (Ctsk), and osteoclast-associated Ig-like receptor (Oscar) genes. In BALB-derived macrophages, PB reduced osteoclastic differentiation by inhibiting the nuclear factor of the activated T cell 1 (NFatC1) pathway [58]. In addition, the C57Bl/6-derived macrophages tolerated elevated doses of PB (up to 2 μM), while in the BALB-derived macrophages, the same dosage was toxic for the cells. The authors hypothesized this opposite action was due to the different genetic backgrounds of the mice. Therefore, in the pre-clinical evaluation of PB as a therapeutic treatment for bone diseases, researchers should consider the influence of genetic background and the different responses to inflammation of the employed animal models.

3.2. *The Antifibrotic Activity of PB*

Fibrosis refers to the formation of new fibrotic connective tissue following damage or trauma in an organ. The deposition of fibrotic tissue leads to a loss of function in the damaged organ. The increase in collagen and in extracellular matrix levels is often related to the presence of an inflammatory process, that, if persistent, leads to impaired tissue function. Due to this correlation, researchers have focused their attention on the possible implication of PB in the treatment of fibrotic tissues. Wei and co-workers investigated the effects of PB both in vitro and in vivo [59]. For the in vivo studies, they used a model of carbon tetrachloride (CCl4)-induced hepatic fibrosis. CCL4 increased the circulating levels of alanine transaminase (ALT), AST, IL6, and TNF α and this upregulation was suppressed by PB administration. In vitro studies, PB reduced collagen deposition and alpha-smooth muscle actin (α-SMA) expression in hepatic stellate cells [59]. The same in vivo model was used by Chen and colleagues to investigate the molecular pathways involved in PB action [60]. Here, PB inactivated the epidermal growth factor receptor (EGFR) as well as the signal transducer and activator of transcription 3 (STAT3) signaling. With the same mouse model, a few years ago, Chen and co-workers showed how the antifibrotic activity of PB was connected to its antioxidant properties: PB increased the enzymatic activity of SOD, GSH, and GPx, leading to a reduction of collagen III deposition and α-SMA protein expression [61]. Similar findings were obtained in thioacetamide-induced hepatic fibrosis, where PB inhibited the Nf-kB and Akt pathways in the hepatic stellate cells, blocking the collagen deposition [30]. In addition to the α-SMA protein expression, PB downregulated the transforming growth factor (TGF)1-β protein expression in myofibroblasts and hepatic stellate cells [31].

The antifibrotic activity of PB was studied with success in models of lung fibrosis. The Bleomycin-induced lung fibrosis caused tissue structural damage, inflammation, and collagen deposition by fibroblasts [62]. Moreover, PB seems to inhibit the TGF β pathway, as demonstrated in other works [62–64]. A more in-depth analysis clarified the correlation between PB and TGF-β. PB regulated this factor at the epigenetic level by blocking the histone acetyltransferase p300 [62]. As a consequence, all the genes related to collagen production (such as col1a1, col3a1, α-SMA) were not expressed [62]. In addition to the TGF β pathway, PB also modulated mammalian targets of rapamycin (mTOR) and Akt [64].

3.3. *Antibacterial and Antiparasitic Activity of PB*

Over the years, the bacterial evolution and the genetic mutations together with the excessive and abusive usage of antibiotics led to an increase in antibiotic resistance acquired by several pathogenic microorganisms [65,66]. Consequently, there is a pressing need of investigating new antimicrobial agents. In the literature, several reviews regarding the antimicrobial, antimycotic, and antiparasitic effects of PB were written [1,67,68]. Here, we summarized the most recent works that described the molecular mechanisms underlying PB activities.

Several studies demonstrated that PB, in association with antibiotics or antimycotics, allowed reducing drug dosage and increasing their efficacy [69–71]. For example, the synergistic effect of PB with commercial drugs was investigated in *K. pneumoniae* [70]. By combining PB with gentamicin antibiotic, the dosage of gentamicin decreased from 16 μg/mL to 4 μg/mL. PB augmented the efficacy of gentamicin by promoting the intracellular uptake of the drug into the pathogen [70]. Moreover, the efficacy of PB was also studied on methicillin-resistant *S. aureus* MRSA [72]. Dissanayake and colleagues described an antimicrobial mechanism of action of PB in S. *aureus*: PB inhibited DNA gyrase activity, a topoisomerase relevant for the pathogen duplication, by binging the enzyme active site [73]. Interesting results were obtained in the treatment of colistin-resistant *P. aeruginosa* [71]. The usage of colistin antibiotics has been limited since this drug is nephrotoxic. Researchers combined PB with colistin and observed a synergistic effect stopping biofilm formation. In particular, PB increased pathogen susceptibility to colistin by altering the membrane permeability [71]. A similar alteration of membrane permeability following PB treatment was previously reported by Reddy and colleagues in *B. subtilis* [74].

In 2020, Sarkar and colleagues proposed an additional mechanism of action for PB. When *M. tuberculosis* was treated with PB, PB bound and inhibited the activity of thymidylate synthase X (ThyX), an enzyme implicated in bacterial duplication and is necessary for dTMP synthesis, starting from dUMP [75].

The antiparasitic activity of PB was also tested on *Leishmania* [76] and *C. elegans* [77], where PB perturbated the mitochondrial membrane potential and reduced the number of mitochondria, respectively.

A limit of PB usage consists of its poor solubility in water [11]. Rashidzadeh and co-workers synthesized nanoparticles to optimize the PB delivery and its bioavailability for the treatment of malaria. The nanoparticles carrying PB were tested in vivo on mice infected with *P. berghei*. The obtained data showed a better antiplasmodial activity of PB-loaded micelles in comparison with free PB activity [78]. In addition, this study showed the great potential of the nanoparticles as carriers for PB.

4. The Role of PB in Stem Cells and in Cell Senescence

While the anti-inflammatory effect of PB has been extensively studied, the effect of this natural compound on stem cells or other progenitor cell types is still unknown. Several studies have investigated the antitumoral effect of PB [13,79], specifically focusing on PB and cancer stem cell (CSC) biology. CSCs are responsible for the abnormal growth and metastatic processes of tumors, as well as for the resistance to chemotherapeutic agents [80,81]. Among researchers' goals, identifying the signature and eradication of CSCs by using targeted therapy is the biggest challenge. Pan and co-workers investigated the role of PB on human tongue squamous carcinoma cells [82]: PB promoted cell apoptosis by increasing ROS levels and, on the other hand, it downregulated the expression of stemness markers, such as octamer-binding transcription factor 4 (Oct4), sex-determining region Y-box 2 (Sox2), Nanog and Polycomb complex protein Bmi-1 [82]. Aldehyde dehydrogenase (ALDH) is highly expressed in cells including CD 34^+ cells, c-kit^+ cells, $CD133^+$ cells, and lineage-antigen negative (Lin-) cells, and its expression has been well described in normal and cancer precursor cells of various lineages [83]. Interestingly, PB exerted a selective activity against $ALDH1^+$ cells, which is also specifically expressed by a breast CSC subpopulation [84]. Here, PB reduced stem cell proliferation and the formation of the mammosphere. PB treatment inhibited the Wnt/beta-catenin pathway, blocking the epithelial–mesenchymal transition in breast CSCs. These data were also confirmed on prostate CSCs [85], where PB downregulated fibroblast growth factor (FGF)2 expression, as well as Nanog, ALDH1, and Oct4. These results have been confirmed in vivo on orthotopic xenograft nude mice [86]. All these data are in line with the previous findings [82,84,85].

The putative role of PB in modulating the biological properties of stem cells destined for tissue homeostasis is still unknown. Stem cell properties important for regenerative medicine applications. However, they need to be expanded in vitro to obtain a sufficient number of cells for transplantation. During the in vitro culturing, stem cells undergo replicative senescence [87,88], a process that can be triggered by elevated ROS levels [89,90]. In 2013, a protective role of PB against senescence in human amniotic stem cells was proposed [89]. PB, as an antioxidant agent, reduced senescence by inhibiting NADPH oxidase 4 (Nox4) activity and ROS production, as well as by enhancing stem cell proliferation [91]. PB at 2 μM showed a protective effect, while at higher doses it was cytotoxic [91]. The anti-senescent role of PB was also investigated on hair follicle dermal papilla (DP) cells [92]. DP cells are responsible for hair growth by secreting growth regulatory factors, such as FGF-7/keratinocyte growth factor (KGF) and insulin-like growth factor-1 (IGF-1). Moreover, DP cells seem to play a key role in the development of androgenic alopecia by secreting TGF-β, which negatively regulates hair follicle development [93]. PB treatment promoted DP cell proliferation and downregulated the expression of 5α-reductase type II (SRD5A2), an enzyme that works in senescence and hair follicle development [92]. Thus, PB could be implied in the treatment of androgenic alopecia.

All the reviewed effects of PB are summarized in the table below (Table 1).

Table 1. Summary of the biological effects of PB.

	Cytoprotective Activity		
Biological Effect	**In Vitro**	**In Vivo**	**Ref.**
↑ Cell viability ↑ SOD, CAT, GST, GPx activity ↑ Nrf-2 expression ↓ Cas3, 8, 9	✓		[21–24]
↓ Bax ↑ Bcl2	✓	✓	[26,27]
↑ Mitochondrial membrane potential ↓ Apoptosis	✓		[24,28,29]
↑ Liver regeneration ↑ Autophagy	✓	✓	[30]
↓ LDH, AST, CK ↑ pTEN ↓ PI3K/Akt		✓	[34]
↓ ROS levels ↑ LC3A II/LC3 I ratio	✓		[35]

Table 1. *Cont.*

	Cytoprotective Activity		
Biological Effect	**In Vitro**	**In Vivo**	**Ref.**
	Anti-inflammatory activity		
↑ ROS levels Inactivating Nf-kB	✓		[19,20,37]
↑ Cell viability ↓ IL6, IL1 β, NO		✓	[18,20,39]
Blocking PKM2 ↓ IL1 β, NO, HMGB1	✓		[38]
↑ iNOS, IL1 α, G-CSF, IL12, MCP1, MCP5 Inactivating NLRP3-induced inflammasome	✓		[28,40,42]
↑ GSH, GPx	✓		[26]
Inactivating IRAK1	✓		[43]
↓ ROS levels, lipid peroxidation ↓ iNOS, IL6, IL18, TNF α ↑ Nrf2	✓		[21,45]
↓ RANKL ↑ OPG	✓		[52]
Inhibiting NIF ↑ Runx2, OCN, OPN, ALP → Decreasing osteoporosis	✓	✓	[24,57]
	Antifibrotic activity		
↓ ALT, AST, IL6, TNF α ↑ SOD, GSH, GPx activity ↓ TGF1- β, Collagen III, α-SMA	✓	✓	[30,31,59,61,62]
Inactivating EGFR/STAT3 signaling		✓	[60]
Inhibiting TGF1- β/mTOR/Akt pathways	✓	✓	[64]
	Antibacterial and antiparasitic activity		
↑ Intracellular uptake of gentamicin	✓		[70]
Inhibiting DNA gyrase activity	✓		[73]
↑ Membrane permeability and drug effect	✓		[71,74]
Inhibiting ThyX activity	✓		[75]
↓ Mitochondrial number Altering mitochondrial membrane potential	✓		[76,77]
↓ Plasmodial infection and propagation	✓	✓	[78]
	Stem cells and cell senescence		
↑ ROS levels [1] ↓ Oct4, Sox2, Nanog, BMI-1 [1]	✓		[82]
↓ Proliferation [1] ↓ FGF2, Oct4, Nanog, ALDH1 [1]	✓	✓	[84–86]
↓ ROS levels [2] ↑ Proliferation [2] Inhibiting NOX4 activity	✓		[91]
↑ Proliferation [2] ↓ SRD5A2 → Decreasing senescence [2]	✓		[92]

[1] Cancer stem cells. [2] Normal stem cells. SOD, Superoxide dismutase; CAT, Catalase; GST, Glutathione S-transferase; GPx, Glutathione peroxidase; Nrf-2, Nuclear factor erythroid 2-related factor; Cas 3, 8, 9, Caspase 3, 8, 9; Bax, Bcl-2 associated X-protein; Bcl2, B-cell lymphoma 2; LDH, Lactate dehydrogenase; AST, Aspartate aminotransferase; CK, Creatine kinase; pTEN, Phosphatase and tensin homolog; PI3K/Akt, Phosphoinositide-3-kinase/protein kinase B (Akt); ROS, reactive oxygen species; LC3A II, Microtubule-associated protein light chain 3 II; Nf-kB, Nuclear factor-kB; IL6, Interleukin 6; IL1 β, Interleukin 1 β; NO, Nitric oxide; PKM2, Pyruvate kinases M2; HMGB1, High mobility group box 1; iNOS, Inducible nitric oxide synthase; IL1 α, Interleukin 1 α; G-CSF, Granulocyte-colony stimulating factors; IL12, Interleukin 12; MCP1, Monocyte chemoattractant protein 1; MCP5, Monocyte chemoattractant protein 5; NLRP3, NRL family pyrin domain containing 3; GSH, Glutathione; IRAK1, Interleukin-1-receptor-associated kinase; IL18, Interleukin 18; TNF α, Tumor necrosis factor α; RANKL, RANK-ligand; OPG, Osteoprotegerin; NIF, Neutrophil inhibitory factor; Runx2, Runt-related transcription factor 2; OCN, Osteocalcin; OPN, Osteopontin; ALP, Alkaline phosphatase; ALT, Alanine transaminase; TGF1-β, Transforming growth factor 1-β; α-SMA, α-smooth muscle actin; EGFR/STAT3, Epidermal growth factor receptor/Signal transducer and activator of transcription 3; mTOR, Mammalian target of rapamycin; ThyX, Thymidylate synthase; Oct4, Octamer-binding transcription factor 4; Sox2, Sex determining region Y-box 2; Nanog, Homeobox protein Nanog; ALDH1, Aldehyde dehydrogenase 1; NOX4, NADPH oxidase 4; SRD5A2, 5α-reductase type II. ↑ means "increase in", ↓ means "decrease in", → means "imply".

5. Discussion

Natural quinones, occurring not only in plants but also in animals, have garnered attention due to their pharmacological properties and potential therapeutic significance. Drugs having quinone moiety such as anthracyclines have been used in cancer therapy revolutionizing the clinical practice. Plumbagin (PB), a biocompound occurring in many plant families which showed many interesting biological properties (Figure 1), is a member of the naphthoquinones, the most prominent type of quinones [1].

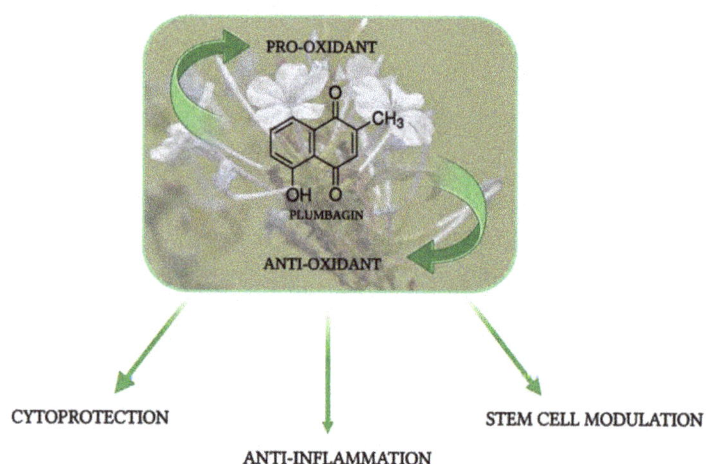

Figure 1. Biological activities related to Plumbagin. (Created with BioRender.com).

On relevance, being a plant-derived small compound, PB should not be specifically delivered, but a simple supplementation and administration can be effective to produce changes in cells. Beneficial effects can be found in how PB indirectly influences the immune system mainly by owning protective effects against many cell types. The PB capacity of promoting the downregulation of inflammation in different pathological models [18–20] is a result not only of its intrinsic anti-inflammatory activity but also of its antimicrobial and antifibrotic effects, in particular, in the pathologies that involve an infectious pathophysiological background.

The influence of PB on stem cell features [91] should be subjected to investigation due to the relevance of clinical research on these cells. If stem cell homeostasis is impaired, many tissues will continue to suffer chronic inflammation insults. The oxidative stress balance appears to be critical in vivo to maintain stem cell properties and retain them in vitro. Small molecules and antioxidants are usually added to stem cell media to improve ex-vivo culture conditions, for maintaining proliferation and prolonging the stemness. On the contrary, CSCs can be a target for many pro-oxidant molecules to avoid metastasis development and drug resistance. Fortunately, the scientific community is presenting encouraging evidence in searching for both these respective effects of PB on normal and cancer stem cells.

Since, the antimicrobial effects have been deeply demonstrated [1,67], not only the biomedical field can take advantage of PB, but it can also be a promising tool if applied in other health relative fields, such as in food packaging safety.

Lastly, considering the antimicrobial effect and wondering about the stem cell modulation, PB may be soon become a valid molecule to be investigated in the regenerative medicine field. In this context, PB may be employed for biomaterial functionalization, thus evolving into a useful tool for drug delivery in future biomedical clinical applications.

Of course, there are some limitations in the use and supplementation of phytochemicals like PB, even though researchers usually and successfully showed a broad spectrum of applications and advantages for inflammatory-based pathologies. As with many phytochemicals, PB has a double face, as pro-oxidant and anti-oxidant, thus its effects on cells and tissues have to be carefully considered, in terms of dosage and in vitro conditions (pH, media composition, oxidative stress, and oxygen levels), because they can determine an unexpected biological effect by revealing concealed mechanisms of action. In this context, it will also be important, before any translational application in the clinical field, to find bioavailable as well as safe doses without complications for the patients. In the future, further chemical modifications of the PB may allow some wanted specificities for targeting single cell types or increase its bioavailability for in vivo studies.

Author Contributions: Conceptualization, P.M.; methodology, G.P. and P.M.; investigation, G.P.; writing—original draft preparation, G.P. and P.M.; writing—review and editing, P.M., L.B., F.F., F.A. and S.C.; visualization, G.P.; supervision, L.B., F.F., F.A. and S.C. All authors have read and agreed to the published version of the manuscript.

Funding: This research received no external funding.

Institutional Review Board Statement: Not applicable.

Informed Consent Statement: Not applicable.

Data Availability Statement: Not applicable.

Acknowledgments: J.M. Garg is the author of the plant photo, available under CC license, which was adapted for creating Figure 1.

Conflicts of Interest: The authors declare no conflict of interest.

References

1. Padhye, S.; Dandawate, P.; Yusufi, M.; Ahmad, A.; Sarkar, F.H. Perspectives on medicinal properties of plumbagin and its analogs. *Med. Res. Rev.* **2012**, *32*, 1131–1158. [CrossRef]
2. Aziz, M.H.; Dreckschmidt, N.E.; Verma, A.K. Plumbagin, a medicinal plant-derived naphthoquinone, is a novel inhibitor of the growth and invasion of hormone-refractory prostate cancer. *Cancer Res.* **2008**, *68*, 9024–9032. [CrossRef]
3. Shivani, D.; Venkatesh; D'Souza, R.; Shenoy, B.D.; Udupi, R.H.; Udupa, N. Niosomal delivery of plumbagin ester for better anti-fertility activity. *Indian Drugs* **2002**, *39*, 163–165.
4. Ahmad, I.; Mehmood, Z.; Mohammad, F.; Ahmad, S. Antimicrobial potency and synergistic activity of five traditionally used Indian medicinal plants. *J. Med. Aromat. Plant Sci.* **2000**, *23*, 173–176.
5. Jeyachandran, R.; Mahesh, A.; Cindrella, L.; Sudhakar, S.; Pazhanichamy, K. Antibacterial activity of plumbagin and root extracts of *Plumbago zeylanica* L. *Acta Biol. Crac. Ser. Bot.* **2009**, *51*, 17–22.
6. Oyedapo, O.O. Studies on Bioactivity of the Root Extract of Plumbago zeylanica. *Int. J. Pharmacogn.* **1996**, *34*, 365–369. [CrossRef]
7. Panichayupakaranant, P.; Ahmad, M.I. Plumbagin and its role in chronic diseases. *Adv. Exp. Med. Biol.* **2016**, *929*, 229–246. [CrossRef]
8. Shukla, B.; Saxena, S.; Usmani, S.; Kushwaha, P. Phytochemistry and pharmacological studies of *Plumbago zeylanica* L.: A medicinal plant review. *Clin. Phytosci.* **2021**, *7*, 34. [CrossRef]
9. Hafeez, B.B.; Zhong, W.; Fischer, J.W.; Mustafa, A.; Shi, X.; Meske, L.; Hong, H.; Cai, W.; Havighurst, T.; Kim, K.; et al. Plumbagin, a medicinal plant (Plumbago zeylanica)-derived 1,4-naphthoquinone, inhibits growth and metastasis of human prostate cancer PC-3M-luciferase cells in an orthotopic xenograft mouse model. *Mol. Oncol.* **2013**, *7*, 428–439. [CrossRef]
10. Yadav, A.M.; Bagade, M.M.; Ghumnani, S.; Raman, S.; Saha, B.; Kubatzky, K.F.; Ashma, R. The phytochemical plumbagin reciprocally modulates osteoblasts and osteoclasts. *Biol. Chem.* **2021**, *403*, 211–229. [CrossRef]
11. Hsieh, Y.J.; Lin, L.C.; Tsai, T.H. Measurement and pharmacokinetic study of plumbagin in a conscious freely moving rat using liquid chromatography/tandem mass spectrometry. *J. Chromatogr. B Analyt. Technol. Biomed. Life Sci.* **2006**, *844*, 1–5. [CrossRef] [PubMed]
12. Nair, H.A.; Snima, K.S.; Kamath, R.C.; Nair, S.V.; Lakshmanan, V.K. Plumbagin nanoparticles induce dose and pH dependent toxicity on prostate cancer cells. *Curr. Drug Deliv.* **2015**, *12*, 709–716. [CrossRef] [PubMed]
13. Tripathi, S.K.; Panda, M.; Biswal, B.K. Emerging role of plumbagin: Cytotoxic potential and pharmaceutical relevance towards cancer therapy. *Food Chem. Toxicol.* **2019**, *125*, 566–582. [CrossRef] [PubMed]
14. Thakor, N.; Janathia, B. Plumbagin: A potential candidate for future research and development. *Curr. Pharm. Biotechnol.* **2022**, *23*, 1800–1812. [CrossRef]
15. Sukkasem, N.; Chatuphonprasert, W.; Tatiya-Aphiradee, N.; Jarukamjorn, K. Imbalance of the antioxidative system by plumbagin and Plumbago indica L. extract induces hepatotoxicity in mice. *J. Intercult. Ethnopharmacol.* **2016**, *24*, 137–145. [CrossRef] [PubMed]

16. Avila-Carrasco, L.; Majano, P.; Sánchez-Toméro, J.A.; Selgas, R.; López-Cabrera, M.; Aguilera, A.; González Mateo, G. Natural plants compounds as modulators of epithelial-to-mesenchymal transition. *Front. Pharmacol.* **2019**, *10*, 715. [CrossRef] [PubMed]
17. Yin, Z.; Zhang, J.; Chen, L.; Guo, Q.; Yang, B.; Zhang, W.; Kang, W. Anticancer effects and mechanisms of action of plumbagin: Review of research advances. *Biomed. Res. Int.* **2020**, *2020*, 6940953. [CrossRef]
18. Luo, P.; Wong, Y.F.; Ge, L.; Zhang, Z.F.; Liu, Y.; Liu, L.; Zhou, H. Anti-inflammatory and analgesic effect of plumbagin through inhibition of nuclear factor-κB activation. *J. Pharmacol. Exp. Ther.* **2010**, *335*, 735–742. [CrossRef]
19. Wang, T.; Wu, F.; Jin, Z.; Zhai, Z.; Wang, Y.; Tu, B.; Yan, W.; Tang, T. Plumbagin inhibits LPS-induced inflammation through the inactivation of the nuclear factor-kappa B and mitogen activated protein kinase signaling pathways in RAW 264.7 cells. *Food Chem. Toxicol.* **2014**, *64*, 177–183. [CrossRef]
20. Checker, R.; Patwardhan, R.S.; Sharma, D.; Menon, J.; Thoh, M.; Sandur, S.K.; Sainis, K.B.; Poduval, T.B. Plumbagin, a vitamin K3 analogue, abrogates lipopolysaccharide-induced oxidative stress, inflammation and endotoxic shock via NF-kappaB suppression. *Inflammation* **2014**, *37*, 542–554. [CrossRef]
21. Guo, Y.X.; Liu, L.; Yan, D.Z.; Guo, J.P. Plumbagin prevents osteoarthritis in human chondrocytes through Nrf-2 activation. *Mol. Med. Rep.* **2017**, *15*, 2333–2338. [CrossRef] [PubMed]
22. Kuan-Hong, W.; Bai-Zhou, L. Plumbagin protects against hydrogen peroxide-induced neurotoxicity by modulating NF-κB and Nrf-2. *Arch. Med. Sci.* **2018**, *14*, 1112–1118. [CrossRef] [PubMed]
23. Zhang, W.; Cheng, L.; Hou, Y.; Si, M.; Zhao, Y.P.; Nie, L. Plumbagin protects against spinal cord injury-induced oxidative stress and inflammation in wistar rats through Nrf-2 upregulation. *Drug Res.* **2015**, *65*, 495–499. [CrossRef] [PubMed]
24. Zhang, S.; Li, D.; Yang, J.Y.; Yan, T.B. Plumbagin protects against glucocorticoid-induced osteoporosis through Nrf-2 pathway. *Cell Stress Chaperones* **2015**, *20*, 621–629. [CrossRef]
25. Chu, H.; Yu, H.; Ren, D.; Zhu, K.; Huang, H. Plumbagin exerts protective effects in nucleus pulposus cells by attenuating hydrogen peroxide-induced oxidative stress, inflammation and apoptosis through NF-κB and Nrf-2. *Int. J. Mol. Med.* **2016**, *37*, 1669–1676. [CrossRef]
26. Zaki, A.M.; El-Tanbouly, D.M.; Abdelsalam, R.M.; Zaki, H.F. Plumbagin ameliorates hepatic ischemia-reperfusion injury in rats: Role of high mobility group box 1 in inflammation, oxidative stress and apoptosis. *Biomed. Pharmacother.* **2018**, *106*, 785–793. [CrossRef]
27. Chen, X.J.; Zhang, J.G.; Wu, L. Plumbagin inhibits neuronal apoptosis, intimal hyperplasia and also suppresses TNF-α/NF-κB pathway induced inflammation and matrix metalloproteinase-2/9 expression in rat cerebral ischemia. *Saudi J. Biol. Sci.* **2018**, *25*, 1033–1039. [CrossRef]
28. Zhang, Q.; Zhao, S.; Zheng, W.; Fu, H.; Wu, T.; Hu, F. Plumbagin attenuated oxygen-glucose deprivation/reoxygenation-induced injury in human SH-SY5Y cells by inhibiting NOX4-derived ROS-activated NLRP3 inflammasome. *Biosci. Biotechnol. Biochem.* **2020**, *84*, 134–142. [CrossRef]
29. Marrazzo, P.; Angeloni, C.; Hrelia, S. Combined treatment with three natural antioxidants enhances neuroprotection in a SH-SY5Y 3D culture model. *Antioxidants* **2019**, *8*, 420. [CrossRef]
30. Wang, H.; Zhang, H.; Zhang, Y.; Wang, D.; Cheng, X.; Yang, F.; Zhang, Q.; Xue, Z.; Li, Y.; Zhang, L.; et al. Plumbagin protects liver against fulminant hepatic failure and chronic liver fibrosis via inhibiting inflammation and collagen production. *Oncotarget* **2016**, *7*, 82864–82875. [CrossRef]
31. Pan, P.H.; Wang, Y.Y.; Lin, S.Y.; Liao, S.L.; Chen, Y.F.; Huang, W.C.; Chen, C.J.; Chen, W.Y. Plumbagin ameliorates bile duct ligation-induced cholestatic liver injury in rats. *Biomed. Pharmacother.* **2022**, *151*, 113133. [CrossRef] [PubMed]
32. Vaduganathan, M.; Mensah, G.A.; Turco, J.V.; Fuster, V.; Roth, G.A. The global burden of cardiovascular diseases and risk: A compass for future health. *J. Am. Coll. Cardiol.* **2022**, *80*, 2361–2371. [CrossRef] [PubMed]
33. McGowan, J.V.; Chung, R.; Maulik, A.; Piotrowska, I.; MalcolmWalker, J.; Yellon, D.M. Anthracycline chemotherapy and cardiotoxicity. *Cardiovasc. Drugs Ther.* **2017**, *31*, 63–75. [CrossRef] [PubMed]
34. Li, Z.; Chinnathambi, A.; Ali Alharbi, S.; Yin, F. Plumbagin protects the myocardial damage by modulating the cardiac biomarkers, antioxidants, and apoptosis signaling in the doxorubicin-induced cardiotoxicity in rats. *Environ. Toxicol.* **2020**, *35*, 1374–1385. [CrossRef]
35. Zhang, Q.; Fu, H.; Gong, W.; Cao, F.; Wu, T.; Hu, F. Plumbagin protects H9c2 cardiomyocytes against TBHP-induced cytotoxicity by alleviating ROS-induced apoptosis and modulating autophagy. *Exp. Ther. Med.* **2022**, *24*, 501. [CrossRef]
36. Haddad, J.J.; Harb, H.L. L-gamma-Glutamyl-L-cysteinyl-glycine (glutathione; GSH) and GSH-related enzymes in the regulation of pro- and anti-inflammatory cytokines: A signaling transcriptional scenario for redox(y) immunologic sensor(s)? *Mol. Immunol.* **2005**, *42*, 987–1014. [CrossRef]
37. Checker, R.; Sharma, D.; Sandur, S.K.; Subrahmanyam, G.; Krishnan, S.; Poduval, T.B.; Sainis, K.B. Plumbagin inhibits proliferative and inflammatory responses of T cells independent of ROS generation but by modulating intracellular thiols. *J. Cell. Biochem.* **2010**, *110*, 1082–1093. [CrossRef]
38. Zhang, Z.; Deng, W.; Kang, R.; Xie, M.; Billiar, T.; Wang, H.; Cao, L.; Tang, D. Plumbagin protects mice from lethal sepsis by modulating immunometabolism upstream of PKM2. *Mol. Med.* **2016**, *22*, 162–172. [CrossRef]
39. Arruri, V.; Komirishetty, P.; Areti, A.; Dungavath, S.K.N.; Kumar, A. Nrf2 and NF-κB modulation by Plumbagin attenuates functional, behavioural and biochemical deficits in rat model of neuropathic pain. *Pharmacol. Rep.* **2017**, *69*, 625–632. [CrossRef]

40. Messeha, S.S.; Zarmouh, N.O.; Mendonca, P.; Kolta, M.G.; Soliman, K.F.A. The attenuating effects of plumbagin on pro-inflammatory cytokine expression in LPS-activated BV-2 microglial cells. *J. Neuroimmunol.* **2017**, *313*, 129–137. [CrossRef]
41. Chiocchio, I.; Prata, C.; Mandrone, M.; Ricciardiello, F.; Marrazzo, P.; Tomasi, P.; Angeloni, C.; Fiorentini, D.; Malaguti, M.; Poli, F.; et al. Leaves and spiny burs of *Castanea Sativa* from an experimental chestnut grove: Metabolomic analysis and anti-neuroinflammatory activity. *Metabolites* **2020**, *10*, 408. [CrossRef] [PubMed]
42. Marrazzo, P.; Mandrone, M.; Chiocchio, I.; Zambonin, L.; Barbalace, M.C.; Zalambani, C.; Angeloni, C.; Malaguti, M.; Prata, C.; Poli, F.; et al. By-product extracts from *Castanea sativa* counteract hallmarks of neuroinflammation in a microglial model. *Antioxidants* **2023**, *12*, 808. [CrossRef] [PubMed]
43. Mahmoud, I.S.; Hatmal, M.M.; Abuarqoub, D.; Esawi, E.; Zalloum, H.; Wehaibi, S.; Nsairat, H.; Alshaer, W. 1,4-naphthoquinone is a potent inhibitor of IRAK1 kinases and the production of inflammatory cytokines in THP-1 differentiated macrophages. *ACS Omega* **2021**, *6*, 25299–25310. [CrossRef] [PubMed]
44. Jiang, Y. Osteoarthritis year in review 2021: Biology. *Osteoarthr. Cartil.* **2022**, *30*, 207–215. [CrossRef]
45. Marrazzo, P.; O'Leary, C. Repositioning natural antioxidants for therapeutic applications in tissue engineering. *Bioengineering* **2020**, *7*, 104. [CrossRef]
46. Agarwal, S.; Misra, R.; Aggarwal, A. Synovial fluid RANKL and matrix metalloproteinase levels in enthesitis related arthritis subtype of juvenile idiopathic arthritis. *Rheumatol. Int.* **2009**, *29*, 907–911. [CrossRef]
47. Chemel, M.; Le Goff, B.; Brion, R.; Cozic, C.; Berreur, M.; Amiaud, J.; Bougras, G.; Touchais, S.; Blanchard, F.; Heymann, M.F.; et al. Interleukin 34 expression is associated with synovitis severity in rheumatoid arthritis patients. *Ann. Rheum. Dis.* **2012**, *71*, 150–154. [CrossRef]
48. Udagawa, N.; Koide, M.; Nakamura, M.; Nakamichi, Y.; Yamashita, T.; Uehara, S.; Kobayashi, Y.; Furuya, Y.; Yasuda, H.; Fukuda, C.; et al. Osteoclast differentiation by RANKL and OPG signaling pathways. *J. Bone Miner. Metab.* **2021**, *39*, 19–26. [CrossRef]
49. Ainola, M.; Mandelin, J.; Liljeström, M.; Konttinen, Y.T.; Salo, J. Imbalanced expression of RANKL and osteoprotegerin mRNA in pannus tissue of rheumatoid arthritis. *Clin. Exp. Rheumatol.* **2008**, *26*, 240–246.
50. Hensvold, A.H.; Joshua, V.; Li, W.; Larkin, M.; Qureshi, F.; Israelsson, L.; Padyukov, L.; Lundberg, K.; Defranoux, N.; Saevarsdottir, S.; et al. Serum RANKL levels associate with anti- citrullinated protein antibodies in early untreated rheumatoid arthritis and are modulated following methotrexate. *Arthritis Res. Ther.* **2015**, *17*, 239. [CrossRef]
51. Abimannan, T.; Peroumal, D.; Parida, J.R.; Barik, P.K.; Padhan, P.; Devadas, S. Oxidative stress modulates the cytokine response of differentiated Th17 and Th1 cells. *Free Radic. Biol. Med.* **2016**, *99*, 352–363. [CrossRef] [PubMed]
52. Cui, M.Y.; Li, X.; Lei, Y.M.; Xia, L.P.; Lu, J.; Shen, H. Effects of IL-34 on the secretion of RANKL/OPG by fibroblast-like synoviocytes and peripheral blood mononuclear cells in rheumatoid arthritis. *Eur. Cytokine Netw.* **2019**, *30*, 67–73. [CrossRef]
53. Weinstein, R.S. Clinical practice. Glucocorticoid-induced bone disease. *N. Engl. J. Med.* **2011**, *365*, 62–70. [CrossRef] [PubMed]
54. Aya, K.; Alhawagri, M.; Hagen-Stapleton, A.; Kitaura, H.; Kanagawa, O.; Novack, D.V. NF-(kappa)B-inducing kinase controls lymphocyte and osteoclast activities in inflammatory arthritis. *J. Clin. Investig.* **2005**, *115*, 1848–1854. [CrossRef]
55. Yamashita, T.; Yao, Z.; Li, F.; Zhang, Q.; Badell, I.R.; Schwarz, E.M.; Takeshita, S.; Wagner, E.F.; Noda, M.; Matsuo, K.; et al. NF-kappaB p50 and p52 regulate receptor activator of NF-kappaB ligand (RANKL) and tumor necrosis factor-induced osteoclast precursor differentiation by activating c-Fos and NFATc1. *J. Biol. Chem.* **2007**, *282*, 18245–18253. [CrossRef] [PubMed]
56. Yao, Z.; Getting, S.J.; Locke, I.C. Regulation of TNF-induced osteoclast differentiation. *Cells* **2021**, *11*, 132. [CrossRef]
57. Shen, G.; Liu, X.; Lei, W.; Duan, R.; Yao, Z. Plumbagin is a NF-κB-inducing kinase inhibitor with dual anabolic and antiresorptive effects that prevents menopausal-related osteoporosis in mice. *J. Biol. Chem.* **2022**, *298*, 101767. [CrossRef]
58. Sultanli, S.; Ghumnani, S.; Ashma, R.; Kubatzky, K.F. Plumbagin, a biomolecule with (anti)osteoclastic properties. *Int. J. Mol. Sci.* **2021**, *22*, 2779. [CrossRef]
59. Wei, Y.; Huang, M.; Liu, X.; Yuan, Z.; Peng, Y.; Huang, Z.; Duan, X.; Zhao, T. Anti-fibrotic effect of plumbagin on CCl_4-lesioned rats. *Cell. Physiol. Biochem.* **2015**, *35*, 1599–1608. [CrossRef]
60. Chen, S.; Chen, Y.; Chen, B.; Cai, Y.J.; Zou, Z.L.; Wang, J.G.; Lin, Z.; Wang, X.D.; Fu, L.Y.; Hu, Y.R.; et al. Plumbagin ameliorates CCl_4-induced hepatic fibrosis in rats via the epidermal growth factor receptor signaling pathway. *Evid. Based Complement. Altern. Med.* **2015**, *2015*, 645727. [CrossRef]
61. Chen, Y.; Zhao, C.; Liu, X.; Wu, G.; Zhong, J.; Zhao, T.; Li, J.; Lin, Y.; Zhou, Y.; Wei, Y. Plumbagin ameliorates liver fibrosis via a ROS-mediated NF-κB signaling pathway in vitro and in vivo. *Biomed. Pharmacother.* **2019**, *116*, 108923. [CrossRef]
62. Lee, S.Y.; Kim, M.J.; Jang, S.; Lee, G.E.; Hwang, S.Y.; Kwon, Y.; Hong, J.Y.; Sohn, M.H.; Park, S.Y.; Yoon, H.G. Plumbagin suppresses pulmonary fibrosis via inhibition of p300 histone acetyltransferase activity. *J. Med. Food* **2020**, *23*, 633–640. [CrossRef] [PubMed]
63. Mehdizadeh, S.; Taherian, M.; Bayati, P.; Mousavizadeh, K.; Pashangzadeh, S.; Anisian, A.; Mojtabavi, N. Plumbagin attenuates bleomycin-induced lung fibrosis in mice. *Allergy Asthma Clin. Immunol.* **2022**, *18*, 93. [CrossRef] [PubMed]
64. Shi, W.; Fang, Y.; Jiang, Y.; Jiang, S.; Li, Y.; Li, W.; Xu, M.; Aschner, M.; Liu, G. Plumbagin attenuates traumatic tracheal stenosis in rats and inhibits lung fibroblast proliferation and differentiation via TGF-β1/Smad and Akt/mTOR pathways. *Bioengineered* **2021**, *12*, 4475–4488. [CrossRef] [PubMed]
65. Chiș, A.A.; Rus, L.L.; Morgovan, C.; Arseniu, A.M.; Frum, A.; Vonica-Țincu, A.L.; Gligor, F.G.; Mureșan, M.L.; Dobrea, C.M. Microbial resistance to antibiotics and effective antibiotherapy. *Biomedicines* **2022**, *10*, 1121. [CrossRef]

66. Marrazzo, P.; Pizzuti, V.; Zia, S.; Sargenti, A.; Gazzola, D.; Roda, B.; Bonsi, L.; Alviano, F. Microfluidic tools for enhanced characterization of therapeutic stem cells and prediction of their potential antimicrobial secretome. *Antibiotics* **2021**, *10*, 750. [CrossRef]
67. Kapoor, N.; Kandwal, P.; Sharma, G.; Gambhir, L. Redox ticklers and beyond: Naphthoquinone repository in the spotlight against inflammation and associated maladies. *Pharmacol. Res.* **2021**, *174*, 105968. [CrossRef]
68. Singh, A.P.; Sharma, A. Structural insights and pharmaceutical relevance of plumbagin in parasitic disorders: A comprehensive review. *Recent Adv. Antiinfect. Drug Discov.* **2022**, *17*, 187–198. [CrossRef]
69. Hassan, S.T.; Berchová-Bímová, K.; Petráš, J. Plumbagin, a plant-derived compound, exhibits antifungal combinatory effect with amphotericin B against Candida albicans clinical isolates and anti-hepatitis C virus activity. *Phytother. Res.* **2016**, *30*, 1487–1492. [CrossRef]
70. Chen, X.; Yin, L.; Peng, L.; Liang, Y.; Lv, H.; Ma, T. Synergistic effect and mechanism of plumbagin with gentamicin against carbapenem-resistant *Klebsiella pneumoniae*. *Infect. Drug Resist.* **2020**, *13*, 2751–2759. [CrossRef]
71. Wang, Y.; Kong, J.; Zhang, X.; Liu, Y.; Huang, Z.; Yuan, L.; Zhang, Y.; Cao, J.; Chen, L.; Liu, Y.; et al. Plumbagin resurrect colistin susceptible against colistin-resistant *Pseudomonas aeruginosa* in vitro and in vivo. *Front. Microbiol.* **2022**, *13*, 1020652. [CrossRef] [PubMed]
72. Periasamy, H.; Iswarya, S.; Pavithra, N.; Senthilnathan, S.; Gnanamani, A. In vitro antibacterial activity of plumbagin isolated from Plumbago zeylanica L. against methicillin-resistant Staphylococcus aureus. *Lett. Appl. Microbiol.* **2019**, *69*, 41–49. [CrossRef] [PubMed]
73. Dissanayake, D.M.I.H.; Perera, D.D.B.D.; Keerthirathna, L.R.; Heendeniya, S.; Anderson, R.J.; Williams, D.E.; Peiris, L.D.C. Antimicrobial activity of *Plumbago indica* and ligand screening of plumbagin against methicillin-resistant *Staphylococcus aureus*. *J. Biomol. Struct. Dyn.* **2022**, *40*, 3273–3284. [CrossRef]
74. Reddy, P.J.; Ray, S.; Sathe, G.J.; Prasad, T.S.; Rapole, S.; Panda, D.; Srivastava, S. Proteomics analyses of Bacillus subtilis after treatment with plumbagin, a plant-derived naphthoquinone. *OMICS* **2015**, *19*, 12–23. [CrossRef] [PubMed]
75. Sarkar, A.; Ghosh, S.; Shaw, R.; Patra, M.M.; Calcuttawala, F.; Mukherjee, N.; Das Gupta, S.K. Mycobacterium tuberculosis thymidylate synthase (ThyX) is a target for plumbagin, a natural product with antimycobacterial activity. *PLoS ONE* **2020**, *15*, e0228657. [CrossRef] [PubMed]
76. Awasthi, B.P.; Kathuria, M.; Pant, G.; Kumari, N.; Mitra, K. Plumbagin, a plant-derived naphthoquinone metabolite induces mitochondria mediated apoptosis-like cell death in Leishmania donovani: An ultrastructural and physiological study. *Apoptosis* **2016**, *21*, 941–953. [CrossRef] [PubMed]
77. Chaweeborisuit, P.; Suriyonplengsaeng, C.; Suphamungmee, W.; Sobhon, P.; Meemon, K. Nematicidal effect of plumbagin on Caenorhabditis elegans: A model for testing a nematicidal drug. *Z. Naturforsch. C. J. Biosci.* **2016**, *71*, 121–131. [CrossRef]
78. Rashidzadeh, H.; Zamani, P.; Amiri, M.; Hassanzadeh, S.M.; Ramazani, A. Nanoincorporation of plumbagin in micelles increase its in vivo anti-plasmodial properties. *Iran. J. Parasitol.* **2022**, *17*, 202–213. [CrossRef]
79. Roy, A. Plumbagin: A potential anti-cancer compound. *Mini Rev. Med. Chem.* **2021**, *21*, 731–737. [CrossRef]
80. Liu, F.; Kong, X.; Lv, L.; Gao, J. TGF-beta1 acts through miR-155 to down-regulate TP53INP1 in promoting epithelial-mesenchymal transition and cancer stem cell phenotypes. *Cancer Lett.* **2015**, *359*, 288–298. [CrossRef]
81. Donnenberg, V.S.; Donnenberg, A.D. Stem cell state and the epithelial to-mesenchymal transition: Implications for cancer therapy. *J. Clin. Pharmacol.* **2015**, *55*, 603–619. [CrossRef] [PubMed]
82. Pan, S.T.; Qin, Y.; Zhou, Z.W.; He, Z.X.; Zhang, X.; Yang, T.; Yang, Y.X.; Wang, D.; Zhou, S.F.; Qiu, J.X. Plumbagin suppresses epithelial to mesenchymal transition and stemness via inhibiting Nrf2-mediated signaling pathway in human tongue squamous cell carcinoma cells. *Drug Des. Dev. Ther.* **2015**, *9*, 5511–5551. [CrossRef]
83. Ma, I.; Allan, A.L. The role of human aldehyde dehydrogenase in normal and cancer stem cells. *Stem Cell Rev. Rep.* **2011**, *7*, 292–306. [CrossRef] [PubMed]
84. Somasundaram, V.; Hemalatha, S.K.; Pal, K.; Sinha, S.; Nair, A.S.; Mukhopadhyay, D.; Srinivas, P. Selective mode of action of plumbagin through BRCA1 deficient breast cancer stem cells. *BMC Cancer* **2016**, *16*, 336. [CrossRef]
85. Reshma, R.S.; Sreelatha, K.H.; Somasundaram, V.; Satheesh Kumar, S.; Nadhan, R.; Nair, R.S.; Srinivas, P. Plumbagin, a naphthaquinone derivative induces apoptosis in BRCA 1/2 defective castrate resistant prostate cancer cells as well as prostate cancer stem-like cells. *Pharmacol. Res.* **2016**, *105*, 134–145. [CrossRef]
86. Sakunrangsit, N.; Ketchart, W. Plumbagin inhibits cancer stem-like cells, angiogenesis and suppresses cell proliferation and invasion by targeting Wnt/β-catenin pathway in endocrine resistant breast cancer. *Pharmacol. Res.* **2019**, *150*, 104517. [CrossRef] [PubMed]
87. Turinetto, V.; Vitale, E.; Giachino, C. Senescence in human mesenchymal stem cells: Functional changes and implications in stem cell-based therapy. *Int. J. Mol. Sci.* **2016**, *17*, 1164. [CrossRef]
88. Abruzzo, P.M.; Canaider, S.; Pizzuti, V.; Pampanella, L.; Casadei, R.; Facchin, F.; Ventura, C. Herb-Derived Products: Natural tools to delay and counteract stem cell senescence. *Stem Cells Int.* **2020**, *2020*, 8827038. [CrossRef]
89. Marrazzo, P.; Angeloni, C.; Freschi, M.; Lorenzini, A.; Prata, C.; Maraldi, T.; Hrelia, S. Combination of epigallocatechin gallate and sulforaphane counteracts in vitro oxidative stress and delays stemness loss of amniotic fluid stem cells. *Oxid. Med. Cell. Longev.* **2018**, *2018*, 5263985. [CrossRef]

90. Pizzuti, V.; Abruzzo, P.M.; Chatgilialoglu, A.; Zia, S.; Marrazzo, P.; Petrocelli, G.; Zannini, C.; Marchionni, C.; Poggi, P.; Simonazzi, G.; et al. A tailored lipid supplement restored membrane fatty acid composition and ameliorates in vitro biological features of human amniotic epithelial cells. *J. Clin. Med.* **2022**, *11*, 1236. [CrossRef]
91. Guida, M.; Maraldi, T.; Resca, E.; Beretti, F.; Zavatti, M.; Bertoni, L.; La Sala, G.B.; De Pol, A. Inhibition of nuclear Nox4 activity by plumbagin: Effect on proliferative capacity in human amniotic stem cells. *Oxid. Med. Cell. Longev.* **2013**, *2013*, 680816. [CrossRef] [PubMed]
92. Yamada, N.; Miki, K.; Yamaguchi, Y.; Takauji, Y.; Yamakami, Y.; Hossain, M.N.; Ayusawa, D.; Fujii, M. Extract of Plumbago zeylanica enhances the growth of hair follicle dermal papilla cells with down-regulation of 5α-reductase type II. *J. Cosmet. Dermatol.* **2020**, *19*, 3083–3090. [CrossRef] [PubMed]
93. Inui, S.; Itami, S. Molecular basis of androgenetic alopecia: From androgen to paracrine mediators through dermal papilla. *J. Dermatol. Sci.* **2011**, *61*, 1–6. [CrossRef] [PubMed]

Disclaimer/Publisher's Note: The statements, opinions and data contained in all publications are solely those of the individual author(s) and contributor(s) and not of MDPI and/or the editor(s). MDPI and/or the editor(s) disclaim responsibility for any injury to people or property resulting from any ideas, methods, instructions or products referred to in the content.

 life

MDPI

Review

Critical Assessment of the Anti-Inflammatory Potential of Usnic Acid and Its Derivatives—A Review

Wojciech Paździora [1], Irma Podolak [1], Marta Grudzińska [2], Paweł Paśko [2], Karolina Grabowska [1] and Agnieszka Galanty [1,*]

1 Department of Pharmacognosy, Jagiellonian University Medical College, Medyczna 9, 30-688 Kraków, Poland; wpazdziora@interia.pl (W.P.)

2 Department of Food Chemistry and Nutrition, Jagiellonian University Medical College, Medyczna 9, 30-688 Kraków, Poland

* Correspondence: agnieszka.galanty@uj.edu.pl

Abstract: Inflammation is a response of the organism to an external factor that disrupts its natural homeostasis, and it helps to eliminate the cause of tissue injury. However, sometimes the body's response is highly inadequate and the inflammation may become chronic. Thus, the search for novel anti-inflammatory agents is still needed. One of the groups of natural compounds that attract interest in this context is lichen metabolites, with usnic acid (UA) as the most promising candidate. The compound reveals a broad spectrum of pharmacological properties, among which anti-inflammatory properties have been studied both in vitro and in vivo. The aim of this review was to gather and critically evaluate the results of the so-far published data on the anti-inflammatory properties of UA. Despite some limitations and shortcomings of the studies included in this review, it can be concluded that UA has interesting anti-inflammatory potential. Further research should be directed at the (i) elucidation of the molecular mechanism of UA; (ii) verification of its safety; (iii) comparison of the efficacy and toxicity of UA enantiomers; (iv) design of UA derivatives with improved physicochemical properties and pharmacological activity; and (v) use of certain forms or delivery carriers of UA, especially in its topical application.

Keywords: usnic acid; anti-inflammatory; enantioselective

Citation: Paździora, W.; Podolak, I.; Grudzińska, M.; Paśko, P.; Grabowska, K.; Galanty, A. Critical Assessment of the Anti-Inflammatory Potential of Usnic Acid and Its Derivatives—A Review. *Life* **2023**, *13*, 1046. https://doi.org/10.3390/life13041046

Academic Editors: Azahara Rodríguez-Luna, Salvador González and Stefano Manfredini

Received: 10 February 2023
Revised: 19 March 2023
Accepted: 14 April 2023
Published: 19 April 2023

Copyright: © 2023 by the authors. Licensee MDPI, Basel, Switzerland. This article is an open access article distributed under the terms and conditions of the Creative Commons Attribution (CC BY) license (https://creativecommons.org/licenses/by/4.0/).

1. Introduction

Inflammation is a dynamic response of the organism to an external factor that disrupts its natural homeostasis—most commonly apathogenic microorganisms or physical agents. Generally, this process helps to eliminate the cause of tissue injury, but in some diseases, the body's response is highly inadequate. The unfolding chronic inflammation can result in cellular destruction and damage to tissues, or even promote the development of some serious diseases, such as cancer [1]. The visual signs of inflammation predominantly include local redness and swelling, but also pain, heat, and loss of function [2]. These result from a response of the organism to inflammation agents that initiates the sequential process, which comprises the activation of phospholipase A2 followed by the release of arachidonic acid and a number of inflammatory mediators (e.g., proinflammatory TNF-α or IL-1, anti-inflammatory IL-10 or IL-13) [3,4]. These mediators are one of the possible important targets in the search for novel anti-inflammatory drugs.

Nature is an almost inexhaustible source of bioactive compounds that can be considered new drug candidates for the treatment of various disorders, including inflammatory diseases. The classic anti-inflammatory drug used worldwide, aspirin, is derived from salicylic acid, a natural phenolic abundant in *Salix* sp. One of the groups of natural compounds that attract interest in this context is lichen metabolites. Lichens are composite organisms, that are primarily formed by the symbiotic co-existence of algal and/or cyanobacterial units and fungi, with the participation of basidiomycete yeasts and some bacterial communities [5]. A unique feature of lichens is that their metabolism stops under anhydrous conditions and returns to full metabolic activity under more favorable conditions. Despite their not very advanced evolutionary development, they contain primary and secondary metabolites, often with unique structures. One of the most interesting and promising, in terms of the pharmacological potential of lichen secondary metabolites, is usnic acid (2,6-diacetyl-7,9-dihydroxy-8,9b-dimethyl-1,3(2H,9bH)-dibenzofurandione), which is found at a particularly high content (up to 10%) in genera such as *Usnea, Alectoria, Cladonia, Lecanora, Ramalina*, and *Flavocetraria*. Usnic acid (UA) was first isolated in 1844 and, since then, its biological properties have been intensively studied, focusing mainly on antimicrobial, cytotoxic, antioxidant, and anti-inflammatory activities. It should be mentioned that this is a chiral compound (Figure 1). Even though there are many examples, among both synthetic and natural compounds, indicating that chirality can determine the activity observed, the enantioselectivity of usnic acid is still an open question, mainly due to scarce research data having been published so far [5].

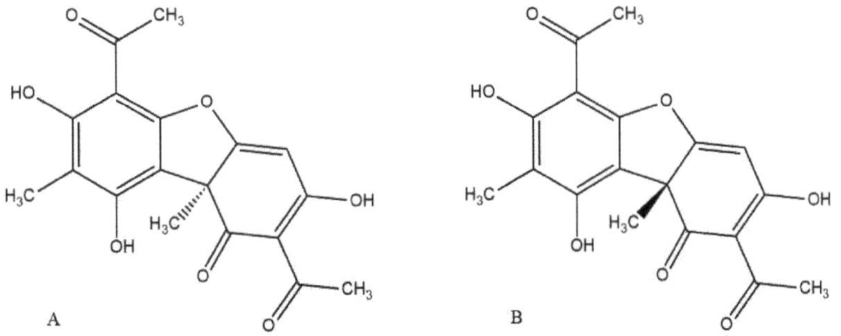

A

B

Figure 1. Structures of (−)-usnic acid (**A**) and (+)-usnic acid (**B**).

None of the recent reviews on the pharmacological activity of usnic acid have specifically focused on its anti-inflammatory potential. Therefore, the present paper summarizes studies published to date on the anti-inflammatory properties of usnic acid in various in vitro and in vivo models and critically assesses the prospects of this compound with the view of using it as a lead structure for further chemical modifications. Furthermore, the enantioselectivity of the action of the compound is also discussed.

2. Materials and Methods

This review was conducted in accordance with the Preferred Reporting Items for Systematic Reviews and Meta-analyses (PRISMA). A literature search was conducted in the PubMed, Google Scholar, and Scopus databases, covering reports up to December 2022. Initially, the search term "usnic acid" was used, but it was too general and gathered papers on all of the different activities of this compound. For example, Scopus found 1313 articles containing this keyword. Accordingly, the following search terms were refined: "usnic acid anti-inflammatory", "usnic acid antiinflammatory", "usnic acid inflammation", and "anti-inflammatory effects of usnic acid". An additional criterion was the English language of the articles. After checking the titles and abstracts of the papers, 72 articles were selected. Then, after a deeper analysis of the full text, 23 duplicates and 26 studies were excluded. Further reports were found by checking the reference lists of previously identified scientific publications. Of the remaining 23 papers, 5 review articles were excluded, resulting in a total of 18 original studies used to prepare this review. The flow chart of the search method is shown in Figure 2.

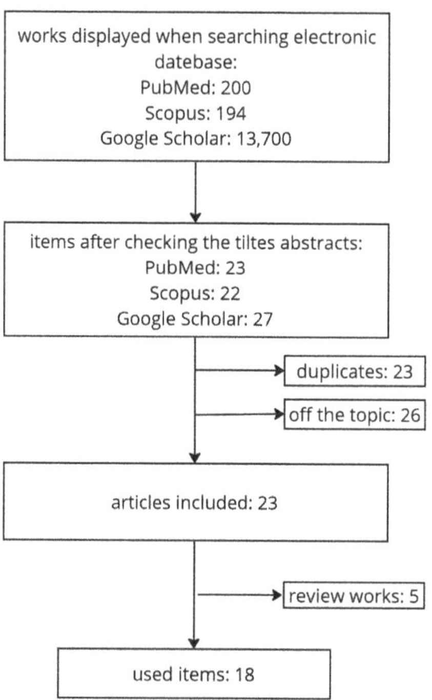

Figure 2. Searching strategy flowchart.

3. Anti-Inflammatory Potential of Usnic Acid

3.1. Results from the In Vitro Studies

Several in vitro studies described the anti-inflammatory activity of usnic acid in an attempt to discover the potential mechanism at the cellular level. The published studies involved experiments on leukocytes or platelets isolated from blood, referring to the production of an eicosanoid inflammatory mediator, but also on RAW 264.7 macrophages stimulated by LPS, where NO or a different cytokine release was measured. Details of the experiments published to date and their results are shown in Table 1.

Kumar and Müller investigated the effect of (+)-UA on leukotriene B4 (LTB4) synthesis from bovine polymorphonuclear leukocytes. The compound was shown to have only a weak inhibitory effect on LTB4 biosynthesis, with an IC_{50} value of 42 $\pm$ 2.2 μM, whereas the values for the reference substances were 0.4 $\pm$ 0.21 μM (nordihydroguaiaretic acid) and 37 $\pm$ 4.6 μM (anthralin) [6]. The in vitro effect of (+)-UA on human plate-type 12(S)-lipoxygenase activity was also verified. However, in the concentration range of up to 100 μg/mL, UA did not inhibit the activity of the enzyme tested [7].

In the study on LPS-stimulated RAW 264.7 macrophages, significant reductions in the TNF-α level and NO production were observed after UA treatment at doses of 0.5–400 μM, with IC_{50} values of 12.8 μM and 4.7 μM, respectively. TNF-α mRNA expression was also inhibited. Western blot assay showed that UA suppressed LPS-induced inducible nitric oxide synthase (iNOS) protein synthesis and NF-κB p65 nuclear translocation in the cells tested. The degradation of I-κBα, a protein that inhibits NF-κB by masking the nuclear localization signals of NF-κB proteins and keeping them sequestered in an inactive state in the cytoplasm, was inhibited [8].

In a study by Huang et al. (2014), in the same cellular model, a similar decrease in the production of pro-inflammatory factors TNF-α, IL-1β, IL-6, and NO after treatment with UA at the concentrations of 1.5 and 10 μg/mL was noted. The observed decrease in the expression of TNF-α mRNA, COX-2 mRNA, and iNOS mRNA confirmed the activity of the compound at the cellular transcriptional and translational levels. At the same time, the mRNA levels of anti-inflammatory IL-10 and anti-inflammatory mediator heme oxygenase-1 (HO-1) increased significantly. Furthermore, a reduction in NF-κB activation was observed. These results indicated a dual effect of UA in reducing inflammation by stimulating the secretion of anti-inflammatory factors and inhibiting pro-inflammatory factors [9].

Table 1. Summary of in vitro anti-inflammatory activity of usnic acid.

In Vitro Model	Experimental Conditions	Effects	Ref.
bovine polymorphonuclear leukocytes (PMNL)	(+)-UA Reference: nordihydroguaiaretic acid, anthralin Groups: Ca-ionophore A23187-stimulated cells Methods: RP-HPLC (inhibition of LTB4 biosynthesis)	• weak inhibitory effect on LTB4 biosynthesis • IC_{50} 42 ± 2.2 μM for UA vs. 0.4 ± 0.21 μM for nordihydroguaiaretic acid vs. 37 ± 4.6 μM for anthralin	[6]
human platelets	(+)-UA (3.33–100 μg/mL) Reference: baicalein (IC_{50} = 24.6 μM) Methods: optical density, RP-HPLC (inhibition of platelet-type 12(S)-LOX)	• no activity of UA up to 100 μg/mL	[7]
RAW 264.7 macrophages	UA (0.5–400 μM) Reference: none Groups: LPS-stimulated cells, untreated cells Methods: Griess reagent (NO), ELISA assay (TNF-α, iNOS, NF-κB, I-κB).	• ↓ TNF-α (dose-dependent effect)—IC_{50} 12.8 μM. • ↓ NO (dose-dependent effect)—IC_{50} 4.7 μM. • ↓ iNOS for 2.5, 5, 10 μM UA. • ↓ NF-κB p65 for 2.5, 5, 10 μM UA. • ↓ I-κB for 2.5, 5, 10 μM UA	[8]
RAW 264.7 cells	UA (1, 5, 10 μg/mL) Reference: dexamethasone 0.5 μg/mL Groups: LPS-stimulated cells, untreated cells (control) Methods: ELISA assay (TNF-α, IL-1β, IL-6, IL-10), Griess reagent (NO), RT-PCR (TNF-α mRNA, COX2 mRNA, iNOS mRNA, HO-1 mRNA), immunocytochemical assay (NF-κB), Western Blot (COX-2, HO-1)	• dose-dependent effect—most effective dose 10 μg/mL UA. • ↓ TNF-α, IL-1β, IL-6, NO, mRNA of TNF-α, mRNA of COX2, mRNA of iNOS, NF-κB. • ↓ HO-1 mRNA (only 1 μg/mL)	[9]
RAW 264.7 cells	(+)-UA, (−)-UA (10, 25 μg/mL) Reference: dexamethasone 0.5 μg/mL Groups: LPS-stimulated cells, untreated cells (control) Methods: ELISA assay (TNF-α, IL-6) Griess reagent (NO), Western Blot (TLR4, cPLA2, COX-1, COX-2).	• ↓ NO for all variants • ↓ IL-6 (only for (+)-UA 25 μg/mL) • no influence on TNF-α production. • ↓ TRL4 for all variants • ↓ cPLA2 for all variant • ↓ COX-1 for all variant ↑ COX-2 (only for (+)-UA 25 μg/mL)	[10]
MCF-7 breast cancer cells	UA (0.623–15,638 μM) Reference: none Groups: untreated (control) Methods: biochemical analysis (MDA, GSH), Griess reagent (NO), ELISA assay (PGE2, IL-2, IL-6, TNF-α), Bio-Plex assay (VEGF), RT-PCR (COX-2, iNOS)	• dose-dependent effect—most effective dose 15,638 μM (group 6) UA • ↓ NO, PGE2, IL-6, TNF-α, VEGF • ↓ COX-2 and iNOS (by 81% in 6 groups compared to control) • ↓ GSH (1,33-fold compared to control) • ↑ MDA (1,62-fold compared to control)	[11]

UA, usnic acid; RP-HPLC, reverse-phase high-performance liquid chromatography; LTB4, leukotriene B4; TNF-α, tumor necrosis factor alpha; iNOS, inducible nitric oxide synthase; NF-κB, nuclear factor kappa B; I-κB, IκB kinase; LPS, lipopolysaccharide; IL-1β, interleukin-1β; IL-6, interleukin-6; IL-10, interleukin-10; NO, nitric oxide; RT-PCR, reverse transcription polymerase chain reaction; COX-2, cyclooxygenase-2; HO-1, heme oxygenase; TRL4, toll-like receptor 4; cPLA2, cytosolic phospholipase A2; COX-1, cyclooxygenase-1; MDA, malondialdehyde; GSH, glutathione; PGE2, prostaglandin E2; IL-2, interleukin-2; VEGF, vascular endothelial growth factor; ↓ decrease; ↑ increase.

In our own studies on LPS-stimulated RAW 264.7 cells, the effects of both UA enantiomers at concentrations of 10 and 25 μg/mL were compared. A significant reduction in NO production was found for both concentrations, irrespective of the enantiomer used. In the case of IL-6, only the 25 μg/mL dose of both enantiomers had a significant effect on its release, whereas TNF-α production decreased only slightly, with no significant differences compared to control cells treated with LPS. In addition, the effect of both enantiomers on the expression of pro-inflammatory proteins: toll-like receptor 4 (TLR4), cytosolic phospholipase A2 (cPLA2), and cyclooxygenases (COX-1, COX-2) was also assessed. The inhibitory effect on TLR4 was observed at UA concentrations of 10 and 25 μg/mL, irrespective of the enantiomer used. Both UA enantiomers significantly and dose-dependently reduced cPLA2 synthesis in comparison to LPS-stimulated macrophages, with the strongest effect observed for (+)-UA at a concentration of 10 μg/mL. A dose-dependent decrease in COX-1 protein levels was observed for both enantiomers, but for (+)-UA only the higher dose made the effect significantly different from LPS-stimulated macrophages. Both UA enantiomers significantly decreased COX-2 protein levels, but for (−)-UA, the effect was dose-independent. Surprisingly, (+)-UA slightly increased COX-2 synthesis at the higher dose. The study showed a slight pro-inflammatory effect of (+)-UA, as the compound increased cPLA2 and COX-2 expression at the higher dose of 25 μg/mL, whereas no such effect was observed for (−)-UA [10].

A recent study tested the effect of UA on a broad panel of cytokines produced by unstimulated human breast cancer MCF-7 cells. In a concentration range of 0.62–15.64 μM, the compound significantly reduced the release of NO, vascular endothelial growth factor (VEGF), prostaglandin E2 (PGE2), cytokines (IL-2, CXCL 10, CXCL8, CCL2 (MCP-1), TNF-α, IL-6) in the cells, in a dose-dependent manner, compared to control cells. The compound also reduced the expression levels of COX-2 and iNOS genes [11].

3.2. *Results from In Vivo Studies*

It is noteworthy that UA was also tested in vivo in several models involving wound-healing, neurodegenerative, or lung diseases. Nevertheless, only one of these studies compared the impact of both enantiomers. Details of the experiments published to date and their results are shown in Table 2.

Table 2. Summary on in vivo anti-inflammatory activity of usnic acid and its derivatives.

In Vivo Model	Experimental Conditions	Effects	Ref.
Induced chronic and acute inflammation in Wistar rats (n = 30)	(+)-UA: 25, 50, 100 mg/kg orally (p.o.) Reference: ibuprofen 100 mg/kg Different groups: untreated control Methods: volume of paw edema, weight of cotton pellets.	• anti-edematous and anti-inflammatory effects of UA • dose-dependent effect, with most effective dose of 100 mg/kg • ↓ paw edema volume • ↓ cotton pellet weight	[12]
	Wound healing models		
Burn wound in male Wistar rats (n = 45)	Collagen film with liposomal UA: 330 mg/4 cm^2, dermal application for 7, 14, and 21 days. Reference: no data Different groups: collagen film, collagen film with empty liposomes. Methods: histological assessment of inflammatory profile, epithelization rates, collagen deposition, mean of myofibroblasts for histological field.	• day 7: moderate neutrophil infiltration over the entire wound surface (UA group) vs. infiltration only at the edges of the wound (other groups). • day 14: ↓ inflammation with high plasma cell infiltration in UA group vs. others. • day 21: slight inflammation in all groups. Content of highly undulating and dense type I and III collagen fibers, ↑ conversion of type III to type I collagen (UA group).	[13]

Table 2. *Cont.*

In Vivo Model	Experimental Conditions	Effects	Ref.
	Wound healing models		
Burn wound in a porcine model (n = 9)	Gelatin-based membranes with liposomal UA: 127.02 mg/7 cm^2, dermal application for 8, 18, and 30 days. Reference: ointment with silver sulfadiazine. Different groups: duoDerme® dressing. Methods: histological assessment of burn healing grading, collagen deposition.	• day 8: severe inflammation (UA group) vs. moderate (other groups). • day 18: granulation tissue neoplasia advanced in all groups; more visible fibroblasts (UA group). • day 30: 100% wound healing (UA and DuoDerme groups) vs. 80% (silver sulfadiazine ointment group).	[14]
Healing of wound in 8-week-old male Wistar rats (n = 64)	SUA: 38.4 mg/L in DMSO, daily dermal application for 21 days. Reference: gentamicin sulfate 0.01%. Different groups: untreated control, pure DMSO. Methods: wound area measured at 3, 7, 10, and 14 days after wounding. Histological assessment, immunohistochemistry analysis (VEGF).	• ↑ wound healing, re-epithelialization, ↓ inflammation (SUA and gentamicin groups). • on day 21, full skin regeneration (SUA and gentamicin groups). • VEGFT highest on day 1 (SUA, gentamicin) and day 3 (no treatment, pure DMSO). • No significant differences between gentamicin and SU.	[15]
	Neurodegenerative diseases models		
Model of cerebral ischemia/reperfusion by 20-min occlusion of the carotid arteries in male Wistar rats (n = 42)	UA: 25 mg/kg in DMSO, intraperitoneally (i.p.), 20 min of ischemia, and 48 h of reperfusion. Reference: no data Different groups: sham-operated, untreated control. Methods: Morris water maze task, spatial training test, spatial probe test, immunohistochemistry analysis (caspase-3, GFAP, Iba-1), biochemical assessment (SOD, GSH, MDA).	• ↑ caspase-3, GFAP, Iba-1 proteins • ↑ SOD and GSH • ↓ MDA	[16]
MPTP-induced Parkinson's disease model in mice C57BL/6 (n = 40)	UA: 5 and 25 mg/kg intraperitoneally (i.p.) used daily for 10 days before MPTP-induced Parkinson's disease. Reference: no data Different groups: sham control, untreated control. Methods: motor performance testing (rota-rod), immunocytochemical and immunochemical tests (Iba-1, GFAP, iNOS).	• ↓ astrocytic GFAP, microglial Iba-1, inducible nitric oxide synthase (iNOS) in the *substantia nigra* in UA group. • dose-dependent effect—most effective dose 25 mg/kg UA	[17]
Aβ1-42-induced Alzheimer's disease model in female mice (n = 81)	(R)-(+)- and (S)-(−)-UA: 25, 50, and 100 mg/kg, orally (p.o.) for 24 days. Reference: donepezil 2 mg/kg. Different groups: naïve, untreated control, sham-operated, Methods: open field test, novel object recognition test, Morris water maze task, Inhibitory-avoidance test, biochemical analysis (SOD, GSH, LOOH, MPO, IL-1β).	• ↑ SOD ((R)-(+)-UA (50 and 100 mg/kg), (S)-(−)-UA (100 mg/kg)) in the hippocampus. • ↑ GSH ((R)-(+)-UA 100 mg/kg) in the hippocampus. • ↓ LOOH, MPO (all variants) in the cerebral cortex and hippocampus. • ↓ IL-1β only in the hippocampus (without (S)-(−)-UA 100 mg/kg). • no effect on TNF-α.	[18]

Table 2. *Cont.*

In Vivo Model	Experimental Conditions	Effects	Ref.
Neurodegenerative diseases models			
Okadaic acid-induced memory impairment in male rats SD (n = 32)	UA derivative No 30 *: 5 and 10 mg/kg, intraperitoneally (i.p.) for 7 days after okadaic acid injection. Reference: no data Different groups: sham-operated, natrium chloratum 0.9%. Methods: Morris water maze task.	• ↑ memory and cognitive abilities of the derivative. • dose-dependent effect—most effective dose 10 mg/kg UA. • ↓ escape latency • no impact on swimming speed.	[19]
Lung diseases models			
LPS-induced acute lung injury (ALI) in mice (n = no data)	UA: 25, 50, or 100 mg/kg used daily for 5 days intratracheal. Reference: dexamethasone 5 mg/kg. Different groups: naïve, untreated control. Methods: histological assessment of BALF, immunochemical analysis (MPO, MDA, TNF-α, IL-6, IL-10, IL-8, MIP-2, GSH, SOD).	• dose-dependent effect—most effective dose 100 mg/kg UA • ↓ mortality (50, 100 mg/kg) • ↓ immune cells in the bronchoalveolar lavage fluid (BALF). • ↓ MPO, MDA, TNF-α, IL-6, IL-10, IL-8, MIP-2 (50, 100 mg/kg). • ↑ GSH and SOD (50, 100 mg/kg)	[20]
Bleomycin-induced lung fibrosis in mice (n = no data)	UA: 25, 50, or 100 mg/kg with bleomycin 15 mg/kg used daily for 21 days i.p. Reference: prednisone acetate 5 mg/kg. Different groups: natrium chloratum 0.9%, untreated control. Methods: histological assessments, immunochemical analysis (TGF-β1, TNF-α, IL-1β,IL-6, SOD, MDA).	• dose-dependent effect—most effective dose 100 mg/kg UA. • ↑ SOD, ↓ MDA (100 mg/kg UA) • ↓ TGF-β1, TNF-α, IL-1β and IL-6 (all doses of UA).	[21]

UA, usnic acid; SUA, sodium usnic acid; DMSO, dimethyl sulfoxide; VEGF, vascular endothelial growth factor; GFAP, glial fibrillary acidic protein; Iba-1, ionized calcium-binding adapter molecule 1; SOD, superoxide dismutase; GSH, glutathione; MDA, malondialdehyde; MPTP, 1-methyl-4-phenyl-1,2,3,6-tetrahydropyridine; iNOS, inducible nitric oxide synthase; LOOH, lipid hydroperoxide; MPO, myeloperoxidase; IL-1β, interleukin-1β; TNF-α, tumor necrosis factor alpha; BALF, bronchoalveolar lavage fluid; IL-6, interleukin-6; IL-10, interleukin-10; IL-8, interleukine-8; MIP-2, macrophage inflammatory protein-2; TGF-β1, transforming growth factor β1, * UA derivative No 30, according to [19]; ↓ decrease; ↑ increase.

In probably the first published study, the anti-inflammatory potential of (+)-UA was evaluated in a rat model of induced chronic and acute inflammation; the compound's activity was comparable to ibuprofen, which was used as a reference substance [12]. After a hiatus of almost a decade, studies exploring the anti-inflammatory potential of UA began to continue, targeting more specific problems, such as dermal inflammation and neurodegenerative- or lung disease-related inflammation.

Hard-to-heal wounds are a major health care problem. Inflammation is one of the natural stages of wound healing, forming an immune barrier against microbes. In many chronic wounds, there is clinically significant wound infection and/or excessive inflammation. The interesting efficacy of UA, in liposome form, in the treatment of burn wounds has been demonstrated in two experiments by the same research group (Table 2). In animals treated with UA, a significant improvement was observed in collagen quality and density [13], but also in granulation tissue and scar repair—better than in the case of the reference compound (sulfadiazine silver ointment) [14]. Despite some information on the allergic potential of UA [22], the authors did not observe such effects, even during the prolonged exposure time (up to 30 days). This could be explained by the use of a liposomal form in the study, which is safer for the body than the direct application of the compound [23].

Some recent studies, probably inspired by the lipophilic properties of UA and the proven ability of (−)-UA to cross the blood–brain barrier in vitro [24], have attempted to verify the exploitation of UA's anti-inflammatory potential in neurodegenerative diseases.

Cerebral ischemia causes oxidative stress, inflammation, and cell apoptosis due to oxygen deficiency. Astrocytes, some of the largest cells in the brain, are capable of producing pro-inflammatory factors under hypoxia, such as glial fibrillar acidic protein (GFAP). This protein is used as an indicator of astrocyte ischemia. Another indicator used to assess microglia activation is ionized calcium-binding adapter protein-1 (Iba-1). A study by Erfani et al. reported that UA significantly reduced the increase in caspase-3, GFAP, and Iba-1 values after cerebral ischemia in rats (Table 2). In addition, UA also revealed antioxidant activity, observed as an increase in superoxide dismutase (SOD) and glutathione synthetase (GSH) activity in hippocampal cells, which may support its anti-inflammatory activity against ischemia [16].

The neuroprotective effect of UA, resulting from its anti-inflammatory properties, was also suggested by the results of another study, with Parkinson's-like brain changes induced by MPTP (1-methyl-4-phenyl-1,2,3,6-tetrahydropyridine) in mice (Table 2). UA suppressed motor dysfunction and effectively attenuated neurodegenerative changes (loss of dopaminergic neurons) in the substantia nigra and striatum. Moreover, the aforementioned markers, astrocytic GFAP and microglia Iba-1, were reduced in UA-treated animals, followed by the reduced activation of inducible NOS (an inflammation-related gene) in the substantia nigra. This confirms the ability of UA to inhibit inflammatory processes in the central nervous system [17].

One hypothesis for Alzheimer's disease is that amyloid-β protein (Aβ) is deposited as amyloid fibers or non-fibrous amorphous aggregates in senile plaques, resulting in impaired neuronal transmission [25]. Aβ1−42 is one of the more cytotoxic amyloid isoforms, whose aggregation in the central nervous system causes neuroinflammation, oxidative stress, and apoptosis of the neuronal cells. Cazarin et al. tested different concentrations of UA enantiomers for the reduction in cognitive deficits, oxidative imbalance, and inflammation after the injection of Aβ1-42 into female mice (Table 2). This compound was chosen because, according to the authors, its structure reveals some similarities to galantamine, a drug used in Alzheimer's disease. This was further supported by the results of their in silico experiment, where both UA enantiomers exhibited a complex-receptor interaction with acetylcholinesterase (AchE), similar to that of galantamine. In addition, both enantiomers revealed nootropic properties, observed as improved learning and memory in animals in various tests, and also reduced the activity of myeloperoxidase (MPO) and lipid hydroperoxides (LOOH) in the cortex and hippocampus and IL-1β levels in the hippocampus, without an effect on TNF-α. Despite the use of two enantiomers, the authors did not discuss differences in their activity [18].

The anti-inflammatory properties of UA have also been assessed for acute lung injury and acute respiratory distress syndrome, inflammatory diseases characterized by lung infiltrates, pulmonary edema, or hypoxemia, but also by a rapid overproduction of pro-inflammatory cytokines and chemokines, with a mortality rate of up to 40%. Both diseases are serious, with a long course for which there is still no effective treatment [26]. Zu-Qing Su et al. investigated the effect of UA on LPS-induced acute lung injury (ALI) in mice (Table 2). The application of UA significantly reduced mortality in mice with ALI, as well as neutrophils, macrophage levels, and the production of the studied cytokines in bronchoalveolar lavage fluid. However, the amount of anti-inflammatory IL-10 in the UA group was lower than in the LPS group. These results may be related to the suppressive effect of UA on neutrophil infiltration, which led to a reduction in the number of neutrophils in the lavage fluid. IL-10, as a counter-regulatory cytokine, is known to be produced more intensively after the increase in TNF-α production induced by LPS, which may also explain the high IL-10 content in the LPS group. Furthermore, the levels of myeloperoxidase (MOP), malondialdehyde (MDA), and H_2O_2 were significantly reduced, while the observed increase in SOD and GSH activities indicated the antioxidant properties of the compound [20]. In an experiment by Huang et al., lung fibrosis was induced with bleomycin in mice, and the impact of UA on selected markers, such as SOD, MDA, transforming growth factor beta 1 (TGF-β1), TNF-α, IL-1β, and IL-6, were investigated.

Bleomycin caused a significant increase in MDA concentration and a decrease in SOD activity in the samples tested. The compound effectively inhibited MDA levels and reversed the bleomycin-induced decrease in SOD activity, and this effect was comparable to that of the reference prednisone acetate. In addition, there was a significant decrease in the expression of the cytokines tested, as recorded for UA, in a dose-dependent manner, thereby reducing inflammation in the lung tissue [21].

4. Anti-Inflammatory Potential of Synthetic Usnic Acid Derivatives In Vitro and In Vivo

Due to the documented anti-inflammatory effects of usnic acid, attempts have also been made to modify its structure in order to obtain synthetic derivatives, with improved physicochemical and anti-inflammatory properties. The studies were mainly conducted within in vitro models on LPS-stimulated cells of various origins, but two in vivo experiments have also been described. Details of the experiments published to date and their results are shown in Table 3.

Table 3. Summary of in vitro anti-inflammatory activity of usnic acid derivatives.

Cellular Model	Experimental Conditions	Effects	Ref.
lymphoma U937 cells	16 derivatives of UA (10 μM) Reference: prednisolone 10 μM groups: LPS-stimulated cells Methods: ELISA assay (IL-1β, TNF-α)	• Compounds No 5f and 5h—the highest scores • ↓ TNF-α by 90.94% and 83.75%, respectively, vs. prednisolone at 60.69% • ↓ IL-1β by 12.4% and 16.74%, respectively, vs. prednisolone at 46.11% • IC_{50} 1. (No 5f) and 1.88 (No 5h) vs. prednisolone 0.52	[27]
lymphoma U937 cells	UA derivatives No 4-13 (10 μM) Reference: dexamethasone 10 μM. Different groups: LPS-stimulated cells, Methods: ELISA assay (IL-1β, TNF-α)	• Compounds No 5, 6, and 13—the highest scores • ↓ TNF-α by 80.1%, 17.3%, and 4.7%, respectively, vs. dexamethasone at 81.4% • ↓ IL-1β by 25.4%, 90.4%, 85.4%, respectively, vs. dexamethasone at 80.5% • IC_{50} from 5.3 ± 0.01 (No 5) to 7.5 ± 0.1 (No 6) vs. dexamethasone from 1.5 ± 0.04 to 2.9 ± 0.05	[28]
microglia BV2 cells	UA derivative No 30 (2.5, 5, 10 μM) Reference: sodium usnate 10 μM Different groups: LPS-stimulated cells Methods: Griess reagent (NO)	• Dose-dependent effect—most effective dose 10 μM • ↓ NO by 41% (10 μM)	[19]

UA, usnic acid; LPS, lipopolysaccharide; TNF-α, tumor necrosis factor alpha; IL-1β, interleukin-1β; NO, nitric oxide. ↓ decrease.

Vanga et al. [27], synthesized sixteen novel (+)-UA-based triazole hybrids and evaluated their in vitro anti-inflammatory potential against TNF-α and IL-1β release in the LPS-stimulated human lymphoma U937 cell line. Four intermediates (Figure 3) of the target synthesis (which were also included in the study) and sixteen synthesized triazole derivatives (Figure 4) showed promising anti-inflammatory activity against TNF-α, with IC_{50} values ranging from 1.40 to 5.70 μM compared to an IC_{50} > 100 μM for the parent compound.

The authors suggest that the triazole ring with an aliphatic side chain may be responsible for the increase in anti-inflammatory activity. Two of the triazole derivatives, 5f and 5h (Figure 4), were the most promising in terms of activity, as their IC_{50} (1.40 and 1.88 μM) were most similar to the values obtained for prednisolone (IC_{50} 0.52 μM), used as a reference drug. Interestingly, two of the intermediates tested (compounds 2a and 2b, Figure 3) showed stronger activity than some of the final triazole derivatives [27].

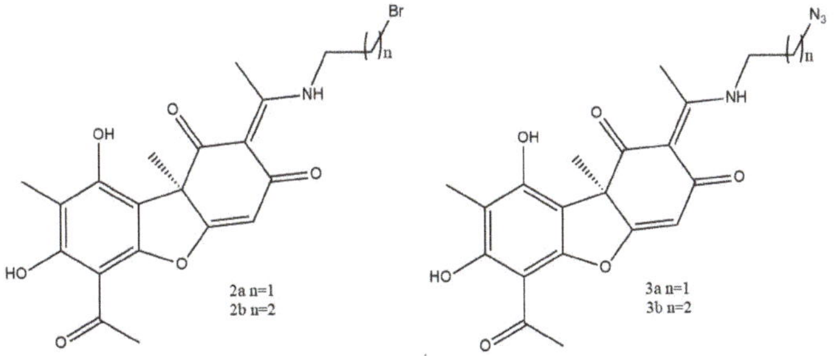

Figure 3. Intermediates of the target UA derivatives synthesis, according to Vanga et al. [27].

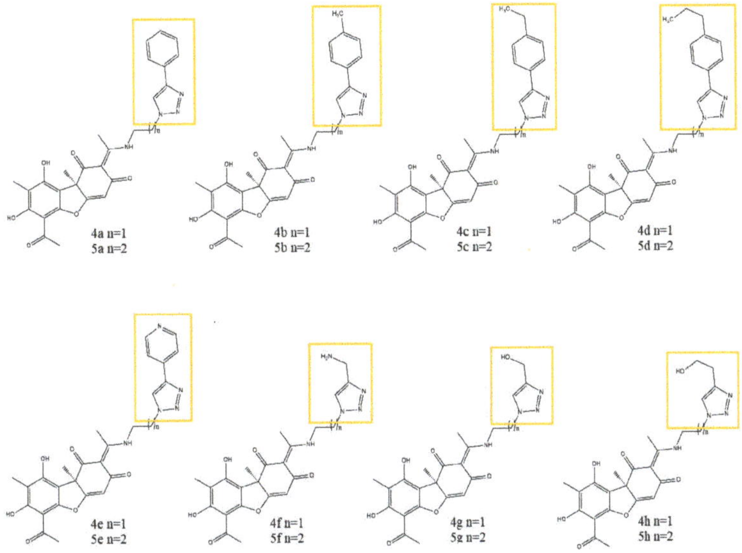

Figure 4. Structures of the synthesized UA triazole derivatives according to Vanga et al. [27], with the modification of the parent structure marked with yellow frames.

In a subsequent study, the same group of authors synthesized ten new (+)-UA imidazolium salts, which were evaluated for the in vitro anti-inflammatory potential of TNF-α and IL-1β on the LPS-stimulated human lymphoma U937 cell line. The three most active synthesized derivatives inhibited the release of TNF-α and IL-1β in 80.1 and 25.4% (compound No. 5, Figure 5); 17.3 and 90.4% (compound No. 6, Figure 5); and 4.7 and 85.5% (compound No. 13, Figure 5), respectively. The values for the reference substance, dexamethasone, were 81.4% and 80.5%, respectively. The IC_{50} values of the three most active compounds (No. 5, No. 6, No. 13) ranged from 5.3 μM to 7.5 μM and were many times lower compared to the parent UA (>100 μM), while for dexamethasone the IC_{50} values were 1.5 and 2.9 μM, respectively. The authors noted that the introduction of an enamine group at the C-2 position of (+)-UA, as present in the derivatives 5, 6, and 13, significantly increased the assessed anti-inflammatory activity, compared to the parent compound. Moreover, these most active derivatives were also characterized by the presence of electron-withdrawing groups in the phenacyl moiety, such as chloro, nitro, and bromo groups, while compounds bearing aromatic or heteroaromatic substituents (4 and 8–12) were significantly less active [28].

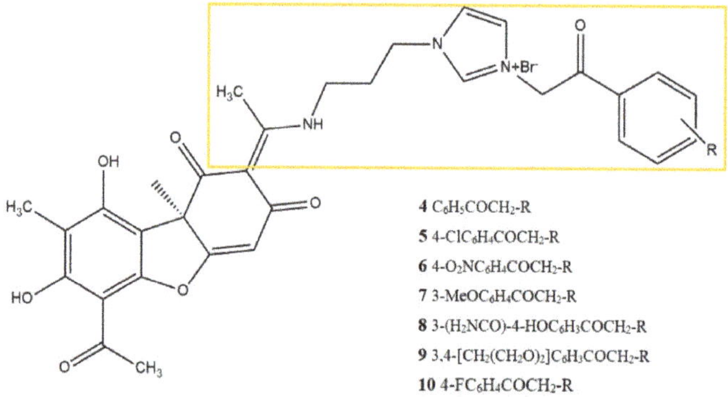

Figure 5. Structures of imidazolium salts of (+)-UA, according to Somasekhar et al. [28], with the modification of parent structure marked with yellow frames.

Another study focused on the modification of the (+)-UA structure, retaining its anti-inflammatory properties, with additional properties to inhibit tau protein aggregation (an important element in the pathogenesis of Alzheimer's disease). Twenty-five enamine derivatives and twenty-five hydrazines and hydrazides of (+)-UA were synthesized, but due to their better water solubility, the sodium salt of usnic acid (sodium usnate, SU, Figure 6A) was used as the reference parent compound. Compound No 30, with a substituted p-benzoic acid group (Figure 6B), appeared to be the most promising in terms of inhibition of tau aggregation.

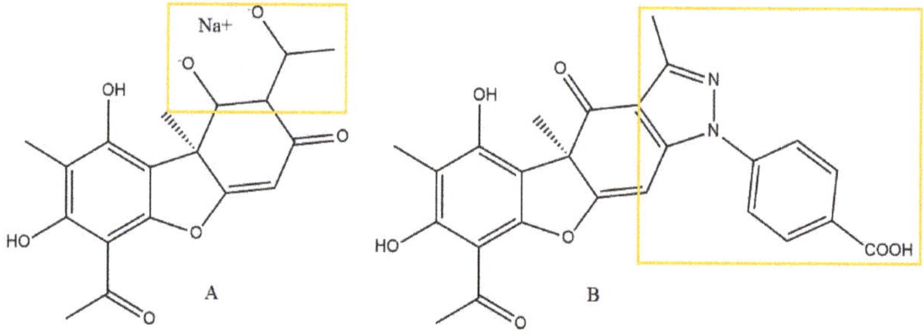

Figure 6. Structures of usnic acid sodium salt (**A**) and compound No 30 (**B**), according to Shi et al. [19], with the modification of parent structure marked with yellow frames.

This compound was also evaluated for the inhibition of LPS-induced nitric oxide release in the BV-2 mouse microglia cell line compared to SU. Interestingly, compound No 30 retained the anti-inflammatory effect of SUA and inhibited NO release by 41%, while it was significantly less toxic to the cells. The authors also assessed the neurotoxic and hepatotoxic potential of this derivative in vitro, and only minor effects on the viability of human neuroblastoma SH-SY5Y and LO2 hepatocytes were observed. Furthermore, in an in vivo Morris water maze test (see Table 2 for details), compound No 30 improved conventional reference spatial memory and cognitive abilities in okadaic acid-induced Alzheimer's disease model rats [19].

Zhiheng Zhang et al. [15], investigated the sodium salt of usnic acid (SUA, Figure 6A), in the healing of an experimentally prepared wound in 8-week-old Wistar rats (see Table 2 for details). After 14 days of the experiment, a significant increase in wound healing activity was observed in the group treated with SUA (38.4 mg/kg), and the reference gentamicin sulfate (GA, 0.01%), compared to the untreated group. Furthermore, after the third day of treatment, the level of VEGF was significantly elevated in the SUA and GA groups, indicating faster skin regeneration processes. Unfortunately, UA alone was not included in the study; therefore, a comparison of the activity of SUA and the parent structure is not possible.

Despite the small number of experiments performed, the derivatives of UA designed so far, even as simple as its sodium salt, clearly demonstrate the utility of the compound's parent structure to enhance its anti-inflammatory potential, both in vitro and in vivo.

5. Limitations of the Studies Included in the Review

Surprisingly few in vitro studies have been carried out so far, and their results still do not answer the question of UA's anti-inflammatory mechanism. Only a general conclusion can be drawn, indicating an effect of the compound on the release and synthesis of inflammatory mediators, while more in-depth mechanistic studies are really needed. Although UA revealed significant anti-inflammatory activity in a relatively low concentration range of 5–25 μg/mL, the effect of dexamethasone used as the reference drug was observed at a much lower dose of 0.5 μg/mL. Furthermore, the control drug was used only in two studies, conducted on LPS-stimulated macrophages, and the results obtained in these studies are contradictory; the activity of UA was similar [9] or much weaker [10] than that of dexamethasone. The other two studies mentioned above did not include the reference drug [8,11].

Despite the interesting effects, the in vivo studies published so far can only be treated as preliminary observations, as a relatively small number of animals were used. The observed effects of usnic acid in reducing inflammation are particularly promising in wound-healing models using the liposomal form of UA, which may reduce the risk of UA's allergic potential.

Although information on the pharmacokinetics of UA is limited, the experimental data suggest its high bioavailability [22], which may justify its potential oral use—for example, in neurodegenerative or lung diseases—as presented in the cited articles. However, in our opinion, the issue of UA toxicity, especially hepatotoxicity [29,30], was not taken into account during these experiments, as the cited studies generally lack information on the effects of UA on the liver or other organs. In one paper, the authors mentioned this problem [18], speculating that the effective UA dose of 25 mg/kg proven in their study was much lower than the toxic doses (<50 mg/kg) reported in some previous toxicological experiments [29]. As the effective dose of UA was 100 mg/kg in some studies included in this review, there is still a question about its safety.

The results obtained also cannot provide direct information on the superiority of one UA enantiomer over the other, as only two studies directly compared the activity of both enantiomers [10,18], while most of the other studies tested only (+)-. However, the significant differences in activity, as well as the small pro-inflammatory effect observed in our study for (+)-UA only, may suggest that this issue requires further research.

6. Conclusions

Although the studies included in this review have some limitations and shortcomings, it can be concluded, without a doubt, that usnic acid has interesting anti-inflammatory potential. The summary of the results of the studies included in the review is presented in Figure 7.

Further research into its action in inflammatory diseases is highly anticipated, particularly directed at the (i) elucidation of the molecular mechanism of UA's anti-inflammatory activity; (ii) verification of UA's hepatotoxic properties, especially at the higher doses used; (iii) comparison of the efficacy and toxicity of UA enantiomers; (iv) design of UA derivatives, with improved physicochemical properties (especially solubility) and pharmacological activity, as well as high safety; and (v) use of certain forms or delivery carriers of UA, especially in its topical application.

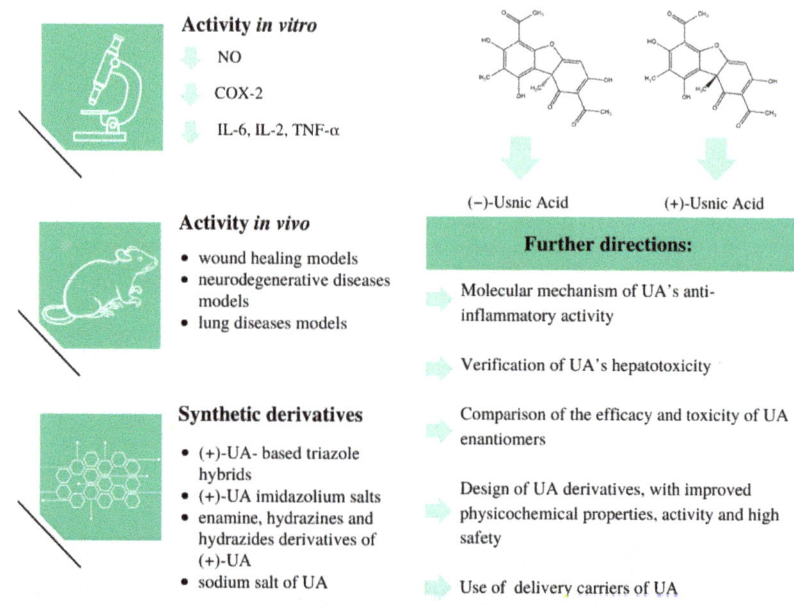

Figure 7. Summary of the anti-inflammatory effects of usnic acid and its derivatives.

Author Contributions: Conceptualization, A.G., W.P. and P.P.; methodology A.G., W.P. and P.P.; writing—original draft preparation, W.P. and A.G.; writing—review and editing, A.G., P.P., K.G. and I.P.; visualization, M.G.; supervision, A.G.; All authors have read and agreed to the published version of the manuscript.

Funding: This research received no external funding.

Institutional Review Board Statement: Not applicable.

Data Availability Statement: Not applicable.

Acknowledgments: The article was prepared within the grant Miniatura no. K/MNT/000250 from the National Science Center in Poland.

Conflicts of Interest: The authors declare no conflict of interest.

References

1. Marchi, S.; Guilbaud, E.; Tait, S.W.; Yamazaki, T.; Galluzzi, L. Mitochondrial control of inflammation. *Nat. Rev. Immunol.* **2023**, *23*, 159–173. [CrossRef] [PubMed]
2. Nunes, C.D.R.; Barreto Arantes, M.; Menezes de Faria Pereira, S.; Leandro da Cruz, L.; de Souza Passos, M.; Pereira de Moraes, L.; Vieira, I.J.C.; Barros de Oliveira, D. Plants as Sources of Anti-Inflammatory Agents. *Molecules* **2020**, *25*, 3726. [CrossRef] [PubMed]
3. de Cássia da Silveira e Sá, R.; Andrade, L.N.; de Sousa, D.P. A review on anti-inflammatory activity of monoterpenes. *Molecules* **2013**, *18*, 1227–1254. [CrossRef] [PubMed]
4. Karpel, E. Mediatory ogólnoustrojowej odpowiedzi zapalnej—Znaczenie w praktyce klinicznej intensywnej terapii. *Anestezjol. Intensywna Ter.* **2001**, *3*, 181–190.
5. Galanty, A.; Paśko, P.; Podolak, I. Enantioselective activity of usnic acid: A comprehensive review and future perspectives. *Phytochem. Rev.* **2019**, *18*, 527–548. [CrossRef]
6. Kumar, S.; Muller, K. Lichen Metabolites. 2. Antiproliferative and cytotoxic activity of gyrophoric, usnic, and diffractaic acid on human keratinocyte growth. *J. Nat. Prod.* **1999**, *62*, 821–823. [CrossRef]
7. Bucar, F.; Schneider, I.; Ogmundsdóttir, H.; Ingólfsdóttir, K. Anti-proliferative lichen compounds with inhibitory activity on 12(S)-HETE production in human platelets. *J. Phytomed.* **2004**, *11*, 602–606. [CrossRef]
8. Jin, J.; Li, C.; He, L. Down-regulatory effect of usnic acid on nuclear factor-κB-dependent tumor necrosis factor-α and inducible nitric oxide synthase expression in lipopolysaccharide-stimulated macrophages RAW 264.7. *Phytother. Res.* **2008**, *22*, 1605–1609. [CrossRef]

9. Huang, Z.; Tao, J.; Ruan, J.; Li, C.; Zheng, G. Anti-inflammatory effects and mechanisms of usnic acid, a compound firstly isolated from lichen *Parmelia saxatilis*. *J. Med. Plant Res.* **2014**, *8*, 197–207.
10. Galanty, A.; Zagrodzki, P.; Gdula-Argasińska, J.; Grabowska, K.; Koczurkiewicz-Adamczyk, P.; Wróbel-Biedrawa, D.; Podolak, I.; Pękala, E.; Paśko, P. A comparative survey of anti-melanoma and anti-inflammatory potential of usnic acid enantiomers—A comprehensive in vitro approach. *Pharmaceuticals* **2021**, *14*, 945. [CrossRef]
11. Yildirim, M.; Degirmenci, U.; Akkapulu, M.; Gungor, M.; Oztornacı, R.O.; Berkoz, M.; Comelekoglu, U.; Yalın, A.E.; Yalın, S. Anti-inflammatory effects of usnic acid in breast cancer. *Russ. J. Bioorg. Chem.* **2022**, *14*, 945. [CrossRef]
12. Vijayakumar, C.S.; Viswanathan, S.; Reddy, M.K.; Parvathavarthini, S.; Kundu, A.B.; Sukumar, E. Anti-inflammatory activity of (+)-usnic acid. *Fitoterapia* **2000**, *71*, 564–568. [CrossRef]
13. Nunes, P.S.; Albuquerque-Júnior, R.L.C.; Cavalcante, D.R.R.; Dantas, M.D.M.; Cardoso, J.C.; Bezerra, M.S.; Souza, J.C.C.; Russo Serafini, M.; Quitans, L.J., Jr.; Bonjardim, L.R.; et al. Collagen-based films containing liposome-loaded usnic acid as dressing for dermal burn healing. *Biomed. Res. Int.* **2011**, *2011*, 761593. [CrossRef]
14. Nunes, P.S.; Rabelo, A.S.; Campos de Souza, J.C.; Vasconcelos Santana, B.; Monteiro Menezes da Silva, T.; Russo Serafini, M.; Dos Passos Menezes, P.; Dos Santos Lima, B.; Cordeiro Cardoso, J.; Santana Alves, J.C.; et al. Gelatin-based membrane containing usnic acid-loaded liposome improves dermal burn healing in a porcine model. *Int. J. Pharm.* **2016**, *513*, 473–482. [CrossRef]
15. Zhang, Z.; Zheng, Y.; Li, Y.; Bai, H.; Ma, T.; Song, X.; Zhao, J.; Gao, L. The effects of sodium usnic acid by topical application on skin wound healing in rats. *Biomed. Pharmacother.* **2018**, *97*, 587–593. [CrossRef]
16. Erfani, S.; Valadbeigi, T.; Aboutaleb, N.; Karimi, N.; Moghimi, A.; Khaksari, M. Usnic acid improves memory impairment after cerebral ischemia/reperfusion injuries by anti-neuroinflammatory, anti-oxidant, and anti-apoptotic properties. *Iran. J. Basic Med. Sci.* **2020**, *23*, 1225–1231. [CrossRef] [PubMed]
17. Lee, S.; Lee, Y.; Ha, S.; Chung, H.Y.; Kim, H.; Hur, J.S.; Lee, J. Anti-inflammatory effects of usnic acid in an MPTP-induced mouse model of Parkinson's disease. *Brain Res. J.* **2020**, *1730*, 146642. [CrossRef]
18. Cazarin, C.A.; Dalmagro, A.P.; Gonçalves, A.E.; Boeing, T.; Mota da Silva, L.; Correa, R.; Klein-Júnior, L.C.; Carlesso Pinto, B.; Savoldi Lorenzett, T.; Uchoa da Costa Sobrinho, T.; et al. Usnic acid enantiomers restore cognitive deficits and neurochemical alterations induced by Aβ1-42 in mice. *Behav. Brain Res.* **2021**, *397*, 112945. [CrossRef] [PubMed]
19. Shi, C.J.; Peng, W.; Zhao, J.H.; Yang, H.J.; Qu, L.L.; Wang, C.; Kong, L.Y.; Wang, X.B. Usnic acid derivatives as tau-aggregation and neuroinflammation inhibitors. *Eur. J. Med. Chem.* **2020**, *187*, 111961. [CrossRef]
20. Su, Z.-Q.; Mo, Z.Z.; Liao, J.-B.; Feng, X.-X.; Liang, Y.-Z.; Zhang, X.; Liu, Y.-H.; Chen, X.-Y.; Chen, Z.-W.; Su, Z.-R.; et al. Usnic acid protects LPS-induced acute lung injury in mice through attenuating inflammatory responses and oxidative stress. *Int. Immunopharmacol.* **2014**, *22*, 371–378. [CrossRef]
21. Huang, X.Q.; Ai, G.X.; Zheng, X.H.; Liao, H.J. Usnic acid ameliorates bleomycin-induced pulmonary fibrosis in mice via inhibition of inflammatory responses and oxidative stress. *Trop. J. Pharm. Res.* **2019**, *18*, 2563–2569.
22. Wang, H.; Xuan, M.; Huang, C.; Wang, C. Advances in research on bioactivity, toxicity, metabolism, and pharmacokinetics of usnic acid in vitro and in vivo. *Molecules* **2022**, *27*, 7469. [CrossRef] [PubMed]
23. da Silva Santos, N.P.; Nascimento, S.C.; Wanderley, M.S.O.; Pontes-Filho, N.T.; da Silva, J.F.; de Castro, C.M.M.B.; Santos-Magalhaes, N.S. Nanoencapsulation of usnic acid: An attempt to improve antitumour activity and reduce hepatotoxicity. *Eur. J. Pharm. Biopharm.* **2006**, *64*, 154–160. [CrossRef]
24. Studzińska-Sroka, E.; Majchrzak-Celińska, A.; Zalewski, P.; Szwajgier, D.; Baranowska-Wójcik, E.; Kaproń, B.; Plech, T.; Żarowski, M.; Cielecka-Piontek, J. Lichen-derived compounds and extracts as biologically active substances with anticancer and neuroprotective properties. *Pharmaceuticals* **2021**, *14*, 1293. [CrossRef]
25. Gaweł, M.; Potulska-Chromik, A. Neurodegenerative diseases: Alzheimer's and Parkinson's disease. *Postępy Nauk Med.* **2015**, *28*, 468–472.
26. Parekh, D.; Dancer, R.C.; Thickett, D.R. Acute lung injury. *Clin. Med.* **2011**, *11*, 615–618. [CrossRef]
27. Vanga, N.R.; Kota, A.; Sistla, R. Synthesis and anti-inflammatory activity of novel triazole hybrids of (+)-usnic acid, the major dibenzofuran metabolite of the lichen *Usnea longissima*. *Mol. Divers.* **2017**, *21*, 273–282. [CrossRef] [PubMed]
28. Somasekhar, T.; Javadi, M.; Sistla, R. Synthesis of novel anti-inflammatory usnic acid-based imidazolium salts. *Eur. Chem. Bull.* **2021**, *10*, 67–72. [CrossRef]
29. Guo, L.; Shi, Q.; Fang, J.L.; Mei, N.; Ali, A.A.; Lewis, S.M.; Frankos, V.H. Review of usnic acid and *Usnea barbata* toxicity. *J. Environ. Health Part C* **2008**, *26*, 317–338. [CrossRef]
30. Croce, N.; Pitaro, M.; Gallo, V.; Antonini, G. Toxicity of Usnic Acid: A Narrative Review. *J. Toxicol.* **2022**, *2022*, 8244340. [CrossRef]

Disclaimer/Publisher's Note: The statements, opinions and data contained in all publications are solely those of the individual author(s) and contributor(s) and not of MDPI and/or the editor(s). MDPI and/or the editor(s) disclaim responsibility for any injury to people or property resulting from any ideas, methods, instructions or products referred to in the content.

Review

Current Update on Role of Hesperidin in Inflammatory Lung Diseases: Chemistry, Pharmacology, and Drug Delivery Approaches

Salman Hosawi [1,2]

[1] Department of Biochemistry, Faculty of Science, King Abdulaziz University, Jeddah 21589, Saudi Arabia; shosawi@kau.edu.sa

[2] Centre for Artificial Intelligence in Precision Medicines, King Abdulaziz University, Jeddah 21589, Saudi Arabia

Abstract: Inflammation is a common feature of many respiratory diseases, such as pneumonia, asthma, pulmonary fibrosis, chronic obstructive pulmonary disease (COPD), lung cancer, acute lung injury, and COVID-19. Flavonoids have demonstrated their anti-inflammatory and antioxidant effects by influencing inflammation at different stages and majorly impacting several respiratory diseases' onset and development. According to current studies, hesperidin, one of the most abundant polyphenols, can inhibit transcription factors or regulatory enzymes essential for controlling inflammation-linked mediators, including nuclear factor-kappa B (NF-κB), Inducible nitric oxide synthase (iNOS), and cyclooxygenase-2 (COX-2). It also improved cellular antioxidant defences by activating the ERK/Nrf2 signalling pathway. Therefore, this review provides the latest studies on the effect of hesperidin in different respiratory diseases, its pharmacokinetic profile, and innovative drug delivery methods.

Keywords: hesperidin; inflammation; respiratory lung diseases; flavonoids

Citation: Hosawi, S. Current Update on Role of Hesperidin in Inflammatory Lung Diseases: Chemistry, Pharmacology, and Drug Delivery Approaches. *Life* **2023**, *13*, 937. https://doi.org/10.3390/life13040937

Academic Editor: Ramón Cacabelos

Received: 9 February 2023
Revised: 21 March 2023
Accepted: 30 March 2023
Published: 3 April 2023

Copyright: © 2023 by the author. Licensee MDPI, Basel, Switzerland. This article is an open access article distributed under the terms and conditions of the Creative Commons Attribution (CC BY) license (https://creativecommons.org/licenses/by/4.0/).

1. Introduction

Airway inflammation has a causative role in the pathophysiology of several significant respiratory diseases, including lung fibrosis, asthma, chronic obstructive pulmonary disease (COPD), acute respiratory distress syndrome (ARDS), pulmonary fibrosis, emphysema, and other respiratory diseases. Inflammation increases due to the production of pro-inflammatory cytokines, which cause mononuclear cells and polymorphonuclear eosinophils neutrophils to migrate into the lung tissue [1,2]. Over the last 10 years, many natural substances or phytochemicals with anti-inflammatory activities have been discovered as potential treatment possibilities. Additionally, drug development efforts focused on these natural compounds' high efficacy and few adverse effects.

Hesperidin is a member of polyphenolic phytochemicals called flavonoids known for their wide range of pharmacological activities, such as anti-oxidative, anti-inflammatory, and anti-neoplastic properties [3,4]. Hesperidin is beneficial in treating and managing respiratory problems, as demonstrated by several clinical investigations [5]. In addition, hesperidin has been found to have a preventative impact against respiratory infections [6,7]. The primary purpose of this study is to provide readers with a thorough grasp of how hesperidin benefits conditions related to the lungs or the respiratory system. The central focus of this review is the significance of hesperidin in managing all lung disorders and the healing process.

2. Search Strategy

We searched PubMed/Medline, Science Direct, Scopus, ProQuest, and Google Scholar as well as Google databases using the following keywords: "hesperidin", "drug delivery", "inflammation", "lung fibrosis", "asthma", "COPD", "acute respiratory distress syndrome (ARDS)", "pulmonary fibrosis", "emphysema", "Pharmacokinetics", "toxicity", "bleomycin", and "oxidative stress.

Relevant English-language papers published up through December 2022 were eligible for inclusion in this analysis. This review comprised every paper that evaluated the impact of hesperidin on inflammatory lung disorders. This assessment excludes articles with minimal information, including letters and comments. In addition, research investigating the impact of naringenin on disorders other than those related to lung inflammation were excluded.

Two reviewers separately retrieved literature by title/abstract to choose eligible articles; studies that did not satisfy the eligibility criteria were eliminated. The entire texts of the shortlisted articles were next scrutinised for eligibility and data extraction. In situations of disagreement, debated articles were considered by writers, who then reached a conclusion.

3. Sources of Hesperidin

Hesperidin is present in citrus fruits, which include lemons, oranges, and several other vegetables and fruits. French scientist Lebreton initially isolated it, using the ellipticity (the spongy inside of the peel) of oranges in 1828 [8]. Citrus maxima, sweet oranges, and unshiu rinds have all been widely used to isolate them [9]. It has also been documented in other citrus species, including *Citrus reticulata Blanco* and *Citrus aurantium* L. *Var. Dulcis* (mandarin or tangerine) [10]. In addition, citrus species have higher concentrations of hesperidin in their albedo, membranes, and pith than in their seeds and juice vesicles. Furthermore, it has been claimed that the hesperidin content in green fruits changes as they mature. Furthermore, there is a significant connection between hesperidin levels and seed germination, demonstrating that light exposure increases the formation of this molecule (Table 1).

Table 1. Natural sources and quantity of hesperidin.

Citrus Fruit Juices	Quantity	References
Grapefruit concentrate juice	1.55 mg/100 mL	[11]
Pure grapefruit juice	0.65 mg/100 mL	[11]
Juice from the concentrate of lemon	24.99 mg/100 mL	[12]
Lemon juice, pure	17.81 mg/100 mL	[12]
Pure juice, lime	13.41 mg/100 mL	[12]
Orange [Blond], concentrate juice	52.68 mg/100 mL	[13]
Pure orange [Blond] juice	25.85 mg/100 mL	[13]
Orange [Blond], concentrate juice	51.30 mg/100 mL	[13]
Orange [Blond] juice, undiluted	43.61 mg/100 mL	[13]
Tangerine, concentrate juice	36.11 mg/100 mL	[14]
Herbs	**Quantity**	**References**
Dried, peppermint	480.65 mg/100 g FW	[15]
Fresh welsh onion	0.02 mg/100 g FW	[16]

4. Chemistry

An aglycone called hesperetin or methyl eriodictyol with an associated disaccharide called rutinose makes up the flavanone glycoside known as hesperidin. Therefore, hesperetin's -7-rutinoside is called hesperidin. One molecule of glucose and one molecule of rhamnose comprise the disaccharide unit (C12H22O10), which may take on either of its two isomeric forms, neohesperidose or rutinose. Chemically, rhamnosyl-(1→6) glucose or 6-O-(6-deoxy-L-mannopyranosyl)-D-glucose make up rutinose. Chemically speaking, neohesperidose is O-L-rhamnosyl-(1→2) glucose, with the two sugar units' arrangement being the sole difference. By partially hydrolyzing hesperetin with diluted acid to produce L-rhamnose and hesperetin-7-Dglucoside, by which the enzyme could break down -Dglucosidase, the location of the sweeteners in hesperidin was established.

As a result, in hesperidin, rhamnose is connected to glucose and glucose is coupled to hesperetin. Chemically speaking, hesperetin ($C_{16}H_{14}O_6$) is a 3,5,7-trihydroxy-4-methoxy flavanone. Hesperidin, following alkaline hydrolysis, gives phloroglucinol and hesperetin acid. Hesperidin is thus 3,5,7-trihydroxy-4-methoxy flavanone-7-(6-L-rhamnopyranosyl-D-glucopyranoside or -7-rutinoside [7] (Figure 1).

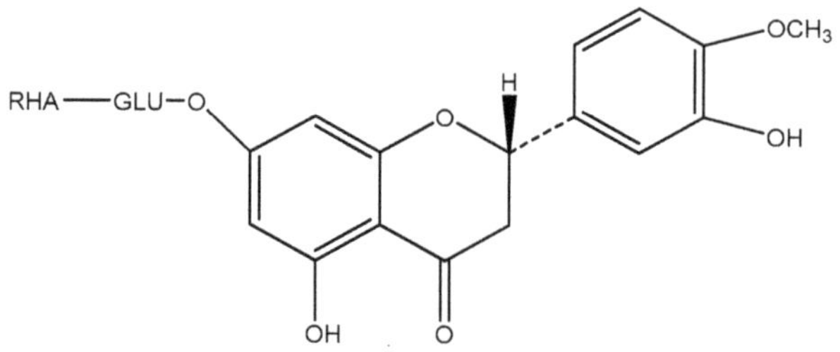

Hesperidin (hesperetin-7-rhamnoglucoside)

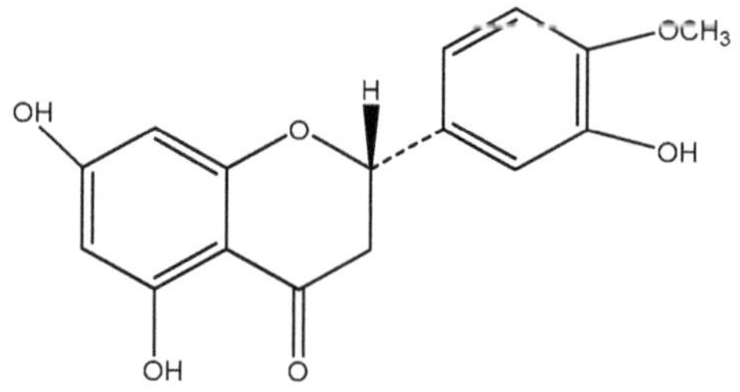

Hesperetin (3, 5, 7-trihydoxy-4-methoxyflavanone)

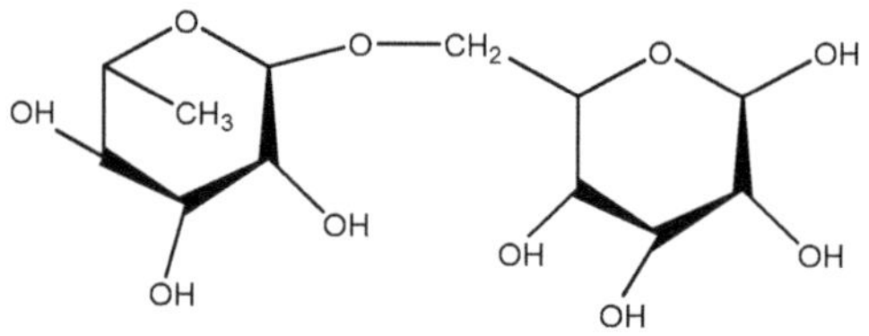

Rutinose [O-α–L-rhamnosyl-(1⟶ 6) glucose}

Figure 1. Structure of hesperetin and its derivatives.

5. Physical Properties

Long, hair-like needles of pure hesperidin are tan or light yellow. It softens at 250 °C and has a melting point between 258 and 262 °C. Its chemical composition is $C_{18}H_{34}O_{15}$, and its molecular weight is 610.57 daltons. Hesperidin is readily soluble in pyridine, providing a clear yellow solution, and in diluted alkali. It is virtually insoluble in acetone and marginally soluble in hot glacial acetic acid. Upon L-rhamnose, hesperetin, aglycone, and D-glucose are all produced in equal amounts by the hydrolysis of hesperidin. By hydrolyzing (±)- and (−)-hesperetin with either diluted sulfuric acid in ethylene glycol or sulfuric acid, an optically active mixture of (±)- and (−)-hesperetin is created, which may be divided by fractional recrystallization. Both share the trait of becoming derivatives. Hesperidin produces the well-controlled acid hydrolysis of the optically active laevorotatory aglycone (−)-hesperetin. This compound then produces L-malic acid upon ozonization, demonstrating that (−)-hesperetin possesses the 2S configuration. The disaccharide's structure and the amount or lack of unpleasantness in the chemical are known to be correlated. Neohesperidosides are quite bitter, but rutinosides have no flavour. Due to its flavonoid rutinoside nature, hesperidin is not bitter in and of itself. While non-bitter rutinosides predominate in oranges and lemons, grapefruit mainly assembles bitter neohesperidosides [17,18].

6. Pharmacokinetics

Healthy white men (human volunteers) aged 25 were given 500 mg of the medication in water and equal volumes of orange juice and grapefruit to test the oral bioavailability of hesperidin from citrus products. It was taken from the digestive tract after the oral route in either form, but continuous urinary collection revealed low bioavailability. Hesperetin, an aglycone, was found in both plasma and urine. Absorbed before urinary excretion, citrus flavanones were assumed to go through glucuronidation [19]. To investigate membrane absorption of hesperidin derivatives, a cultivated layer of Caco-2 cells was used to represent the small bowel epithelium. Hesperidin does not penetrate the Caco-2 single layer, likely because of its lower solubility, but its glycosides do, and this penetration is dose-dependent. The paracellular route had been assumed to be the mechanism for this penetration [20]. The diet has been discovered to affect the rabbits' ability to absorb hesperidin when given orally. When combined with a synthetic ration, it was absorbed, but not when combined with a standard pelleted feed. After oral administration to rats, flavonoids, including hesperetin and hesperidin, had their metabolic M-hydroxyphenyl propionic acid studied and smaller levels of m-coumaric acid in the aglycones were the common metabolic products found in urine. The aglycones were both free and glucuronic acid conjugated. This suggested dehydroxylation, demethoxylation, or demethylation coupled with dehydroxylation occurred after intestinal absorption, producing m—acid hydroxyphenyl propionic. According to research, hesperetin was more quickly absorbed than hesperidin in mice, rabbits, and humans [7]. Human volunteers consumed hesperidin, and a significant change in metabolism was seen. The main urinary metabolite, 3-hydroxy-4-methoxyphenylhydracrylic acid, showed that the breaks of the pyran circle of hesperetin produce hydracrylic acid. Additionally, a minor quantity of hesperetin's glucuronide was found. The bacteria that produce endo-beta-glucosidase, beta-glucosidase, and alpha-mannosidase were shown to convert hesperidin to its hesperetin, aglycone, in the gut. Moreover, it was discovered that the metabolite produced in the human gut had stronger antiplatelet action and cytotoxicity than the parent molecule. The research was recently conducted on the plasma dynamics and urine outflow of hesperetin and naringenin from grapefruit and orange juice. Juices were administered to healthy participants, and plasma and urinary tests were taken and evaluated with HPLC. Hesperetin was shown to be available from the sample juices. However, there were significant inter-individual differences. Since the urine levels varied so much, it was determined that they could not be used as food consumption indicators [21]. Animal tests demonstrated that Daflon-500 mg (a mixture of fractionated flavonoids) almost completely disappeared 96 h after treatment, with no adverse buildup in any specific organ. Rat resorption and excretion research showed that a semi-synthetic derivative of natural hesperidin, a hydrosoluble, 14C-hesperidin methyl chalcone, absorbed 1–2 h after an oral delivery system at a dosage of 10 mg/kg body weight. The intravenous drug injection at the same dosage confirmed the entero-hepatic cycle predicted by the blood kinetic patterns. The blood profiles showed that the medication was bioavailable in a favorable way. Following oral consumption, fecal excretion was higher than urinary excretion, but the two were equivalent after intravenous treatment. Furthermore, excretion mostly happens during the first 24 h after treatment, regardless of the route [22].

7. Toxicity of Hesperidin

Hesperidin is a nontoxic flavonoid, and there are not many studies regarding its negative effects, even in pregnant women. In dosages between 0.3% and 5%, it is also classified as non-chemical. According to clinical studies, phosphorylated hesperidin (100–125 mg/kg/day) is utilized as an antifertility agent, and its antifertility effect is unaffected by trauma, infections, or systemic illnesses. Another clinical trial found that 10% of patients receiving Daflon 500 mg (hesperidin + diosmin) two times daily for six weeks to one year had mild side effects compared to those of the placebo group. Hesperidin may, however, interact with widely used medications such as daunomycin and vincristine, increasing vincristine's absorption (10–50 M) across the BBB. Therefore, this research recommended avoiding hesperidin-containing meals while consuming vinblastin (especially P-glycoprotein substrates) [7]. Phosphorylated hesperidin (4%), readily digested and generating no adverse responses in animal experiments, is safe and non-toxic. Additionally, the oral bioavailability of methyl hesperidin (5%) has not been shown to have any cancerous or mutagenic effects on a rat. Moreover, consumption of hesperidin (0.3–5%) showed no side effects on food consumption, body weight, or food efficiency. Another research found that rats receiving methyl hesperidin (0.3–5%) for 13 weeks did not die, lose weight, or exhibit anomalies in their normal activities [23].

8. Role of Hesperidin in Inflammatory Lung Diseases

8.1. Asthma

Asthma, a primary noncommunicable disease (NCD) that affects both children and adults, is the most common chronic illness in children, and may include coughing, wheezing, and shortness of breath caused by inflammation, tightness in the chest, and constriction of the small airways in the lungs [24,25]. Citrus-farming yields hesperidin, a flavanone derivative of the flavonoid hesperetin and the sugars rutinose in large quantities and at a low cost. Using an allergic asthma model of the mouse, Dajun Wei et al. sought to ascertain if hesperidin inhibits ovalbumin-induced airway inflammation. OVA (ovalbumin) challenged and sensitized mice, resulting in persistent airway inflammation and remodeling. Compared to the mice exposed to OVA, hesperidin dramatically reduced the quantity of infiltration inflammatory responses and bronchoalveolar lavage fluid with Th2 cytokines. Hesperidin also decreased serum OVA-specific IgE levels. The OVA-induced airway hyperresponsiveness (AHR) to breathed methacholine was significantly reduced by hesperidin. According to lung histological examinations utilizing hematoxylin, eosin, and alcian blue-periodic acid-Schiff staining, hesperidin decreased hypersecretion, and inflammatory cell infiltration combined with the mice group was exposed to OVA. Their results explain the immune function of hesperidin in case it protects a mouse asthma model [26]. In another study, Hesperidin-3′-O-Methylether inhibits phosphodiesterase more effectively than hesperidin airway hyperresponsiveness brought on by ovalbumin: inhibition and suppression hesperidin and hesperidin-3′-O-effects methyl ethers on phosphodiesterase suppression and airway hyperresponsiveness (AHR) in an asthma model of the mouse were compared by Yang et al. According to their findings, hesperidin-3′-O-methyl ether significantly decreased the induced pause value and suppressed the total number of inflamed cells, including macrophages, eosinophils, neutrophils, and lymphocytes as well as complete and OVA-specific immunoglobulin (Ig)E levels in serum and BALF. It also increased the total quantity in mice's serum sensitized to the substance and challenged with it. They found that hesperidin-3′-O-methyl ether has a greater therapeutic ratio and is effective in inhibiting phosphodiesterase and suppressing AHR. As a result, the potential for hesperidin-3′-O-methyl ether to be used in treating allergic asthma and COPD may be greater [27]. Hesperidin, a purported Th2 cytokine blocker, benefits asthma in a mouse model of allergic asthma.

Seung-Hyung Kim et al. examined the anti-inflammatory and anti-asthmatic effects of hesperidin on the production of interleukin -17 (IL-17), eotaxin, and -OVA-specific IgE and Th2 cytokines. They looked at how hesperidin affected the generation of Th2 cytokines, OVA-specific IgE, pulmonary eosinophilic infiltration, different immune cell morphologies, and hyperresponsiveness of the airways in a mouse model of asthma. Results showed that hesperidin-treated groups reduced OVA-specific IgE, IL-5, and IL-17, reducing allergic airway inflammation levels, eosinophil infiltration, and AHR. Their findings revealed that hesperidin successfully cures asthma by decreasing the synthesis of eotaxin, OVA-specific IgE, eosinophil, and Th2 cytokines (IL-5) infiltration via the suppression of the GATA-3 transcription factor [24]. Later, Chang et al. investigated the anti-inflammatory effects of hesperidin and its mechanisms in an asthmatic murine model induced by ovalbumin, and looked into how hesperidin may affect the balance of Th1 and Th2 cells. Airway luminal constriction, the formation of airway hyper-responsiveness, an increase in eosinophils in bronchoscopy (BAL) fluid, and an increase in infiltration by lung tissue caused inflammation in cells surrounding airways and blood vessels were all symptoms of AHR. Hesperidin treatment before the last pulmonary OVA exposure significantly reduced all asthmatic symptoms. As a result, their research has shown that hesperidin is essential for improving the pathogenetic pathway of asthma in mice. Their results enhance existing views in comprehension of the immunopharmacological activities of hesperidin and provide unique insight into the immunopharmacological action of hesperidin as it relates to its effects in a mouse asthma model [25,26] (Figure 2).

8.2. COPD

A set of diseases known as a COPD includes chronic bronchitis and emphysema [21]. Hesperidin's ability to reduce oxidative stress and inflammatory reaction in COPD rats may be connected to the PGC-1/NF-B signalling axis/ SIRT1. Shuyun Wang et al. studied the function of hesperidin in a COPD mice model to provide a foundation for using hesperidin. Cigarette smoke extract (CSE) was exposed to the mice to create COPD models. Hesperidin was then administered to the animals. The results demonstrated that inflammatory cell infiltrating and CSE brought on cell death in the mouse lung tissues, but hesperidin efficiently reversed these pathological alterations. Hesperidin may successfully enhance superoxide dismutase (SOD) and catalase (CAT) levels in bronchoalveolar lavage fluid (BLAF) while decreasing IL-6, malondialdehyde (MDA), and IL-8 in BLAF and the lung MPO content in COPD mice. Hesperidin was also shown to consistently increase PGC-1 and SIRT1 expression levels while decreasing the phosphorylation of p65 in COPD mouse models. Hesperidin used in large doses generally had a better impact on COPD mice. It was finally observed that hesperidin reduced inflammation and oxidative stress in CES-induced COPD mice and was linked to the SIRT1/PGC-1/NF-B signaling axis, suggesting a potentially novel approach to the treatment of COPD [27].

8.3. Pulmonary Fibrosis

The injury and scarring of lung tissue lead to the development of the lung ailment pulmonary fibrosis [28]. Hesperidin improves experiments on bleomycin-induced lung fibrosis through suppression of Smad3/MPK/ TGF-beta1 and NF-kB pathways, according to research by Zheng Zhou et al. When bleomycin (BLM) was administered intraperitoneally to Sprague–Dawley (SD) rats, pulmonary fibrosis was induced. Hesperidin was administered to rats, and different lung and BALF parameters were measured. Hesperidin dramatically reduced elevated myeloperoxidase levels, hydroxyproline, and oxide-nitrosative stress in the lung and BALF. It also reduced the expressions of lung Nrf2 and HO-1 that BLM upregulated, and the expressions of IL-1, TNF-α, collagen-1, TGF-β, IL-6, and Smad-3 were upregulated. According to a Western blot examination, hesperidin improved lung AMPK, NF-kB, IB, and PP2C-protein expression. Hesperidin also lessens BLM-induced pulmonary fibrosis by inhibiting the IB/NF-kB,/ TGF-1/MPK, Smad3, and oxide-inflammatory marker (HO-1 and Nrf2) pro-inflammatory marker (IL-1, TNF-α, and IL-6) pathways. This improves the variation of oxide-inflammatory markers (HO-1 and Nrf2) and pro-inflammatory indicators (TNF-α) [29].

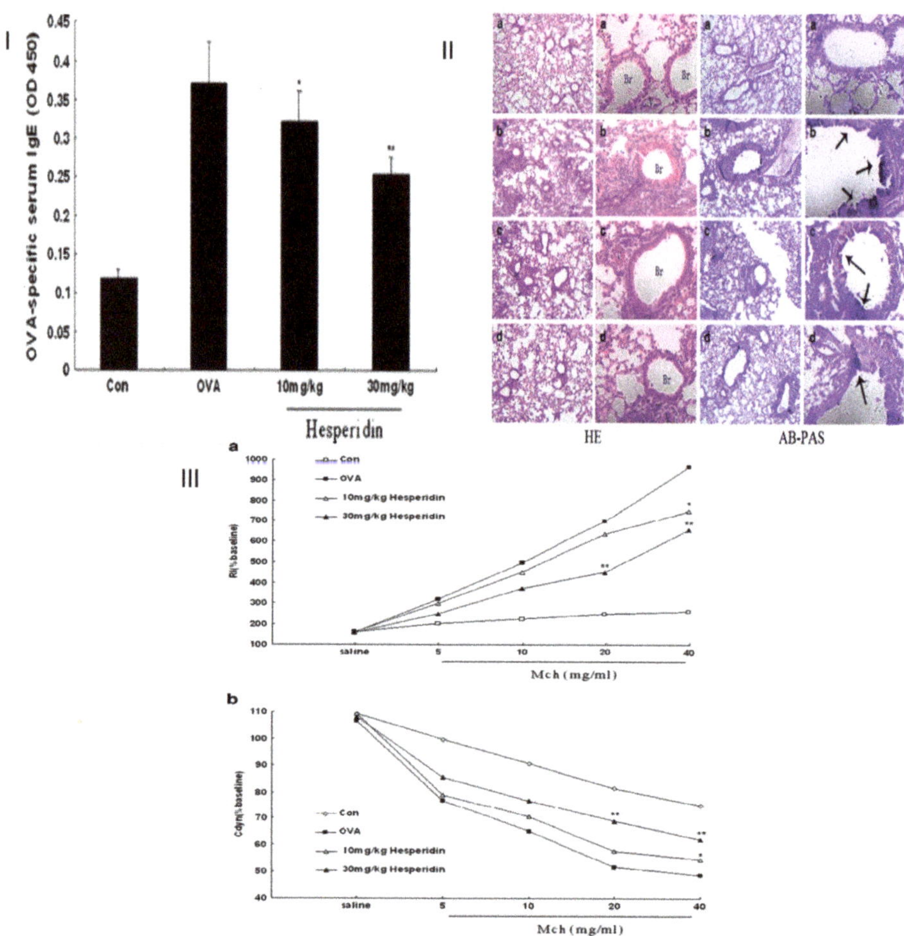

Figure 2. Schematic representation of hesperidin's ability to reduce asthma symptoms in an allergic airway inflammation model. (**I**) Hesperidin's (milligrams per kilogram) impact on serum OVA-specific IgE. (**II**) Representative hematoxylin–eosin and alcian blue-periodic acid-Schiff stained sections of lung from: (a) PBS-challenged mice; (b) OVA-challenged mice; (c) OVA-challenged mice treated with hesperidin (10 mg/kg); (d) OVA-challenged mice treated with hesperidin (30 mg/kg). The left panel is magnified 100×; the right panel is magnified 400×. Br bronchi, V vessel. Arrows indicate areas of alcian blue+cells. (**III**) In response to methacholine, hesperidin therapy decreased RI and restored Cdyn in OVA-challenged mice. Airway hyperresponsiveness was assessed by percentage change from the baseline level of (a) lung resistance (RI, n = 6 mice per treatment group) and (b) dynamic compliance The values represent the mean ± SEM of three independent experiments. * $p < 0.05$, ** $p < 0.01$. vs. OVA [26].

In addition, Jiasen Guo et al. discovered that the natural substance, neohesperidin, suppresses TGF-1/Smad3 signaling and reduces lung fibrosis brought on by bleomycin in rats. They discovered that neo hesperidin reduced TGF-1-induced extracellular matrix formation, myofibroblast differentiation, and fibroblast migration, inhibiting TGF-1's damage to alveolar epithelial cells. Additionally, they acquired in vivo proof that neo hesperidin therapy prevented bleomycin-induced lung damage, even in rats with preexisting pulmonary fibrosis. According to their research, neohesperidin may treat progressive pulmonary fibrosis since it may target a key signaling pathway and profibrogenic responses [30]. Histopathological research of hesperidin as a radioprotector towards radiation-induced lung damage was examined by Gholam Hassan Haddadi et al. Three groups of fifty rats were created. G1: No HES or radiation exposure (sham). G2: Thorax received beta-irradiation. HES was administered to G3 along with orradiation. Results from histopathology after 24 h revealed radiation-induced inflammation and the presence of more inflammatory cells than in G1. Compared to G2, the administration of hesperidin greatly reduced such an impact. When comparing G2 to G1, histopathological analysis revealed a substantial rise in alveolar thickness, mast cells, inflammatory cells, vascular thickness, pulmonary oedema, inflammation, and fibrosis. The research has shown that hesperidin is a strong radioprotector against radiation-induced lung injury in tissue of rats. Compared to G2, hesperidin greatly reduced the mast cells, inflammatory response, and fibrosis. Hesperidin administration reduced radiation fibrosis and radiation pneumonitis in the lung tissue [31].

Hesperidin's effects on idiopathic pulmonary fibrosis were assessed using micro-computed tomography, histopathology, and a bleomycin-rat model. Due to its biochemical, anti-inflammatory, and antioxidant capabilities, Cemile Ayse Görmeli et al. examined the therapeutic potential of hesperidin against pulmonary fibrosis using histological, biochemical, and micro-CT investigations. Based on their research, they proposed that hesperidin's biochemical, anti-inflammatory, and antioxidant characteristics might prevent BLC-induced lung fibrosis. The lungs of BLC-treated rats underwent obvious histological alterations. Infiltration of macrophages and lymphocytes, along with fibroblast growth, were shown to thicken the interalveolar septa in these rats. Histopathological alterations have been less extensive in the BLC+ hesperidin group than in the BLC group. Compared to the BLC group, the hesperidin therapy resulted in lower lipid peroxidation and higher antioxidant status. Additionally, data from histopathology and biochemistry were significantly positively correlated with micro-CT findings. It was also shown to effectively reduce the severity of BLC-induced lung damage, which was employed as a model for IPF [32].

Hesperidin was later studied to see whether it may treat pulmonary fibrosis by preventing lung fibroblast senescence by Di Han et al. They showed that hesperidin could help mice with lung fibrosis brought on by BLC. Hesperidin therapy in vitro and in vivo dramatically downregulated the amount of senescence-associated galactosidase-positive cells by downregulating the expression of senescence signature proteins p53, p21, and p16 as well as the myofibroblast marker-SMA. Hesperidin, meanwhile, may block the IL6/STAT3 signaling pathway. Further research revealed that when the IL-6/STAT3 signaling pathway was blocked in vitro with LMT-28 pre-treatment, the inhibitory impact of hesperidin on fibroblast senescence appeared to fail [33].

Rezaeyan et al. sought to determine the male rat lung tissue damage caused by irradiation and the radioprotective efficiency of hesperidin. As a radiosensitive organ, the lung, radiation doses must consider this. Their findings indicated that oral treatment of hesperidin protected against -irradiation-induced oxidative stress and pulmonary damage in rats. This protection was most likely provided by hesperidin's ability to stabilize membranes and scavenge free radicals, which may have a beneficial impact on respiratory diseases [34] (Figure 3).

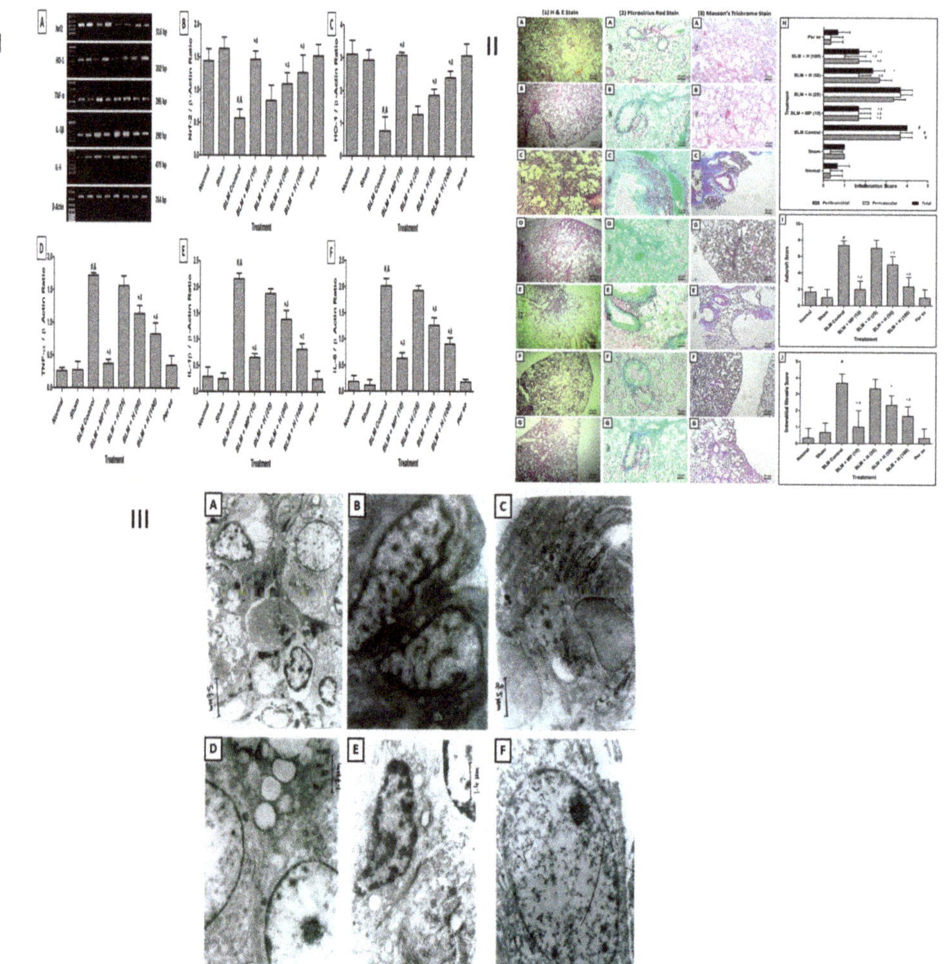

Figure 3. Schematic representation of hesperidin bleomycin-induced experimental pulmonary fibrosis. (**I**) Hesperidin's impact on modifications in rat lung Nrf2, HO-1, TNF-β, IL-1, and IL-6 mRNA expression brought on by BLM. (A) on day 14, quantitative representation of mRNA expression of Nrf2 (B), HO-1 (C), TNF-α (D), IL-1β (E) and IL-6 (F). $p < 0.05$ when compared with: sham group (&), normal group (#), each other (MP or hesperidin) group ($), and BLM control group (*). (**II**) Effect of hesperidin on BLM-induced modifications in the rats' lungs' and airways' histology. This is magnified 400×. Photomicrograph of sections of lungs of normal (A), sham control (B), BLM control (C), MP (10 mg/kg) treated (D), hesperidin (50 mg/kg) treated (E), hesperidin (100 mg/kg) treated (F) and Perse treated (G) rats. Lung H&E staining (1A–1G) on day 14, Lung Picro-Sirius red staining (2A–2G) on day 28 and Lung Masson's trichrome staining (3A–3G) on day 28. Effect of hesperidin on alterations induced by BLM in lung airway inflammation score (H), Ashcroft Score (I) and interstitial fibrosis score (J). $p < 0.05$ when compared with: sham group (&), normal group (#), each other (MP or hesperidin) group ($), and BLM control group (*). (n = 3) (**III**) BLM altered rats' lung ultrastructure, which was inhibited by hesperidin. Photomicrographs of lung from representative animals, normal (at 3474×) (A), sham control (at 11,580×) (B), BLM control (at 7720×) (C), methylprednisolone (10 mg/kg) treated (at (11,580×) (D), hesperidin (100 mg/kg) treated (at 13,510×) (E), and per Se treated rats (at 11,500×) (F) [34].

8.4. Lung Cancer

When lung cells divide too quickly, tumors form and the condition is known as lung cancer [35]. The worldwide burden of lung cancer is high; up to 18% of cancer-related fatalities are attributable to lung tumors. In terms of prevalence, lung cancer ranks second, and experts worldwide have made enormous attempts to treat it. Tumors may damage one's ability to breathe, as well as have the possibility of spreading to other parts of the body [36]. Natural products are a major source of anti-tumour medications. Citrus sinensis Osbeck, a Rutaceae plant lime, contains the flavanone known as hesperidin. It is thought to reduce the viability of cancer cells in culture. However, hesperidin's impact on lung cancer and its possible mechanisms are unknown. It was discovered that pinx1 expression level has a significant role in controlling invasion, cell proliferation, lung cancer senescence, and migration, and is strongly connected to overall survival. More notably, hesperidin markedly enhanced pinx1 protein expression, and pinx1's specific siRNA prevented hesperidin's protective effects. Additionally, it was determined that hesperidin is effective in vivo. The results demonstrated that hesperidin is a promising therapy for halting lung cancer development [35].

The carotenoid cryptoxanthin and the flavonoid hesperidin have inhibitory activity on carcinogenesis in various organs. Takuji Tanaka and colleagues, by using juice from a satsuma mandarin (MJs), created citrus juices with high levels of β-cryptoxanthin and hesperidin. Hesperidin and cryptoxanthin are present in significant concentrations. They showed that cationized peroxidase (CHRP) and MJs inhibit medically induced mouse lung, rat colon, and rat tongue carcinogenesis. Rats' liver, tongue and colon detoxifying enzyme activity were increased after CHRP gavage.

Additionally, the target tissues' production of pro-inflammatory enzymes and cytokines might be reduced by CHRP and MJs. Their research shows that orally ingested CHRP and MJs block the growth of epithelial neoplasms in mouse colon, lung, and tongue cancers by acting on various biological processes. Further, they concluded that CHRP and MJs might be used as cancer chemopreventive agents against the growth of tongue, colon, and lung cancer due to their ability to control proliferation and the mRNA expression of many cytokines and inflammatory enzymes, as well as their ability to prevent chemically induced detoxification enzymes and carcinogenesis [36].

Citrus unshiu Marc juice (MJ) and MJ5, especially MJ5 and MJ2, completely inhibited colon cancerogenesis in rats brought on by chemicals, according to Hiroyuki Kohno et al.'s study on the β-cryptoxanthin and flavonoids-rich mandarin juice and inhibitory impact on the compound 4-(methylnitrosamino)-1-(3-pyridyl)-1-butanone. MJ's therapy reduced the number of lung neoplasms with no statistically significant differences. MJs, particularly MJ5, immunohistochemically decreased lung cancers with a PCNA-positive index (proliferating cell nuclear antigen). Hyperplastic alveolar cell lesions were unaffected. According to their results, MJ5 can chemoprevention 4-(methyl-nitrosamine)-1-(3-pyridyl)-1-butanone (NNK)-induced mouse lung carcinogenesis [37].

Later, Sattu Kamaraj et al. investigated the anticancer and antioxidant hesperidin's involvement in benzo (a) pyrene-induced lung cancer in mice. Hesperidin's ability to prevent lung cancer caused by B(a)P exposure in Swiss albino mice was investigated in this research. B(a)P administration to mice caused an increase in lung-specific tumour marker and lipid peroxides (LPO), serum marker enzymes, gamma-glutamyl transpeptidase (GGT), 5′nucleotidase (5′ND), lactate dehydrogenase (LDH), carcinoembryonic antigen (CEA), and aryl hydrocarbon hydroxylase (AHH), along with a concurrent reduced in the levels of tissue antioxidants such as superoxide dismutase (SOD), Vitamins E and C, peroxidase (gpx), reduced glutathione (GSH), and glutathione catalase (CAT). Treatment with hesperidin dramatically reduced these changes, demonstrating a strong anticancer impact on lung cancer. Additionally, proliferating cell nuclear antigen (PCNA) immunofluorescent and histological investigation supported hesperidin's antiproliferative action. The results support hesperidin's chemopreventive ability against chemically generated lung disease in mice [38].

8.5. Acute Lung Injury or ARDS

Acute lung injury (ALI) and the more severe immediate respiratory distress syndrome are pulmonary manifestations of an acute systemic inflammation process, including hypoxemia, pulmonary infiltrates, and oedema (ARDS) [39]. It was shown by Xin-xin Liu et al. that high-mobility group box 1 (HMGB1) release might be decreased to lessen the acute lung damage caused by lipopolysaccharide in mice. The investigation aimed to determine how HMGB1 is related to HDN-induced immunoregulation of ALI. LPS was administered intravenously to male BALB/c mice, where it caused ALI. Reduced lung wet-to-dry weight ratio increase, total cells, neutrophils, myeloperoxidase (MPO) activity, macrophages, and lower histological lung damage are signs that HDN substantially protected rats against LPS-induced ALI. The production of chemokines and pro-inflammatory cytokines, such as monocyte chemoattractant protein-1, interleukin-6 (IL-6), and, as tumour necrosis factor (TNF), were significantly reduced by HDN pre-treatment in the meanwhile (MCP-1). Additionally, HDN pre-treatment significantly reduced macrophage infiltration and lowered both HMGB1 release and expression in vivo and in vitro. Moreover, exogenous HMGB1 used intranasally might cause lung damage, which was likewise mitigated by HDN treatment. According to the findings, HDN treatment again shields mice against ALI-induced LPS by reducing TNF and IL-6 production. Additionally, they discovered that HDN might prevent the synthesis of MCP-1 and the infiltration of macrophages, inhibiting the expression and release of HMGB1 [37].

Hesperidin (HES) modulates local immunological responses in the lung in reaction to in vivo acute lung inflammation caused by LPS, according to research by Chia-chouyeh et al regarding the immunomodulation of both in vivo and in vitro endotoxin-induced acute lung injury. Mice were given intratracheal lipopolysaccharide, and hesperidin administration reduced the production of IL-6, KC, MCP-1, IL-12, MIP-2, and IL-1; the LPS caused this. Additionally, it increased IL-4 and IL-10 production. Hesperidin dramatically reduced inos expression, nitric oxide generation and proteins, and total leukocyte counts. HES reduced the production of THP-1 cells and IL-8 on A549 cells, IL-1 THP-1 cells IL-1, and VCAM-1 on A549 cells, all of which have an impact on the function of cell adhesion. IB and MAPK pathways stimulate NF-B and AP-1, which then regulate the repression of those molecules. Because HES blocks those pathways, the production of IL-8 and VCAM-1, IL-6, and CAM-1 is reduced. According to the research, HES significantly modulated the immune system in an important clinical ARDS model. However, further research is needed to assess the potential therapeutic value of HES in the supplementary treatment of ARDS [40].

Ana Beatriz Fariasde Souza et al. postulated the results of hesperidin treatment in a possible mechanism of acute lung inflammation both in vitro and in vivo, where they assessed cell survival and production of reactive oxygen species in macrophages using various hesperidin doses. Hesperidin, they found, did not affect cell survival, but it did decrease intracellular ROS generation in cells treated with lipopolysaccharide (LPS). They also examined the effects of hesperidin in vivo in mice given the measles virus (MV). Animals given MV had greater macrophage, lymphocyte, and neutrophil counts in their bronchoalveolar lavage fluid, which suggested an inflammatory response. Although the antioxidant enzyme activity was reduced, MV caused oxidative damage and elevated myeloperoxidase activity in the lung tissue. The findings show that pretreatment with hesperidin may protect infected mice's lungs to respiratory support by regulating the inflammatory process and redox disequilibrium, and may work to prevent MV harm [41] (Figure 4).

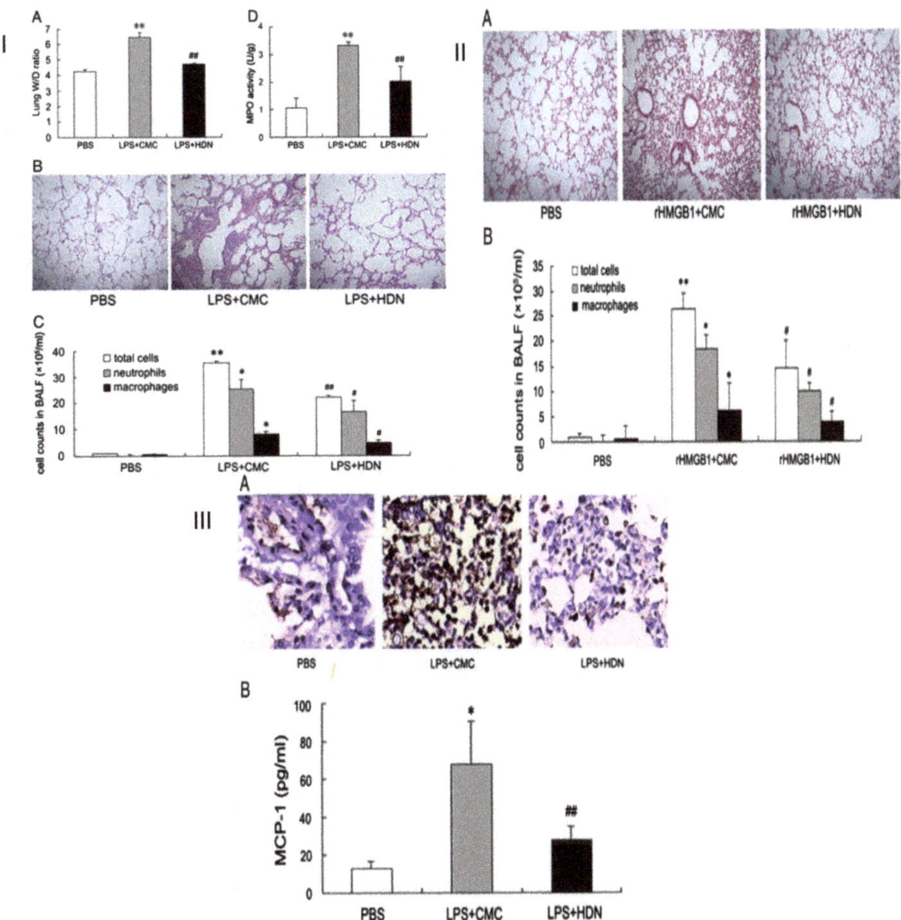

Figure 4. Schematic representation of hesperidin reduces the acute lung damage brought on by lipopolysaccharide in mice. (**I**) Pre-treatment with hesperidin reduces acute lung damage brought on by LPS. (A) The lung wet to dry weight (W/D) ratio determined 24 h after LPS challenge. (B) Hematoxylin and eosin staining of lung specimens 24 h after LPS administration (H&E staining, original magnification × 100). (C) The cell counts in BALF 24 h after LPS administration. (D) The MPO activity in lung tissues 24 h after LPS challenge. ** $p < 0.01$ versus the PBS group, * $p < 0.05$ versus the PBS group; ## $p < 0.01$ versus the LPS + CMC group, # $p < 0.05$ versus the LPS + CMC group. (**II**) Pre-treatment with hesperidin reduces HMGB1 release and expression brought on by LPS. (A) Hematoxylin and eosin staining of lung specimens collected 24 h after rHMGB1 exposure (H&E staining, original magnification × 100). (B) The cell counts in BALF collected 4 h and 24 h after rHMGB1 exposure. The values are presented as means ± SD (n = 6–8 in each group). ** $p < 0.01$ versus the PBS group, * $p < 0.05$ versus the PBS group; # $p < 0.05$ versus the rHMGB1 + CMC group (**III**) In LPS-induced ALI, hesperidin pre-treatment reduces macrophage infiltration. (A) Lung samples were collected 24 h after LPS administration for immunohistochemistry staining of macrophage (original magnification × 400). (B) BALF samples were collected at 4 h to determine the level of MCP-1. Data are expressed as means ± SD (n = 6–8 in each group). * $p < 0.05$ versus the PBS group; ## $p < 0.01$ versus the LPS + CMC group [42].

8.6. Pneumonia

The air sacs in one or both lungs might become inflamed due to pneumonia. The air sacs may enlarge with liquid or pus (purulent material), which might cause fever, breathing problems, cough, and chills that produce pus [41,43]. Numerous species, such as fungi, viruses, and bacteria, may cause pneumonia. Hassan Haddadi et al. investigated the radioprotective effects of hesperidin on radiation-induced lung injury. Three groups of fifty rats were created to prevent radiation-induced lung damage in rats. G1: no hesperidin or radiation exposure (sham). G2: Thorax received beta-irradiation. Hesperidin was administered to G3 along with irradiation. A considerable rise in inflammatory cells, mast cells, vascular thickness, inflammation, alveolar thickness, pulmonary oedema, and fibrosis were seen after an eight-week histopathological assessment. Hesperidin administration reduced radiation fibrosis and radiation pneumonitis in the lung tissue. As a result, the research demonstrated that hesperidin is a powerful radioprotector in radiation-induced lung injury in tissue rats [31].

8.7. COVID-19

The viral disease known as coronavirus is caused by the SARS-CoV-2 virus (COVID-19) [42,44]. Canrongwu and others, by using computational approaches, analyzed the pharmaceutical targets for SARS-CoV-2 and found promising medications. Several naturally occurring substances and their derivatives with anti-inflammatory properties also demonstrated a binding affinity high too, 2-hydroxy-3,4-seco-friedelolactone-27-oic acid, with several derivatives of andrographolide, chrysin-7-O-glucuronide from isodecortinol, Cassine xylocarp, and cerevisterol from hesperidin, neohesperidin, kitchen side I, and deacetylcentapicrin from *Citrus aurantium*, *Viola diffusa*, and the *Swertia* plant species. The findings imply that these small-molecule substances may be suitable 3clpro inhibitors and useful in managing SARS-CoV-2 [45] (Table 2).

Table 2. Summarizes the role pharmacological role of Hesperidin in various respiratory lung diseases.

Respiratory Diseases	Model	Aim	Findings	References
Asthma	OVA-challenged mice	Allergic asthma model of the mouse. Hesperidin downregulated ovalbumin-induced inflammation of the airways.	Hesperidin suppressed inflammatory cell infiltration and mucus hypersecretion, decreasing OVA-specific IgE levels.	[26]
	OVA-induced lung eosinophilia and mucus hypersecretion in a mouse model of asthma	Hesperidin-3′-O-Methylether inhibits phosphodiesterase more effectively than hesperidin airway hyperresponsiveness brought on by ovalbumin: inhibition and suppression	It decreased the blood levels of OVA-specific immunoglobulin (Ig)E and the total number of macrophages, macrophages, neutrophils, lymphocytes, and eosinophils.	[27]
	Eotaxin, IL-17, and OVA-specific IgE in vivo asthma model mice	Hesperidin, a prospective Th2 cytokine antagonist, has anti-asthmatic effects in a mouse model of asthma.	HPN efficiently cures asthma by inhibiting the transcription factor GATA-3, which lowers the development of the eotaxin and Th2 cytokines (IL-5).	[46]

Table 2. *Cont.*

Respiratory Diseases	Model	Aim	Findings	References
	Asthmatic mouse model induced by ovalbum	Hesperidin's anti-inflammatory properties and its underlying mechanisms in a mouse model of asthma induced by ovalbumin	Decreased serum OVA-specific immunoglobulin (Ig)E levels, total inflammatory cell counts, macrophages, lymphocytes, neutrophils, and eosinophils, considerably reducing all asthmatic symptoms.	[26]
COPD	Cigarette smoke-exposed mice	Hesperidin's ability to reduce oxidative stress and inflammatory reactions in mice having COPD may be connected to the SIRT1/PGC-1/NF-B signalling axis.	In mice with COPD caused by CES, hesperidin reduced the inflammatory and oxidative stress responses.	[47]
Pulmonary Fibrosis	Sprague–Dawley rats	By inhibiting the TGF-beta1/Smad3/AMPK, kappa alpha/NF-kappa B, and bleomycin-induced experimental lung fibrosis pathways, hesperidin reduces the severity of the condition.	Hesperidin reduces the effects of BLM-induced IPF by inhibiting the IB/NF-B, TGF-1/AMPK/Smad3 and oxide-inflammatory marker (HO-1and Nrf2) pro-inflammatory marker (TNF-, IL-6 and IL-1,) pathways.	[29]
	Mice	Neohesperidin reduces bleomycin-induced lung fibrosis in rats and suppresses TGF-1/Smad3 signalling.	Lowered the formation of extracellular matrix, fibroblast migration, and myofibroblast differentiation caused by TGF-1.	[30]
	Tissue damage in the lung of male rats by induced radiation	Analyzing hesperidin's radioprotective effect on acute radiation damage in rat lung tissue	In rats, oral treatment of HES was observed to protect against irradiation-induced pulmonary damage. This protective action against inflammatory diseases is likely a result of HES's capacity to scavenge free radicals and stabilise membranes.	[34]
	Radiation-induced lung injury of male rats	Hesperidin as a radioprotector regarding radiation-induced lung damage in rat	HES dramatically reduced lung tissue fibrosis, inflammation, and mast cell proliferation. It also reduced radiation fibrosis and radiation pneumonitis.	[31]
	Bleomycin-rat model.	Hesperidin's effects on pulmonary fibrosis were assessed using micro-computed tomography, histopathology, and a bleomycin-rat model.	Due to its anti-inflammatory, chemical, and antioxidant capabilities, HP therapy causes lung fibrosis while decreasing lipid peroxidation and raising antioxidant status.	[29]

Table 2. *Cont.*

Respiratory Diseases	Model	Aim	Findings	References
	Bleomycin-induced pulmonary fibrosis in mice.	Hesperidin reduces pulmonary fibrosis by inhibiting lung fibroblast senescence through the IL-6/STAT3 signaling pathway.	Hesperidin decreased senescence-associated-galactosidase-positive cells both in vivo and in vitro, and it can block the IL6/STAT3 signalling pathway.	[33]
Lung cancer	Inhibit cancer cell viability in vitro	Through pinx1, hesperidin prevents lung cancer both in vitro and in vivo.	Hesperidin dramatically enhanced pinx1 protein levels, and blocking pinx1 with a targeted siRNA prevented hesperidin's protective effects.	[35]
	Rodent model of lung cancer	Citrus juices and pulp with high levels of hesperidin and cryptoxanthin prevent cancer	HPs are a promising cancer chemopreventive agent against the formation of tongue, colon, and lung cancer because they inhibit chemically induced carcinogenesis by detoxifying enzymes, controlling the proliferation and mRNA expression of various cytokines and inflammatory enzymes.	[36]
	Lung initiated with 4-(methylnitrosamino)-1-(3-pyridyl)-1-butanone (NNK) in male A/J mice	Cryptoxanthin and hesperidin-rich mandarin juice's inhibitory effects on 4-(methylnitrosamino)-1-(3-pyridyl) mouse pulmonary tumorigenesis brought on by 1-butanone	HP decreased the PCNA-positive index in lung cancers while not affecting the PCNA index in lesions with hyperplastic alveolar cells.	[48]
	Swiss albino mice	Hesperidin's ability to fight cancer and benzo(a)pyrene-induced lung cancer in mice	Lipid peroxides (LPO), carcinoembryonic antigen (CEA), a lung-specific tumour marker, and the serum marker aryl hydrocarbon hydroxylase (AHH) and lactate dehydrogenase were all enhanced by HP. Conversely, these modifications exhibiting a strong anticancer impact in lung cancer were dramatically decreased by hesperidin.	[38]
Acute lung injury	Male BALB/c mice	By preventing the release of HMGB1, hesperidin reduces the acute lung damage brought on by lipopolysaccharide in mice.	HP inhibits the invasion of neutrophils and the synthesis of MCP-1, preventing the transcription and release of HMGB1. Along with neutrophils, macrophages, and myeloperoxidase (MPO) activity, HP cells lowered the elevation of the lung wet-to-dry weight ratio, and therefore lessened lung histological damage	[37]

Table 2. *Cont.*

Respiratory Diseases	Model	Aim	Findings	References
	In vivo and in vitro model of acute lung inflammation	Effects of hesperidin treatment in a model system of acute respiratory inflammation both in vitro and in vivo.	By regulating the chronic inflammation and redox imbalance, hesperidin may protect the lungs of mice exposed to mechanical ventilation and may even work to stop MV harm.	[41]
	LPS mice in vivo and cell line in vitro	Hesperidin's in vivo and in vitro innate immunity of endotoxin-induced severe lung damage.	HES inhibits the inflammatory pathways and suppresses the production of ICAM-1, IL-8, IL-1, TNF, IL-6, IL-12, and VCAM-1 in THP-1 and A549 cells.	[40]
	Mouse model of acute lung injury	In a mouse model of acute lung injury, nasal administration of chitosan/ nanoparticles of hesperidin suppresses the cytokine storm syndrome.	In a mouse model of chronic lung illness, nasal administration of HPD/NPS lowers CSS and ALI/ARDS, indicating that anti-inflammatory nanoparticle-based therapeutic approaches may be employed to lessen CSS and ALI in people with inflammatory lung damage.	[49]
Pneumonia	Radiation-induced lung injury male rats	A histopathological study on hesperidin as a radioprotector against radiation-induced lung damage in rats.	HES dramatically reduced lung tissue fibrosis, inflammation, and mast cell proliferation. It also reduced radiation fibrosis and radiation pneumonitis.	[31]
COVID-19	Computational models	Using computational approaches, the medical targets for SARS-CoV-2 are analyzed, and promising medications are found.	One group of andrographolide compounds, hesperidin, showed a strong affinity.	[45]

9. Drug Delivery in Respiratory Diseases

Ultimately, nanotechnology could be the most creative and successful strategy to increase bioavailability. When some substances are covered with a coating of nanoparticles, their permeability and the amount of that substance that may enter the bloodstream are boosted [50]. Healthy cells seem to be only marginally harmed by the nanoparticle covering. Nanotechnology research has also been conducted to improve hesperidin's bioavailability. The antimicrobial zinc oxide nanostructures influenced by in silico and hesperidin direct comparison among antiviral phenolics for anti-SARS-CoV-2 purposes were studied by Gouda H. Attia et al. These natural medications and dietary supplements may be more affordable, accessible, safe, and have fewer adverse effects. This research compares ten phenolic antiviral drugs in silico against SARS-CoV-2 and identifies the potent metabolite of natural sources. Then, as a reducing agent, these metabolites were used to create zinc oxide nanoparticles (ZnO NPS). Every tested substance exhibited the anticipated anti-SARS-CoV-2 action. Hesperidin had a high docking score, so they isolated it from the peels of orange and used conventional spectroscopic research to establish its structure.

Moreover, IR, UV, XRD, and TEM were used to characterize the production of zinc oxide nanoparticles of hesperidin. Hepatitis A, an example of an RNA virus, was used to test the in vitro antibacterial activities of hesperidin and ZNO NPS. Hesperidin demonstrated an antiviral effect against HAV, although ZnO NPS did so more effectively. Therefore, more research against SARS-CoV-2 is necessary before using hesperidin and its linked ZnO nanoparticles as a viable therapeutic [51].

An effective biocompatible, phagocytosis, anti-cancer, anti-inflammatory-inducing model was examined by Ghassan M. Sulaiman et al. A medication delivery method based on hesperidin encapsulated on gold nanoparticles was explored. Using a chemical synthesis process, hesperidin packed on gold nanoparticles (Hsp-aunps) was created. The synthesis of Hsp-aunps was confirmed using various characterization methods, including Zeta potential measurement, XRD, FESEM, FTIR, EDX, TEM, and particle size analysis. The MTT and crystal violet tests evaluated the cytotoxic impact of Hsp-aunps breast cancer cells of the human line. Compared to a human breast epithelial cell line with normal proliferation, the findings showed that the treated cells' growth was significantly inhibited (HBL-100). Acridine orange-propidium iodide dual stained test was also used for fluorescence microscopy apoptosis determination. The in vivo experiment aimed to assess the hazard of Hsp-aunps in rats. The indicators for kidney and liver function were evaluated. For the examined indicators, no statistically significant changes were discovered. After receiving therapy with Hsp-aunps, histological pictures of the lung, liver, spleen, and kidney revealed no obvious damages or abnormalities. Hsp-aunps improved macrophages' functional efficacy against mice carrying ehrlich ascites tumour cells. It was also determined if bone marrow-derived cells treated with Hsp-aunps produced the pro-inflammatory cytokines. The findings showed that treatment with Hsp-aunps greatly reduced the release of IL-6, TNF and IL-1 [52].

Later, Hua Jin et al. conducted a study titled "Nasal Distribution of Hesperidin/Chitosan Nanoparticles Decreases Cytokine Storm Syndrome (CSS) in an Animal Model of Acute Lung Injury." It showed that CSS, also known as cytokine storm, is linked to a high mortality rate in individuals with (ARDS) and lung injury (ALI). However, no proven therapies for ALI or ALI/ARDS caused by CSS exist. Therefore, there is still a pressing need to create potent medications and treatment approaches to combat CSS and ALI/ARDS. Due to its capacity to enhance medication delivery to the lungs, inhaled drug delivery techniques provide a potential option for managing respiratory disease.

Another potential approach in the battle against ALI/ARDS is increasing the nasal mucosa intake of weakly water-soluble medications with low mucosa solubility to a therapeutically useful level. Hesperidin-loaded chitosan nanoparticles (HPD/NPS) were created in this study for nasal administration of the anti-inflammatory HPD chemical to inflamed lungs. Compared to free HPD, HPD/NPS showed improved cellular absorption in the inflammatory milieu both in vitro and in vivo. Compared to free HPD, the HPD/NPS significantly lowered inflammatory cytokine levels and restricted vascular permeability in an animal model of inflammation lung disease, preventing lung damage. Their research shows that nasal HPD/NP administration lowers CSS in a chronic lung illness rat model. That anti-inflammatory nanoparticle-based therapeutic approaches may be employed to lower CSS and ALI in individuals with inflammatory lung damage [49].

10. Conclusions

In conclusion, hesperidin is a therapeutic, naturally occurring anti-inflammatory flavonoid that is abundantly present in various plants and vegetables. In review, we showed that hesperidin has anti-inflammatory, antiapoptotic, and protective roles in lung cancer, pulmonary fibrosis, COPD, pneumonia, and COVID-19. In addition, we reviewed hesperidin's chemistry, pharmacokinetics, and preclinical pharmacological features, all of which might be useful in planning future clinical studies. Due to the significance of innovative drug delivery methods, we also suggested novel administration formulations and nanotechnology-based drug delivery systems, which may be given greater attention in hesperidin research in the future. Based on sufficient data, hesperidin may likely be seen as an alternate flavonoid medication for respiratory disorders. Studies using experimental animal models show that hesperidin significantly reduced organ damage and lung dysfunction.

Additionally, these results imply that it is possible to use the information from hesperidin-based animal research to develop a treatment approach for human lung disorders, namely pulmonary fibrosis and lung cancer. Furthermore, hesperidin may be used in experimental models of various respiratory conditions such as COPD, ALI, and asthma. Interestingly, these lung injuries have several characteristics, including tissue remodeling, oxidative stress, and inflammation, which are crucial therapeutic targets for hesperidin-mediated pulmonary protection. Furthermore, hesperidin's nontoxicity and natural status as a substance with an outstanding safety record are further benefits. However, further research is necessary to evaluate hesperidin at different phases of illness development. In addition, its mechanism(s) leading to considerable protection still needs to be clarified to offer specific recommendations for assessing hesperidin in patients with respiratory disorders. Finally, as our understanding of the hesperidin processes grows, various therapeutic modalities may be used to improve hesperidin's effectiveness in treating human respiratory illness.

Funding: This research received no external funding.

Institutional Review Board Statement: Not applicable.

Informed Consent Statement: Not applicable.

Data Availability Statement: No data was used for the research described in the article.

Conflicts of Interest: The author declares no conflict of interest.

References

1. Aboussouan, L.S. Respiratory disorders in neurologic diseases. *Clevel. Clin. J. Med.* **2005**, *72*, 511–520. [CrossRef] [PubMed]
2. Benditt, J.O. Pathophysiology of Neuromuscular Respiratory Diseases. *Clin. Chest Med.* **2018**, *39*, 297–308. [CrossRef]
3. Birnkrant, D.J. The assessment and management of the respiratory complications of pediatric neuromuscular diseases. *Clin. Pediatr.* **2002**, *41*, 301–308. [CrossRef]
4. Boentert, M.; Wenninger, S.; Sansone, V.A. Respiratory involvement in neuromuscular disorders. *Curr. Opin. Neurol.* **2017**, *30*, 529–537. [CrossRef] [PubMed]
5. Aksu, E.H.; Kandemir, F.M.; Küçükler, S. Ameliorative effect of hesperidin on streptozotocin-diabetes mellitus-induced testicular DNA damage and sperm quality degradation in Sprague-Dawley rats. *J. Food Biochem.* **2021**, *45*, e13938. [CrossRef] [PubMed]
6. Ferreira de Oliveira, J.M.P.; Santos, C.; Fernandes, E. Therapeutic potential of hesperidin and its aglycone hesperetin: Cell cycle regulation and apoptosis induction in cancer models. *Phytomed. Int. J. Phytother. Phytopharm.* **2020**, *73*, 152887. [CrossRef]
7. Garg, A.; Garg, S.; Zaneveld, L.J.; Singla, A.K. Chemistry and pharmacology of the Citrus bioflavonoid hesperidin. *Phytother. Res. PTR* **2001**, *15*, 655–669. [CrossRef]
8. Kanaze, F.I.; Gabrieli, C.; Kokkalou, E.; Georgarakis, M.; Niopas, I. Simultaneous reversed-phase high-performance liquid chromatographic method for the determination of diosmin, hesperidin and naringin in different citrus fruit juices and pharmaceutical formulations. *J. Pharm. Biomed. Anal.* **2003**, *33*, 243–249. [CrossRef]
9. Inoue, T.; Tsubaki, S.; Ogawa, K.; Onishi, K.; Azuma, J.-I. Isolation of hesperidin from peels of thinned Citrus unshiu fruits by microwave-assisted extraction. *Food Chem.* **2010**, *123*, 542–547. [CrossRef]
10. Rouseff, R.L.; Martin, S.F.; Youtsey, C.O. Quantitative survey of narirutin, naringin, hesperidin, and neohesperidin in citrus. *J. Agric. Food Chem.* **1987**, *35*, 1027–1030. [CrossRef]
11. Suzuki, H.; Asakawa, A.; Kawamura, N.; Yagi, T.; Inui, A. Hesperidin potentiates ghrelin signaling. *Recent Pat. Food Nutr. Agric.* **2014**, *6*, 60–63. [CrossRef]
12. Xiong, H.; Wang, J.; Ran, Q.; Lou, G.; Peng, C.; Gan, Q.; Hu, J.; Sun, J.; Yao, R.; Huang, Q. Hesperidin: A Therapeutic Agent for Obesity. *Drug Des. Dev. Ther.* **2019**, *13*, 3855–3866. [CrossRef]
13. Tadros, F.J.; Andrade, J.M. Impact of hesperidin in 100% orange juice on chronic disease biomarkers: A narrative systematic review and gap analysis. *Crit. Rev. Food Sci. Nutr.* **2022**, *62*, 8335–8354. [CrossRef]
14. Mas-Capdevila, A.; Teichenne, J.; Domenech-Coca, C.; Caimari, A.; Del Bas, J.M.; Escoté, X.; Crescenti, A. Effect of Hesperidin on Cardiovascular Disease Risk Factors: The Role of Intestinal Microbiota on Hesperidin Bioavailability. *Nutrients* **2020**, *12*, 1488. [CrossRef]

15. Sroka, Z.; Fecka, I.; Cisowski, W. Antiradical and anti-H_2O_2 properties of polyphenolic compounds from an aqueous peppermint extract. *Z. Naturforsch. C J. Biosci.* **2005**, *60*, 826–832. [CrossRef]
16. Ćirić, A.; Prosen, H.; Jelikić-Stankov, M.; Đurđević, P. Evaluation of matrix effect in determination of some bioflavonoids in food samples by LC-MS/MS method. *Talanta* **2012**, *99*, 780–790. [CrossRef]
17. Pyrzynska, K. Hesperidin: A Review on Extraction Methods, Stability and Biological Activities. *Nutrients* **2022**, *14*, 2387. [CrossRef]
18. Majumdar, S.; Srirangam, R. Solubility, stability, physicochemical characteristics and in vitro ocular tissue permeability of hesperidin: A natural bioflavonoid. *Pharm. Res.* **2009**, *26*, 1217–1225. [CrossRef]
19. Ameer, B.; Weintraub, R.A.; Johnson, J.V.; Yost, R.A.; Rouseff, R.L. Flavanone absorption after naringin, hesperidin, and citrus administration. *Clin. Pharmacol. Ther.* **1996**, *60*, 34–40. [CrossRef]
20. Kim, M.; Kometani, T.; Okada, S.; Shimuzu, M. Permeation of Hesperidin Glycosides across Caco-2 Cell Monolayers Via the Paracellular Pathway. *Biosci. Biotechnol. Biochem.* **1999**, *63*, 2183–2188. [CrossRef]
21. Erlund, I.; Meririnne, E.; Alfthan, G.; Aro, A. Plasma Kinetics and Urinary Excretion of the Flavanones Naringenin and Hesperetin in Humans after Ingestion of Orange Juice and Grapefruit Juice. *J. Nutr.* **2001**, *131*, 235–241. [CrossRef] [PubMed]
22. Chanal, J.L.; Cousse, H.; Sicart, M.T.; Bonnaud, B.; Marignan, R. Absorption and elimination of (14C) hesperidin methylchalcone in the rat. *Eur. J. Drug Metab. Pharmacokinet.* **1981**, *6*, 171–177. [CrossRef] [PubMed]
23. Devi, K.P.; Rajavel, T.; Nabavi, S.F.; Setzer, W.N.; Ahmadi, A.; Mansouri, K.; Nabavi, S.M. Hesperidin: A promising anticancer agent from nature. *Ind. Crops Prod.* **2015**, *76*, 582–589. [CrossRef]
24. Granner, D.K. The role of glucocorticoid hormones as biological amplifiers. *Monogr. Endocrinol.* **1979**, *12*, 593–611. [CrossRef] [PubMed]
25. Birnhuber, A.; Biasin, V.; Schnoegl, D.; Marsh, L.M.; Kwapiszewska, G. Transcription factor Fra-2 and its emerging role in matrix deposition, proliferation and inflammation in chronic lung diseases. *Cell. Signal.* **2019**, *64*, 109408. [CrossRef]
26. Wei, D.; Ci, X.; Chu, X.; Wei, M.; Hua, S.; Deng, X. Hesperidin Suppresses Ovalbumin-Induced Airway Inflammation in a Mouse Allergic Asthma Model. *Inflammation* **2012**, *35*, 114–121. [CrossRef]
27. Yang, Y.-L.; Hsu, H.-T.; Wang, K.-H.; Wang, C.-S.; Chen, C.-M.; Ko, W.-C. Hesperidin-3′-*O*-Methylether Is More Potent than Hesperidin in Phosphodiesterase Inhibition and Suppression of Ovalbumin-Induced Airway Hyperresponsiveness. *Evid.-Based Complement. Altern. Med.* **2012**, *2012*, 908562. [CrossRef]
28. Wilson, M.S.; Wynn, T.A. Pulmonary fibrosis: Pathogenesis, etiology and regulation. *Mucosal Immunol.* **2009**, *2*, 103–121. [CrossRef]
29. Zhou, Z.; Kandhare, A.D.; Kandhare, A.A.; Bodhankar, S.L. Hesperidin ameliorates bleomycin-induced experimental pulmonary fibrosis via inhibition of TGF-β1/Smad3/AMPK and IκBalpha/NF-κB pathways. *Excli J.* **2019**, *18*, 723–745. [CrossRef]
30. Guo, J.; Fang, Y.; Jiang, F.; Li, L.; Zhou, H.; Xu, X.; Ning, W. Neohesperidin inhibits TGF-β1/Smad3 signaling and alleviates bleomycin-induced pulmonary fibrosis in mice. *Eur. J. Pharmacol.* **2019**, *864*, 172712. [CrossRef]
31. Haddadi, G.H.; Rezaeyan, A.; Mosleh-Shirazi, M.A.; Hosseinzadeh, M.; Fardid, R.; Najafi, M.; Salajegheh, A. Hesperidin as Radioprotector against Radiation-Induced Lung Damage in Rat: A Histopathological Study. *J. Med. Phys.* **2017**, *42*, 25–32. [CrossRef]
32. Gormeli, C.; Saraç, K.; Ciftci, O.; Timurkaan, N.; Malkoç, S. The effects of hesperidin on idiopathic pulmonary fibrosis evaluated by histopathologial-biochemical and micro-computed tomography examinations in a bleomycin-rat model. *Biomed. Res.* **2016**, *27*, 737–742.
33. Waters, D.W.; Blokland, K.E.C.; Pathinayake, P.S.; Wei, L.; Schuliga, M.; Jaffar, J.; Westall, G.P.; Hansbro, P.M.; Prele, C.M.; Mutsaers, S.E.; et al. STAT3 Regulates the Onset of Oxidant-Induced Senescence in Lung Fibroblasts. *Am. J. Respir. Cell Mol. Biol.* **2019**, *61*, 61–73. [CrossRef]
34. Rezaeyan, A.; Fardid, R.; Haddadi, G.H.; Takhshid, M.A.; Hosseinzadeh, M.; Najafi, M.; Salajegheh, A. Evaluating Radioprotective Effect of Hesperidin on Acute Radiation Damage in the Lung Tissue of Rats. *J. Biomed. Phys. Eng.* **2016**, *6*, 165–174.
35. Yao, Y.; Lin, M.; Liu, Z.; Liu, M.; Zhang, S.; Zhang, Y. Hesperidin Inhibits Lung Cancer In Vitro and In Vivo through PinX1. *Front. Pharm.* **2022**, *13*, 918665. [CrossRef]
36. Tanaka, T.; Tanaka, T.; Tanaka, M.; Kuno, T. Cancer Chemoprevention by Citrus Pulp and Juices Containing High Amounts of *β*-Cryptoxanthin and Hesperidin. *J. Biomed. Biotechnol.* **2012**, *2012*, 516981. [CrossRef]
37. Liu, X.-X.; Yu, D.-D.; Chen, M.-J.; Sun, T.; Li, G.; Huang, W.-J.; Nie, H.; Wang, C.; Zhang, Y.-X.; Gong, Q.; et al. Hesperidin ameliorates lipopolysaccharide-induced acute lung injury in mice by inhibiting HMGB1 release. *Int. Immunopharmacol.* **2015**, *25*, 370–376. [CrossRef]
38. Kamaraj, S.; Ramakrishnan, G.; Anandakumar, P.; Jagan, S.; Devaki, T. Antioxidant and anticancer efficacy of hesperidin in benzo(a)pyrene induced lung carcinogenesis in mice. *Investig. New Drugs* **2009**, *27*, 214–222. [CrossRef]
39. Matuschak, G.M.; Lechner, A.J. Acute lung injury and the acute respiratory distress syndrome: Pathophysiology and treatment. *Mo. Med.* **2010**, *107*, 252–258.
40. Yeh, C.-C.; Kao, S.-J.; Lin, C.-C.; Wang, S.-D.; Liu, C.-J.; Kao, S.-T. The immunomodulation of endotoxin-induced acute lung injury by hesperidin in vivo and in vitro. *Life Sci.* **2007**, *80*, 1821–1831. [CrossRef]
41. de Souza, A.B.F.; de Matos, N.A.; de Castro, T.F.; de Costa, G.P.; Oliveira, L.A.M.; de Nogueira, K.O.P.C.; Ribeiro, I.M.L.; Talvani, A.; Cangussú, S.D.; de Menezes, R.C.A.; et al. Effects in vitro and in vivo of hesperidin administration in an experimental model of acute lung inflammation. *Free Radic. Biol. Med.* **2022**, *180*, 253–262. [CrossRef] [PubMed]

42. Sharma, A.; Ahmad Farouk, I.; Lal, S.K. COVID-19: A Review on the Novel Coronavirus Disease Evolution, Transmission, Detection, Control and Prevention. *Viruses* **2021**, *13*, 202. [CrossRef] [PubMed]
43. Mani, C.S. Acute Pneumonia and Its Complications. *Princ. Pract. Pediatr. Infect. Dis.* **2018**, *18*, 238.
44. Lai, C.C.; Shih, T.P.; Ko, W.C.; Tang, H.J.; Hsueh, P.R. Severe acute respiratory syndrome coronavirus 2 (SARS-CoV-2) and coronavirus disease-2019 (COVID-19): The epidemic and the challenges. *Int. J. Antimicrob. Agents* **2020**, *55*, 105924. [CrossRef] [PubMed]
45. Wu, C.; Liu, Y.; Yang, Y.; Zhang, P.; Zhong, W.; Wang, Y.; Wang, Q.; Xu, Y.; Li, M.; Li, X.; et al. Analysis of therapeutic targets for SARS-CoV-2 and discovery of potential drugs by computational methods. *Acta Pharm. Sin. B* **2020**, *10*, 766–788. [CrossRef]
46. Kim, S.-H.; Kim, B.-K.; Lee, Y.-C. Antiasthmatic Effects of Hesperidin, a Potential Th2 Cytokine Antagonist, in a Mouse Model of Allergic Asthma. *Mediat. Inflamm.* **2011**, *2011*, 485402. [CrossRef]
47. Wang, S.; He, N.; Xing, H.; Sun, Y.; Ding, J.; Liu, L. Function of hesperidin alleviating inflammation and oxidative stress responses in COPD mice might be related to SIRT1/PGC-1α/NF-κB signaling axis. *J. Recept. Signal Transduct.* **2020**, *40*, 388–394. [CrossRef]
48. Kohno, H.; Taima, M.; Sumida, T.; Azuma, Y.; Ogawa, H.; Tanaka, T. Inhibitory effect of mandarin juice rich in β-cryptoxanthin and hesperidin on 4-(methylnitrosamino)-1-(3-pyridyl)-1-butanone-induced pulmonary tumorigenesis in mice. *Cancer Lett.* **2001**, *174*, 141–150. [CrossRef]
49. Jin, H.; Zhao, Z.; Lan, Q.; Zhou, H.; Mai, Z.; Wang, Y.; Ding, X.; Zhang, W.; Pi, J.; Evans, C.E.; et al. Nasal Delivery of Hesperidin/Chitosan Nanoparticles Suppresses Cytokine Storm Syndrome in a Mouse Model of Acute Lung Injury. *Front. Pharm.* **2020**, *11*, 592238. [CrossRef]
50. Gunasekaran, T.; Haile, T.; Nigusse, T.; Dhanaraju, M.D. Nanotechnology: An effective tool for enhancing bioavailability and bioactivity of phytomedicine. *Asian Pac. J. Trop. Biomed.* **2014**, *4*, S1–S7. [CrossRef]
51. Attia, G.H.; Moemen, Y.S.; Youns, M.; Ibrahim, A.M.; Abdou, R.; El Raey, M.A. Antiviral zinc oxide nanoparticles mediated by hesperidin and in silico comparison study between antiviral phenolics as anti-SARS-CoV-2. *Colloids Surf. B Biointerfaces* **2021**, *203*, 111724. [CrossRef]
52. Sulaiman, G.M.; Waheeb, H.M.; Jabir, M.S.; Khazaal, S.H.; Dewir, Y.H.; Naidoo, Y. Hesperidin Loaded on Gold Nanoparticles as a Drug Delivery System for a Successful Biocompatible, Anti-Cancer, Anti-Inflammatory and Phagocytosis Inducer Model. *Sci. Rep.* **2020**, *10*, 9362. [CrossRef]

Disclaimer/Publisher's Note: The statements, opinions and data contained in all publications are solely those of the individual author(s) and contributor(s) and not of MDPI and/or the editor(s). MDPI and/or the editor(s) disclaim responsibility for any injury to people or property resulting from any ideas, methods, instructions or products referred to in the content.

 life

Review

Anti-Inflammatory Properties of Plants from Serbian Traditional Medicine

Katarina Radovanović [1], Neda Gavarić [1] and Milica Aćimović [2,*]

[1] Department of Pharmacy, Faculty of Medicine, University of Novi Sad, Hajduk Veljkova 3, 21000 Novi Sad, Serbia
[2] Department of Vegetable and Alternative Crops, Institute of Field and Vegetable Crops Novi Sad, Maksima Gorkog 30, 21000 Novi Sad, Serbia
* Correspondence: acimovicbabicmilica@gmail.com

Abstract: Inflammation is a natural protective response of the human body to a variety of hostile agents and noxious stimuli. Standard anti-inflammatory therapy includes drugs whose usage is associated with a number of side effects. Since ancient times, natural compounds have been used for the treatment of inflammation. Traditionally, the use of medicinal plants is considered safe, inexpensive, and widely acceptable. In Serbia, traditional medicine, based on the strong belief in the power of medicinal herbs, is the widespread form of treatment. This is supported by the fact that Serbia is classified as one of 158 world centers of biodiversity, which confirms that this country is a treasure of medicinal herbs. Some of the most used herbs for the treatment of inflammations of various causes in Serbian tradition are yarrow, common agrimony, couch grass, onion, garlic, marshmallow, common birch, calendula, liquorice, walnut, St. John's wort, chamomile, peppermint, white willow, sage, and many others. The biological activity and anti-inflammatory effect of selected plants are attributed to different groups of secondary biomolecules such as flavonoids, phenolic acids, sterols, terpenoids, sesquiterpenes, and tannins. This paper provides an overview of plants with traditional anti-inflammatory use in Serbia with reference to available studies that examined this effect. Plants used in traditional medicine could be a powerful source for the development of new remedies. Therefore intensive research on the bioactive potential of medicinal plants in each region should be the focus of scientists around the world.

Keywords: traditional medicine; anti-inflammatory properties; ethnopharmacology

Citation: Radovanović, K.; Gavarić, N.; Aćimović, M. Anti-Inflammatory Properties of Plants from Serbian Traditional Medicine. *Life* **2023**, *13*, 874. https://doi.org/10.3390/life13040874

Academic Editors: Azahara Rodríguez-Luna, Salvador González and Jianfeng Xu

Received: 20 February 2023
Revised: 15 March 2023
Accepted: 24 March 2023
Published: 24 March 2023

Copyright: © 2023 by the authors. Licensee MDPI, Basel, Switzerland. This article is an open access article distributed under the terms and conditions of the Creative Commons Attribution (CC BY) license (https://creativecommons.org/licenses/by/4.0/).

1. Introduction

Inflammation is the body's automatic protective defence reaction to tissue injury or the invasion of foreign factors (toxins and pathogens) [1]. A controlled inflammatory response is an important beneficial process that is part of the maintained normal homeostasis of tissue [2]. The duration and extent of the inflammatory response are of key importance for its outcome and consequences. The process of acute inflammation includes phagocytosis, apoptosis, or activations of pro-inflammatory mediators that leads to the clearance of injurious stimuli and restore normal physiology [3]. However, chronic inflammation is not a useful process and it causes various pathological disorders including Alzheimer's disease, cancer, rheumatoid arthritis, type 2 diabetes, and obesity, as well as a cardiovascular and pulmonary disease [4]. Chronic disorders are regarded as a leading cause of death globally, with 60% of these deaths due to chronic inflammatory diseases [5]. These inflammatory changes are induced by cytokines and other inflammatory mediators. Cytokines are classified into two major categories: pro-inflammatory and anti-inflammatory cytokines. Several cytokines including interleukin (IL)-1, tumour necrosis factor-alpha (TNF-a), IL-6 and IL-8, and chemokines such as granulocyte colony-stimulating factor (G-CSF) and granulocyte-macro-phage colony-stimulating factor (GM-CSF) play a key role in acute inflammatory reactions [6].

A conventional therapeutic option for the treatment of inflammation and associated pain is nonsteroidal anti-inflammatory drugs, but their use is associated with a multitude of unwanted effects. For this reason, new research is focused on the search for safe natural substitutes for conventional anti-inflammatory drugs. The latest research has determined numerous pharmacological targets including cytokines, chemokines, transcription factors, complement activation pathways, eicosanoids, reactive oxygen species (ROS), and reactive nitrogen species (RNS) [4,7].

Since ancient times, people have relied on medicinal plants in the treatment of various health disorders, including inflammation and its complications. These actions are attributed to the complex chemical composition and the presence of secondary biomolecules in plants such as phenolic compounds, flavonoids, saponins, and sesquiterpenes [8]. The challenge of science is to find the exact chemical compound responsible for the observed pharmacological effects [9]. Moreover, it is important to recognize the side effects and potential interactions between medicinal plants and other synthetic drugs [10].

Bearing in mind that for centuries all nations have developed their traditional medicine based on the plants that grow in their environment and that precisely traditional methods of treatment have become very popular in recent years, and ubiquitous in everyday life, the aim of this paper was to review the plants with anti-inflammatory properties that are used in Serbian traditional medicine. Many ethnopharmacological studies indicate the use of certain plants for the treatment of diseases related to inflammatory processes, and in this review, we have selected 15 plants for which the anti-inflammatory potential has been confirmed by studies.

2. Materials and Methods

After a detailed selection of plants used in the Serbian tradition based on written sources (books) by local authors [11–13], the scientific literature that examines the named plants was searched in various scientific databases such as Google Scholar, PubMed, Science Direct, and Web of Science. The keywords include a combination of Latin or English names of selected plants with the words "anti-inflammatory", "phytochemistry", and "pharmacology". The content of selected articles was evaluated to determine their suitability for our topic. Mainly, articles in the English language, published from 2000 to 2023, with full text available were included. The exception is a few articles (books) in Serbian and Russian. These references are published before 2000 [11–16]. However, they have been exclusively included because of relevance as they provide a detailed list of all plants used in traditional Serbian medicine and were used to select plants with anti-inflammatory activity. The search was performed from November 2022 to the end of January 2023.

3. Anti-Inflammatory Plants from Serbian Traditional Medicine

The use of medicinal plants in Serbia for the treatment of diseases related to inflammatory processes has a long history. The most frequent, in Serbia, inflammation treated with plants, is located at the mucous membrane, the upper respiratory tract, the gastrointestinal tract, and the skin [11,17]. There are several studies dealing with the ethnopharmacology approach of medicinal plant application among Serbian people, mainly in rural areas [18–20]. However, plants used as anti-inflammatories in Serbian traditional medicine belong to different families: Asteraceae (*Achillea millefolium, Calendula officinalis, Matricaria chamomilla*), Rosaceae (*Agrimonia eupatoria*), Poaceae (*Agropyrum repens*), Liliaceae (*Allium cepa, A. sativum*), Malvaceae (*Althaea officinilas*), Betulaceae (*Betula pendula*), Fabaceae (*Glycyrrhiza glabra*), Juglandaceae (*Juglans regia*), Hypericaceae (*Hypericum perforatum*), Lamiaceae (*Mentha piperita, Salvia officinalis*), and Salicaceae (*Salix alba*) (Figure 1).

Figure 1. Plants used as anti-inflammatory in Serbian traditional medicine: (**a**) Achillea millefolium; (**b**) Agrimonia eupatoria; (**c**) Agropyrum repens; (**d**) Allium cepa; (**e**) Allium sativum; (**f**) Althaea officinilas; (**g**) Betula pendula; (**h**) Calendula officinalis; (**i**) Glycyrrhiza glabra; (**j**) Juglans regia; (**k**) Hypericum perforatum; (**l**) Matricaria chamomilla; (**m**) Mentha piperita; (**n**) Salix alba; (**o**) Salvia officinalis.

3.1. Achillea millefolium L., Asteraceae (Eng. Yarrow, Srb. Hajdučka Trava)

Yarrow is one of the most famous and most commonly used plants in traditional Serbian medicine. The internal and external use of this plant in which people have great confidence is widespread. In the form of poultices and ointments, it is used by people to treat various inflammatory and injured skin conditions. This action was confirmed by in vitro and in vivo studies in which extracts of yarrow showed a significant anti-inflammatory effect as well as a healing effect. In addition to independent use, it is also used as part of complex herbal mixtures for various medicinal purposes [13,14,21]. The data obtained in one in vivo study showed that the oil yarrow extracts had a significant anti-inflammatory property. The application of tested oil extracts on the artificially irritated skin of volunteers demonstrated the ability to re-establish their optimal skin parameters to the values measured prior to the irritation. The topical anti-inflammatory activity is attributed to the sesquiterpenes being caused by their inhibition of the arachidonic acid metabolism [22]. Internally, yarrow is part of tea mixtures intended for the treatment of asthma as chronic inflammation of the respiratory tract [14]. One study aimed to evaluate the ethanolic extract of *A. millefolium* flower for antitussive and anti-asthmatic potential through animal experimental models. The results of the study revealed the potent antitussive and antiasthmatic activities of *A. millefolium* flower extract [23].

3.2. *Agrimonia eupatoria L., Rosaceae (Eng. Common Agrimony, Srb. Petrovac)*

Common agrimony, church steeples, or sticklewort is another important herb of traditional medicine with numerous uses. Aerial parts of this plant are used internally in the form of tea or externally in the form of baths. It is known to be used for painful joints and inflammatory diseases of the mouth and throat [13,14]. Anti-inflammatory effects are well studied and confirmed in a lot of in vitro and in vivo studies. Experiments showed that *A. eupatoria* exerts an immunoprotective effect and decreases the levels of pro-inflammatory cytokines while increasing those of anti-inflammatory cytokines [24,25]. Antioxidant, anti-inflammatory, and peripheral analgesic activities were observed for *A. eupatoria* infusion and polyphenol-enriched fraction. Based on the results of this study, it was concluded that the traditional use of the *A. eupatoria* infusion as an antioxidant and anti-inflammatory is justified and suggests that its polyphenols (isoquercetin, tiliroside, and kaempferol O-acetyl-hexosyl-O-rhamnoside) contribute to this activity and should be considered as lead molecules for designing new pharmacophores [26,27].

3.3. *Agropyrum repens L., Poaceae (Eng. Couch Grass, Srb. Pirevina)*

Couch grass is a very common perennial species of grass native to most parts of Europe. For medicinal purposes, the rhizome is used to make tea for the treatment of various inflammatory conditions. Traditional use has been recorded for inflammation of the bladder, bronchitis, arthritis, and rheumatism [13,14]. One study showed that oral administration of ethanol extract of rhizomes of *A. repens* induced moderate inhibition of carrageenan foot oedema of the rat hind-paw compared to indomethacin. In the other study, the cream containing dry couch grass extract was tested for allergic contact dermatitis in rats. The results showed that the anti-inflammatory effects of the couch grass cream were comparable to the standard glucocorticoid cream activity [28].

3.4. *Allium cepa L., Liliaceae (Eng. Onion, Srb. Crni Luk)*

Onion is widely used in Serbia both for therapeutic purposes and as a spice and part of traditional cuisine. The whole plant is edible, but the bulbs that grow underground are most commonly used. It is used in fresh and dried form as a spice in food and in the form of teas or poultices for the treatment of various diseases, including inflammatory conditions such as headache, common cold, arthritis, and asthma. Externally, onion juice or juice is used in the treatment of skin inflammation, purulent wounds, burns, frostbite, and insect bites [13,14]. Numerous modern research has confirmed the justification of the use of onion as an anti-inflammatory agent. It was reported that the anti-inflammatory properties of *Allium* species are due to the presence of effective compounds such as tannin, flavonoids, anthocyanin, saponin, etc. Thiosulfinates and cepaenes from onion showed anti-inflammatory properties mediated through the inhibition of chemotaxis of human polymorphonuclear leukocytes. Furthermore, it has been established that cepaenes inhibit cyclooxygenase (COX) and lipoxygenase (LOX) enzymes. Quercetin, a well-known constituent of onion decreased the production of inflammatory cytokines such as IL-1a, IL-4, and TNF-a and inhibited the proliferation and activity of lymphocytes. These effects have been confirmed through several studies on animals and humans [29–31].

3.5. *Allium sativum L., Liliaceae (Eng. Garlic, Srb. Beli Luk)*

Garlic is another valuable plant from the genus *Allium* which is very widespread. Among people, garlic is a medicine for all ailments, in which there is a great and unshakable trust. This belief often goes so far that garlic is still used today in some households in the countryside not only as a preventative but also as a protective agent against "evil spirits" and other dangerous "invisibles", for fortune-telling, recovery, spells, and magic. The primordial belief in the medicinal, protective, and magical power of garlic left a deep mark on the material and spiritual life of the people. Garlic is eaten almost regularly and is added as an ingredient to various dishes. During epidemics of typhus, cholera, plague, dysentery, flu, and in general, whenever there was a great plague of infectious diseases, garlic was always recommended and used daily as a preventive and curative tool [13,14]. Garlic extracts and their related phytochemicals have been reported to possess anti-inflammatory activity in numerous studies [32]. Allicin, the main constituent of garlic, demonstrated a defensive mechanism against pathogens by its ability to enhance the activity of immune cells and influence signaling pathways associated with these immune cells. Moreover, allicin works on T-cell lymphocytes by inhibiting the SDF1α chemokine which is associated with the weakness of the dynamic structure of the actin cytoskeleton in addition to this, it leads to inhibit the transendothelial migration of neutrophils [33]. Another report indicated that thiacremonone (a sulfur compound isolated from garlic) prevents neuroinflammation and amyloidogenesis by blocking the nuclear factor kappa-light-chain-enhancer of activated B cells (NF-κB) activity, and for that reason can be used to treat neurodegenerative disorders related to inflammation [34].

3.6. *Althaea officinilas L., Malvaceae (Eng. Marshmallow, Srb. Beli Slez)*

Marshmallow is a widespread plant in Serbia. All parts of the plant can be used for treatment, but the root, which is the richest in active principle, is used most often. Macerate of white marshmallow is used as an auxiliary mucus agent for inflammation of the respiratory tract and gastroenteritis, and as compresses for inflammation of the skin. Cultivated plants are generally used to obtain plant raw materials because they are of better quality with a higher content of active principles [13,14]. The in vitro experiments on human monocytic cell line THP-1 showed a significant anti-oxidant and anti-inflammatory activity of root extracts of *A. officinalis.* The investigated preparation showed the ability to ameliorate the migratory capacity of macrophages. These anti-inflammatory effects were comparable to or even better than diclofenac [35].

3.7. *Betula pendula Roth., Betulaceae (Eng. Common Birch, Srb. Bela Breza)*

Common birch is widely distributed in temperate and northern climate zones. It is used in the traditional medicine of numerous countries, and its positive effects on human health have been known since ancient times. It belongs to the group of medicinal plants. Numerous studies on the chemical composition and activities of birch isolates aim to confirm their biological effects and use in traditional medicine. Birch leaf is one of the main ingredients of tea mixtures for the treatment of inflammation of the urinary tract acting as a diuretic and does not irritate the renal parenchyma. Birch tar is also used in folk medicine, a substance derived from the dry distillation of the wood, bark, and roots of the birch tree. It is used in dermatology, especially in eczema as a condition of chronic inflammation of the skin. In addition to tea and tar, birch sap is also used, a colorless liquid that oozes in the spring from cut birch trees [13,14]. One study investigates the influence of the aqueous extract of *Betula pendula* on primary human lymphocytes in comparison to the synthetic anti-arthritis drug methotrexate in vitro on human peripheral blood mononuclear cells (PBMC). These results provide a strong rational base for the widespread use of the leaf extract of *Betula pendula* in the treatment of immune disorders such as rheumatoid arthritis, through the reduction of proliferating inflammatory lymphocytes [36].

3.8. *Calendula officinalis L. Asteraceae (Eng. Calendula, Srb. Neven)*

Calendula is native to the Mediterranean region but it is widely cultivated all over Serbia. It has a very wide application in traditional medicine. It is used in the form of tea, as an addition to salads and other dishes, and externally in the form of oils, ointments, compresses, or rinsing teas. For curative purposes, inflorescences are used and rarely is the above-ground part of the plant in bloom. Traditionally, many beneficial effects are attributed to this plant. It is used as an anti-inflammatory and a remedy for healing wounds and skin disorders. Calendula ointments, suspensions, or tinctures are used topically for treating acne, reducing inflammation, controlling bleeding, and soothing irritated tissue [13,14]. Numerous studies confirm the anti-inflammatory effect of this plant applied topically. One in vitro study assessed the anti-inflammatory potential of calendula oil using lipopolysaccharide (LPS)-stimulated macrophages, as an in vitro model of inflammation. Scientists investigated the ability of a commercial calendula flower extract to inhibit NO production on macrophages exposed to LPS. The obtained results showed a dose-dependent NO inhibition of up to 50%, presenting a safety profile, thus, reinforcing the anti-inflammatory activity of calendula flower extract. In conclusion, the results of this study support the usefulness of Calendula oil in the treatment of injured skin and for conditions or diseases for which NOS contributes to the pathophysiology, such as contact dermatitis, vitiligo, rosacea, melasma, and psoriasis [37]. A study on rats provided evidence that *Calendula officinalis* presented anti-inflammatory and antibacterial activities as well as the capability of stimulating fibroplasia and angiogenesis. Calendula extracts showed a positive effect on the inflammatory and proliferative phases of the healing process of cutaneous wounds in rats [38]. In addition, the topical application of *C. officinalis* ointment has helped to prevent dermatitis and pain, thus, reducing the incidence rate of skipped radiation treatments in randomized trials [39].

3.9. *Glycyrrhiza glabra L., Fabaceae (Eng. Liquorice, Srb. Sladić)*

Liquorice is a famous medicinal plant worldwide. The root is the most used part of this plant with an extremely sweet flavor and pleasant odor. Liquorice has had an important place in traditional Serbian medicine since ancient times. In monastery hospitals, tea made from a mixture of liquorice root, rhizomes, and barley was used as a universal remedy for reconvalescence. Modern research has also confirmed its beneficial effect on the liver, so licorice root is a common ingredient in detox tea mixes. The active ingredients show anti-inflammatory action and have a beneficial effect on spasms and pain relief. Liquorice is part of several tea mixtures with therapeutic usages, namely *Species pectorales, Species diureticae,* and *Species urologicae*) and it is also part of a tea mix for the pediatric population. Black sugar is a product of liquorice made from the aqueous extract by steaming to dryness, which contains around 25% glycyrrhizin [13,14]. Many studies have investigated the anti-inflammatory effect of this plant and the results have confirmed the justification of its use in tradition. Some studies concluded that glycyrrhetinic acid and aqueous extract of liquorice possess strong anti-inflammatory activity, which was comparable with diclofenac [40,41]. Additionally, it was further recommended that the activity of anti-inflammatory formulations such as famotidine or diclofenac can be further enhanced through the addition of liquorice aqueous extract [41]. A lot of studies evaluated the impact of *G. glabra* and its bioactive components on different mechanisms of inflammation. Results showed inhibition of proinflammatory cytokine through inhibition of LPS-induced IL-1β, IL-6, IL-8, and TNF-α responses of macrophages. Furthermore, one study showed that extract of this plant inhibits serum levels of TNF-α and reduces antigen induce arthritis symptoms in mice [42].

3.10. *Juglans regia* L. *Juglandaceae* (*Eng. Walnut, Srb. Orah*)

Walnut is an ancient plant that is cultivated and grows wild in Serbia. In addition to the nutritional value of the fruit, the leaf and pericarp of the young fruit are traditionally used in healing. Externally, walnut-leaf tea is used to rinse the skin and mucous membranes in various inflammatory processes. Walnut tea is drunk orally for inflammation of the mucous membrane of the digestive organs. Furthermore, walnut leaves are added to tea mixtures for improving the resistance of the body's immunity [13,14]. The justification of traditional use has been confirmed by research. One study evaluated the antitussive, antioxidant, and anti-inflammatory effects of a walnut extract rich in bioactive compounds, using a citric acid-induced cough model in rats. Walnut septum showed significant antitussive and anti-inflammatory activities [43]. The ethanolic extracts of *J. regia* leaves exhibited potent anti-inflammatory activity comparable with indomethacin against carrageenan-induced hind paw edema model in mice without inducing any gastric side effects [44].

3.11. *Hypericum perforatum* L. *Hypericaceae* (*Eng. St. John's Wort, Srb. Kantarion*)

St. John's wort is a very common plant in the flora of Serbia. The aerial part of the plant is used for a variety of external and internal uses. Traditional written sources mention the use of St. John's wort in the form of an infusion or tincture as an anti-inflammatory, styptic, and antiseptic agent. St. John's wort oil, which is made as a macerate with sunflower or olive oil, is a well-known and valued remedy for healing wounds and burns [45]. An important note with the traditional use of the preparation is that the chemical compounds in the composition cause photosensitization, so precaution must be taken when exposed to UV light during treatment [15]. Imaninum is traditional antibacterial preparation for external application based on *H. perforatum* used for the treatment of fresh wounds, burns, and ulcers [13,14]. Imaninum is described as a dark brown powder obtained by boiling ground aerial parts of plant material (without stem) in 10% NaOH (1:10). The NaOH is removed and the cooking is repeated 5–6 times, always with a new amount of NaOH. The herbal residue that remains after cooking is acidified with HCl to an acidic reaction and then grind to a fine powder. In addition, if necessary, chlorophyll and pigments are separated from this preparation. It is used as an external agent in the form of solutions, ointments, and powders for the treatment of patients with fresh and infected wounds, burns, ulcers, abscesses, mastitis, carbuncles, boils, etc. They are also used for acute rhinitis, pharyngitis, laryngitis, and sinusitis [16].

Many in vitro and in vivo studies have justified the traditional use of St. John's wort in the therapy of inflammatory skin disorders by proving that the lipophilic extract as well as the pure individual components of its composition possess notable anti-inflammatory potential [46]. The study on rats showed that *H. perforatum* decreased levels of enzymes associated with colonic inflammation [47]. Results from a study on rats demonstrated that *H. perforatum* exhibits antiedematogenic and antinociceptive properties, which may be of value for the management of inflammatory painful conditions. However, the side effect is gastric irritation [48].

3.12. *Matricaria chamomilla* L. *Asteraceae* (*Eng. Chamomile, Srb. Kamilica*)

Medicinal properties of chamomile have been known since ancient times: in folk medicine, chamomile is considered a "panacea"—a cure for all diseases, given that its primary medicinal properties are antiseptic and anti-inflammatory, whether it is infections or inflammatory processes on the skin, mucous membranes of the mouth and throat or mucous membranes of the respiratory organs, digestive organs, or urogenital system [13,14]. Results from many studies confirmed traditional uses. The result from one study on animals showed that the volatile essential oil and non-volatile components (the aqueous extract and flower-water of chamomile) could significantly inhibit pedal swelling induced by carrageenan in rats, and ear swelling induced by xylol in mice [49]. Moreover, extracts provoked the increase of celiac capillary vessel permeability induced by HAC (glacial acetic acid dissolved in normal saline) in mice and the concentration increase of PGE2 and NO during pedal swelling induced by carrageenan in rats, as well as heterogeneity passive skin allergy in mice's ear and the inching reaction caused by dextran in mice [50,51]. As expected, the essential oil had the most remarkable anti-inflammatory and antiallergic effects [49]. An animal study showed that bisabolol from chamomile reduces inflammation and fever and has a favourable effect as adjuvant therapy in arthritis [49,52].

3.13. *Mentha piperita L. Lamiaceae (Eng. Peppermint, Srb. Nana)*

The peppermint leaf has multiple uses in Serbian tradition. Mint is a common ingredient in various tea blends, including tea blends for children which confirms the people's trust in this plant. In addition to consumption for medicinal purposes, mint tea is also drunk as a beverage due to its pleasant taste. Peppermint has gained the trust of folk medicine as an effective and safe tool with no restrictions for any type of use in all age populations. Later, based on the evidence, this precious plant was given generally recognized as safe (GRAS) status by the Food and Drug Administration (FDA). The essential oil of this plant is widely used for various purposes. Externally, it is used as part of the preparation for rubbing against rheumatic and neuralgic pains. By the way, the largest quantities of essential oil are used in the food, cosmetic, and alcoholic beverage industries. It is also used for the extraction of menthol [13,14]. The essential oil and extract of *M. piperita* were evaluated for their in vitro and in vivo anti-inflammatory activity. The investigations showed that the oil of *M. piperita* exerted significant anti-inflammatory activity, without inducing any apparent acute toxicity or gastric damage as compared to indomethacin, as the reference drug [53,54].

3.14. *Salix alba L. Salicaceae (Eng. White Willow, Srb. Bela Vrba)*

White willow has been widely used in traditional Serbian medicine since ancient times. In the treatment, the young bark is used, which is peeled in early spring. Willow bark decoction is used for colds, flu, and rheumatic diseases due to its analgesic, antipyretic, and anti-inflammatory effects. In the form of oral rinses, a decoction of willow bark is recommended for the treatment of inflammatory conditions of the mucous membranes. Until the synthetic production of salicylic acid was perfected, this plant and related willow species were long used as a raw material from which this acid was obtained [13,14]. Contemporary research has confirmed the experiences of folk medicine. In addition to the content of salicylic acid also, the presence of flavonoids is responsible for the anti-inflammatory effects of willow bark extracts, so to achieve a healing effect in the case of lower back pain relief, much lower doses are needed than for aspirin-based treatment [55]. This fact is very important from the aspect of safety of use. Experimental animal models showed that *S. alba* possesses an anti-inflammatory effect in xylene-induced ear oedema or carrageenan-induced paw oedema [56,57].

3.15. *Salvia officinalis L. Lamiaceae (Eng. Sage, Srb. Žalfija)*

Sage is an ethereal, luxurious, medicinal, and Mediterranean plant that has been cultivated for centuries for its healing properties. The most serious diseases were treated with sage before the era of antibiotics, so it could be said that it was the only salvation in those cases. It was prepared and used in different ways. It was used fresh and dry, it was pressed and essential oil was made from it, sage tea was brewed, it was chewed fresh for toothaches and diseased gums, and bandaged on wounds and injuries received in battles. Its exceptional medicinal properties, given that it is one of the strongest natural antibiotics, antimycotics, and antiseptics, rank sage even today in the first place, as the queen among medicinal herbs. In Serbia, people have great confidence in this plant as one of the most important medicinal plants [13,14]. Pharmacological studies have shown that *S. officinalis* has anti-inflammatory and antinociceptive effects. For example, it has been shown that this plant helps to control neuropathic pain in chemotherapy-induced peripheral neuropathy [58]. Among different extracts of *S. officinalis*, the chloroform one shows more anti-inflammatory action, while the methanolic extract and essential oil demonstrate low action [59]. Flavonoids and terpenes are the compounds that most likely contribute to the anti-inflammatory and antinociceptive actions of the herb. One study reported that flavonoids extracted from *S. officinalis* reduce inflammation in the mouse carrageenan model and induce an analgesic effect in a dose-dependent manner [60]. Investigation of individual constituents of *S. officinalis* showed that topical application of rosmarinic acid inhibits epidermal inflammation [61]. Manool, carnosol, and ursolic acid are terpenes/terpenoids with anti-inflammatory potential [62]. The anti-inflammatory action of ursolic acid is significantly more potent than that of indomethacin [59]. This proven action of *S. officinalis* constituents may be responsible for its high value as an anti-inflammatory agent [63].

4. Scientific Data on Anti-Inflammatory Potential of Selected Plants

All the above plants have a long history of traditional use as medicinal agents by many peoples around the world. That application is founded and confirmed by experience about the beneficial effect of their application. Modern research tends to investigate and document the justification of their use, explain the mechanisms of action, and isolate the active principles. Various plant-derived compounds inhibit inflammation through a reduction in the levels of several cytokines including IL-1β, IL-6, and TNF-α, and the suppression of COX-2, prostaglandins, and nitric oxide (NO) release. Active organosulphur compounds in garlic primarily ajoene, alliin, and allicin work by reducing levels of pro-inflammatory cytokines while increasing levels of anti-inflammatory interleukins. Different subclasses of flavonoids have been shown to suppress inflammatory molecules such as TNF-α, IL-1, IL-6, IL-17, and IFN-γ, which are secreted through the activation of several signaling pathways, predominantly the NF-κB pathway [9]. The ultimate goal of all research is to provide knowledge for the safe and effective use of the whole plant as an herbal remedy, as well as to find new raw materials for the isolation of bioactive substances for direct use as drugs or starting substances for further chemical modification to improve activity and/or reduce toxicity. Nonetheless, herbal preparations can be used as adjuvant therapy in addition to conventional therapy. In favor of plants and all the advantages of their use, it is worth noting that molecular diversity is a valuable advantage in relation to synthetic molecules because it allows acting on more different molecular mechanisms with lower doses and fewer adverse effects.

An overview of the scientific literature on plants with traditional use in Serbia in the treatment of inflammation is given in Table 1.

Table 1. Mini-review of scientific data about the anti-inflammatory effect of plants from Serbian tradition.

Name of Herb	Part Used	Phytoconstituents	Scientific Data	Ref.
Achillea millefolium	aerial parts	alkaloids, glucoside, choline, essential oils, salicylic acid, sesquiterpenoids, dicaffeoylquinic acids, luteolin, apigenin	• down-regulating the expression iNOS • inhibition of the inflammation-related proteases, namely, HNE and MMP-2 and MMP-9 • macrophage activation modulating agents • in vivo study on humans showed the ability to re-establish optimal pH and hydration of the skin to the values measured prior to the irritation • significant antitussive and antiasthmatic activity on mice	[21–23,64]

Table 1. *Cont.*

Name of Herb	Part Used	Phytoconstituents	Scientific Data	Ref.
Agrimonia eupatoria	aerial parts	carbohydrates, tannins, terpenoids, flavonoids, agrimony lactone, glycosides and oils; polysaccharides, triterpenoids, silicic acid, salicylic acid, nicotinamide complex, thiamine, and vitamin K.	• inhibiting inflammatory cytokine (IL-1β, IL-4, IL-6) and INF-β production • inhibition of NO and PGE2 production • significantly reduced carrageenan-induced paw oedema • decreases the levels of pro-inflammatory cytokines while increasing those of anti-inflammatory cytokines. • improving human markers of lipid metabolism, oxidative status, and inflammation after one month's consumption in healthy volunteers	[24–28]
Agropyrum repens	rhizome	carbohydrates, mucilages, saponins, essential oils	• moderate inhibition of carrageenan-induced foot oedema of the rat • anti-inflammatory activity impacting plasma lipid peroxidation parameters MDA, DC, and catalase activity	[28]
Allium cepa	bulb	Phenolic acids, thiosulphates, and flavonoids	• modulating COX-mediated prostaglandin production • reducing lung inflammatory cytokines such as IL-4, 5, and 13 and T helper 2 • significantly reduced total WBC and lung inflammatory cells such as neutrophil, eosinophil, and monocyte counts, but led to a significant increase in lymphocyte counts in asthmatic Wistar rats	[30,31,65]
Allium sativum	bulb	Sulphur compounds, enzymes, amino acids, minerals	• modulating leucocyte cell proliferation and cytokine production • inhibiting Th1 and inflammatory cytokines and upregulating IL-10 production	[32,66–68]
Althaea officinilas	root, leaf, flower	starch, pectins, saccharose, mucilage, flavonoids, caffeic acid, p-coumaric acid, isoquercitrin, coumarins, phytosterols, tannins, amino acids	• in vitro Phytohustil® and root extracts of A. officinalis were able to protect human MΦ against H_2O_2-induced cytotoxicity and H_2O_2-induced ROS production. • inhibition of the LPS-induced release of TNF-α as well as of IL6 in MΦ. • results of a pilot active-controlled trial revealed that topical use of 1% ointment contained ethanolic extract in petrolatum base which has a higher efficacy in children with atopic dermatitis in comparison to topical hydrocortisone 1%.	[35,69]
Betula pendula	leaves and leaf buds	flavonoids, tannins, resins, essential oils	• in vitro reduction of proliferating inflammatory lymphocytes • in vivo study on rats showed a significant effect of tablets with dense extract on carrageenin-induced inflammation	[36,70]

Table 1. *Cont.*

Name of Herb	Part Used	Phytoconstituents	Scientific Data	Ref.
Calendula officinalis	inflorescence	triterpenoids, flavonoids, coumarins, quinones, essential oils, carotenoids, and amino acids.	• dose-dependent NO inhibition reinforcing the anti-inflammatory activity of calendula flower extract. • potent anti-inflammatory response extract may be mediated by the inhibition of proinflammatory cytokines and Cox-2 and subsequent prostaglandin synthesis. • methanolic extract of flowers showed the most potent inhibition of TPA-a-induced inflammation in mice • antioedematous effect in croton oil-induced mouse oedema • cream containing calendula extract has been reported to be effective in dextran and burn oedemas as well as in acute lymphoedema in rats. • reducing inflammatory bone resorption in an experimental rat periodontitis • in vivo on rats aqueous flower extract demonstrated anti-inflammatory and antibacterial activities as well as the capability of stimulating fibroplasia and angiogenesis • phase III Randomized Tria showed that calendula-based ointment was statistically significantly more effective than trolamine in preventing acute dermatitis grade 2 or higher during adjuvant postoperative breast irradiation.	[38,39,71–74]
Glycyrrhiza glabra	root	triperpenic saponins, sterols, flavonoids	• the hydroalcoholic extract showed a maximum inhibitory action on carrageenan-induced paw oedema at the dose of 200 mg/kg and inhibited the leukocyte migration in a dose-dependent manner. The anti-inflammatory activity was comparable to indomethacin • methanol extract was evaluated for the COX-2 inhibitory activity using Cayman COX (ovine) inhibitory screening assay. A few molecules (dominantly glycyrrhizic acid) showed potent COX-2 inhibitory activity which may be beneficial as anti-inflammatory agents • glycyrrhizin exhibited steroid-like anti-inflammatory activity, comparable with hydrocortisone, due to inhibition of phospholipase A2 activity, glycyrrhizic acid inhibited cyclooxygenase activity and prostaglandin formation (specifically PGE2), as well as indirectly inhibiting platelet aggregation • outcomes of one in vitro study show that licoflavanone can decrease iNOS and COX2 expression levels in LPS-stimulated RAW 264.7 cells, interfering with the inflammatory cascade mediated by NO and PGE2	[42,75–77]

Table 1. *Cont.*

Name of Herb	Part Used	Phytoconstituents	Scientific Data	Ref.
Juglans regia	leaves	tannins, naphthoquinone derivatives, flavonoids	• in vivo on mice aqueous and ethanolic extracts showed activity against acute and chronic inflammation • potent anti-inflammatory activity comparable with indomethacin against carrageenan-induced hind paw edema model in mice without inducing any gastric side effects • citric acid-induced cough model in rats showed that walnut septum expressed potent antitussive and anti-inflammatory activities.	[43,44,78]
Hypericum perforatum	aerial part	hyperforin, hypericin, flavonoids, tannins	• inhibition of the production of PGE2on mouse macrophage cells • the compounds isolated from H. perforatum induced a dose-dependent reduction in oedema in mice • the results from a study on rats provide evidence for the usage of Oleum Hyperici as an anti-inflammatory and gastroprotective agent	[79–81]
Matricaria chamomilla	flowers	essential oils, sesquiterpene lactones, coumarins, mucilages	• reduction of NO production and induction of anti-inflammatory cytokine production (IL-10) • significantly inhibiting swelling of mouse ears caused by xylene, pedal swelling caused by carrageenan in rats, and the increase of celiac capillary vessel permeability in mice • inhibitory effect on the increase in PGE2 and NO levels in rat pedal edema caused by carrageenan • study on rats showed that chamomile extract prevented a significant increase in serum levels of TNF-α, CRP, IL-6, and fibrinogen. • synergic anti-inflammatory effects with diclofenac and indomethacin on carrageenan-induced paw inflammation and stomach damage in rats	[49,51,52]
Mentha piperita	leaves	essential oils, phenolics, flavonoids, tannins	• potent anti-inflammatory activity in the croton oil-induced mouse ear oedema model • inhibitory effect on the production of NO and PGE2	[53,82]
Salix alba	bark	phenolic glycosides, flavonoids, tannins, aromatic aldehydes, and acids	• in vivo tests of the methanolic and aqueous extracts of the barks showed a strong effect on carrageenan-induced paw oedema and xylene-induced ear edema • in vivo on mice methanolic extracts exhibited a dose-dependent analgesic property with more potency than the standard drug aspirin in all tested doses, as well as inhibited the paw edema by interruption of the arachidonic acid metabolism and shows inhibition of the inflammation greater than the inhibitory effect of the aspirin	[55,56]

Table 1. *Cont.*

Name of Herb	Part Used	Phytoconstituents	Scientific Data	Ref.
Salvia officinalis	leaves	essential oil, tannins, diterpenes, triterpenes, flavonoids	• the chloroform extracts showed strong anti-inflammatory properties on croton oil-induced ear oedema in mice after topical application. • in mice, arnosol and ursolic acid/oleanolic acid inhibited the inflammatory phase of formalin and the nociception and mechanical allodynia induced by cinnamaldehyde	[59,62,63]

COX2—cyclooxygenase-2; CRP—C-Reactive Protein; DC—diene conjugates; HNE—neutrophil elastase; IL-6—Interleukin 6; NO—nitric oxide; INF-β—interferon-β; iNOS—inducible nitric oxide synthase; LPS—lipopolyccharide; MDA—malondialdehyde; MMP—matrix metalloproteinases; MΦ—*Macrophages;* PGE2—prostaglandine E2; RAW 264.7—macrophage cell line; ROS—reactive oxygen species; TNF-α—Tumor Necrosis Factor-α; TPA—12-O-tetradecanoylphorbol-13-acetate; WBC—white blood cells.

Medicinal plants could be a powerful source of raw material for the pharmaceutical industry and synthesis of new remedies, considering that every nation has its own heritage in ethnobotany and ethnopharmacology, so traditional medicine is key important in the processes of plant-based medicines products [83,84]. Taking into account the diversity of the plants, which is conditioned by numerous factors such as geographical, pedological and climatic, intensive research on the bioactive potential of medicinal plants in each region should be the focus of scientists around the world [85,86]. The medicinal plants with anti-inflammatory properties used among Serbian people could be promising candidates for further research and identifying new bioactive potentials, especially in combination with modern green technologies [87–89].

5. Conclusions

People in Serbia have relied on healing with various herbs since ancient times. Most of the plants from the natural treasures of Serbia found traditional use. In the therapy of inflammation, plants from Asteraceae and Lamiaceae families are most frequently used. The easy availability of numerous herbs and empirically proven effectiveness with high safety are the reason for their popularity in tradition. Today, through in vitro and in vivo tests, the justification for the use of traditional plants is confirmed. With modern knowledge, the best possibilities for using plant potential and their safe application are reached. Where the limitations of conventional medicine are due to side effects, herbal treatment could gain its full recognition due to fewer side effects, complex composition, and synergistic action of individual components. In summary, all those facts open the possibility of using the natural plants' treasure for the purpose of isolating the active principles and further clinical trials which would relive the reveal benefits and limitations of their application as well as the potential synergistic effect with conventional therapy.

Author Contributions: Conceptualization, N.G.; investigation, K.R.; writing—original draft preparation, K.R.; writing—review and editing, N.G.; visualization, K.R.; supervision, M.A.; project administration, M.A.; funding acquisition, M.A. All authors have read and agreed to the published version of the manuscript.

Funding: This research received no external funding.

Institutional Review Board Statement: Not applicable.

Informed Consent Statement: Not applicable.

Data Availability Statement: Not applicable.

Acknowledgments: This research was supported by the Ministry of Science, Technological Development and Innovation of the Republic of Serbia (grants 451-03-68/2020-14/200114 and 451-03-47/2023-01/200032).

Conflicts of Interest: The authors declare no conflict of interest.

References

1. Chen, L.; Deng, H.; Cui, H.; Fang, J.; Zuo, Z.; Deng, J.; Li, Y.; Wang, X.; Zhao, L. Inflammatory responses and inflammation-associated diseases in organs. *Oncotarget* **2017**, *9*, 7204–7218. [CrossRef] [PubMed]
2. Janssen, W.J.; Henson, P.M. Cellular regulation of the inflammatory response. *Toxicol. Pathol.* **2012**, *40*, 166–173. [CrossRef] [PubMed]
3. Varela, M.L.; Mogildea, M.; Moreno, I.; Lopes, A. Acute inflammation and metabolism. *Inflammation* **2018**, *41*, 1115–1127. [CrossRef] [PubMed]
4. Tasneem, S.; Liu, B.; Li, B.; Choudhary, M.I.; Wang, W. Molecular pharmacology of inflammation: Medicinal plants as anti-inflammatory agents. *Pharmacol. Res.* **2019**, *139*, 126–140. [CrossRef]
5. Pahwa, R.; Goyal, A.; Jialal, I. Chronic Inflammation. In *StatPearls*; StatPearls Publishing: Treasure Island, FL, USA, 2022.
6. Abdulkhaleq, L.; Assi, M.; Abdullah, R.; Zamri-Saad, M.; Taufiq-Yap, Y.; Hezmee, M. The crucial roles of inflammatory mediators in inflammation: A review. *Vet. World* **2018**, *11*, 627. [CrossRef]
7. Khan, M.; Ali, S.; Al Azzawi, T.N.; Saqib, S.; Ullah, F.; Ayaz, A.; Zaman, W. The key roles of ROS and RNS as a signaling molecule in plant–microbe interactions. *Antioxidants* **2023**, *12*, 268. [CrossRef]
8. Abidullah, S.; Rauf, A.; Khan, S.W.; Ayaz, A.; Liaquat, F.; Saqib, S. A comprehensive review on distribution, paharmacological uses and biological activities of Argyrolobium roseum (*Cambess.*) Jaub. & Spach. *Sheng Tai Xue Bao* **2022**, *42*, 198–205.
9. Ginwala, R.; Bhavsar, R.; Chigbu, D.G.I.; Jain, P.; Khan, Z.K. Potential role of flavonoids in treating chronic inflammatory diseases with a special focus on the anti-inflammatory activity of apigenin. *Antioxidants* **2019**, *8*, 35. [CrossRef]
10. Zhang, L.; Virgous, C.; Si, H. Synergistic anti-inflammatory effects and mechanisms of combined phytochemicals. *J. Nutr. Biochem.* **2019**, *69*, 19–30. [CrossRef]
11. Pelagić, V. *Pelagić's National Teacher*; Sloboda: Belgrade, Serbia, 1974. (In Serbian)
12. Tucakov, J. *Healing with Plants—Phytotherapy*; Kultura: Belgrade, Serbia, 1971. (In Serbian)
13. Sarić, M.R. *Medicinal Plants of SR Serbia*; Srpska Akademija Nauka i Umetnosti: Belgrade, Serbia, 1989. (In Serbian)
14. Gostuški, R. *Healing with Medicinal Plants*; Narodna knjiga: Belgrade, Serbia, 1967. (In Serbian)
15. Araya, O.S.; Ford, E.J. An investigation of the type of photosensitization caused by the ingestion of St John's Wort (*Hypericum perforatum*) by calves. *J. Comp. Pathol.* **1981**, *91*, 135–141. [CrossRef]
16. Dobrotiko, V.G.; Aizenman, B.E.; Schvaiger, S.E.; Zelepucha, S.I.; Mandrik, T.P. *Animicrobial Substances Plants*; Academy of Sciences of Ukraine: Kiev, Ukraine, 1958. (In Russian)
17. Dajić-Stevanović, Z.; Petrović, M.; Aćić, S. Ethnobotanical knowledge and traditional use of plants in Serbia in relation to sustainable rural development. In *Ethnobotany and Biocultural Ddiversities in the Balkans*; Pieroni, A., Quave, C.L., Eds.; Springer: New York, NY, USA, 2014; pp. 229–252.
18. Živković, J.; Ilić, M.; Šavikin, K.; Zdunić, G.; Ilić, A.; Stojković, D. Traditional Use of Medicinal Plants in South-Eastern Serbia (Pčinja District): Ethnopharmacological Investigation on the Current Status and Comparison with Half a Century Old Data. *Front. Pharmacol.* **2020**, *11*, 1020. [CrossRef]
19. Matejić, J.S.; Stefanović, N.; Ivković, M.; Živanović, N.; Marin, P.D.; Džamić, A.M. Traditional uses of autochthonous medicinal and ritual plants and other remedies for health in Eastern and South-Eastern Serbia. *J. Ethnopharmacol.* **2020**, *261*, 113186. [CrossRef]
20. Zlatković, B.; Bogosavljević, S.; Radivojević, A.; Pavlović, M. Traditional use of the native medicinal plant resource of Mt. Rtanj (Eastern Serbia): Ethnobotanical evaluation and comparison. *J. Ethnopharmacol.* **2014**, *151*, 704–713. [CrossRef]
21. Ali, S.I.; Gopalakrishnan, B.; Venkatesalu, V. Pharmacognosy, phytochemistry and pharmacological properties of *Achillea millefolium* L.: A review. *Phytother. Res.* **2017**, *31*, 1140–1161. [CrossRef]
22. Tadić, V.; Arsić, I.; Zvezdanović, J.; Zugić, A.; Cvetković, D.; Pavkov, S. The estimation of the traditionally used yarrow (*Achillea millefolium* L. Asteraceae) oil extracts with anti-inflamatory potential in topical application. *J. Ethnopharmacol.* **2017**, *199*, 138–148. [CrossRef]
23. Choudhary, G.P. Pharmacological evaluation for antitussive and antiasthmatic potential of *Achillea millefolium* flower extracts. *Respir. Care* **2022**, *67*, 3771721.
24. Malheiros, J.; Simões, D.M.; Figueirinha, A.; Cotrim, M.D.; Fonseca, D.A. *Agrimonia eupatoria* L.: An integrative perspective on ethnomedicinal use, phenolic composition and pharmacological activity. *J. Ethnopharmacol.* **2022**, *296*, 115498. [CrossRef]
25. Huzio, N.; Grytsyk, A.; Raal, A.; Grytsyk, L.; Koshovyi, O. Phytochemical and pharmacological research in *Agrimonia eupatoria* L. herb extract with anti-inflammatory and hepatoprotective properties. *Plants* **2022**, *11*, 2371. [CrossRef]
26. Paluch, Z.; Biriczova, L.; Pallag, G.; Marques, E.C.; Vargova, N.; Kmoníčková, E. The therapeutic effects of *Agrimonia eupatoria* L. *Physiol. Res.* **2020**, *69*, S555. [CrossRef]
27. Santos, T.N.; Costa, G.; Ferreira, J.P.; Liberal, J.; Francisco, V.; Paranhos, A.; Cruz, M.T.; Castelo-Branco, M.; Figueiredo, I.V.; Batista, M.T. Antioxidant, anti-inflammatory, and analgesic activities of *Agrimonia eupatoria* L. infusion. *Evid.-Based Complement. Altern. Med.* **2017**, *2017*, 8309894. [CrossRef]
28. Al-Snafi, A.E. Chemical constituents and pharmacological importance of *Agropyron repens*—A review. *Res. J. Pharmacol. Toxicol.* **2015**, *1*, 37–41.
29. Kumar, K.S.; Bhowmik, D.; Chiranjib, B.; Tiwari, P. *Allium cepa*: A traditional medicinal herb and its health benefits. *J. Chem. Pharmaceut. Res.* **2010**, *2*, 283–291.

30. Marefati, N.; Ghorani, V.; Shakeri, F.; Boskabady, M.; Kianian, F.; Rezaee, R.; Boskabady, M.H. A review of anti-inflammatory, antioxidant, and immunomodulatory effects of *Allium cepa* and its main constituents. *Pharm. Biol.* **2021**, *59*, 285–300. [CrossRef] [PubMed]
31. Amin, M.; Putra, K.S.; Amin, I.F.; Earlia, N.; Maulina, D.; Lukiati, B.; Lestari, U. Quercetin: The bioactive compound from *Allium cepa* L. as anti-inflammation based on in silico screening. *Biol. Med. Nat. Prod. Chem.* **2018**, *7*, 27–31. [CrossRef]
32. El-Saber, B.G.; Beshbishy, M.A.; Wasef, G.L.; Elewa, Y.H.; Al-Sagan, A.A.; Abd El-Hack, M.E.; Taha, A.E.; Abd-Elhakim, Y.M.; Devkota, H.P. Chemical constituents and pharmacological activities of garlic (*Allium sativum* L.): A review. *Nutrients* **2020**, *12*, 872. [CrossRef]
33. Sela, U.; Ganor, S.; Hecht, I.; Brill, A.; Miron, T.; Rabinkov, A.; Wilchek, M.; Mirelman, D.; Lider, O.; Hershkoviz, R. Allicin inhibits SDF-1α-induced T cell interactions with fibronectin and endothelial cells by down-regulating cytoskeleton rearrangement, Pyk-2 phosphorylation and VLA-4 expression. *Immunology* **2004**, *111*, 391–399. [CrossRef]
34. Jin, P.; Kim, J.A.; Choi, D.Y.; Lee, Y.J.; Jung, H.S.; Hong, J.T. Anti-inflammatory and anti-amyloidogenic effects of a small molecule, 2, 4-bis (p-hydroxyphenyl)-2-butenal in Tg2576 Alzheimer's disease mice model. *J. Neuroinflamm.* **2013**, *10*, 767. [CrossRef]
35. Bonaterra, G.A.; Bronischewski, K.; Hunold, P.; Schwarzbach, H.; Heinrich, E.U.; Fink, C.; Aziz-Kalbhenn, H.; Müller, J.; Kinscherf, R. Anti-inflammatory and anti-oxidative effects of Phytohustil® and root extract of *Althaea officinalis* L. on macrophages in vitro. *Front. Pharmacol.* **2020**, *11*, 290. [CrossRef]
36. Gründemann, C.; Gruber, C.W.; Hertrampf, A.; Zehl, M.; Kopp, B.; Huber, R. An aqueous birch leaf extract of *Betula pendula* inhibits the growth and cell division of inflammatory lymphocytes. *J. Ethnopharmacol.* **2011**, *136*, 444–451. [CrossRef]
37. Silva, D.; Ferreira, M.S.; Sousa-Lobo, J.M.; Cruz, M.T.; Almeida, I.F. Anti-inflammatory activity of *Calendula officinalis* L. Flower extract. *Cosmetics* **2021**, *8*, 31. [CrossRef]
38. Parente, L.M.L.; de Souza Lino, R., Jr.; Tresvenzol, L.M.F.; Vinaud, M.C.; de Paula, J.R.; Paulo, N.M. Wound healing and anti-inflammatory effect in animal models of *Calendula officinalis* L. growing in Brazil. *Evid.-Based Complement. Altern. Med.* **2012**, *2012*, 375671. [CrossRef]
39. Pommier, P.; Gomez, F.; Sunyach, M.; D'hombres, A.; Carrie, C.; Montbarbon, X. Phase III randomized trial of *Calendula officinalis* compared with trolamine for the prevention of acute dermatitis during irradiation for breast cancer. *J. Clin. Oncol.* **2004**, *22*, 1447–1453. [CrossRef]
40. Wang, H.; Li, R.; Rao, Y.; Liu, S.; Hu, C.; Zhang, Y.; Meng, L.; Wu, Q.; Ouyang, Q.; Liang, H.; et al. Enhancement of the Bioavailability and Anti-Inflammatory Activity of Glycyrrhetinic Acid via Novel Soluplus®—A Glycyrrhetinic Acid Solid Dispersion. *Pharmaceutics* **2022**, *14*, 1797. [CrossRef]
41. Aly, A.M.; Al-Alousi, L.; Salem, H.A. Licorice: A possible anti-inflammatory and anti-ulcer drug. *AAPS PharmSciTech* **2005**, *6*, E74–E82. [CrossRef]
42. Bisht, D.; Rashid, M.; Arya, R.K.K.; Kumar, D.; Chaudhary, S.K.; Rana, V.S.; Sethiya, N.K. Revisiting liquorice (*Glycyrrhiza glabra* L.) as anti-inflammatory, antivirals and immunomodulators: Potential pharmacological applications with mechanistic insight. *Phytomed. Plus* **2022**, *2*, 100206. [CrossRef]
43. Fizeșan, I.; Rusu, M.E.; Georgiu, C.; Pop, A.; Ștefan, M.G.; Muntean, D.M.; Mirel, S.; Vostinaru, O.; Kiss, B.; Popa, D.S. Antitussive, antioxidant, and anti-inflammatory effects of a walnut (*Juglans regia* L.) septum extract rich in bioactive compounds. *Antioxidants* **2021**, *10*, 119. [CrossRef]
44. Erdemoglu, N.; Küpeli, E.; Yeşilada, E. Anti-inflammatory and antinociceptive activity assessment of plants used as remedy in Turkish folk medicine. *J. Ethnopharmacol.* **2003**, *89*, 123–129. [CrossRef]
45. Arsić, I.; Žugić, A.; Tadić, V.; Tasić-Kostov, M.; Mišić, D.; Primorac, M.; Runjaić-Antić, D. Estimation of dermatological application of creams with St. John's Wort oil extracts. *Molecules* **2011**, *17*, 275–294. [CrossRef]
46. Huang, N.; Rizshsky, L.; Hauck, C.; Nikolau, B.J.; Murphy, P.A.; Birt, D.F. Identification of anti-inflammatory constituents in Hypericum perforatum and Hypericum gentianoides extracts using RAW 264.7 mouse macrophages. *Phytochemistry* **2011**, *72*, 2015–2023. [CrossRef]
47. Barnes, J.; Anderson, L.A.; Phillipson, J.D. St John's wort (*Hypericum perforatum* L.): A review of its chemistry, pharmacology and clinical properties. *J. Pharm. Pharmacol.* **2001**, *53*, 583–600. [CrossRef]
48. Abdel-Salam, O.M. Anti-inflammatory, antinociceptive, and gastric effects of *Hypericum perforatum* in rats. *Sci. World J.* **2005**, *5*, 586–595. [CrossRef] [PubMed]
49. Wu, Y.N.; Xu, Y.; Yao, L. Anti-inflammatory and anti-allergic effects of German chamomile (*Matricaria chamomilla* L.). *J. Essent. Oil Bear. Plants* **2012**, *15*, 75–83. [CrossRef]
50. Ortiz, M.I.; Cariño-Cortés, R.; Ponce-Monter, H.A.; González-García, M.P.; Castañeda-Hernández, G.; Salinas-Caballero, M. Synergistic interaction of *Matricaria chamomilla* extract with diclofenac and indomethacin on carrageenan-induced aw inflammation in rats. *Drug Dev. Res.* **2017**, *78*, 360–367. [CrossRef] [PubMed]
51. Nargesi, S.; Moayeri, A.; Ghorbani, A.; Seifinejad, Y.; Shirzadpour, E.; Amraei, M. The effects of *Matricaria chamomilla* L. hydroalcoholic extract on atherosclerotic plaques, antioxidant activity, lipid profile and inflammatory indicators in rats. *Biomed. Res. Ther.* **2018**, *5*, 2752–2761. [CrossRef]
52. El Mihyaoui, A.; Esteves da Silva, J.C.; Charfi, S.; Candela Castillo, M.E.; Lamarti, A.; Arnao, M.B. Chamomile (*Matricaria chamomilla* L.): A review of ethnomedicinal use, phytochemistry and pharmacological uses. *Life* **2022**, *12*, 479. [CrossRef]

53. Mahendran, G.; Rahman, L.U. Ethnomedicinal, phytochemical and pharmacological updates on peppermint (*Mentha× piperita* L.)—A review. *Phytother. Res.* **2020**, *34*, 2088–2139. [CrossRef]
54. Li, Y.; Liu, Y.; Ma, A.; Bao, Y.; Wang, M.; Sun, Z. In vitro antiviral, anti-inflammatory, and antioxidant activities of the ethanol extract of *Mentha piperita* L. *Food Sci. Biotechnol.* **2017**, *26*, 1675–1683. [CrossRef]
55. Gyawali, R.; Bhattarai, P.; Dhakal, S.; Jha, B.; Sharma, S.; Poudel, P.N. Analgesic and anti-inflammatory properties of *Salix alba* Linn. and *Calotropis procera* (Aiton) Dryand. *Int. J. Pharm. Biol. Arch.* **2013**, *4*, 873–877.
56. Roumili, I.; Mayouf, N.; Charef, N.; Arrar, L.; Baghiani, A. HPLC analysis, acute toxicity and anti-inflammatory effects of *Salix alba* L. barks extracts on experimental animal models. *Indian J. Exp. Biol.* **2022**, *60*, 842–850.
57. Tawfeek, N.; Mahmoud, M.F.; Hamdan, D.I.; Sobeh, M.; Farrag, N.; Wink, M.; El-Shazly, A.M. Phytochemistry, pharmacology and medicinal uses of plants of the genus Salix: An updated review. *Front. Pharmacol.* **2021**, *12*, 593856. [CrossRef]
58. Abad, N.A.A.; Nouri, M.H.K.; Tavakkoli, F. Effect of *Salvia officinalis* hydroalcoholic extract on vincristine-induced neuropathy in mice. *Chin. J. Nat. Med.* **2011**, *9*, 354–358.
59. Baricevic, D.; Sosa, S.; Della Loggia, R.; Tubaro, A.; Simonovska, B.; Krasna, A.; Zupancic, A. Topical anti-inflammatory activity of *Salvia officinalis* L. leaves: The relevance of ursolic acid. *J. Ethnopharmacol.* **2001**, *75*, 125–132. [CrossRef]
60. Mansourabadi, A.M.; Sadeghi, H.M.; Razavi, N.; Rezvani, E. Anti-inflammatory and analgesic properties of salvigenin, *Salvia officinalis* flavonoid extracted. *Adv. Herb. Med.* **2015**, *1*, 31–41.
61. Osakabe, N.; Yasuda, A.; Natsume, M.; Yoshikawa, T. Rosmarinic acid inhibits epidermal inflammatory responses: Anticarcinogenic effect of *Perilla frutescens* extract in the murine two-stage skin model. *Carcinogenesis* **2004**, *25*, 549–557. [CrossRef]
62. Rodrigues, M.R.A.; Kanazawa, L.K.S.; das Neves, T.L.M.; da Silva, C.F.; Horst, H.; Pizzolatti, M.G.; Santos, A.R.S.; Baggio, C.H.; de Paula Werner, M.F. Antinociceptive and anti-inflammatory potential of extract and isolated compounds from the leaves of *Salvia officinalis* in mice. *J. Ethnopharmacol.* **2012**, *139*, 519–526. [CrossRef]
63. Ghorbani, A.; Esmaeilizadeh, M. Pharmacological properties of *Salvia officinalis* and its components. *J. Tradit. Complement. Med.* **2017**, *7*, 433–440. [CrossRef]
64. Akram, M. Minireview on *Achillea millefolium* Linn. *J. Membr. Biol.* **2013**, *246*, 661–663. [CrossRef]
65. Upadhyay, R.K. Nutraceutical, pharmaceutical and therapeutic uses of *Allium cepa*: A review. *Int. J. Green Pharm.* **2016**, *10*, S46–S64.
66. Hussein, H.J.; Hameed, I.H.; Hadi, M.Y. A review: Anti-microbial, anti-inflammatory effect and cardiovascular effects of garlic: *Allium sativum*. *Res. J. Pharm. Technol.* **2017**, *10*, 4069–4078. [CrossRef]
67. Jayanthi, M.; Dhar, M.; Jayanthi, M. Anti-inflammatory effects of *Allium sativum* (garlic) in experimental rats. *Biomedicine* **2011**, *31*, 84–89.
68. Sadeghi, M.; Miroliaei, M.; Fateminasab, F.; Moradi, M. Screening cyclooxygenase-2 inhibitors from *Allium sativum* L. compounds: In silico approach. *J. Mol. Model.* **2022**, *28*, 24. [CrossRef] [PubMed]
69. Naseri, V.; Chavoshzadeh, Z.; Mizani, A.; Daneshfard, B.; Ghaffari, F.; Abbas-Mohammadi, M.; Gachkar, L.; Kamalinejad, M.; Hajati, R.J.; Bahaeddin, Z.; et al. Effect of topical marshmallow (*Althaea officinalis*) on atopic dermatitis in children: A pilot double-blind active-controlled clinical trial of an in-silico-analyzed phytomedicine. *Phytother. Res.* **2021**, *35*, 1389–1398. [CrossRef]
70. Chumak, O.; Bezrukavyi, Y.; Maloshtan, L.; Maloshtan, A. Studies of the anti-inflammatory and diuretic properties of tablets with dense extract of *Betula pendula* leaves. *Dan. Sci. J.* **2019**, *1*, 46–48.
71. Muley, B.; Khadabadi, S.; Banarase, N. Phytochemical constituents and pharmacological activities of *Calendula officinalis* Linn (Asteraceae): A review. *Trop. J. Pharm. Res.* **2009**, *8*, 5. [CrossRef]
72. Givol, O.; Kornhaber, R.; Visentin, D.; Cleary, M.; Haik, J.; Harats, M. A systematic review of *Calendula officinalis* extract for wound healing. *Wound Repair Regen.* **2019**, *27*, 548–561. [CrossRef]
73. Preethi, K.C.; Kuttan, G.; Kuttan, R. Anti-inflammatory activity of flower extract of *Calendula officinalis* Linn. and its possible mechanism of action. *Indian J. Exp. Biol.* **2009**, *47*, 113–120.
74. Alexandre, J.T.M.; Sousa, L.H.T.; Lisboa, M.R.P.; Furlaneto, F.A.; do Val, D.R.; Marques, M.; Vasconcelos, H.C.; de Melo, I.M.; Leitão, R.; Brito, G.A.C.; et al. Anti-inflammatory and antiresorptive effects of Calendula officinalis on inflammatory bone loss in rats. *Clin. Oral. Investig.* **2018**, *22*, 2175–2185. [CrossRef]
75. Al-Snafi, A.E. *Glycyrrhiza glabra*: A phytochemical and pharmacological review. *IOSR J. Pharm.* **2018**, *8*, 1–17.
76. Frattaruolo, L.; Carullo, G.; Brindisi, M.; Mazzotta, S.; Bellissimo, L.; Rago, V.; Curcio, R.; Dolce, V.; Aiello, F.; Cappello, A.R. Antioxidant and anti-inflammatory activities of flavanones from *Glycyrrhiza glabra* L. (licorice) leaf phytocomplexes: Identification of licoflavanone as a modulator of NF-kB/MAPK pathway. *Antioxidants* **2019**, *8*, 186. [CrossRef]
77. Thakur, A.; Raj, P. Pharmacological perspective of *Glycyrrhiza glabra* Linn: A mini-review. *J. Anal. Pharm. Res.* **2017**, *5*, 00156. [CrossRef]
78. Hosseinzadeh, H.; Zarei, H.; Taghiabadi, E. Antinociceptive, anti-inflammatory and acute toxicity effects of *Juglans regia* L. leaves in mice. *Iran. Red Crescent Med. J.* **2011**, *13*, 27–33.
79. Hammer, K.D.; Hillwig, M.L.; Solco, A.K.; Dixon, P.M.; Delate, K.; Murphy, P.A.; Wurtele, E.S.; Birt, D.F. Inhibition of prostaglandin E2 production by anti-inflammatory *Hypericum perforatum* extracts and constituents in RAW264. 7 mouse macrophage cells. *J. Agric. Food Chem.* **2007**, *55*, 7323–7331. [CrossRef]
80. Zdunić, G.; Gođevac, D.; Milenković, M.; Vučićević, D.; Šavikin, K.; Menković, N.; Petrović, S. Evaluation of *Hypericum perforatum* oil extracts for an antiinflammatory and gastroprotective activity in rats. *Phytother. Res.* **2009**, *23*, 1559–1564. [CrossRef]

81. Sosa, S.; Pace, R.; Bornanciny, A.; Morazzoni, P.; Riva, A.; Tubaro, A.; Loggia, R.D. Topical anti-inflammatory activity of extracts and compounds from *Hypericum perforatum* L. *J. Pharm. Pharmacol.* **2007**, *59*, 703–709. [CrossRef]
82. Sun, Z.; Wang, H.; Wang, J.; Zhou, L.; Yang, P. Chemical composition and anti-inflammatory, cytotoxic and antioxidant activities of essential oil from leaves of *Mentha piperita* grown in China. *PLoS ONE* **2014**, *9*, e114767. [CrossRef]
83. Moshi, M.J. Current and future prospects of integrating traditional and alternative medicine in the management of diseases in Tanzania. *Tanzan Health Res. Bull.* **2005**, *7*, 159–167. [CrossRef]
84. Riordan, A.; Schofield, J. Beyond biomedicine: Traditional medicine as cultural heritage. *Int. J. Herit. Stud.* **2015**, *21*, 280–299. [CrossRef]
85. Kujawska, M.; Hilgert, N.I.; Keller, H.A.; Gil, G. Medicinal plant diversity and inter-cultural interactions between indigenous Guarani, Criollos and Polish migrants in the Subtropics of Argentina. *PLoS ONE* **2017**, *12*, e0169373. [CrossRef]
86. Chen, S.L.; Yu, H.; Luo, H.M.; Wu, Q.; Li, C.F.; Steinmetz, A. Conservation and sustainable use of medicinal plants: Problems, progress, and prospects. *Chin. Med.* **2016**, *11*, 37. [CrossRef]
87. Nastić, N.; Švarc-Gajić, J.; Delerue-Matos, C.; Barroso, M.F.; Soares, C.; Moreira, M.M.; Morais, S.; Mašković, P.; Gaurina Srček, V.; Slivac, I.; et al. Subcritical water extraction as an environmentally-friendly technique to recover bioactive compounds from traditional Serbian medicinal plants. *Ind. Crops Prod.* **2018**, *111*, 579–589. [CrossRef]
88. Šeregelj, V.; Šovljanski, O.; Švarc-Gajić, J.; Cvanić, T.; Ranitović, A.; Vulić, J.; Aćimović, M. Modern green approaches for obtaining *Satureja kitaibelii* Wierzb. ex Heuff extracts with enhanced biological activity. *J. Serb. Chem. Soc.* **2022**, *87*, 1359–1365. [CrossRef]
89. Radovanović, K.; Gavarić, N.; Švarc-Gajić, J.; Brezo-Borjan, T.; Zlatković, B.; Lončar, B.; Aćimović, M. Subcritical water extraction as an effective technique for the isolation of phenolic compounds of *Achillea* species. *Processes* **2023**, *11*, 86. [CrossRef]

Disclaimer/Publisher's Note: The statements, opinions and data contained in all publications are solely those of the individual author(s) and contributor(s) and not of MDPI and/or the editor(s). MDPI and/or the editor(s) disclaim responsibility for any injury to people or property resulting from any ideas, methods, instructions or products referred to in the content.

 *life*

MDPI

Review

The Use of Medicinal Plants in Blood Vessel Diseases: The Influence of Gender

Guglielmina Froldi

Department of Pharmaceutical and Pharmacological Sciences, University of Padova, 35131 Padova, Italy; g.froldi@unipd.it; Tel.: +39-049-827-5092; Fax: +39-049-827-5093

Abstract: Data available in the literature on the use of herbal products to treat inflammation-related vascular diseases were considered in this study, while also assessing the influence of gender. To this end, the articles published in PubMed over the past 10 years that described the use of plant extracts in randomized clinical trials studying the effectiveness in vascular pathologies were analyzed. The difference in efficacy of plant-derived preparations in female and male subjects was always considered when reporting. The safety profiles of the selected plants were described, reporting unwanted effects in humans and also by searching the WHO database (VigiBase®). The medicinal plants considered were *Allium sativum*, *Campomanesia xanthocarpa*, *Sechium edule*, *Terminalia chebula*. Additionally, an innovative type of preparation consisting of plant-derived nanovesicles was also reported.

Keywords: vascular dysfunction; endothelium; plant extracts; gender; safety; botanicals; garlic

Citation: Froldi, G. The Use of Medicinal Plants in Blood Vessel Diseases: The Influence of Gender. *Life* **2023**, *13*, 866. https://doi.org/10.3390/life13040866

Academic Editors: Azahara Rodríguez-Luna and Salvador González

Received: 21 February 2023
Revised: 17 March 2023
Accepted: 23 March 2023
Published: 23 March 2023

Copyright: © 2023 by the author. Licensee MDPI, Basel, Switzerland. This article is an open access article distributed under the terms and conditions of the Creative Commons Attribution (CC BY) license (https://creativecommons.org/licenses/by/4.0/).

1. Introduction

Inflammation is a well-known biological response of the organism to chemical and physical injuries that help the healing of tissues [1]. However, the inflammatory reaction can be harmful when it is excessively great, causing acute organ failure, or when too persistent, triggering chronic systemic inflammation [2]. In general, the inflammatory status is manifested by an increase in serum C-reactive protein (CRP) level and the release of pro-inflammatory cytokines, such as interleukin-6 (IL-6) and tumor necrosis factor-α (TNF-α) [3]. Arterial and venous vessels are directly involved in inflammatory progression with alterations in endothelial permeability, causing liquid and protein leakage and tissue edema formation. TNF-α increases intracellular calcium and regulates myosin light chain kinase and RhoA, which disrupts endothelial junctions, reducing barrier function and enabling leukocyte transmigration [4,5]. Subjects with high levels of homocysteine, cholesterol, and triglycerides show a greater risk of stroke by approximately 50% than those with normal values [6]. Furthermore, the correction of hyperhomocysteinemia involves a reduction in stroke risk from 34% to 70% [7]. Among the factors that affect the amount of homocysteine present in the blood, are various physiological factors, such as age, sex, and body mass index [8,9]. In fact, women generally have lower levels than men [10]; therefore, homocysteine may be a possible factor responsible for gender differences in atherosclerosis and coronary artery disease [11].

In the blood vessel, acute and chronic inflammation results in endothelial dysfunction and vascular remodeling, affecting cardiovascular function. Several endogenous modulators play a role in vascular inflammatory processes; in particular, an increase in advanced glycation end-products (AGEs) causes activation of inflammatory pathways, oxidative stress, and procoagulant activity, leading to endothelial dysfunction [12,13]. Fishman et al. suggested that AGEs and their receptors may be useful biomarkers of the presence and severity of coronary artery disease [12]. Other inflammatory modulators are proinflammatory cytokines, such as TNF-α and interleukins (ILs) [14]. IL-1α and IL-1β are responsible for the disruption effects on the endothelial barrier, while nuclear factor kappa-light-chain-enhancer of activated B cells (NF-kB) and the mitogen-activated protein kinase (MAPK) signaling cause pro-atherogenic effects [15]. Moreover, the vascular adhesion protein-1 (VAP-1), also known as amine oxidase copper-containing 3 (AOC3), is a pro-inflammatory modulator with a greater expression in endothelial cells during inflammation [16,17]. VAP-1 is an ectoenzyme that catalyzes the oxidative deamination of primary amines and produces hydrogen peroxide, ammonium, and aldehydes, regulates leukocyte extravasation, and causes vascular damage and atherosclerosis [16,17]. Furthermore, endothelial dysfunction is mediated by the activation of enzymes such as heparinase and metalloproteinases (MMPs), which cleave glycoproteins anchored to the endothelial glycocalyx, which are activated by pro-inflammatory cytokines and reactive oxygen species (ROS) [18]. Furthermore, monosodium urate and cholesterol crystals, and islet amyloid polypeptides can damage the phagolysosome membrane and promote persistent activation of the NLRP3 inflammasome, also known as NALP3 or cryopyrin, causing severe inflammatory diseases such as gout, atherosclerosis, and type 2 diabetes mellitus [19–21]. Recently, extracellular vehicles such as platelet and endothelial microparticles have been reported to be involved in vascular regulation, including inflammatory and thrombotic homeostasis [22,23]. Therefore, vascular inflammation is clearly a complex condition that has several modulators.

The literature findings suggest that the treatment of inflammation with targeted drugs may promote the regression of vascular dysfunction [4,24,25]. However, non-steroid and steroid anti-inflammatory drugs have not shown protective effects against arterial stiffening; however, some promising results have been obtained with the use of selective inhibitors of IL-1β, IL-6, and TNF-α [4,25]. Unfortunately, the use of these inhibitors can cause the appearance of significant side effects, and their risk/benefit ratio remains to be further determined. In this context, plant-derived products with potential activity in the treatment of vascular dysfunction may be of interest. For this purpose, this study focused on the literature delineating potential plant products of utility in the treatment of cardiovascular diseases related to inflammatory damage. Particular attention was paid to the identification of differences in response to herbal medicines in women versus men.

2. Sex Differences in Vascular Function

Few studies considered the sex difference related to inflammation and metabolic and cardiovascular diseases. The general opinion is that men tend to have a worse risk factor profile than women, although this relationship changes with advancing age [26]. Differences in the prevalence of cardiovascular diseases were observed in premenopausal women compared to men of the same age; clinical data suggest that women are protected against various cardiovascular diseases primarily during the fertile period of life [27–30]. In fact, estrogens increase vascular $NO^\bullet$ signaling, improving vasodilatation and insulin responsiveness, protecting against diabetes mellitus [30]. However, the relative risk of cardiovascular disease morbidity and mortality in subjects affected by diabetes mellitus ranges from 1 to 3 in men and from 2 to 5 in women, related to those without diabetes [31]. Furthermore, clinical studies failed to demonstrate that hormone replacement therapy (HRT) in postmenopausal women could improve cardiovascular outcomes [30]. It can also be observed that men and women have different lifestyle risk factors, such as male behavior that may include more frequently cigarette smoking, alcohol abuse, higher intake of red meat, and lower fruit and vegetable consumption [32]. Effectively, in part, behavior differences could explain why women live longer than men [33]. Focus has been on the relationship between the Western lifestyle and chronic metabolic inflammatory diseases and also on finding preventive approaches [34]. A meta-analysis that considered a total of 462,194 participants demonstrated that high ingestion of fruits and vegetables (flavonoids) was inversely associated with the risk of total cardiovascular disease mortality [35]. Recently, Parmenter et al. (2022) found that in older women, a higher intake of habitual dietary flavonoids, mainly black tea, is associated with less extensive abdominal aortic calcification [36]. Previous authors have also shown that high black tea consumption is associated with lower coronary artery and abdominal aorta calcifications [37–39].

3. Materials and Methods

This study considered articles identified using a PubMed search strategy up to January 2023. The terms included in the search query were 'blood vessel', 'inflammation', and 'plant-derived compound'. The inclusion criteria for the search were as follows: 1. Published in the last 10 years, 2. Full texts, 3. English language. Of a total of 16,395 articles included, some were excluded according to other inclusion criteria, such as 'endothelium' and 'female', excluding the term 'review' (Figure S1). A total of 20 articles was reviewed according to their relevance to the selected topic. Additionally, other references were examined also through Google Scholar to find additional relevant articles, including in vitro and in vivo studies to facilitate the explanation of the pharmacological mechanisms. The safety profiles of the selected plants were described, along with the unwanted effects reported in humans, also by searching VigiBase®, the WHO global database of possible side effects of medicinal products [40].

4. Results

The selected medicinal plants evaluated in this review were *Allium sativum*, *Campomanesia xanthocarpa*, *Sechium edule*, and *Terminalia chebula*. Additionally, an innovative type of preparation consisting of plant-derived nanovesicles was considered. Table 1 reports a synthesis of the clinical trials that detected the efficacy of the medicinal plants considered in this review.

4.1. *Allium sativum L.*

4.1.1. Botanical Characteristics

Garlic (Amaryllidaceae) is a herbaceous plant whose bulbs are often used to flavor different types of food, widely known for its typical smell and taste [41]. Its alimentary use is widespread throughout the world, and its medicinal use is well-known in popular medicine as crude drug, standardized extracts, and also as a food supplement [42]. The bulbs are harvested in late spring and early summer, then dried in the shade at 40 °C, to enable storage [41]. Several traditional uses are known, such as antimicrobial, diuretic, vermifuge, adjuvant in the prevention of atherosclerosis and the relief of the common cold [43,44].

4.1.2. Phytoconstituents and Preclinical Activity

Among the most peculiar constituents, there are several sulfur compounds, including alliin, an odorless compound, which in turn is transformed by the alliinase enzyme into allicin that has the typical garlic smell [45–48]. Allicin is a diallyl thiosulfinate considered one of the most active components, although other sulfur compounds provide garlic properties, such as ajoene, allyl propyl disulfide, diallyl trisulfide, and S-allylcysteine [49]. Additionally, garlic contains saponins, flavonoids, vitamins, and minerals [41,50]. Numerous in vitro and in vivo studies suggested its efficacy in several human diseases [51,52]. Garlic powder, aged garlic, and garlic oil have shown antiplatelet and anticoagulant effects by interfering with cyclooxygenase-mediated thromboxane synthesis [49,53]. Garlic extracts showed antioxidant property, decreased expression of vascular endothelial growth factor (VEGF), hypoxia-inducible factor 1 alpha (HIF-1α), inducible nitric oxide synthase (iNOS), and metalloproteinase (MMP)-9 [54–56]. Furthermore, garlic prevents the expression of inflammatory cytokines such as IL-6 and monocyte chemoattractant protein-1 (MCP-1) in lipopolysaccharide (LPS)-stimulated 3T3-L1 adipocytes [57].

Table 1. Effectiveness of herbal preparations used for the treatment of vascular diseases administered orally to human subjects.

Natural Products	Clinical Trials	Participants	Dosage	Outcomes	Refs.
Allium sativum (garlic)	a. R, single-blind, PC b. R, DB, controlled c. R, DB, PC d. R, DB, PC	a. 50 pregnant subjects b. 44 pregnant subjects c. 92 obese subjects d. 91 T2DM subjects	a. 8 weeks 800 mg day garlic (1 mg allicin) b. 9 weeks 400 mg day garlic (1 mg allicin) c. 3 months 400 mg day garlic extract (2% allicin) d. 4 weeks 500 mg twice day garlic (2–3 mg allicin)	a. Reduces systolic blood pressure, total cholesterol b. Reduces CRP, increases GSH c. Decreases CRP, PAI-1, TG and LDL-C. Increases TAC d. Improves visual acuity	[46,58–62]
Campomanesia xanthocarpa (guavirova)	a. R, DB, PC b. R, DB, PC c. R, DB, pilot	a. 33 hypercholesterolemic subjects b. 156 hypercholesterolemic subjects c. 23 healthy subjects	a. 3 months 250 or 500 mg day dried leaves b. 3 months 500, 750 and 1000 mg day dried leaves c. 1000 mg day dried leaves	a. Decreases TG, LDL b. Decreases TG, LDL, CRP, oxidative stress. Increases $NO^\bullet$ c. Antiplatelet activity	[63–65]
Sechium edule (chayote)	ND	ND	ND	ND	ND
Terminalia chebula (black myrobalan)	a. R, DB, PC b. R, DB, PC	a. 56 subjects with metabolic syndrome b. 60 T2DM subjects	a. 12 weeks 250 or 500 mg twice daily aqueous fruit extract b. 12 weeks 250 or 500 mg twice daily aqueous fruit extract	a. and b. Improves endothelial function, increases $NO^\bullet$, GSH, HDL, decreases CRP, HbA1c, MDA, TG, LDL, VLDL	[66–68]
Plant-derived nanovesicles	a. Open-label	20 healthy subjects	3 months 1000 mg day *Citrus limon* EVs	Decreases waist circumference in women	[69]

CRP: C-reactive protein; DB: double-blind; EVs: extracellular vesicles; GSH: glutathione; PC: placebo-controlled; T2DM: type 2 diabetes mellitus; HbA1c: glycosylated hemoglobin A1c; HDL: high-density lipoproteins; LDL-C, low density lipoprotein cholesterol; MDA: malondialdehyde; PAI-1: plasminogen activator inhibitor 1; PC: protein carbonyls; R: randomized trial; TG, triglycerides; TAC: total antioxidant capacity; TC, total cholesterol; HDL-C, high density lipoprotein cholesterol. ND: no documented clinical trials.

4.1.3. Therapeutic Efficacy: Clinical Trials

Garlic is a very well-known plant that is used around the world. Studies on garlic preparations have mainly tested hypocholesterolemic, antihypertensive, antimicrobial, and, also, antitumor activities [54–56]. A randomized placebo-controlled trial (RCT) enrolled a total of 100 pregnant women at high risk of pre-eclampsia who were treated during the third trimester of pregnancy for 8 weeks with 800 mg garlic tablets (dried garlic powder containing 1 mg allicin, and ajoene) per day or placebo [60]. The treatment prevented the increase of total cholesterol and reduced hypertension [60]. Another RCT conducted in 44 pregnant women treated with 400 mg garlic tablet (equal to 400 mg garlic and 1 mg allicin) for 9 weeks showed reduced serum CRP levels and increased plasma glutathione in the treated women compared to the untreated group [61]. Another study showed that garlic supplementation (400 mg/day, garlic extract 2% allicin) positively modifies endothelial biomarkers of cardiovascular risk, suggesting that treatment can reduce chronic inflammation in obese individuals of both sexes [58]. Furthermore, one trial also suggested that 500 mg granulated garlic powder (2–3 mg allicin) can be considered as an adjuvant treatment in patients with diabetic macular edema [62]. Recently, the acute efficacy of 180 mg fermented garlic extract enriched with 7 mg inorganic nitrite (NO_2^-) in healthy women and men showed a significant decrease in both systolic and diastolic pressure 30 min after ingestion of the product [70].

Systematic reviews and meta-analysis evaluating the effects of garlic supplementation have generally reported an improvement of lipid profile and insulin-resistance; however, the low quality of the trials does not permit at the moment the real effectiveness of garlic preparations to be defined [71,72]. Overall, the data available in the literature support the use of garlic extracts in the treatment of vascular diseases mainly with an atherosclerotic basis, without apparent differences between the two sexes. However, additional RCTs with standardized extracts are required to confirm the therapeutic use of garlic in cardiovascular diseases.

4.1.4. Safety

According to toxicity data from experimental studies and medicinal use, garlic preparations are generally considered safe in the usual dosage regimen. Female and male rats treated with 300 and 600 mg/kg day of an aqueous garlic bulb extract for 21 days showed changes in weight growth, biological parameters, and histological structures [73]. In humans, common side effects included odor and skin rash, and gastrointestinal upset [60,74,75]. In VigiBase® *Allium sativum* has 232 reports of potential side effects from all countries, mainly Europe (31%), in patients of both sexes (female 49%, male 47%, unknown 5%), in adults and older people [40]. Figure 1 shows the types and percentages of reported side effects. Among these, there are mainly gastrointestinal disorders (18%, such as vomiting, gastrointestinal pain, diarrhea, etc.), general disorders (13%, such as drug interaction, asthenis, etc.), nervous system disorders (11%, such as dizziness, hemorrhages, skin disorders, etc.) and various others (Figure 1). The large number of side effects related to garlic use may depend on the very high consumption of this plant worldwide. In general, the use of garlic even for curative purposes is considered safe in humans, avoiding use in atopic subjects [51,76,77]. The medicinal use of garlic is not recommended during pregnancy and breastfeeding due to the absence of clinical evidence showing both efficacy and non-toxicity [44]. Furthermore, it is also not recommended in patients being treated with antiplatelet and anticoagulant drugs due to the increased risk of bleeding [44,75].

4.1.5. Future Needs

The use of garlic preparations is widespread both in the diet and for healing purposes. To optimize its medical use, it is essential to define the type of formulation and the titer of sulfur derivatives, the dose, and the mode of administration, to standardize treatments and compare data from different clinical trials. Certainly, double-blind randomized clinical trials, with a sufficient number of subjects of both sexes are needed, to validate the use of garlic in therapy in the treatment of vascular diseases. Otherwise, unfortunately, it would remain only a traditional use, without valid evidence of effectiveness, the use on the basis of scientific evidence being renounced.

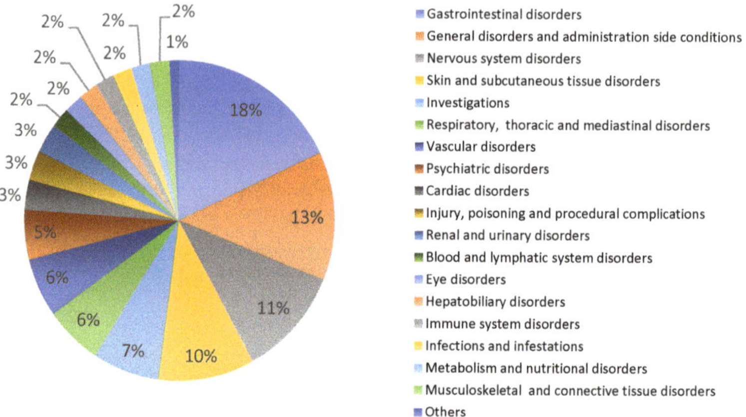

Figure 1. *Allium sativum*: potential side effects reported in VigiBasis®.

4.2. *Campomanesia xanthocarpa Berg.*

4.2.1. Botanical Characteristics

Campomanesia xanthocarpa (Myrtaceae) is a semi-deciduous tree, commonly known as "guavirova", which grows in Brazil, Argentina, Paraguay, and Uruguay. It has edible fruits with a succulent pulp and a sweet flavor [78]. Leaves are traditionally used in herbal teas to treat inflammatory, urinary, rheumatic diseases, high blood pressure, and high cholesterol [78,79].

4.2.2. Phytoconstituents and Preclinical Activity

Leaves contain phenolic compounds, such as chlorogenic, gallic, ellagic and rosmarinic acids, glycosylated flavanols mainly of quercetin and myricetin, and pro-anthocyanidins [80–84]. Alkaloid theobromine (3,7-dimethyl-xanthine) was also identified in an aqueous infusion of leaves [83]. Acute administration of an aqueous extract showed a dose-dependent hypotensive effect in rats, by inhibiting the renin–angiotensin system through the block of the angiotensin II type 1 receptor (AT1R) and of calcium currents, as well as by K_{ATP} channel activation [83,84]. Furthermore, several studies reported the antioxidant activity of the fruits and leaves [80,82,85].

4.2.3. Therapeutic Efficacy: Clinical Trials

The authors studied *Campomanesia xanthocarpa* on inflammatory processes, oxidative stress, and lipid biomarkers of hypercholesterolemia, showing a decrease in total cholesterol and LDL levels in treated hypercholesterolemic subjects [63,64]. A small trial in 33 hypercholesterolemic subjects treated with 250 and 500 mg capsules that contained dried *Campomanesia xanthocarpa* leaves for 90 days revealed a significant reduction in total cholesterol and LDL levels in hypercholesterolemic subjects with total cholesterol >240 mg/dL (n = 22) [63]. In addition, another trial involving a larger number of subjects treated for 90 days with 500, 750, and 1000 mg of dried encapsulated leaves also demonstrated anticholesterolemic activity [64]. These authors also suggested that this treatment attenuates oxidative stress and pro-inflammatory reactions, improving blood flow and endothelial function [64]. Furthermore, healthy subjects were treated with 1000 mg of powdered *Campomanesia xanthocarpa* leaves (n = 8) and compared to those treated with 100 mg of acetylsalicylic acid (ASA, n = 7), or 500 mg of *Campomanesia xanthocarpa* plus 50 mg of ASA

($n = 7$). The authors showed that *Campomanesia xanthocarpa* leaves have antiplatelet activity when administered at 1000 mg for 5 days alone, or at 500 mg with low doses of ASA [65].

4.2.4. Safety

The extract of *Campomanesia xanthocarpa* leaf administered to male rats at 300 mg/kg iv caused cardiac depression with a dramatic drop in blood pressure and animal death [84]. In contrast, a 5000 mg/kg ethanol leaf extract administered orally to five male and five female mice did not show toxicity [81]. Genotoxic effects were observed after treatment with an aqueous leaf extract administered to male rats at 1000 mg/kg [82]; this observation should be taken into account in future studies. No adverse reactions were reported for the use of *Campomanesia xanthocarpa* (guavirova) in VigiBase® [40]. No information is available in the literature on its safety during pregnancy or breastfeeding in humans.

4.2.5. Future Needs

Preparations with dry leaves of *Campomanesia xanthocarpa* seem to have interesting hypocholesterolemic and antiplatelet activities, potentially useful in vascular diseases. However, no specific active compounds were identified in the formulations administered to the subjects, and therefore there is no reference compound for titration. Data from clinical trials are very limited, and there is certainly a need for double-blind randomized clinical trials with a sufficient number of subjects of both sexes to define the clinical usefulness of this medicinal plant.

4.3. Sechium edule (Jacq.) Sw.

4.3.1. Botanical Characteristics

Sechium edule (Cucurbitaceae) is a perennial herbaceous climbing plant cultivated mainly by Asian and Latino-American populations for food use, in particular for Chayote fruits [86–88]. Likewise, fruits, roots, and leaves are known in traditional medicine against kidney stones, as a diuretic and antihypertensive [89–93]. Furthermore, alcoholic extracts showed a very good antimicrobial efficacy against all strains of multi-resistant Staphylococci and Enterococci [94]. Moreover, various leaf and seed preparations have shown remarkable antioxidant activity [95]. Several investigations in different animal models, such as rats, mice, and dogs, defined the capacity of this plant to reduce blood pressure [86,89,91,96]. The hydroalcoholic extract and the acetone fraction obtained from the roots of *Sechium edule* showed antihypertensive activity by a relaxant effect on blood vessels [91,97].

4.3.2. Phytoconstituents and Preclinical Activity

Leaf, seed, stem, and also the fruit *Sechium edule* are rich in various bioactive components, as well as flavonoids, phenolics, vitamin C, and carotenoids [94,95,98]. The leaves contained the highest concentration of luteolin glycosides, while the most significant concentration of apigenin derivatives (C-glycosidic and O-glycosidic bonds) was found in the root extract [92]. Trans-cinnamic acid, phenylacetic acid, and α-linolenic acid were identified in the leaf extract [99]. Fruits can contain bitter principles called cucurbitacins [86,100,101].

In isolated aorta rings without endothelium, a hydro-alcoholic root extract caused a concentration-dependent vasorelaxation of angiotensin II-induced vasocontraction [91]. Furthermore, the authors reported in vivo antihypertensive effects in mice treated with angiotensin II [91]. The distinctive components of the highest active fraction were identified as cinnamic compounds, such as cinnamic acid methyl ester [91,97]. An aqueous leaf extract administered at 200 mg/kg showed nephroprotective activity against various types of chemically induced renal damage in rats [102]. The extract used at 100–200 mg/kg showed anti-inflammatory activity reducing levels of TGF-β, TNF-α, and ICAM-1 [103,104]. In rats fed with a high-fat diet, *Sechium edule* shoots can prevent hepatic steatosis and attenuate fatty tissue by inhibiting lipogenic enzymes and stimulating lipolysis by upregulating AMP-activating protein kinase (AMPK) [105]. The same authors also showed that the shoot extracts inhibited the expressions of fatty acid synthase and HMG-CoA reductase in rats, while also isolated caffeic acid and hesperetin, the main characteristic components of *Sechium edule* shoots, prevented hepatic lipid accumulation [105]. The acetone fraction of the hydro-alcoholic extract of *Sechium edule* roots administered to female mice at 10 mg/kg per day, orally, for 10 weeks was able to control hypertension, as well as the oxidative and inflammatory status in the kidneys, as efficiently as losartan, returning mice to normotensive levels [97,103]. Furthermore, the acetonic fraction was more effective than losartan in preventing liver and kidney damage. Therefore, the fraction was able to control endothelial dysfunction and related diseases [97,103].

4.3.3. Therapeutic Efficacy: Clinical Trials

As far as was found in the literature, no clinical studies have been conducted in patients using this plant as a single treatment. A clinical trial studied a commercial antioxidant supplement containing three components, including *Sechium edule*, showing an improvement of the hemorheology in alcoholics [106]. Based on available data, clinical studies are required on the use of *Sechium edule* in hypertension, diabetes mellitus, obesity, and, in general, in vascural-related diseases.

4.3.4. Safety

Acute toxicity was tested in rats and mice that received a single oral dose of 2000 mg/kg of an aqueous leaf extract. Treated animals showed no change in the normal behavior pattern and no evidence of toxicity and mortality [102]. Negative effects on humans were not documented. In VigiBase® there are no reports of potential side effects related to *Sechium edule* [40], and the literature does not provide data on the safety of medicinal use during pregnancy or breastfeeding.

4.3.5. Future Needs

Although traditional use and preclinical data suggest great interest in the use of *Sechium edule* in vascular diseases, the total absence of clinical studies strongly limits its use. Therefore, proper clinical trials are desirable.

4.4. Terminalia chebula Retz.

4.4.1. Botanical Characteristics

Terminalia chebula (Combretaceae), also known as black myrobalan, is a deciduous tree that grows up to 30 m, widely known in India and Southern Asia for its use in Ayurvedic medicine [107,108]. Fruits are used in traditional medicine to treat various diseases, such as used as a laxative, stomachic, tonic, and antispasmodic [107,108].

4.4.2. Phytoconsituents and Preclinical Activity

The main components of the fruit are phenolic compounds, such as hydrolysable tannins and flavonoids, saccharides, such as D-glucose, D-fructose, and saccharose [67]. The aqueous extract of *Terminalia chebula* fruits contains chebulagic acid, chebulinic acid, and other low molecular weight hydrolysable tannins [68]. The leaves contain polyphenols such as punicalin, punicalagin, terflavins B, C, and D [108].

The antioxidant activity of *Terminalia chebula* has been reported in vivo and in vitro assays [108–111]. A study described a significant decrease in glucose level in normal and alloxan-induced diabetic rats four hours after oral administration of a methanolic fruit extract (100 mg/kg) [110]. Similarly, *Terminalia chebula* fruit extract at a concentration of 200 mg/kg administered for 30 days significantly reduced blood glucose, glycosylated hemoglobin, urea, and creatinine levels in diabetic rats [112]. Furthermore, a cardioprotective effect of an ethanolic extract of fruits (500 mg/kg) was described in rats [113].

4.4.3. Therapeutic Efficacy: Clinical Trials

Few trials have been reported in the literature on the potential therapeutic use of *Terminalia chebula*. A 12-week prospective trial showed that an aqueous extract of *Terminalia chebula* fruits administered at 250 mg and 500 mg, twice daily, significantly improved endothelial function, systemic inflammation, and lipid profile in 60 subjects with type 2 diabetes mellitus of either gender, compared to placebo treatment [66]. *Terminalia chebula* extract significantly increased $NO^{\bullet}$ and GSH levels, reducing oxidative stress, malondialdehyde, and CRP levels [66]. Previously, the same authors reported similar beneficial effects in 56 patients of either gender with metabolic syndrome [68]. In this study, *Terminalia chebula* reduced malondialdehyde levels and increased glutathione levels, improving antioxidant status. Furthermore, the treatment significantly decreased total cholesterol, triglycerides, and low-density lipoprotein cholesterol, and increased high-density lipoprotein cholesterol, while the placebo did not have a significant effect on endothelial function or any of the other clinical parameters [68].

4.4.4. Safety

Treatment with *Terminalia chebula* extracts was well tolerated with very few side effects; in fact, very few patients experienced dyspepsia [66]. Other authors reported increased libido, dry mouth, colic, and confusion [114]. In general, the fruit preparations are well-tolerated and do not adversely affect health [67,108]. In VigiBase®, *Terminalia chebula* has five reports of potential side effects, such as gastrointestinal disorders (i.e., diarrhea and gastrointestinal pain), and other general disorders [40].

4.4.5. Future Needs

The few clinical studies that have evaluated the efficacy of *Terminalia chebula* fruit extracts suggest a potential use in the treatment of vascular dysfunction in diabetes mellitus and/or metabolic syndrome. Unfortunately, data are limited, and other studies are certainly necessary to determine the efficacy and safety of this medicinal plant.

4.5. Plant-Derived Nanovesicles

4.5.1. General Characteristics

In recent years, various investigations have suggested that plant cells, through an exosome-like process, may release nanosized particles, which are involved in plant cell-cell communication [115,116]. Furthermore, various studies suggest that plant-derived nanovesicles may also play a role in the properties of medicinal plants in human diseases, mainly based on their biological cargo [117–120]. It is assumed that the interaction between vegetal extracellular vesicles and mammalian cells may have beneficial effects through antioxidant and anti-inflammatory activities [121,122]. Plants produce nanovesicles in response to numerous biotic and abiotic environmental stresses, including pathogen infection and attack. Plant nanovesicles carry a wide variety of molecules, including proteins, lipids, miRNAs, vitamins, and various plant metabolites [123].

4.5.2. In Vitro and In Vivo Studies

Lemon and strawberry-derived nanovesicles showed antioxidant activity in mesenchymal stem cells [124,125]. The authors verified the potential anti-osteoporotic effects of apple-derived nanovesicles using MC3T3-E1 cells, inhibiting osteoporosis by promoting osteoblastogenesis in osteoblastic MC3T3-E1 cells, by regulating the BMP2/Smad1 pathway [126]. Lemon-derived nanovesicles have been found to be rich in citric acid and vitamin C, which have a significant protective effect on oxidative stress in mesenchymal stromal cells [124].

Few studies have been carried out in animal models and humans. Oral administration of grape exosome-like nanoparticles showed beneficial effects in dextran sulfate sodium (DSS)-induced experimental colitis in mice, via induction of intestinal stem cells [120]. Similarly, broccoli-derived nanoparticles administered orally protected against various types of mice colitis, through activation of adenosine monophosphate-activated protein kinase (AMPK) in dendritic cells [127]. Additionally, ginger-derived nanoparticles protected mice from alcohol-induced liver injury, activating nuclear factor erythroid 2-related factor 2 (Nrf2) and inhibiting ROS production [128].

4.5.3. Therapeutic Efficacy: Clinical Trials

Very few trials considered the administration of plant-derived nanovesicles in the treatment of human diseases. In a prospective open-label study, 20 healthy volunteers (9 women and 11 men) were treated with a commercial preparation of extracellular vesicles from *Citrus limon* L., administered at 1000 mg daily for 3 months [69]. A decrease in waist circumference was found in women after 4 and 12 weeks of treatment, while no significant reduction was detected in men [69]. In the same study, the authors also observed a significant reduction in low-density lipoproteins (LDL) [69]. Significant correlations were also found in the stratified analysis between alkaline phosphatase enzymes (ALP) and glucose for women and between ALP and LDL for men [69]. A phase 1 clinical trial is currently underway studying the ability of a grape exosome preparation, administered orally for 35 days, to act as an anti-inflammatory agent against oral mucositis during radiation and chemotherapy treatment for head and neck tumors (NCT01668849) [129]. Another clinical study with ginger and aloe-derived exosomes studying the ability to mitigate insulin resistance and chronic inflammation in patients diagnosed with polycystic ovary syndrome was withdrawn because the investigator left the university before study approval (NCT03493984).

4.5.4. Safety

As reported in the literature, there are no toxicity studies or reports of undesirable effects related to human administration of these types of preparations. The use of plant-derived extracellular vesicles represents a new and very interesting approach in the treatment of diseases; however, other studies are needed to explore the advantages and, also, the disadvantages of plant-derived nanovesicles in therapy [129].

4.5.5. Future Needs

Plant-derived nanovesicles are certainly an innovative type of plant preparation, which also has considerable industrial implications. However, there are many aspects to be validated, starting from the techniques for obtaining and the definition of the constituents, up to the possible uses in the prevention or treatment of human pathologies.

5. Conclusions

Based on the data collected, it can be observed that clinical studies concerning the use of products of plant origin in the treatment of human pathologies and, in particular, in cardiovascular diseases are few and consider only small groups of subjects. Furthermore, the studies generally do not examine the differences in treatment response comparing the female or male gender. In studies in which the efficacy of the products used were reported separately, women versus men, it was not possible to obtain evidence of the difference in efficacy because the number of subjects enrolled in the trials was too small to perform any statistical estimation.

Among the plants considered, garlic has been the most studied and there are several data on its effectiveness in the treatment of vascular-related disorders. However, the available data are insufficient to validate the pharmacological use of garlic preparations for any of the conditions under consideration. Additional research that recruits more patients is desirable. Some plants, such as *Campomanesia xanthocarpa*, *Sechium edule*, and *Terminalia chebula*, have been proposed for their potential use in vascular problems in diabetic or hypertensive subjects. Finally, a new type of innovative preparation based on plant-derived extracellular vesicles has been suggested, but this is only an idea that still requires long investigation. Importantly, greater attention must be paid in carrying out clinical trials with the aim of obtaining a personalized use of plant products, noting the differences in the effectiveness between women and men. Improved consideration of gender-based medicine is required to improve the efficacy of therapeutic interventions and reduce adverse reactions.

Supplementary Materials: The following supporting information can be downloaded at: https://www.mdpi.com/article/10.3390/life13040866/s1, Figure S1: Flowchart of criteria used in PubMed research.

Funding: This research received no external funding.

Institutional Review Board Statement: Not applicable.

Informed Consent Statement: Not applicable.

Data Availability Statement: Not applicable.

Conflicts of Interest: The author declares no conflict of interest.

References

1. Chen, L.; Deng, H.; Cui, H.; Fang, J.; Zuo, Z.; Deng, J.; Li, Y.; Wang, X.; Zhao, L. Inflammatory responses and inflammation-associated diseases in organs. *Oncotarget* **2018**, *9*, 7204–7218. [CrossRef] [PubMed]
2. Furman, D.; Campisi, J.; Verdin, E.; Carrera-Bastos, P.; Targ, C.; Franceschi, C.; Ferrucci, L.; Gilroy, D.W.; Fasano, A.; Miller, G.W.; et al. Chronic inflammation in the etiology of disease across the life span. *Nat. Med.* **2019**, *25*, 1822–1832. [CrossRef] [PubMed]
3. Sproston, N.R.; Ashworth, J.J. Role of C-reactive protein at sites of inflammation and infection. *Front. Immunol.* **2018**, *9*, 754. [CrossRef] [PubMed]
4. Zanoli, L.; Briet, M.; Empana, J.P.; Cunha, P.G.; Maki-Petaja, K.M.; Protogerou, A.D.; Tedgui, A.; Touyz, R.M.; Schiffrin, E.L.; Spronck, B.; et al. Vascular consequences of inflammation: A position statement from the ESH working group onvascular structure and function and the ARTERY Society. *J. Hypertens.* **2020**, *38*, 1682–1698. [CrossRef] [PubMed]
5. Shen, Q.; Rigor, R.R.; Pivetti, C.D.; Wu, M.H.; Yuan, S.Y. Myosin light chain kinase in microvascular endothelial barrier function. *Cardiovasc. Res.* **2010**, *87*, 272–280. [CrossRef]
6. Hao, L.; Chen, L.M.; Sai, X.Y.; Liu, Z.F.; Yang, G.; Yan, R.Z.; Wang, L.L.; Fu, C.Y.; Xu, X.; Cheng, Z.Z.; et al. Synergistic effects of elevated homocysteine level and abnormal blood lipids on the onset of stroke. *Neural Regen. Res.* **2013**, *8*, 2923–2931. [CrossRef]
7. Spence, J.D. Stroke Prevention: A Lifetime of Lessons. *Stroke* **2020**, *51*, 2255–2262. [CrossRef]
8. Zhu, W.; Huang, X.; Li, M.; Neubauer, H. Elevated plasma homocysteine in obese schoolchildren with early atherosclerosis. *Eur. J. Pediatr.* **2006**, *165*, 326–331. [CrossRef]
9. Brattström, L.; Lindgren, A.; Israelsson, B.; Andersson, A.; Hultberg, B. Homocysteine and cysteine: Determinants of plasma levels in middle-aged and elderly subjects. *J. Intern. Med.* **1994**, *236*, 633–641. [CrossRef]
10. Cohen, E.; Margalit, I.; Shochat, T.; Goldberg, E.; Krause, I. Gender differences in homocysteine concentrations, a population-based cross-sectional study. *Nutr. Metab. Cardiovasc. Dis.* **2019**, *29*, 9–14. [CrossRef]
11. Xu, R.; Huang, F.; Wang, Y.; Liu, Q.; Lv, Y.; Zhang, Q. Gender- and age-related differences in homocysteine concentration: A cross-sectional study of the general population of China. *Sci. Rep.* **2020**, *10*, 17401. [CrossRef] [PubMed]
12. Fishman, S.L.; Sonmez, H.; Basman, C.; Singh, V.; Poretsky, L. The role of advanced glycation end-products in the development of coronary artery disease in patients with and without diabetes mellitus: A review. *Mol. Med.* **2018**, *24*, 59. [CrossRef] [PubMed]
13. Froldi, G.; Ragazzi, E. Selected plant-derived polyphenols as potential therapeutic agents for peripheral artery disease: Molecular mechanisms, efficacy and safety. *Molecules* **2022**, *27*, 7110. [CrossRef]
14. Kany, S.; Vollrath, J.T.; Relja, B. Cytokines in inflammatory disease. *Int. J. Mol. Sci.* **2019**, *20*, 6008. [CrossRef] [PubMed]
15. Martínez, G.J.; Celermajer, D.S.; Patel, S. The NLRP3 inflammasome and the emerging role of colchicine to inhibit atherosclerosis-associated inflammation. *Atherosclerosis* **2018**, *269*, 262–271. [CrossRef]
16. Salmi, M.; Jalkanen, S. Vascular adhesion protein-1: A cell surface amine oxidase in translation. *Antioxid. Redox Signal.* **2019**, *30*, 314–332. [CrossRef] [PubMed]
17. Smith, D.J.; Salmi, M.; Bono, P.; Hellman, J.; Leu, T.; Jalkanen, S. Cloning of vascular adhesion protein i reveals a novel multifunctional adhesion molecule. *J. Exp. Med.* **1998**, *188*, 17–27. [CrossRef]
18. Dogné, S.; Flamion, B. Endothelial glycocalyx impairment in disease: Focus on hyaluronan shedding. *Am. J. Pathol.* **2020**, *190*, 768–780. [CrossRef]
19. Martinon, F.; Pétrilli, V.; Mayor, A.; Tardivel, A.; Tschopp, J. Gout-associated uric acid crystals activate the NALP3 inflammasome. *Nature* **2006**, *440*, 237–241. [CrossRef]
20. Duewell, P.; Kono, H.; Rayner, K.J.; Sirois, C.M.; Vladimer, G.; Bauernfeind, F.G.; Abela, G.S.; Franchi, L.; Nũez, G.; Schnurr, M.; et al. NLRP3 inflammasomes are required for atherogenesis and activated by cholesterol crystals. *Nature* **2010**, *464*, 1357–1361. [CrossRef]

21. Fusco, R.; Siracusa, R.; Genovese, T.; Cuzzocrea, S.; Di Paola, R. Focus on the role of NLRP3 inflammasome in diseases. *Int. J. Mol. Sci.* **2020**, *21*, 4223. [CrossRef] [PubMed]
22. Puhm, F.; Boilard, E.; MacHlus, K.R. Platelet extracellular vesicles; beyond the blood. *Arter. Thromb. Vasc. Biol.* **2021**, *41*, 87–96. [CrossRef]
23. Lugo-Gavidia, L.M.; Burger, D.; Matthews, V.B.; Nolde, J.M.; Galindo Kiuchi, M.; Carnagarin, R.; Kannenkeril, D.; Chan, J.; Joyson, A.; Herat, L.Y.; et al. Role of microparticles in cardiovascular disease: Implications for endothelial dysfunction, thrombosis, and inflammation. *Hypertension* **2021**, *77*, 1825–1844. [CrossRef] [PubMed]
24. Steven, S.; Frenis, K.; Oelze, M.; Kalinovic, S.; Kuntic, M.; Jimenez, M.T.B.; Vujacic-Mirski, K.; Helmstädter, J.; Kröller-Schön, S.; Münzel, T.; et al. Vascular inflammation and oxidative stress: Major triggers for cardiovascular disease. *Oxidative Med. Cell. Longev.* **2019**, *2019*, 7092151. [CrossRef]
25. Engelen, S.E.; Robinson, A.J.B.; Zurke, Y.X.; Monaco, C. Therapeutic strategies targeting inflammation and immunity in atherosclerosis: How to proceed? *Nat. Rev. Cardiol.* **2022**, *19*, 522–542. [CrossRef]
26. Man, J.J.; Beckman, J.A.; Jaffe, I.Z. Sex as a biological variable in atherosclerosis. *Circ. Res.* **2020**, *126*, 1297–1319. [CrossRef]
27. Pucci, G.; Alcidi, R.; Tap, L.; Battista, F.; Mattace-Raso, F.; Schillaci, G. Sex- and gender-related prevalence, cardiovascular risk and therapeutic approach in metabolic syndrome: A review of the literature. *Pharmacol. Res.* **2017**, *120*, 34–42. [CrossRef]
28. Rathod, K.S.; Kapil, V.; Velmurugan, S.; Khambata, R.S.; Siddique, U.; Khan, S.; Van Eijl, S.; Gee, L.C.; Bansal, J.; Pitrola, K.; et al. Accelerated resolution of inflammation underlies sex differences in inflammatory responses in humans. *J. Clin. Investig.* **2017**, *127*, 169–182. [CrossRef]
29. Barrett-Connor, E. Sex differences in coronary heart disease why are women so superior? The 1995 Ancel keys lecture. *Circulation* **1997**, *95*, 252–264. [CrossRef]
30. Murphy, E. Estrogen signaling and cardiovascular disease. *Circ. Res.* **2011**, *109*, 687–696. [CrossRef]
31. Rivellese, A.A.; Riccardi, G.; Vaccaro, O. Cardiovascular risk in women with diabetes. *Nutr. Metab. Cardiovasc. Dis.* **2010**, *20*, 474–480. [CrossRef] [PubMed]
32. Ng, R.; Sutradhar, R.; Yao, Z.; Wodchis, W.P.; Rosella, L.C. Smoking, drinking, diet and physical activity-modifiable lifestyle risk factors and their associations with age to first chronic disease. *Int. J. Epidemiol.* **2020**, *49*, 113–130. [CrossRef] [PubMed]
33. Zarulli, V.; Barthold Jones, J.A.; Oksuzyan, A.; Lindahl-Jacobsen, R.; Christensen, K.; Vaupel, J.W. Women live longer than men even during severe famines and epidemics. *Proc. Natl. Acad. Sci. USA* **2018**, *115*, E832–E840. [CrossRef] [PubMed]
34. Christ, A.; Latz, E. The Western lifestyle has lasting effects on metaflammation. *Nat. Rev. Immunol.* **2019**, *19*, 267–268. [CrossRef]
35. Mazidi, M.; Katsiki, N.; Banach, M. A greater flavonoid intake is associated with lower total and cause-specific mortality: A meta-analysis of cohort studies. *Nutrients* **2020**, *12*, 2350. [CrossRef] [PubMed]
36. Parmenter, B.H.; Bondonno, C.P.; Murray, K.; Schousboe, J.T.; Croft, K.; Prince, R.L.; Hodgson, J.M.; Bondonno, N.P.; Lewis, J.R. Higher habitual dietary flavonoid intake associates with less extensive abdominal aortic calcification in a cohort of older women. *Arter. Thromb. Vasc. Biol.* **2022**, *42*, 1482–1494. [CrossRef] [PubMed]
37. Geleijnse, J.; Launer, L.; Hofman, A.; Pols, H.; Witteman, J. Tea flavonoids may protect against atherosclerosis. The Rotterdam study. *Arch. Intern. Med.* **1999**, *159*, 2170–2174. [CrossRef] [PubMed]
38. Miller, P.; Zhao, D.; Frazier-Wood, A.; Michos, E.; Averill, M.; Sandfort, V.; Burke, G.; Polak, J.; Lima, J.; Post, W.; et al. Associations between coffee, tea, and caffeine intake with coronary artery calcification and cardiovascular events. *Am. J. Med.* **2017**, *130*, 188–197. [CrossRef] [PubMed]
39. Reis, J.P.; Loria, C.M.; Steffen, L.M.; Zhou, X.; Van Horn, L.; Siscovick, D.S.; Jacobs, D.R.; Carr, J.J. Coffee, decaffeinated coffee, caffeine, and tea consumption in young adulthood and atherosclerosis later in life. The CARDIA study. *Arter. Thromb. Vasc. Biol.* **2010**, *30*, 2059–2066. [CrossRef]
40. WHO Global Database VigiBase. Available online: https://www.vigiaccess.org/ (accessed on 14 February 2023).
41. Sethi, N.; Kaura, S.; Dilbaghi, N.; Parle, M.; Pal, M. Garlic: A pungent wonder from nature. *Int. Res. J. Pharm.* **2014**, *5*, 523–529. [CrossRef]
42. Song, K.; Milner, J.A. Recent advances on the nutritional effects associated with the use of garlic as a supplement. Historical perspective on the use of garlic. *J. Nutr.* **2001**, *131*, 1054S–1057S. [CrossRef]
43. Slusarenko, A.J.; Patel, A.; Portz, D. Control of plant diseases by natural products: Allicin from garlic as a case study. *Eur. J. Plant Pathol.* **2008**, *121*, 313–322. [CrossRef]
44. European Medicines Agency (EMA)/Committee on Herbal Medicinal Products (HMPC). European Union herbal monograph *Allium sativum* L. *Eur. Med. Agency* **2017**, *31*, 1–7.
45. Weiner, L.; Shin, I.; Shimon, L.J.W.; Miron, T.; Wilchek, M.; Mirelman, D.; Frolow, F.; Rabinkov, A. Thiol-disulfide organization in alliin lyase (alliinase) from garlic (*Allium sativum*). *Protein Sci.* **2009**, *18*, 196–205. [CrossRef] [PubMed]
46. Beshbishy, A.; Wasef, L.; Elewa, Y.; Al-Sagan, A.; Abd El-Hack, M.; Taha, A.; Abd-Elhakim, Y. Chemical constituents and pharmacological activities of garlic (*Allium sativum* L.): A review. *Nutrients* **2020**, *12*, 872.
47. Rahman, M.S. Allicin and other functional active components in garlic: Health benefits and bioavailability. *Int. J. Food Prop.* **2007**, *10*, 245–268. [CrossRef]
48. Rana, S.V.; Pal, R.; Vaiphei, K.; Sharma, S.K.; Ola, R.P. Garlic in health and disease. *Nutr. Res. Rev.* **2011**, *24*, 60–71. [CrossRef]
49. Lamponi, S. Bioactive natural compounds with antiplatelet and anticoagulant activity and their potential role in the treatment of thrombotic disorders. *Life* **2021**, *11*, 1095. [CrossRef]

50. Martins, N.; Petropoulos, S.; Ferreira, I.C.F.R. Chemical composition and bioactive compounds of garlic (*Allium sativum* L.) as affected by pre- and post-harvest conditions: A review. *Food Chem.* **2016**, *211*, 41–50. [CrossRef]
51. Bayan, L.; Koulivand, P.H.; Gorji, A. Garlic: A review of potential therapeutic effects. *Avicenna J. Phytomed.* **2014**, *4*, 1.
52. Ansary, J.; Forbes-Hernández, T.Y.; Gil, E.; Cianciosi, D.; Zhang, J.; Elexpuru-Zabaleta, M.; Simal-Gandara, J.; Giampieri, F.; Battino, M. Potential health benefit of garlic based on human intervention studies: A brief overview. *Antioxidants* **2020**, *9*, 619. [CrossRef] [PubMed]
53. Chang, H.S.; Yamato, O.; Yamasaki, M.; Maede, Y. Modulatory influence of sodium 2-propenyl thiosulfate from garlic on cyclooxygenase activity in canine platelets: Possible mechanism for the anti-aggregatory effect. *Prostaglandins Leukot. Essent. Fat. Acids* **2005**, *72*, 351–355. [CrossRef] [PubMed]
54. Shiju, T.M.; Rajkumar, R.; Rajesh, N.G.; Viswanathan, P. Aqueous extract of *Allium sativum* L bulbs offer nephroprotection by attenuating vascular endothelial growth factor and extracellular signal-regulated kinase-1 expression in diabetic rats. *Indian J. Exp. Biol.* **2013**, *51*, 139–148. [PubMed]
55. Orozco-Ibarra, M.; Muñoz-Sánchez, J.; Zavala-Medina, M.E.; Pineda, B.; Magaña-Maldonado, R.; Vázquez-Contreras, E.; Maldonado, P.D.; Pedraza-Chaverri, J.; Chánez-Cárdenas, M.E. Aged garlic extract and S-allylcysteine prevent apoptotic cell death in a chemical hypoxia model. *Biol. Res.* **2016**, *49*, 7. [CrossRef] [PubMed]
56. Shin, I.S.; Hong, J.; Jeon, C.M.; Shin, N.R.; Kwon, O.K.; Kim, H.S.; Kim, J.C.; Oh, S.R.; Ahn, K.S. Diallyl-disulfide, an organosulfur compound of garlic, attenuates airway inflammation via activation of the Nrf-2/HO-1 pathway and NF-kappaB suppression. *Food Chem. Toxicol.* **2013**, *62*, 506–513. [CrossRef] [PubMed]
57. Quintero-Fabián, S.; Ortuño-Sahagún, D.; Vázquez-Carrera, M.; López-Roa, R.I. Alliin, a garlic (*Allium sativum*) compound, prevents LPS-induced inflammation in 3T3-L1 adipocytes. *Mediat. Inflamm.* **2013**, *2013*, 381815. [CrossRef]
58. Szulińska, M.; Kręgielska-Narożna, M.; Świątek, J.; Styś, P.; Kuźnar-Kamińska, B.; Jakubowski, H.; Walkowiak, J.; Bogdański, P. Garlic extract favorably modifies markers of endothelial function in obese patients—Randomized double blind placebo-controlled nutritional intervention. *Biomed. Pharmacother.* **2018**, *102*, 792–797. [CrossRef]
59. Atkin, M.; Laight, D.; Cummings, M.H. The effects of garlic extract upon endothelial function, vascular inflammation, oxidative stress and insulin resistance in adults with type 2 diabetes at high cardiovascular risk. A pilot double blind randomized placebo controlled trial. *J. Diabetes Its Complicat.* **2016**, *30*, 723–727. [CrossRef]
60. Ziaei, S.; Hantoshzadeh, S.; Rezasoltani, P.; Lamyian, M. The effect of garlic tablet on plasma lipids and platelet aggregation in nulliparous pregnants at high risk of preeclampsia. *Eur. J. Obstet. Gynecol. Reprod. Biol.* **2001**, *99*, 201–206. [CrossRef]
61. Aalami-Harandi, R.; Karamali, M.; Asemi, Z. The favorable effects of garlic intake on metabolic profiles, hs-CRP, biomarkers of oxidative stress and pregnancy outcomes in pregnant women at risk for pre-eclampsia: Randomized, double-blind, placebo-controlled trial. *J. Matern.-Fetal Neonatal Med.* **2015**, *28*, 2020–2027. [CrossRef]
62. Afarid, M.; Sadeghi, E.; Johari, M.; Namvar, E.; Sanie-Jahromi, F. Evaluation of the effect of garlic tablet as a complementary treatment for patients with diabetic retinopathy. *J. Diabetes Res.* **2022**, *2022*, 6620661. [CrossRef] [PubMed]
63. Klafke, J.Z.; da Silva, M.A.; Panigas, T.F.; Belli, K.C.; de Oliveira, M.F.; Barichello, M.M.; Rigo, F.K.; Rossato, M.F.; dos Santos, A.R.S.; Pizzolatti, M.G.; et al. Effects of *Campomanesia xanthocarpa* on biochemical, hematological and oxidative stress parameters in hypercholesterolemic patients. *J. Ethnopharmacol.* **2010**, *127*, 299–305. [CrossRef] [PubMed]
64. Viecili, P.R.N.; Borges, D.O.; Kirsten, K.; Malheiros, J.; Viecili, E.; Melo, R.D.; Trevisan, G.; da Silva, M.A.; Bochi, G.V.; Moresco, R.N.; et al. Effects of *Campomanesia xanthocarpa* on inflammatory processes, oxidative stress, endothelial dysfunction and lipid biomarkers in hypercholesterolemic individuals. *Atherosclerosis* **2014**, *234*, 85–92. [CrossRef] [PubMed]
65. Otero, J.S.; Hirsch, G.E.; Klafke, J.Z.; Porto, F.G.; de Almeida, A.S.; Nascimento, S.; Schmidt, A.; da Silva, B.; Pereira, R.L.D.; Jaskulski, M.; et al. Inhibitory effect of *Campomanesia xanthocarpa* in platelet aggregation: Comparison and synergism with acetylsalicylic acid. *Thromb. Res.* **2017**, *154*, 42–49. [CrossRef]
66. Pingali, U.; Sukumaran, D.; Nutalapati, C. Effect of an aqueous extract of *Terminalia chebula* on endothelial dysfunction, systemic inflammation, and lipid profile in type 2 diabetes mellitus: A randomized double-blind, placebo-controlled clinical study. *Phytother. Res.* **2020**, *34*, 3226–3235. [CrossRef]
67. Lopez, H.L.; Habowski, S.M.; Sandrock, J.E.; Raub, B.; Kedia, A.; Bruno, E.J.; Ziegenfuss, T.N. Effects of dietary supplementation with a standardized aqueous extract of *Terminalia chebula* fruit (AyuFlex®) on joint mobility, comfort, and functional capacity in healthy overweight subjects: A randomized placebo-controlled clinical trial. *BMC Complement. Altern. Med.* **2017**, *17*, 475. [CrossRef]
68. Kishore, K.K.; Kishan, P.V.; Ramakanth, G.S.H.; Chandrasekhar, N.; Pinhali, U. A Study of *Terminalia chebula* extract on endothelial dysfunction and biomarkers of oxidative stress in patients with metabolic syndrome. *Eur. J. Biomed. Pharm. Sci.* **2016**, *3*, 181–188.
69. Raimondo, S.; Nikolic, D.; Conigliaro, A.; Giavaresi, G.; Lo Sasso, B.; Giglio, R.V.; Chianetta, R.; Manno, M.; Raccosta, S.; Corleone, V.; et al. Preliminary results of citraves™ effects on low density lipoprotein cholesterol and waist circumference in healthy subjects after 12 weeks: A pilot open-label study. *Metabolites* **2021**, *11*, 276. [CrossRef]
70. Baik, J.S.; Min, J.H.; Ju, S.M.; Ahn, J.H.; Ko, S.H.; Chon, H.S.; Kim, M.S.; Shin, Y. Il Effects of fermented garlic extract containing nitric oxide metabolites on blood flow in healthy participants: A randomized controlled trial. *Nutrients* **2022**, *14*, 5238. [CrossRef]
71. Osadnik, T.; Goławski, M.; Lewandowski, P.; Morze, J.; Osadnik, K.; Pawlas, N.; Lejawa, M.; Jakubiak, G.K.; Mazur, A.; Schwingschackl, L.; et al. A network meta-analysis on the comparative effect of nutraceuticals on lipid profile in adults. *Pharmacol. Res.* **2022**, *183*, 106402. [CrossRef]

72. Gyawali, D.; Vohra, R.; Orme-Johnson, D.; Ramaratnam, S.; Schneider, R.H. A systematic review and meta-analysis of Ayurvedic herbal preparations for hypercholesterolemia. *Medicina* **2021**, *57*, 546. [CrossRef] [PubMed]
73. Fehri, B.; Aiache, J.M.; Korbi, S.; Monkni, M.; Ben Said, M.; Memmi, A.; Hizaoui, B.; Boukef, K. Toxic effects induced by the repeat administration of *Allium sativum* L. *J. Pharm. Belg.* **1991**, *46*, 363–374.
74. Mulrow, C.; Lawrence, V.; Ackermann, R.; Ramirez, G.; Morbidoni, L.; Aguilar, C.; Arterburn, J.; Block, E.; Chiquette, E.; Gardener, C.; et al. Garlic: Effects on Cardiovascular Risks and Disease, Protective Effects against Cancer, and Clinical Adverse Effects: Summary. Available online: https://www.ncbi.nlm.nih.gov/books/NBK11910/ (accessed on 16 January 2023).
75. Tattelman, E. Health effects of garlic. *Am. Fam. Physician* **2005**, *72*, 103–106. [PubMed]
76. Borrelli, F.; Capasso, R.; Izzo, A.A. Garlic (*Allium sativum* L.): Adverse effects and drug interactions in humans. *Mol. Nutr. Food Res.* **2007**, *51*, 1386–1397. [CrossRef] [PubMed]
77. Arreola, R.; Quintero-Fabián, S.; Lopez-Roa, R.I.; Flores-Gutierrez, E.O.; Reyes-Grajeda, J.P.; Carrera-Quintanar, L.; Ortuno-Sahagun, D. Immunomodulation and anti-inflammatory effects of garlic compounds. *J. Immunol. Res.* **2015**, *2015*, 401630. [CrossRef]
78. Lorenzi, H. Arvores Brasileiras. In *Manual de Identificação e Cultivo de Plantas Arbóreas Nativas do Brasil*; Editora Plantarum Ltda: São Paulo, Brazil, 1992; p. 256.
79. Bunchen, S. Conhecimento etnobotânico sobre as plantas medicinais utilizadas pela comunidade do Bairro Cidade Alta, município de Videira, Santa Catarina, Brasil. *Unoesc Ciência–ACBS Joaçaba* **2011**, *2*, 129–140.
80. de Oliveira Raphaelli, C.; dos Santos Pereira, E.; Camargo, T.M.; Ribeiro, J.A.; Pereira, M.C.; Vinholes, J.; Dalmazo, G.O.; Vizzotto, M.; Nora, L. Biological activity and chemical composition of fruits, seeds and leaves of guabirobeira (*Campomanesia xanthocarpa* O. Berg–Myrtaceae): A review. *Food Biosci.* **2021**, *40*, 100899. [CrossRef]
81. Markman, B.E.O.; Bacchi, E.M.; Kato, E.T.M. Antiulcerogenic effects of *Campomanesia xanthocarpa*. *J. Ethnopharmacol.* **2004**, *94*, 55–57. [CrossRef]
82. De Sousa, J.A.; da Prado, L.S.; Alderete, B.L.; Boaretto, F.B.M.; Allgayer, M.C.; Miguel, F.M.; De Sousa, J.T.; Marroni, N.P.; Lemes, M.L.B.; Corrêa, D.S.; et al. Toxicological aspects of *Campomanesia xanthocarpa* Berg. associated with its phytochemical profile. *J. Toxicol. Environ. Health—Part A Curr. Issues* **2019**, *82*, 62–74. [CrossRef]
83. Sant'anna, L.S.; Merlugo, L.; Ehle, C.S.; Limberger, J.; Fernandes, M.B.; Santos, M.C.; Mendez, A.S.L.; Paula, F.R.; Moreira, C.M. Chemical composition and hypotensive effect of *Campomanesia xanthocarpa*. *Evid.-Based Complement. Altern. Med.* **2017**, *2017*, 1591762. [CrossRef] [PubMed]
84. de Morais, I.B.M.; Silva, D.B.; Carollo, C.A.; Ferreira-Neto, M.L.; Fidelis-de-Oliveira, P.; Bispo-da-Silva, L.B. Hypotensive activity of *Campomanesia xanthocarpa* leaf extract: Beyond angiotensin II type 1 receptor blockage. *Nat. Prod. Res.* **2021**, *35*, 4798–4802. [CrossRef] [PubMed]
85. Klafke, J.Z.; Pereira, R.L.D.; Hirsch, G.E.; Parisi, M.M.; Porto, F.G.; de Almeida, A.S.; Rubin, F.H.; Schmidt, A.; Beutler, H.; Nascimento, S.; et al. Study of oxidative and inflammatory parameters in LDLr-KO mice treated with a hypercholesterolemic diet: Comparison between the use of *Campomanesia xanthocarpa* and acetylsalicylic acid. *Phytomedicine* **2016**, *23*, 1227–1234. [CrossRef] [PubMed]
86. Cadena-Iñiguez, J.; Arévalo-Galarza, L.; Avendaño-Arrazate, C.H.; Soto-Hernández, M.; del Ruiz-Posadas, L.M.; Santiago-Osorio, E.; Acosta-Ramos, M.; Cisneros-Solano, V.M.; Aguirre-Medina, J.F.; Ochoa-Martínez, D. Production, genetics, postharvest management and pharmacological characteristics of *Sechium edule* (Jacq.) Sw. *Fresh Prod.* **2007**, *1*, 41–53.
87. Booth, S.; Bressani, R.; Johns, T. Nutrient content of selected indigenous leafy vegetables consumed by the Kekchi people of Alta Verapaz, Guatemala. *J. Food Compos. Anal.* **1992**, *5*, 25–34. [CrossRef]
88. Cook, O.F. *The Chayote: A Tropical Vegetable*; US Department of Agriculture, Division of Botany: Washington, DC, USA, 1901; Volume 18, pp. 1–31.
89. Ibarra-Alvarado, C.; Rojas, A.; Mendoza, S.; Bah, M.; Gutiérrez, D.M.; Hernández-Sandoval, L.; Martínez, M. Vasoactive and antioxidant activities of plants used in Mexican traditional medicine for the treatment of cardiovascular diseases. *Pharm. Biol.* **2010**, *48*, 732–739. [CrossRef]
90. Nunes, M.G.S.; Bernardino, A.; Martins, R.D. Use of medicinal plants by people with hypertension. *Rev. Rede Enferm. Nordeste* **2015**, *16*, 775. [CrossRef]
91. Lombardo-Earl, G.; Roman-Ramos, R.; Zamilpa, A.; Herrera-Ruiz, M.; Rosas-Salgado, G.; Tortoriello, J.; Jiménez-Ferrer, E. Extracts and fractions from edible roots of *Sechium edule* (Jacq.) Sw. with antihypertensive activity. *Evid.-Based Complement. Altern. Med.* **2014**, *2014*, 594326. [CrossRef]
92. Siciliano, T.; De Tommasi, N.; Morelli, I.; Braca, A. Study of flavonoids of *Sechium edule* (Jacq) Swartz (Cucurbitaceae) different edible organs by liquid chromatography photodiode array mass spectrometry. *J. Agric. Food Chem.* **2004**, *52*, 6510–6515. [CrossRef]
93. de A. Ribeiro, R.; de Barros, F.; de Melo, M.M.R.F.; Muniz, C.; Chieia, S.; das Graças Wanderley, M.; Gomes, C.; Trolin, G. Acute diuretic effects in conscious rats produced by some medicinal plants used in the state of São Paulo, Brasil. *J. Ethnopharmacol.* **1988**, *24*, 19–29. [CrossRef]
94. Ordoñez, A.A.L.; Gomez, J.D.; Cudmani, N.M.; Vattuone, M.A.; Isla, M.I. Antimicrobial activity of nine extracts of *Sechium edule* (Jacq.) Swartz. *Microb. Ecol. Health Dis.* **2003**, *15*, 33–39. [CrossRef]
95. Ordoñez, A.A.L.; Gomez, J.D.; Vattuone, M.A.; Isla, M.I. Antioxidant activities of *Sechium edule* (Jacq.) Swartz extracts. *Food Chem.* **2006**, *97*, 452–458. [CrossRef]

96. Gordon, E.A.; Guppy, L.J.; Nelson, M. The antihypertensive effects of the Jamaican Cho-Cho (*Sechium edule*). *West Indian Med. J.* **2000**, *49*, 27–31. [PubMed]
97. Trejo-Moreno, C.; Castro-Martínez, G.; Méndez-Martínez, M.; Jiménez-Ferrer, J.E.; Pedraza-Chaverri, J.; Arrellín, G.; Zamilpa, A.; Medina-Campos, O.N.; Lombardo-Earl, G.; Barrita-Cruz, G.J.; et al. Acetone fraction from *Sechium edule* (Jacq.) S.w. edible roots exhibits anti-endothelial dysfunction activity. *J. Ethnopharmacol.* **2018**, *220*, 75–86. [CrossRef] [PubMed]
98. Vieira, E.F.; Pinho, O.; Ferreira, I.M.P.L.V.O.; Delerue-Matos, C. Chayote (*Sechium edule*): A review of nutritional composition, bioactivities and potential applications. *Food Chem.* **2019**, *275*, 557–568. [CrossRef]
99. Ragasa, C.Y.; Biona, K.; Shen, C.C. Chemical constituents of *Sechium edule* (Jacq.) Swartz. *Der Pharma Chem.* **2014**, *6*, 251–255.
100. Cadena-Iñiguez, J.; de la Luz Riviello-Flores, M.; Marcos Soto-Hernández, R.; del Mar Ruiz-Posadas, L.; Gómez-Merino, F.C.; Aguiñiga Sanchez, I.; Arévalo-Galarza, L. Functionally active metabolites in two cultivars of chayote (*Sechium edule* (Jacq.) Swartz). *Acta Hortic.* **2019**, *1256*, 231–237. [CrossRef]
101. Huerta-Reyes, M.; Tavera-Hernández, R.; Alvarado-Sansininea, J.J.; Jiménez-Estrada, M. Selected species of the Cucurbitaceae family used in Mexico for the treatment of diabetes mellitus. *Molecules* **2022**, *27*, 3440. [CrossRef]
102. Firdous Mumtaz, S.M.; Paul, S.; Bag, A.K. Effect of *Sechium edule* on chemical induced kidney damage in experimental animals. *Bangladesh J. Pharmacol.* **2013**, *8*, 28–35. [CrossRef]
103. Trejo-Moreno, C.; Castro-Martínez, G.; Méndez-Martínez, M.; Jiménez-Ferrer, J.E.; Pedraza-Chaverri, J.; Arrellín, G.; Zamilpa-Álvarez, A.; Medina-Campos, O.N.; Lombardo-Earl, G.; Barrita-Cruz, G.J.; et al. Data of the effects of acetone fraction from *Sechium edule* (Jacq.) S.w. edible roots in the kidney of endothelial dysfunction induced mice. *Data Brief* **2018**, *18*, 448–453. [CrossRef]
104. Castañeda, R.; Cáceres, A.; Cruz, S.M.; Aceituno, J.A.; Marroquín, E.S.; Barrios Sosa, A.C.; Strangman, W.K.; Williamson, R.T. Nephroprotective plant species used in traditional Mayan Medicine for renal-associated diseases. *J. Ethnopharmacol.* **2023**, *301*, 115755. [CrossRef]
105. Yang, M.Y.; Chan, K.C.; Lee, Y.J.; Chang, X.Z.; Wu, C.H.; Wang, C.J. *Sechium edule* shoot extracts and active components improve obesity and a fatty liver that involved reducing hepatic lipogenesis and adipogenesis in high-fat-diet-fed rats. *J. Agric. Food Chem.* **2015**, *63*, 4587–4596. [CrossRef]
106. Marotta, F.; Safran, P.; Tajiri, H.; Princess, G.; Anzulovic, H.; Idéo, G.M.; Rouge, A.; Seal, M.G.; Idéo, G.M. Improvement of hemorheological abnormalities in alcoholics by an oral antioxidant. *Hepato-Gastroenterol.* **2001**, *48*, 511–517.
107. Bulbul, M.R.H.; Chowdhury, M.N.U.; Naima, T.A.; Sami, S.A.; Imtiaj, M.S.; Huda, N.; Uddin, M.G. A comprehensive review on the diverse pharmacological perspectives of *Terminalia chebula* Retz. *Heliyon* **2022**, *8*, e10220. [CrossRef] [PubMed]
108. Bag, A.; Bhattacharyya, S.K.; Chattopadhyay, R.R. The development of *Terminalia chebula* Retz. (Combretaceae) in clinical research. *Asian Pac. J. Trop. Biomed.* **2013**, *3*, 244–252. [CrossRef] [PubMed]
109. Suchalatha, S.; Srinivasulu, C.; Devi, S. Antioxidant activity of ethanolic extract of *Terminalia chebula* fruit against isoproterenol-induced oxidative stress in rats. *Indian J. Biochem. Biophys.* **2005**, *42*, 246–249. [PubMed]
110. Sabu, M.C.; Kuttan, R. Anti-diabetic activity of medicinal plants and its relationship with their antioxidant property. *J. Ethnopharmacol.* **2002**, *81*, 155–160. [CrossRef]
111. Lee, H.S.; Nam, H.W.; Kyoung, H.K.; Lee, H.; Jun, W.; Lee, K.W. Antioxidant effects of aqueous extract of *Terminalia chebula* in vivo and in vitro. *Biol. Pharm. Bull.* **2005**, *28*, 1639–1644. [CrossRef] [PubMed]
112. Senthilkumar, G.P.; Subramanian, S.P. Biochemical studies on the effect of *Terminalia chebula* on the levels of glycoproteins in streptozotocin-induced experimental diabetes in rats. *J. Appl. Biomed.* **2008**, *6*, 105–115. [CrossRef]
113. Suchalatha, S.; Devi, C.S.S. Protective effect of *Terminalia chebula* against experimental myocardial injury induced by isoproterenol. *Indian J. Exp. Biol.* **2004**, *42*, 174–178.
114. Banazadeh, M.; Mehrabani, M.; Banazadeh, N.; Dabaghzadeh, F.; Shahabi, F. Evaluating the effect of black myrobalan on cognitive, positive, and negative symptoms in patients with chronic schizophrenia: A randomized, double-blind, placebo-controlled trial. *Phytother. Res.* **2022**, *36*, 543–550. [CrossRef]
115. Yamada, Y.; Benichou, G.; Cosimi, A.B.; Kawai, T.; Cosimi, B.A.; Kawai, T. Apoplastic exosome-like vesicles? A new way of protein secretion in plants. *Plant Signal. Behav.* **2012**, *7*, 544–546. [CrossRef]
116. An, Q.; Hückelhoven, R.; Kogel, K.H.; van Bel, A.J.E. Multivesicular bodies participate in a cell wall-associated defence response in barley leaves attacked by the pathogenic powdery mildew fungus. *Cell. Microbiol.* **2006**, *8*, 1009–1019. [CrossRef] [PubMed]
117. Karamanidou, T.; Tsouknidas, A. Plant-derived extracellular vesicles as therapeutic nanocarriers. *Int. J. Mol. Sci.* **2022**, *23*, 191. [CrossRef] [PubMed]
118. Zuzarte, M.; Vitorino, C.; Salgueiro, L.; Girão, H. Plant nanovesicles for essential oil delivery. *Pharmaceutics* **2022**, *14*, 2581. [CrossRef]
119. Kim, S.Q.; Kim, K.H. Emergence of edible plant-derived nanovesicles as functional food components and nanocarriers for therapeutics delivery: Potentials in human health and disease. *Cells* **2022**, *11*, 2232. [CrossRef]
120. Ju, S.; Mu, J.; Dokland, T.; Zhuang, X.; Wang, Q.; Jiang, H.; Xiang, X.; Deng, Z.B.; Wang, B.; Zhang, L.; et al. Grape exosome-like nanoparticles induce intestinal stem cells and protect mice from DSS-induced colitis. *Mol. Ther.* **2013**, *21*, 1345–1357. [CrossRef]
121. Mu, J.; Zhuang, X.; Wang, Q.; Jiang, H.; Deng, Z.B.; Wang, B.; Zhang, L.; Kakar, S.; Jun, Y.; Miller, D.; et al. Interspecies communication between plant and mouse gut host cells through edible plant derived exosome-like nanoparticles. *Mol. Nutr. Food Res.* **2014**, *58*, 1561–1573. [CrossRef]

122. Nemati, M.; Singh, B.; Mir, R.A.; Nemati, M.; Babaei, A.; Ahmadi, M.; Rasmi, Y.; Golezani, A.G.; Rezaie, J. Plant-derived extracellular vesicles: A novel nanomedicine approach with advantages and challenges. *Cell Commun. Signal.* **2022**, *20*, 69. [CrossRef]
123. Wang, Y.; Wang, J.; Ma, J.; Zhou, Y.; Lu, R. Focusing on future applications and current challenges of plant derived extracellular vesicles. *Pharmaceuticals* **2022**, *15*, 708. [CrossRef]
124. Baldini, N.; Torreggiani, E.; Roncuzzi, L.; Perut, F.; Zini, N.; Avnet, S. Exosome-like nanovesicles isolated from *Citrus limon* L. exert antioxidative effect. *Curr. Pharm. Biotechnol.* **2018**, *19*, 877–885. [CrossRef]
125. Perut, F.; Roncuzzi, L.; Avnet, S.; Massa, A.; Zini, N.; Sabbadini, S.; Giampieri, F.; Mezzetti, B.; Baldini, N. Strawberry-derived exosome-like nanoparticles prevent oxidative stress in human mesenchymal stromal cells. *Biomolecules* **2021**, *11*, 87. [CrossRef]
126. Sim, Y.; Seo, H.-J.; Kim, D.-H.; Lee, S.-H.; Kwon, J.; Kwun, I.-S.; Jung, C.; Kim, J.-I.; Lim, J.-H.; Kim, D.-K.; et al. The effect of apple-derived nanovesicles on the osteoblastogenesis of osteoblastic MC3T3-E1 cells. *J. Med. Food* **2023**, *26*, 49–58. [CrossRef] [PubMed]
127. Deng, Z.; Rong, Y.; Teng, Y.; Mu, J.; Zhuang, X.; Tseng, M.; Samykutty, A.; Zhang, L.; Yan, J.; Miller, D.; et al. Broccoli-derived nanoparticle inhibits mouse colitis by activating dendritic cell AMP-activated protein kinase. *Mol. Ther.* **2017**, *25*, 1641–1654. [CrossRef] [PubMed]
128. Zhuang, X.; Deng, Z.B.; Mu, J.; Zhang, L.; Yan, J.; Miller, D.; Feng, W.; McClain, C.J.; Zhang, H.G. Ginger-derived nanoparticles protect against alcohol-induced liver damage. *J. Extracell. Vesicles* **2015**, *4*, 28713. [CrossRef] [PubMed]
129. Kim, J.; Li, S.; Zhang, S.; Wang, J. Plant-derived exosome-like nanoparticles and their therapeutic activities. *Asian J. Pharm. Sci.* **2022**, *17*, 53–69. [CrossRef] [PubMed]

Disclaimer/Publisher's Note: The statements, opinions and data contained in all publications are solely those of the individual author(s) and contributor(s) and not of MDPI and/or the editor(s). MDPI and/or the editor(s) disclaim responsibility for any injury to people or property resulting from any ideas, methods, instructions or products referred to in the content.

life

MDPI

Review

Neuroprotective Potential of Biflavone Ginkgetin: A Review

İ. İrem Tatlı Çankaya [1], Hari Prasad Devkota [2], Gokhan Zengin [3] and Dunja Šamec [4,*]

1 Department of Pharmaceutical Botany, Faculty of Pharmacy, Hacettepe University, 06100 Ankara, Turkey
2 Graduate School of Pharmaceutical Sciences, Kumamoto University, Chuo-ku, Kumamoto 862-0973, Japan
3 Department of Biology, Science Faculty, Selcuk University, 42130 Konya, Turkey
4 Department of Food Technology, University Center Koprivnica, University North, 48000 Koprivnica, Croatia
* Correspondence: dsamec@unin.hr

Abstract: Neurological disorders are becoming more common, and there is an intense search for molecules that can help treat them. Several natural components, especially those from the flavonoid group, have shown promising results. Ginkgetin is the first known biflavonoid, a flavonoid dimer isolated from ginkgo (*Ginkgo biloba* L.). Later, its occurrence was discovered in more than 20 different plant species, most of which are known for their use in traditional medicine. Herein we have summarized the data on the neuroprotective potential of ginkgetin. There is evidence of protection against neuronal damage caused by ischemic strokes, neurotumors, Alzheimer's disease (AD), and Parkinson's disease (PD). Beneficial effects in ischemic strokes have been demonstrated in animal studies in which injection of ginkgetin before or after onset of the stoke showed protection from neuronal damage. AD protection has been the most studied to date. Possible mechanisms include inhibition of reactive oxygen species, inhibition of β-secretase, inhibition of Aβ fibril formation, amelioration of inflammation, and antimicrobial activity. Ginkgetin has also shown positive effects on the relief of PD symptoms in animal studies. Most of the available data are from in vitro or in vivo animal studies, where ginkgetin showed promising results, and further clinical studies should be conducted.

Keywords: Alzheimer's disease; biflavonoids; ginkgetin; neuroprotection; ginkgo

Citation: Tatlı Çankaya, İ.İ.; Devkota, H.P.; Zengin, G.; Šamec, D. Neuroprotective Potential of Biflavone Ginkgetin: A Review. *Life* **2023**, *13*, 562. https://doi.org/10.3390/life13020562

Academic Editors: Salvador González and Azahara Rodríguez-Luna

Received: 19 December 2022
Revised: 9 February 2023
Accepted: 11 February 2023
Published: 16 February 2023

Copyright: © 2023 by the authors. Licensee MDPI, Basel, Switzerland. This article is an open access article distributed under the terms and conditions of the Creative Commons Attribution (CC BY) license (https://creativecommons.org/licenses/by/4.0/).

1. Introduction

Since 1840, human life expectancy has increased at a rate of nearly 2.5 years per decade, and this trend has continued to this day [1]. According to the World Health Organization [2], by 2030, 1 in 6 people in the world will be 60 or older. At that time, the proportion of the population aged 60 and over is estimated to increase from 1 billion in 2020 to 1.4 billion. By 2050, the global population aged 60 and older will double (2.1 billion). Similarly, the number of people aged 80 or older is estimated to triple between 2020 and 2050, reaching 426 million. On a biological level, aging is a complex process in which a variety of molecules and cellular damage accumulate, leading over time to a gradual decline in physical performance and cognitive functions, as well as an increased risk of disease. Older people are more susceptible to various chronic diseases, especially diseases of the central nervous system, such as strokes, epilepsy, Parkinson's disease (PD), Alzheimer's disease (AD), neuropathy, and other dementias [3]. Neurological disorders are disorders of the nervous system and can affect the activity and physiology of the brain, spinal cord, and nerves. They occur in 5% to 55% of people who are aged 55 and older and are associated with a high risk of adverse health effects, including mortality, disability, and hospitalization [3]. Therefore, scientists have made considerable efforts to understand the pathophysiology of these disorders and to develop effective prevention methods and therapies. However, the reported compounds/active ingredients, which are mostly synthetic, are not considered to be very reliable and therapeutically effective due to their complexity and off-target problems [4]. On the other hand, several natural products may prove to be viable preventive therapeutics to fill the large gap in the treatment of neurological disorders [4].

One of the groups of natural products associated with neuroprotective properties is the flavonoids [5–7], a large and diverse group of specialized plant metabolites characterized by a 15-carbon flavone backbone (C6-C3-C6) with two benzene rings (A and B) linked by a trinuclear pyran ring (C) [8]. Flavonoids can be mainly divided into six groups: flavones, flavonols, flavan-3-ols, flavanonols, flavanones, isoflavones, and anthocyanins. They can be present in free form in plants, but are more often glycosylated, methylated, acetylated, prenylated, or polymerized [9]. The pattern of conjugation, glycosylation, or methylation is responsible for the different chemical and biological properties of these compounds [10]. Flavonoid dimers known as biflavonoids are formed by two linked flavonoid monomers and consist of flavone-flavone, flavane-flavane, flavane-flavone subunits, and in rare cases, dimers of chalcones and isoflavones. Today, nearly 600 different biflavonoids are known to occur in ferns, bryophytes, angiosperms and gymnosperms [11,12]. They are often found in plants used in traditional medicine and are considered responsible factors in the health benefits of these plants [13]. Biflavonoids possess diverse biological activities including therapeutic potential against neurodegenerative diseases [14].

The first biflavonoid isolated was ginkgetin from the yellow leaves of ginkgo (*Ginkgo biloba* L.) (Figure 1a). Chemically, it is a 7,4′-dimethyl ether derivative of the 3′,8″-dimer of the apigenin, known as amentoflavone. Thus, ginkgetin consists of apigenin and apigenin 7,4′-dimethyl ether.

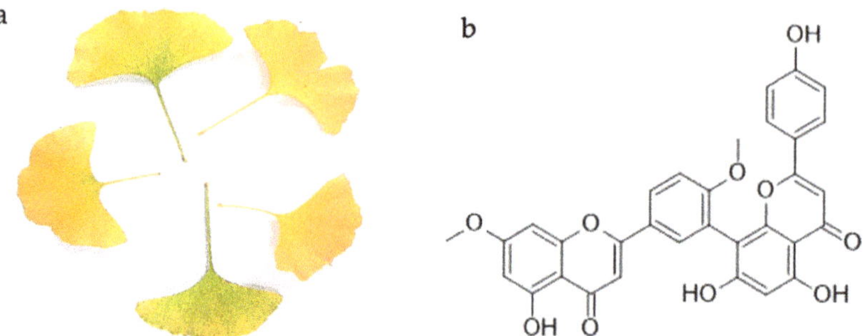

Figure 1. Yellow leaves of *Ginkgo biloba* L. (**a**) and the chemical structure of ginkgetin (**b**).

Ginkgetin is a compound found in ginkgo whose standardized extract (EGb 761) has been used for many years as a supportive therapy and to prevent cognitive impairment [15]. Ginkgo extract can slow the progression of memory loss in AD, usually at a high dose of 240 mg or more per day [16,17], and may have supportive and/or protective effects in the treatment of PD [18]. It is not entirely clear which molecules from the extracts contribute to this activity. Recently, natural products, especially polyphenols, have been intensively studied as potential neuroprotective molecules. One of these molecules is ginkgetin, but as far as we know, there is no review paper summarizing the research to date. Therefore, the review aimed to summarize the data on the potential of ginkgetin in the treatment of neurodegenerative diseases in order to highlight the neuroprotective properties of ginkgetin.

2. Ginkgetin

Ginkgetin (Figure 1b) is a flavonoid dimer, a 7,4′-dimethyl ether derivative of the apigenin dimer amentoflavone. It is the first isolated biflavonoid obtained in the form of a yellow powder from the leaves of ginkgo (Figure 1a) and the first biflavonoid whose structure was described. To date, its occurrence has been confirmed in other ginkgo plant parts [13,19], as well as in more than 20 other plant species [20]. The list of plant species in which the presence of ginkgetin was detected is given in Table 1. It should be noted that the presence of ginkgetin in mosses and liverworts has not been yet reported, and in ferns and fern allies it has been reported only in *Sellaginela* sp. It is commonly found in the conifers, cycads and allies group, and in flowering plants. So far, it has not been found in plants commonly used as food, but rather in plants used in traditional medicine.

Table 1. List of plant species with associated division in which the presence of ginkgetin has been reported.

Division	Species
Thallophyta Unicellular to large algae, fungi, lichens	data not available
Bryophyta Mosses and liverwords	data not available
Pteridophyta Ferns and fern allies	*Selaginella doederleinii* [21] *Selaginella moellendorffii* [22] *Selaginella sinensis* [23]
Gymnosperms Conifers, cycads and allies	*Araucaria angustifolia* [24] *Cephalotaxus drupacea* [25] *Cephalotaxus fortunei* var. *alpina* [26] *Cephalotaxus harringtonia* var. *harringtonia* [27] *Cephalotaxus koreana* [28] *Cephalotaxus sinensis* [29] *Cycas media* [30] *Chamaecyparis obtusa* [31] *Ginkgo biloba* [32–34] *Metasequoia glyptostroboides* [35] *Taxus baccata* [35] *Taxus chinesis* [36] *Taxus cuspidata* [37] *Taxus mairei* [38] *Taxus media* [38] *Torreya nucifera* [39]
Angiosperms Flowering plants	*Capparis spinosa* [40] *Celaenodendron mexicanum* [41] *Cyperus rotundus* [42] *Elateriospermum tapos* [43] *Gaultheria yunnanensis* [44] *Houttuynia cordata* [45]

Most of the plants listed in Table 1 have been used in the traditional medicine systems of various cultures, suggesting that ginkgetin also may have biological activity. As the first known biflavonoid, its biological activity has been studied over the last 30 years. Research shows its potential in treating various inflammation-related diseases such as cancer, cardiovascular disease, inflammation caused by viruses and bacteria, and neurodegenerative disorders [20] (Figure 2).

Most commonly, its anticancer activity has been studied. Recently, Adnan et al. [20] summarized that ginkgetin combats cancer progression by various mechanisms such as arresting the cell cycle, inducing apoptosis, stimulating autophagy, and targeting many deregulated signaling pathways such as JAK / STAT and MAPKs in the colon, lung, prostate, osteosarcoma, breast, leukemia, cervical, medulloblastoma, ovarian, neck, and kidney cell lines. In animal studies, ginkgetin inhibited tumor growth in xenotransplanted nude mice, down-regulated p-STAT3Tyr705 and survivin in tumor tissues [46] and decreased tumor size and weight without apparent toxicity [47]. Ginkgetin may also enhance the therapeutic effects of cisplatin [48] and 5-fluorouracil [49].

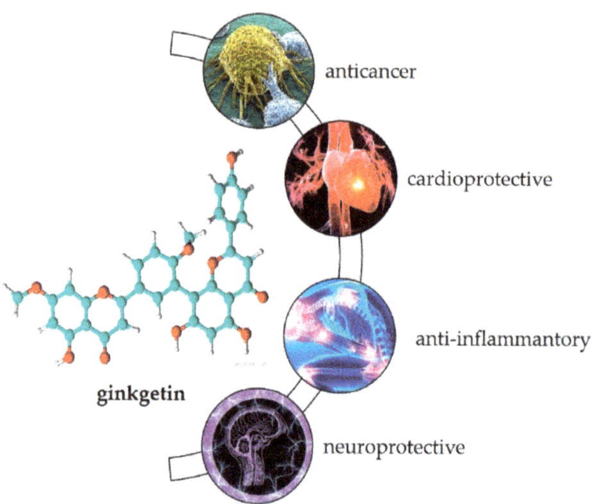

Figure 2. Biological activity of ginkgetin.

It may also be useful for the treatment of cardiovascular disease. Cell-based studies showed its potential as an inhibitor of TRPV4-mediated proatherogenic processes in macrophages [50]. In addition, ginkgetin showed inhibitory effects on human thrombin, an important serine protease that regulates the blood coagulation cascade and processes of thrombosis [51], and pancreatic lipase, an important target that regulates lipid uptake [52]. Animal studies showed its beneficial effects in preventing adipogenesis [53], local vascular damage associated with atherosclerosis [54], and ischemic reperfusion injury [55].

Due to its anti-inflammatory effects, in vitro studies have shown that ginkgetin can be used in the treatment of inflammation-related diseases such as airway inflammation [56] and diabetic nephropathy [57]. It may also be useful as an antiviral [32], antibacterial [30] and antiparasitic [58] agent and has gained attention in recent years as a target for the treatment of SARS-CoV-2 infection [59–61].

In this review, we address in more detail the potential role of gingketin in the treatment of neurodegenerative diseases.

3. Neurodegenerative Diseases

Neurodegenerative disease is a general term for several diseases mainly characterized by a progressive loss of structure or function of neurons that worsens over time [62]. They can be genetic or caused by a tumor, stroke, toxins, viruses, etc., and affect millions of people around the world. Neurodegenerative disorders are caused by various conditions such as abnormal protein dynamics with defective protein aggregation and degradation and aggregation, impaired bioenergetics and mitochondrial dysfunction, excessive free radical formation leading tooxidative stress, and exposure to environntal toxicants such as heavy metals and pesticides [63] (Figure 3).

Neuroprotection is defined as the ability of a specific molecule to prevent neuronal cell death by interfering with and inhibiting the pathogenetic cascade that leads to cell dysfunction and eventual death [64]. AD and PD are the most common neurodegenerative diseases, but others can cause serious problems for individuals and society, and lead to significant healthcare costs. These diseases are known that these diseases cause irreversible cognitive dysfunctions in individuals. It is therefore extremely important for an effective treatment strategy to slow down the prognosis of neurodegenerative diseases by diagnosing them at the earliest possible stage. AD, PD, Huntington's disease (HD), and other neurodegenerative disorders share common features at the cellular and subcellular levels, and utilize similar molecular signaling pathways that can lead to inflammation, apoptosisnecroptosis, etc. These diseases are the consequence of misfolding and dysfunctional trafficking of proteins (Figure 3), mitochondrial dysfunction, oxidative stress, and/or environmental factors [63].

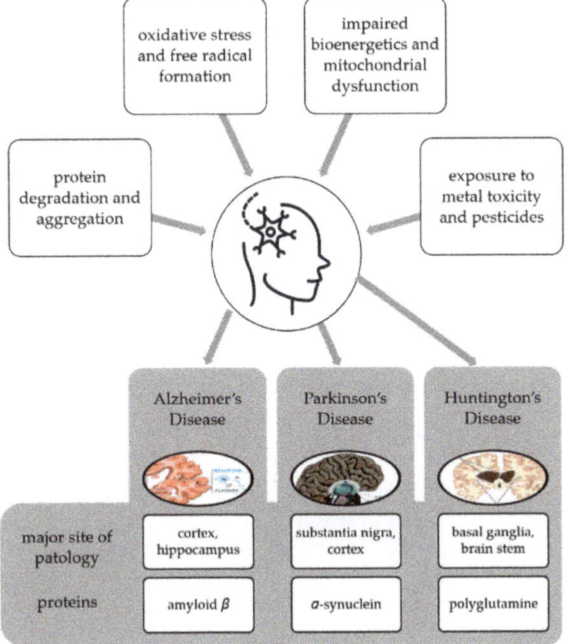

Figure 3. Multifactorial conditions causing neurodegenerative diseases and examples of diseases with the main site of pathology and proteins whose degradation and aggregation cause the pathology.

AD is a highly complex disorder characterized by severe synaptic losses and neuronal death, especially in regions with cognitive functions such as the cerebral cortex, hippocampus, entorhinal cortex, and ventral striatum [65]. Generally, in an average of 10 years, the stage of mild cognitive impairment passes to the advanced stage of AD, and the patient is lost in a completely helpless state at the end of this period. Due to the long duration of the disease and the fact that it affects the vital structures that determine who we are, it creates a great emotional and financial burden on patients' relatives and society [66]. Since the pioneering work of Alois Alzheimer in 1907, neuropathologists have identified amyloid plaques and NFTs in the brains of patients in autopsy examinations and stated that these pathologies cause the disease [67]. Amyloid plaques have been found to be extracellular deposits of amyloid-beta (Aβ) found in the brain parenchyma and cerebral blood vessels. The NFTs observed in the cell were found to consist of hyperphosphorylated tau protein associated with microtubules and clustered in helical filaments [68]. Additional pathological data for amyloid plaques and NFTs can be listed as intracellular granulovacuolar degeneration, decrease in the number of synapses, cholinergic cell losses in Meynert's basal nucleus, and astroglial activation [69]. Studies conducted to understand AD indicate that the disease arises as a result of complex interactions of many genetic, epigenetic, and environmental factors [69,70]. The main histopathological findings observed in the brain parenchyma of the patients have extracellularly located amyloid plaques, neurofibrillary structures consisting of intracellular tau protein clusters, glial activation, and traces of inflammation [71]. Based on these symptoms, many mechanisms have been proposed for the pathogenesis of the disease. The main ones can be listed as the amyloid cascade hypothesis, cholinergic damage hypothesis, neuronal cytoskeleton hypothesis, and oxidative stress hypothesis [72,73]. Other more debatable AD hypotheses are: inflammatory hypothesis, vascular hypothesis, cholesterol hypothesis, metal hypothesis, and cell cycle hypothesis [74].

PD is a progressive neurodegenerative disease that causes involuntary or uncontrollable movements such as tremors, stiffness, balance and coordination problems [75]. Symptoms usually begin insidiously and worsen over time. Degeneration of neurons in the compacta part of substantia nigra and the presence of Lewy bodies in its cytoplasm are the classic pathological findings of the disease [76]. Over the years of PD progression, a picture of dementia can develop that can be severe and debilitating, overshadowing the movement disorder of the disease. Dementia is defined as the presence of impairments in more than one cognitive domain, such as attention, memory, language, executive functioning, practice, and visuo-spatial functioning [77]. These losses reflect a marked decline from previous levels, and this decline is severe enough to interfere with daily, occupational, and social life. PD dementia is a mild or moderate dementia that begins insidiously, progresses slowly, affects some areas of cognitive function, especially executive function, and often develops psychosis during its course. The mechanisms underlying the pathogenesis of PD are still unclear, but there are several proposed mechanisms, such as those related to mitochondrial dysfunction, oxidative stress, ubiquitin-proteosome system, neuroinflammation, excitotoxicity, iron ion accumulation, and genetic issues [78].

HD is an autosomal dominant genetic disorder mainly characterized by progressive motor dysfunction, cognitive decline, and behavioral symptoms [79]. Amyotrophic lateral sclerosis (ALS) is a fatal late-onset neurodegenerative disorder that is characterized by a progressive loss of motor neurons of the CNS leading to muscle weakness, wasting, and spasticity. Patients develop progressive muscle weakness along with fasciculation and hyperreflexia. Mild cognitive deficits and frontotemporal dementia (FTD) are common. FTD is characterized by progressive deficits in executive function, behavior, and language.

Neurodegenerative diseases also share some common pathological features such as the accumulation of characteristic proteins in insoluble aggregates within and/or between neurons and the loss of synapses and death of neurons [80]. These proteins include β-amyloid (Aβ) of senile plaques and tau of neurofibrillary tangles (NFTs) in AD, α-synuclein (α-syn) of Lewy bodies (LBs) and Lewy neurites in PD, polyglutamine (PolyQ)-rich huntingtin inclusions in HD, TDP-43 aggregates in ALS, and TDP-43 aggregates and tau in FTD (Figure 3). In line with the above-mentioned explanations, neurodegenerative diseases occur with the folding and proteasomal disorders of certain proteins due to environmental or genetic reasons, followed by active glial cells secreting various mediators, including proinflammatory cytokines [80]. This whole process repeats each other in a vicious circle, causing apoptosis and developing neuroinflammation. This mechanism is the underlying cause of all diseases. Therefore, it is necessary to develop treatment strategies against neuroinflammation. These may include the use of natural sources or their secondary metabolites with anti-inflammatory effects.

4. Ginkgetin for the Treatment of Neurodegenerative Diseases

4.1. Oxidative Stress Mediation

Oxidative stress is a common trigger that can be associated with the development of neurodegenerative diseases, so compounds with antioxidant activity can be considered beneficial for the development of these diseases. Although the physiological cause of aging is not fully known, the free radical theory states that increasing oxidative stress of aging and aging-related diseases plays a fundamental role in this process by causing cellular degeneration. The increase in the number of free radicals observed in age-related neurodegenerative diseases and the fact that neurons are more sensitive to this damage have both been determined to be important characteristics. Therefore, it is thought that free radical production has an important role in the development and progression of neurological diseases [81]. Neurons are more susceptible to free radical damage [82] for certain reasons shown in Figure 4.

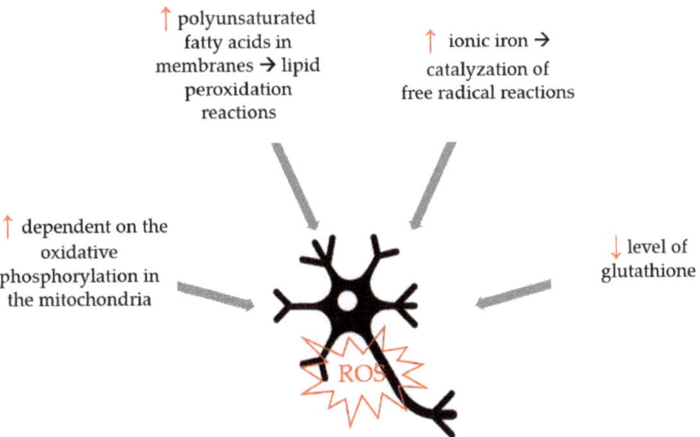

Figure 4. Neurons are more susceptible to free radical damage because of differences in some parameters and biological functions. Higher parameters are marked with ↑, lower parameters with ↓.

Flavonoids are widely recognized as a molecule with good antioxidant activity and beneficial effects in the treatment of neurological disorders [5–83]. Li et al. [84] measured the antioxidant activity of ginkgo leaves from plants grown in different locations and reported that ginkgetin content in the leaves resulted in stronger antioxidant activity and concluded that ginkgetin together with isoginkgetin were the two most active constituents with a strong relationship with antioxidant activity. Several cell-based studies have shown that ginkgetin plays a role in oxidative stress and that ginkgetin can protect fibroblasts from UVB-induced cytotoxicity [85], alleviate oxidative stress induced by H/R injury [86], inhibited NO production from lipopolysaccharide (LPS)-induced RAW 264.7 cells [87] and reduced oxidative stress caused by hyperglycemia [57]. However, reports about the antioxidant activity of ginkgetin are contradictory. Bedir et al. [40] who compared the antioxidant activity of 29 compounds isolated from *G. biloba* reported that ginkgetin was the least potent antioxidant after amentoflavone, showing only 19% inhibition at a concentration of 62.5 μg/mL. In the same study, monomeric biflavonoids had an IC_{50} of less than 10 μg/mL, clearly indicating that ginkgetin itself is not a molecule with antioxidant potential. Kang et al. [14] tested protective effect of biflavonoids on H_2O_2-induced cell death in SH-SY5Y (triple subcloned cell line derived from SK-N-SH neuroblastoma) and showed that all biflavonoids tested, including ginkgetin, significantly reduced H_2O_2-induced cell death. Furthermore, they tested the antioxidant activity of using the well- known 1,1-diphenyl-2-picrylhydrazyl (DPPH) assay, and none of the nine biflavones showed radical scavenging activity at concentrations up to 100 μM. These results suggest that the neuroprotection of biflavonoids may be mediated by direct blockade of cell death cascades rather than by their antioxidant activity. Therefore, they investigated the neuroprotective effects of biflavonoids against the cytotoxic insult induced by staurosporine. Staurosporine is known to mediate apoptosis via the caspase-dependent mitochondrial pathway. Ginkgetin significantly reduced cell death induced by staurosporine at a concentration of 10 μM. They therefore suggest that neuroprotection by biflavonoids is mediated in part, if not completely, by direct blockade of the signaling events that lead to apoptosis during cellular stress. Jeong et al. [31] investigated the neuroprotective effects of four biflavonoids, including ginkgetin, using mouse HT22 hippocampal cells, a model system for studying glutamate-induced oxidative stress. They reported that ginkgetin can protect HT22 neuronal cells from glutamate-induced oxidative damage by preserving the activities of antioxidant enzymes and/or inhibiting ERK1/2 activation.

This example shows that commonly used methods for determining antioxidant activity, such as DPPH, are not ideal for predicting the ability of compounds to reduce oxidative stress at the cellular level and even more so at the tissue or organism level.

4.2. Protection against Neuronal Injury Caused by Ischemic Stroke

Ample evidence has supported the role of neuroinflammation in the development of neurological disorders. Inflammatory components such as astrocytes, microglia, the complement system, and cytokines have been associated with neuroinflammation in the CNS. In particular, inflammatory cytokines have been found to play a central role in neuroinflammation pathway as a large numberof studies have reported abnormally elevated levels of interleukin-1β (IL-1β) and tumor necrosis factor (TNF) in AD and PD patients (reviewd by [88]).

A common method to study the antineuroinflammatory potential of certain molecules is to use animal models exposed to neuronal injury caused by ischemic stroke. Most ischemic strokes occur in the middle cerebral artery territory, so many of the animal models of stroke that have been developed have focused on this artery. In the intraluminal monofilament model of middle cerebral artery occlusion (MCAO), a surgical suture is inserted into the external carotid artery and advanced into the internal carotid artery (ICA) until the tip occludes the origin of the MCA, resulting in interruption of blood flow and subsequent cerebral infarction in the MCA area [89]. This model has been used also for the study of the effects of ginkgetin (Figure 5.).

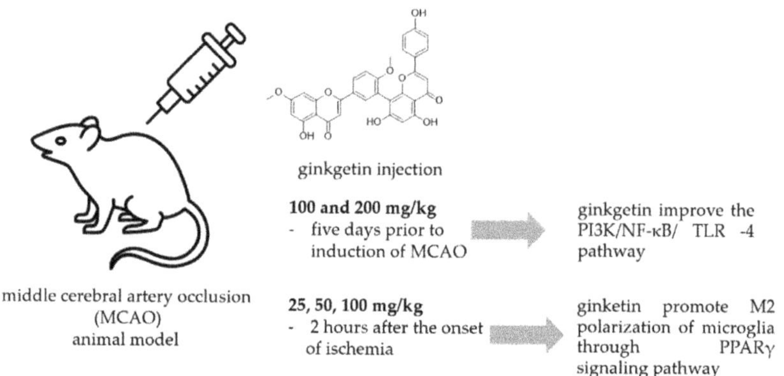

Figure 5. Ginkgetin protection against neuronal injury caused by ischemic stroke.

In a study by Xu et al. [90] animals received ginkgetin at concentrations of 100 and 200 mg/kg i.p. five days prior to induction of MCAO, and they investigated the effect of ginkgetin against stroke. Treatment with ginkgetin attenuated the increased neurological score and decreased the water content in the brain. Ginkgetin-treated rats showed that the levels of pro-inflammatory cytokines NF-κB, IL-1β, and TNF-α were significantly decreased in brain tissue. The authors concluded that ginkgetin aglycone improved the PI3K/NF-κB/ TLR-4 inflammatory pathway. Tian et al. [91] used a transient MCAO procedure to establish the cerebral ischemia/reperfusion model (IR) in rats. Ginkgetin was injected at doses of 25, 50, and 100 mg/kg 2 hours after the onset of ischemia and its administration markedly reduced the volume of cerebral infarction and neurologic deficits. It also reduced the number of apoptotic cells, decreased the amount of cleaved caspase-3 and Bax, and increased the amount of Bcl-2 in rats exposed to IR injury in a dose-dependent manner. In addition, high-dose ginkgetin treatment (100 mg/kg) significantly increased the phosphorylations of Akt and mTOR. Blocking PI3K by LY294002 significantly decreased the antiapoptotic effect and reduced both Akt and mTOR phosphorylation levels. According to the authors, ginkgetin counteracts cerebral IR-induced injury by inhibiting apoptosis in rats, and this effect was attenuated by activation of the PI3K/Akt/mTOR pathway. The same experiment procedure and the same ginkgetin concentration was used by Pan et al. [92] who reported that ginkgetin attenuated I/R-induced autophagy activation, pyramidal neuron death in cerebral I/R, and reduced I/R-induced upregulation of p53. They concluded that ginkgetin can attenuate cerebral ischemia/reperfusion-induced autophagy and apoptosis by inhibiting the NF-κB/p53 pathway. Some other researchers [93] used oxygen glucose deprivation (OGD) cellular and MCAO animal models to study neuroprotective activity of ginkgetin reported that ginkgetin treatment converted microglia from M1 type to M2 type and inhibited neuroinflammation. Detailed study of the neuroprotective mechanism suggested that ginkgetin can inhibit neuroinflammation by promoting M2 polarization of microglia through PPARγ signaling pathway thus promoting recovery of neurological functions in an ischemic stroke.

4.3. Activity against Neurotumors

Different *in vitro* and *in vivo* studies showed that ginkgetin combats cancer progression by arresting the cell cycle, inducing *via* apoptosis, stimulating autophagy, and targeting many deregulated signaling pathways such as JAK/STAT and MAPKs (reviewed by Adnan et al. [20]). Ginkgetin was also studied as a potential agent for the treatment of neurotumores by Ye et. al. [26], who investigated the potential of natural products in the treatment of medulloblastoma (MB), a form of malignant brain tumor that occurs predominantly in infants and children and in which approximately 25% is due to upregulation of the canonical Wnt pathway, with mutations mainly in CTNNB1. They screened for antagonists of Wnt signaling from 600 natural compounds and identified ginkgetin as a potential molecule that showed marked cytotoxicity. Ginkgetin efficiently induced G2/M phase arrest in Daoy cells, reduced the expression of Wnt target genes, including Axin2, CyclinD1, and Survivin in MB cells, and decreased the phosphorylation level of β-catenin. They concluded that ginkgetin is a novel inhibitor of Wnt signaling and, as such, warrants further exploration as a promising candidate against medulloblastoma.

4.4. Protective Effect against Alzheimers' Disease

AD is caused by multiple mechanisms such as excessive accumulation of extracellular amyloid-beta 42 (Aβ42) plaques, intracellular hyperphosphorylated tau neurofibril tangles in the brain, oxidative stress due to mitochondrial dysfunction, and/or genetic as well as environmental factors [94]. Aggregation and accumulation of amyloid-β plaques and tau proteins in the brain are central features in the pathophysiology of AD and are therefore the focus of most research investigating potential therapeutics for this neurodegenerative disease [95]. Kang et al. [14] investigated whether biflavones showed protective effects against Aβ-induced cytotoxicity using SH-SY5Y (triple subcloned cell line derived from SK-N-SH neuroblastoma) cells and found that ginkgetin showed protective effects at 2 μM, with an inhibition percentage of 43.6%. In the same study they tested protective effects against neuronal cell death induced by a DNA-damaging agent, etoposide, but gingetin did not show a protective effect.

Amyloid-β-42 (Aβ42) is proteolytic derivative of the large transmembrane protein amyloid precursor protein (APP) and it plays an early and important role in all cases of AD [96]. Thus, blocking Aβ42 production by specific inhibition of key proteases required for Aβ42 formation is a major focus of AD therapy research. β-Secretase, the aspartic protease that generates the N-terminus of Aβ42, has become a major target and researchers are focused on discovering its inhibitors [96]. Sasaki et al. [97] examined the activity of twenty-one bioflavonoids against β-secretase and ginkgetin showed a significant inhibitory effect with an IC_{50} value of 4.18 μM. The authors indicated that the importance of the position of hydroxyl groups in two apigenin molecules for the inhibition of β-secretase and the presence of hydroxyl groups in the C3′ and C8″ position might enhance the inhibitory effects. Ullah et al. [98] reviewed β-secretase inhibitors from plant sources and, among them, ginkgetin was a significant inhibitor with a low IC_{50} value. In an in silico study performed by Grewal et al. [99], ginkgetin showed a good binding potential on N-methyl-D-aspartate glutamate receptor (NMDA) and beta secretase-1 (BACE-1), and was suggested as a neuroprotective agent. In another study conducted by Choi et al. [100], eight amentoflavone-like bioflavonoids were tested to inhibit amyloid-beta fibrillation and to disaggregate amyloid-beta fibrils. In the study, the IC_{50} value of ginkgetin was 4.92 μM in the inhibition of Aβ fibrils assay. In the same study, ginkgetin exhibited a disaggregation effect on Aβ fibrils with the IC_{50} value of 6.81 μM.

Zeng et al. [101] studied ginkgetin therapeutic potential against AD using a transgenic mouse model of AD, PS1dE9/APPS mice. Prior to the onset of AD-type neuropathology, mice were randomly assigned to four diet groups: ginkgetin group, curcumin group, normal diet group, with wild-type littermates used as a control group. All animals were fed with the above diets for 9 months. The mean daily food consumption of the mice was 0.08– 0.12 g/g body weight and the corresponding daily ginkgetin and curcumin were about 200 and 80 mg/kg/day based on a previous report indicating lack of toxicity. The equivalent consumption in a 60 kg human is about 0.91 g/day for ginkgetin and 0.35 g/day for curcuminas. In their experiments they showed that ginkgetin effectively reduced the Aβ levels in the brain and blood, decreased cerebral microhemorrhage, lowered astrogliosis, and ameliorated inflammation in APP/PS1 transgenic model, which indicates in vivo therapeutic potential of ginkgetin against AD. However, as the authors stated, pathophysiology mechanisms of Aβ clearance need further research.

It has been noted that the development of amyloid-β plaques occurs about 10–20 years before the manifestation of AD symptoms, thus the earlier interventions are necessary to address presymptomatic AD. Studies suggesting that amyloid-β peptides may play a role in innate immunity as antimicrobial peptides indicate that the buildup of amyloid-β plaques may be a response to the presence of viruses and bacteria [94]. This has led to the establishment of the antimicrobial hypothesis for AD and the use of antimicrobial and antiviral drugs as potential therapeutics targeting the root cause of AD. Biflavonoids are in the focus of the sciences as a potent antimicrobial, expecially antiviral agents [13,94] where ginkgetin stands out as a compound with antiviral capabilities against herpes simplex virus type 1 (HSV-1) and type 2 (HSV-2) [25], cytomegalovirus (HCMV) [25], influenza A virus [32], and SARS-CoV-2 virus [59,60,102]. Ginkgetin also shows antifungal activity against *Alternaria alternate, Cladosporium oxysporum,* and *Fusarium culmorum* [35], and antibacterial activity against *Streptococcus suis* [103] (Figure 6).

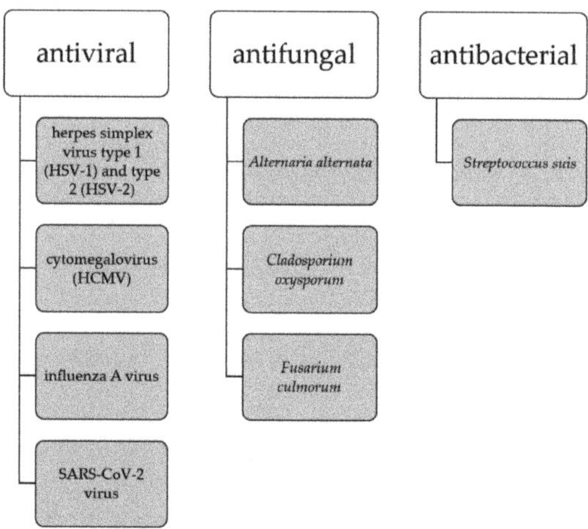

Figure 6. Antimicrobial activity of ginkgetin.

At this point, researchers have not linked any specific bacterium or virus alone to the development of AD. Thus, a number of viruses and bacteria may be involved in the progression of neurodegenerative diseases independently or simultaneously with other pathogens. Given the good antimicrobial, especially antiviral, activity of ginkgetin, its potential role in mechanisms related to the antimicrobial hypothesis for AD is worthy of future research.

Considering all these results, it is obvious that ginkgetin has potential for the treatment of AD, but further studies should be performed to confirm this activity, especially in a clinical trial. Possible mechanisms of ginkgetin related to the protection of AD are shown in Figure 7.

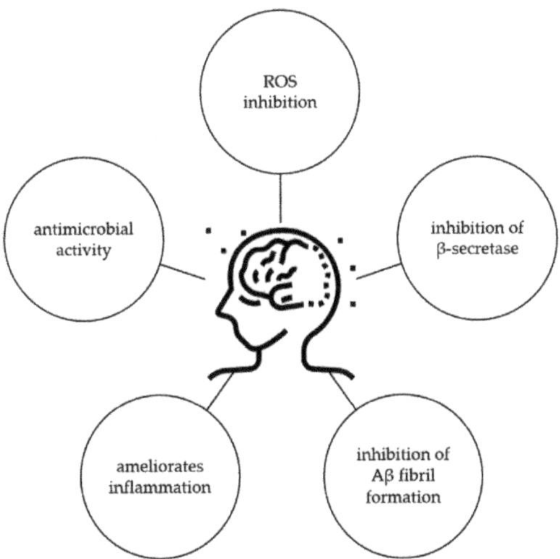

Figure 7. Biological activity of ginkgetin related to protection against AD.

4.5. *Protective Effect against Parkinson' Disease*

PD is reported to be the second most common neurodegenerative disorder after AD [104]. Therefore, great efforts have been made to search for new molecules that could be effective in the treatment of PD. Although there are several in silico and in vitro models for finding new active molecules, our literature search did not yield any results related to ginkgetin. The animal model commonly used is a model where PD is induced by 1-methyl-4-phenyl-1,2,3,6-tetrahydropyridine (MPTP), which is the gold standard for researchers in order to induce all aspects of PD hallmarks in animal model of the disease [104]. Wang et al. [105] investigated the neuroprotective ability of ginkgetin in vivo in a model of PD induced by MPTP. Animals received ginkgetin (80 mg/100 g body weight) via the stomach for 5 days and then were injected intraperitoneally with MPTP (20 mg/kg) once daily for 5 days. The authors showed that ginkgetin significantly improved sensorimotor coordination in a mouse model PD by dramatically inhibiting the decline in tyrosine hydroxylase expression in the substantia nigra and superoxide dismutase activity in the striatum. They reported that ginkgetin can strongly chelate iron ions, thereby inhibiting the increase in intracellular labile iron pool by downregulating L-ferritin and upregulating transferrin receptor 1, suggesting that the neuroprotective mechanism of ginkgetin against neurological damage induced by MPTP is via the regulation of iron homeostasis. In another animal study [106], mice were treated with 1-methyl-4-phenyl-1,2,3,6-tetrahydropyridine (MPTP) (25 mg/kg) and probenecid (250 mg/kg) for five consecutive days to induce PD. Ginkgetin (5, 10, 20 mg/kg) and bromocriptine (10 mg/kg), which is used to treat PD, were administered orally to PD mice for 26 days, including a five-day pretreatment period. The authors reported that in MPTP-induced PD mice, movements and muscle functions improved with ginkgetin. The number of tyrosine hydroxylase-positive cells was reduced and later recovered without degeneration. The level of glial fibrillary acidic protein (GFAP) decreased, while the level of brain-derived neurotrophic factor (BDNF) increased significantly after treatment with ginkgetin. In summary, the authors concluded that ginkgetin effectively protects dopaminergic neurons by reducing oxidative damage, activating microglia, and increasing neurotrophic potential, indicating that it is a potential candidate for the treatment of PD.

5. Conclusions

Ginkgetin is the first biflavonoid isolated from ginkgo, after which it was named. All biflavonoids belong to the flavonoid group, well-studied specialized metabolites from plants, but they are much less studied compared to monomeric flavonoids. In this review, we have summarized the available data on the neuroprotective potential of ginkgetin. The available data are in vitro studies or in vivo animal studies, and as far as we know, there have been no clinical studies performed as yet. There is evidence of protection against neuronal damage caused by ischemic stroke, neurotumors, AD, and PD, but further studies and clinical trials should explain the mechanisms of action and the effective and safe concentration of ginkgetin for clinical use. The focus of future research should be primarily on the potential to cross the blood-brain barrier, as there is currently a lack of information in this regard.

Author Contributions: Conceptualization, D.Š.; writing—original draft preparation, D.Š., İ.İ.T.Ç., H.P.D. and G.Z.; writing—review and editing, D.Š., İ.İ.T.Ç., H.P.D. and G.Z.; visualization, D.Š.; supervision, D.Š.; project administration, D.Š.; funding acquisition, D.Š. All authors have read and agreed to the published version of the manuscript.

Funding: This work has been supported by Croatian Science Foundation project "Biflavonoids role in plants: *Ginkgo biloba* L. as a model system" under the project no. UIP-2019-04-1018.

Institutional Review Board Statement: Not applicable.

Informed Consent Statement: Not applicable.

Data Availability Statement: Not applicable.

Conflicts of Interest: The authors declare no conflict of interest.

References

1. Vaupel, J.W.; Villavicencio, F.; Bergeron-Boucher, M.-P. Demographic Perspectives on the Rise of Longevity. *Proc. Natl. Acad. Sci. USA* **2021**, *118*, e2019536118. [CrossRef] [PubMed]
2. World Health Organization. *WHO Decade of Healthy Ageing: Baseline Report*; World Health Organization: Geneva, Switzerland, 2020; ISBN 9789240017900.
3. Callixte, K.-T.; Clet, T.B.; Jacques, D.; Faustin, Y.; François, D.J.; Maturin, T.-T. The Pattern of Neurological Diseases in Elderly People in Outpatient Consultations in Sub-Saharan Africa. *BMC Res. Notes* **2015**, *8*, 159. [CrossRef]
4. Bhat, B.A.; Almilaibary, A.; Mir, R.A.; Aljarallah, B.M.; Mir, W.R.; Ahmad, F.; Mir, M.A. Natural Therapeutics in Aid of Treating Alzheimer's Disease: A Green Gateway Toward Ending Quest for Treating Neurological Disorders. *Front. Neurosci.* **2022**, *16*, 884345. [CrossRef] [PubMed]
5. Ahmed, T.; Javed, S.; Javed, S.; Tariq, A.; Šamec, D.; Tejada, S.; Nabavi, S.F.; Braidy, N.; Nabavi, S.M. Resveratrol and Alzheimer's Disease: Mechanistic Insights. *Mol. Neurobiol.* **2017**, *54*, 2622–2635. [CrossRef] [PubMed]
6. Nabavi, S.F.; Khan, H.; D'onofrio, G.; Šamec, D.; Shirooie, S.; Dehpour, A.R.; Argüelles, S.; Habtemariam, S.; Sobarzo-Sanchez, E. Apigenin as Neuroprotective Agent: Of Mice and Men. *Pharmacol. Res.* **2017**, *128*, 359–365. [CrossRef]
7. Nabavi, S.M.; Šamec, D.; Tomczyk, M.; Milella, L.; Russo, D.; Habtemariam, S.; Suntar, I.; Rastrelli, L.; Daglia, M.; Xiao, J.; et al. Flavonoid Biosynthetic Pathways in Plants: Versatile Targets for Metabolic Engineering. *Biotechnol. Adv.* **2020**, *38*, 107316. [CrossRef] [PubMed]
8. Ayaz, M.; Sadiq, A.; Junaid, M.; Ullah, F.; Ovais, M.; Ullah, I.; Ahmed, J.; Shahid, M. Flavonoids as Prospective Neuroprotectants and Their Therapeutic Propensity in Aging Associated Neurological Disorders. *Front. Aging Neurosci.* **2019**, *11*, 155. [CrossRef]
9. Dias, M.C.; Pinto, D.C.G.A.; Silva, A.M.S. Plant Flavonoids: Chemical Characteristics and Biological Activity. *Molecules* **2021**, *26*, 5377. [CrossRef]

10. Spagnuolo, C.; Moccia, S.; Russo, G.L. Anti-Inflammatory Effects of Flavonoids in Neurodegenerative Disorders. *Eur. J. Med. Chem.* **2018**, *153*, 105–115. [CrossRef]
11. Šamec, D.; Pierz, V.; Srividya, N.; Wüst, M.; Lange, B.M. Assessing Chemical Diversity in *Psilotum nudum* (L.) Beauv., a Pantropical Whisk Fern That Has Lost Many of Its Fern-like Characters. *Front. Plant Sci.* **2019**, *10*, 868. [CrossRef]
12. He, X.; Yang, F.; Huang, X. Proceedings of Chemistry, Pharmacology, Pharmacokinetics and Synthesis of Biflavonoids. *Molecules* **2021**, *26*, 6088. [CrossRef] [PubMed]
13. Šamec, D.; Karalija, E.; Dahija, S.; Hassan, S.T.S. Biflavonoids: Important Contributions to the Health Benefits of Ginkgo (*Ginkgo biloba* L.). *Plants* **2022**, *11*, 1381. [CrossRef]
14. Kang, S.S.; Lee, J.Y.; Choi, Y.K.; Song, S.S.; Kim, J.S.; Jeon, S.J.; Han, Y.N.; Son, K.H.; Han, B.H. Neuroprotective Effects of Naturally Occurring Biflavonoids. *Bioorg. Med. Chem. Lett.* **2005**, *15*, 3588–3591. [CrossRef]
15. Nowak, A.; Kojder, K.; Zielonka-Brzezicka, J.; Wróbel, J.; Bosiacki, M.; Fabiańska, M.; Wróbel, M.; Sołek-Pastuszka, J.; Klimowicz, A. The Use of *Ginkgo Biloba* L. as a Neuroprotective Agent in the Alzheimer's Disease. *Front. Pharmacol.* **2021**, *12*, 775034. [CrossRef] [PubMed]
16. Napryeyenko, O.; Sonnik, G.; Tartakovsky, I. Efficacy and Tolerability of Ginkgo Biloba Extract EGb 761® by Type of Dementia: Analyses of a Randomised Controlled Trial. *J. Neurol. Sci.* **2009**, *283*, 224–229. [CrossRef] [PubMed]
17. Thancharoen, O.; Limwattananon, C.; Waleekhachonloet, O.; Rattanachotphanit, T.; Limwattananon, P.; Limpawattana, P. Ginkgo Biloba Extract (EGb761), Cholinesterase Inhibitors, and Memantine for the Treatment of Mild-to-Moderate Alzheimer's Disease: A Network Meta-Analysis. *Drugs Aging* **2019**, *36*, 435–452. [CrossRef]
18. Tanaka, K.; Galduroz, R.S.; Gobbi, L.; Galduroz, J. Ginkgo Biloba Extract in an Animal Model of Parkinson's Disease: A Systematic Review. *Curr. Neuropharmacol.* **2013**, *11*, 430–435. [CrossRef]
19. Kovač Tomas, M.; Jurčević, I.; Šamec, D. Tissue-Specific Profiling of Biflavonoids in Ginkgo (*Ginkgo Biloba* L.). *Plants* **2022**, *12*, 147. [CrossRef]
20. Adnan, M.; Rasul, A.; Hussain, G.; Shah, M.A.; Zahoor, M.K.; Anwar, H.; Sarfraz, I.; Riaz, A.; Manzoor, M.; Adem, Ş.; et al. Ginkgetin: A Natural Biflavone with Versatile Pharmacological Activities. *Food Chem. Toxicol.* **2020**, *145*, 111642. [CrossRef]
21. Wang, G.; Yao, S.; Zhang, X.-X.; Song, H. Rapid Screening and Structural Characterization of Antioxidants from the Extract of Selaginella Doederleinii Hieron with DPPH-UPLC-Q-TOF/MS Method. *Int. J. Anal. Chem.* **2015**, *2015*, 849769. [CrossRef]
22. Cao, Y.; Tan, N.-H.; Chen, J.-J.; Zeng, G.-Z.; Ma, Y.-B.; Wu, Y.-P.; Yan, H.; Yang, J.; Lu, L.-F.; Wang, Q. Bioactive Flavones and Biflavones from Selaginella Moellendorffii Hieron. *Fitoterapia* **2010**, *81*, 253–258. [CrossRef] [PubMed]
23. Zhang, Y.; Shi, S.; Wang, Y.; Huang, K. Target-Guided Isolation and Purification of Antioxidants from Selaginella Sinensis by Offline Coupling of DPPH-HPLC and HSCCC Experiments. *J. Chromatogr. B* **2011**, *879*, 191–196. [CrossRef] [PubMed]
24. Yamaguchi, L.F.; Kato, M.J.; Di Mascio, P. Biflavonoids from Araucaria Angustifolia Protect against DNA UV-Induced Damage. *Phytochemistry* **2009**, *70*, 615–620. [CrossRef] [PubMed]
25. Hayashi, K.; Hayashi, T.; Morita, N. Mechanism of Action of the Antiherpesvirus Biflavone Ginkgetin. *Antimicrob. Agents Chemother.* **1992**, *36*, 1890–1893. [CrossRef]
26. Ye, Z.-N.; Yu, M.-Y.; Kong, L.-M.; Wang, W.-H.; Yang, Y.-F.; Liu, J.-Q.; Qiu, M.-H.; Li, Y. Biflavone Ginkgetin, a Novel Wnt Inhibitor, Suppresses the Growth of Medulloblastoma. *Nat. Prod. Bioprospect.* **2015**, *5*, 91–97. [CrossRef]
27. Mendiratta (Nee Chugh), A.; Dayal, R.; Bartley, J.P.; Smith, G. A Phenylpropanoid and Biflavonoids from the Needles of Cephalotaxus Harringtonia Var. Harringtonia. *Nat. Prod. Commun.* **2017**, *12*, 1777–1778. [CrossRef]
28. Lee, M.K.; Lim, S.W.; Yang, H.; Sung, S.H.; Lee, H.-S.; Park, M.J.; Kim, Y.C. Osteoblast Differentiation Stimulating Activity of Biflavonoids from Cephalotaxus Koreana. *Bioorg. Med. Chem. Lett.* **2006**, *16*, 2850–2854. [CrossRef] [PubMed]
29. Li, W.; Deng, Y.; Dai, R.; Yu, Y.; Saeed, M.K.; Li, L.; Meng, W.; Zhang, X. Chromatographic Fingerprint Analysis of Cephalotaxus Sinensis from Various Sources by High-Performance Liquid Chromatography–Diodearray Detection–Electrospray Ionization-Tandem Mass Spectrometry. *J. Pharm. Biomed. Anal.* **2007**, *45*, 38–46. [CrossRef]
30. Attallah, N.G.M.; Al-Fakhrany, O.M.; Elekhnawy, E.; Hussein, I.A.; Shaldam, M.A.; Altwaijry, N.; Alqahtani, M.J.; Negm, W.A. Anti-Biofilm and Antibacterial Activities of Cycas Media R. Br Secondary Metabolites: In Silico, In Vitro, and In Vivo Approaches. *Antibiotics* **2022**, *11*, 993. [CrossRef] [PubMed]
31. Jeong, E.J.; Hwang, L.; Lee, M.; Lee, K.Y.; Ahn, M.-J.; Sung, S.H. Neuroprotective Biflavonoids of Chamaecyparis Obtusa Leaves against Glutamate-Induced Oxidative Stress in HT22 Hippocampal Cells. *Food Chem. Toxicol.* **2014**, *64*, 397–402. [CrossRef]
32. Miki, K.; Nagai, T.; Suzuki, K.; Tsujimura, R.; Koyama, K.; Kinoshita, K.; Furuhata, K.; Yamada, H.; Takahashi, K. Anti-Influenza Virus Activity of Biflavonoids. *Bioorg. Med. Chem. Lett.* **2007**, *17*, 772–775. [CrossRef] [PubMed]
33. Li, M.; Li, B.; Xia, Z.M.; Tian, Y.; Zhang, D.; Rui, W.J.; Dong, J.X.; Xiao, F.J. Anticancer Effects of Five Biflavonoids from *Ginkgo biloba* L. Male Flowers In Vitro. *Molecules* **2019**, *24*, 1496. [CrossRef] [PubMed]
34. Shen, N.; Liu, Y.; Cui, Y.; Xin, H. Large-Scale Targetedly Isolation of Biflavonoids with High Purity from Industrial Waste Ginkgo Biloba Exocarp Using Two-Dimensional Chromatography Coupled with Macroporous Adsorption Resin Enrichment. *Ind. Crops Prod.* **2022**, *175*, 114264. [CrossRef]
35. Krauze-Baranowska, M.; Wiwart, M. Antifungal Activity of Biflavones from Taxus Baccata and Ginkgo Biloba. *Z. Für Naturforsch. C* **2003**, *58*, 65–69. [CrossRef]
36. Meng, A.; Li, J.; Pu, S. Chemical Constituents of Leaves of Taxus Chinensis. *Chem. Nat. Compd.* **2018**, *54*, 841–845. [CrossRef]

37. Choi, S.-K.; Oh, H.-M.; Lee, S.-K.; Jeong, D.G.; Ryu, S.E.; Son, K.-H.; Han, D.C.; Sung, N.-D.; Baek, N.-I.; Kwon, B.-M. Biflavonoids Inhibited Phosphatase of Regenerating Liver-3 (PRL-3). *Nat. Prod. Res.* **2006**, *20*, 341–346. [CrossRef]
38. Hao, J.; Guo, H.; Shi, X.; Wang, Y.; Wan, Q.; Song, Y.-B.; Zhang, L.; Dong, M.; Shen, C. Comparative Proteomic Analyses of Two Taxus Species (Taxus Media and Taxus Mairei) Reveals Variations in the Metabolisms Associated with Paclitaxel and Other Metabolites. *Plant Cell. Physiol.* **2017**, *58*, 1878–1890. [CrossRef]
39. Ryu, Y.B.; Jeong, H.J.; Kim, J.H.; Kim, Y.M.; Park, J.-Y.; Kim, D.; Naguyen, T.T.H.; Park, S.-J.; Chang, J.S.; Park, K.H. Biflavonoids from Torreya Nucifera Displaying SARS-CoV 3CLpro Inhibition. *Bioorg. Med. Chem.* **2010**, *18*, 7940–7947. [CrossRef]
40. Bedir, E.; Tatli, I.I.; Khan, R.A.; Zhao, J.; Takamatsu, S.; Walker, L.A.; Goldman, P.; Khan, I.A. Biologically Active Secondary Metabolites from Ginkgo Biloba. *J. Agric. Food Chem.* **2002**, *50*, 3150–3155. [CrossRef]
41. Castañeda, P.; Garcia, M.R.; Hernandez, B.E.; Torres, B.A.; Anaya, A.L.; Mata, R. Effects of Some Compounds Isolated From-Celaenodendron Mexicanum Standl (Euphorbiaceae) on Seeds and Phytopathogenic Fungi. *J. Chem. Ecol.* **1992**, *18*, 1025–1037. [CrossRef]
42. Zhou, Z.; Fu, C. A New Flavanone and Other Constituents from the Rhizomes of Cyperus Rotundus and Their Antioxidant Activities. *Chem. Nat. Compd.* **2013**, *48*, 963–965. [CrossRef]
43. Pattamadilok, D.; Suttisri, R. Seco-Terpenoids and Other Constituents from Elateriospermum Tapos. *J. Nat. Prod.* **2008**, *71*, 292–294. [CrossRef] [PubMed]
44. Li, J.; Li, F.; Lu, Y.-Y.; Su, X.-J.; Huang, C.-P.; Lu, X.-W. A New Dilactone from the Seeds of Gaultheria Yunnanensis. *Fitoterapia* **2010**, *81*, 35–37. [CrossRef] [PubMed]
45. Li, D.; Liu, J.-P.; Han, X.; Wang, Y.-F.; Wang, C.-H.; Li, Z.; Wang, G.-C. Chemical Constituents of the Whole Plants of Houttuynia Cordata. *Chem. Nat. Compd.* **2017**, *53*, 365–367. [CrossRef]
46. Jeon, Y.J.; Jung, S.; Yun, J.; Lee, C.W.; Choi, J.; Lee, Y.; Han, D.C.; Kwon, B. Ginkgetin Inhibits the Growth of DU −145 Prostate Cancer Cells through Inhibition of Signal Transducer and Activator of Transcription 3 Activity. *Cancer Sci.* **2015**, *106*, 413–420. [CrossRef]
47. Lou, J.-S.; Bi, W.-C.; Chan, G.K.L.; Jin, Y.; Wong, C.-W.; Zhou, Z.-Y.; Wang, H.-Y.; Yao, P.; Dong, T.T.X.; Tsim, K.W.K. Ginkgetin Induces Autophagic Cell Death through P62/SQSTM1-Mediated Autolysosome Formation and Redox Setting in Non-Small Cell Lung Cancer. *Oncotarget* **2017**, *8*, 93131–93148. [CrossRef]
48. Lou, J.-S.; Zhao, L.-P.; Huang, Z.-H.; Chen, X.-Y.; Xu, J.-T.; TAI, W.C.-S.; Tsim, K.W.K.; Chen, Y.-T.; Xie, T. Ginkgetin Derived from Ginkgo Biloba Leaves Enhances the Therapeutic Effect of Cisplatin via Ferroptosis-Mediated Disruption of the Nrf2/HO-1 Axis in EGFR Wild-Type Non-Small-Cell Lung Cancer. *Phytomedicine* **2021**, *80*, 153370. [CrossRef]
49. Hu, W.H.; Chan, G.K.L.; Duan, R.; Wang, H.Y.; Kong, X.P.; Dong, T.T.X.; Tsim, K.W.K. Tsim Synergy of Ginkgetin and Resveratrol in Suppressing VEGF-Induced Angiogenesis: A Therapy in Treating Colorectal Cancer. *Cancers* **2019**, *11*, 1828. [CrossRef]
50. Rahaman, S.O.; Alharbi, M.O.; Dutta, B.; Goswami, R. Identification and Functional Characterization of a Biflavone as a Novel Inhibitor of TRPV4-dependent Proatherogenic Processes in Macrophages. *FASEB J.* **2020**, *34*, 1. [CrossRef]
51. Chen, T.-R.; Wei, L.-H.; Guan, X.-Q.; Huang, C.; Liu, Z.-Y.; Wang, F.-J.; Hou, J.; Jin, Q.; Liu, Y.-F.; Wen, P.-H.; et al. Biflavones from Ginkgo Biloba as Inhibitors of Human Thrombin. *Bioorg. Chem.* **2019**, *92*, 103199. [CrossRef]
52. Liu, P.-K.; Weng, Z.-M.; Ge, G.-B.; Li, H.-L.; Ding, L.-L.; Dai, Z.-R.; Hou, X.-D.; Leng, Y.-H.; Yu, Y.; Hou, J. Biflavones from Ginkgo Biloba as Novel Pancreatic Lipase Inhibitors: Inhibition Potentials and Mechanism. *Int. J. Biol. Macromol.* **2018**, *118*, 2216–2223. [CrossRef] [PubMed]
53. Cho, Y.-L.; Park, J.-G.; Kang, H.J.; Kim, W.; Cho, M.J.; Jang, J.-H.; Kwon, M.-G.; Kim, S.; Lee, S.-H.; Lee, J.; et al. Ginkgetin, a Biflavone from Ginkgo Biloba Leaves, Prevents Adipogenesis through STAT5-Mediated PPARγ and C/EBPα Regulation. *Pharmacol. Res.* **2019**, *139*, 325–336. [CrossRef] [PubMed]
54. Lian, N.; Tong, J.; Li, W.; Wu, J.; Li, Y. Ginkgetin Ameliorates Experimental Atherosclerosis in Rats. *Biomed. Pharmacother.* **2018**, *102*, 510–516. [CrossRef] [PubMed]
55. Zhang, L.; Liu, J.; Geng, T. Ginkgetin Aglycone Attenuates the Apoptosis and Inflammation Response through Nuclear Factor-kB Signaling Pathway in Ischemic-reperfusion Injury. *J. Cell. Biochem.* **2019**, *120*, 8078–8085. [CrossRef]
56. Tao, Z.; Jin, W.; Ao, M.; Zhai, S.; Xu, H.; Yu, L. Evaluation of the Anti-Inflammatory Properties of the Active Constituents in Ginkgo Biloba for the Treatment of Pulmonary Diseases. *Food Funct.* **2019**, *10*, 2209–2220. [CrossRef]
57. Wei, L.; Jian, P.; Erjiong, H.; Qihan, Z. Ginkgetin Alleviates High Glucose-evoked Mesangial Cell Oxidative Stress Injury, Inflammation, and Extracellular Matrix (ECM) Deposition in an AMPK/MTOR-mediated Autophagy Axis. *Chem. Biol. Drug Des.* **2021**, *98*, 620–630. [CrossRef]
58. Weniger, B.; Vonthron-Sénécheau, C.; Kaiser, M.; Brun, R.; Anton, R. Comparative Antiplasmodial, Leishmanicidal and Antitrypanosomal Activities of Several Biflavonoids. *Phytomedicine* **2006**, *13*, 176–180. [CrossRef]
59. Ghosh, R.; Chakraborty, A.; Biswas, A.; Chowdhuri, S. Computer Aided Identification of Potential SARS CoV-2 Main Protease Inhibitors from Diterpenoids and Biflavonoids of Torreya Nucifera Leaves. *J. Biomol. Struct. Dyn.* **2022**, *40*, 2647–2662. [CrossRef]
60. Rana, S.; Kumar, P.; Sharma, A.; Sharma, S.; Giri, R.S.; Ghosh, K. Identification of Naturally Occurring Antiviral Molecules for SARS-CoV-2 Mitigation. *Open. COVID J.* **2021**, *1*, 38–46. [CrossRef]
61. Dey, D.; Hossain, R.; Biswas, P.; Paul, P.; Islam, M.A.; Ema, T.I.; Gain, B.K.; Hasan, M.M.; Bibi, S.; Islam, M.T.; et al. Amentoflavone Derivatives Significantly Act towards the Main Protease (3CLPRO/MPRO) of SARS-CoV-2: In Silico Admet Profiling, Molecular Docking, Molecular Dynamics Simulation, Network Pharmacology. *Mol. Divers.* **2022**, 1–15. [CrossRef]

62. Vajda, F.J.E. Neuroprotection and Neurodegenerative Disease. *J. Clin. Neurosci.* **2002**, *9*, 4–8. [CrossRef] [PubMed]
63. Sheikh, S.; Safia; Haque, E.; Mir, S.S. Neurodegenerative Diseases: Multifactorial Conformational Diseases and Their Therapeutic Interventions. *J. Neurodegener. Dis.* **2013**, *2013*, 563481. [CrossRef] [PubMed]
64. Faden, A.I.; Stoica, B. Neuroprotection. *Arch. Neurol.* **2007**, *64*, 794. [CrossRef] [PubMed]
65. Lane, C.A.; Hardy, J.; Schott, J.M. Alzheimer's Disease. *Eur. J. Neurol.* **2018**, *25*, 59–70. [CrossRef] [PubMed]
66. Wimo, A.; Jönsson, L.; Bond, J.; Prince, M.; Winblad, B. The Worldwide Economic Impact of Dementia 2010. *Alzheimer's Dement.* **2013**, *9*, 1–11.e3. [CrossRef] [PubMed]
67. Ramirez-Bermudez, J. Alzheimer's Disease: Critical Notes on the History of a Medical Concept. *Arch. Med. Res.* **2012**, *43*, 595–599. [CrossRef] [PubMed]
68. Anand, R.; Gill, K.D.; Mahdi, A.A. Therapeutics of Alzheimer's Disease: Past, Present and Future. *Neuropharmacology* **2014**, *76*, 27–50. [CrossRef]
69. Bertram, L.; Lill, C.M.; Tanzi, R.E. The Genetics of Alzheimer Disease: Back to the Future. *Neuron* **2010**, *68*, 270–281. [CrossRef]
70. Day, J.J.; Sweatt, J.D. Epigenetic Mechanisms in Cognition. *Neuron* **2011**, *70*, 813–829. [CrossRef]
71. Terry, R.D. Some Unanswered Questions about the Mechanisms and Etiology of Alzheimer's Disease. *Dan. Med. Bull.* **1985**, *32* (Suppl. S1), 22–24.
72. Christen, Y. Oxidative Stress and Alzheimer Disease. *Am. J. Clin. Nutr.* **2000**, *71*, 621S–629S. [CrossRef] [PubMed]
73. Golde, T.E. Disease Modifying Therapy for AD? *J. Neurochem.* **2006**, *99*, 689–707. [CrossRef]
74. Hroudová, J.; Singh, N.; Fišar, Z.; Ghosh, K.K. Progress in Drug Development for Alzheimer's Disease: An Overview in Relation to Mitochondrial Energy Metabolism. *Eur. J. Med. Chem.* **2016**, *121*, 774–784. [CrossRef] [PubMed]
75. Gamber, K.M. Animal Models of Parkinson's Disease: New Models Provide Greater Translational and Predictive Value. *Biotechniques* **2016**, *61*, 210–211. [CrossRef]
76. Kin, K.; Yasuhara, T.; Kameda, M.; Date, I. Animal Models for Parkinson's Disease Research: Trends in the 2000s. *Int. J. Mol. Sci.* **2019**, *20*, 5402. [CrossRef] [PubMed]
77. Duong, S.; Patel, T.; Chang, F. Dementia. *Can. Pharm. J. Rev. Pharm. Du Can.* **2017**, *150*, 118–129. [CrossRef] [PubMed]
78. Schapira, A.H.; Jenner, P. Etiology and Pathogenesis of Parkinson's Disease. *Mov. Disord.* **2011**, *26*, 1049–1055. [CrossRef]
79. Colpo, G.D.; Ribeiro, F.M.; Rocha, N.P.; Teixeira, A.L. Animal Models for the Study of Human Neurodegenerative Diseases. In *Animal Models for the Study of Human Disease*; Elsevier: Amsterdam, The Netherlands, 2017; pp. 1109–1129.
80. Tutar, Y.; Zgur, A.; Tutar, L. Role of Protein Aggregation in Neurodegenerative Diseases. In *Neurodegenerative Diseases*; InTech: London, UK, 2013. [CrossRef]
81. Huang, Y. Molecular and Cellular Mechanisms of Apolipoprotein E4 Neurotoxicity and Potential Therapeutic Strategies. *Curr. Opin. Drug. Discov. Devel.* **2006**, *9*, 627–641.
82. Uttara, B.; Singh, A.; Zamboni, P.; Mahajan, R. Oxidative Stress and Neurodegenerative Diseases: A Review of Upstream and Downstream Antioxidant Therapeutic Options. *Curr. Neuropharmacol.* **2009**, *7*, 65–74. [CrossRef]
83. Dourado, N.S.; Souza, C.D.S.; de Almeida, M.M.A.; Bispo da Silva, A.; dos Santos, B.L.; Silva, V.D.A.; De Assis, A.M.; da Silva, J.S.; Souza, D.O.; Costa, M.D.F.D.; et al. Neuroimmunomodulatory and Neuroprotective Effects of the Flavonoid Apigenin in in Vitro Models of Neuroinflammation Associated With Alzheimer's Disease. *Front. Aging Neurosci.* **2020**, *12*, 119. [CrossRef]
84. Li, L.; Zhang, M.X.; Wang, X.Y.; Yang, Y.L.; Gong, X.; Wang, C.C.; Xu, J.F.; Li, M.H. Assessment of Components of Gingko Biloba Leaves Collected from Different Regions of China That Contribute to Its Antioxidant Effects for Improved Quality Monitoring. *Food Sci. Technol.* **2021**, *41*, 676–683. [CrossRef]
85. Kim, S.J. Effect of Biflavones of Ginkgo Biloba against UVB-Induced Cytotoxicity in Vitro. *J. Dermatol.* **2001**, *28*, 193–199. [CrossRef]
86. Liu, X.; Bian, H.; Dou, Q.-L.; Huang, X.-W.; Tao, W.-Y.; Liu, W.-H.; Li, N.; Zhang, W.-W. Ginkgetin Alleviates Inflammation, Oxidative Stress, and Apoptosis Induced by Hypoxia/Reoxygenation in H9C2 Cells via Caspase-3 Dependent Pathway. *Biomed Res. Int.* **2020**, *2020*, 1928410. [CrossRef]
87. Cheon, B.S.; Kim, Y.H.; Son, K.S.; Chang, H.W.; Kang, S.S.; Kim, H.P. Effects of Prenylated Flavonoids and Biflavonoids on Lipopolysaccharide-Induced Nitric Oxide Production from the Mouse Macrophage Cell Line RAW 264.7. *Planta Med.* **2000**, *66*, 596–600. [CrossRef]
88. Alam, Q.; Zubair Alam, M.; Mushtaq, G.; Damanhouri, G.A.; Rasool, M.; Amjad Kamal, M.; Haque, A. Inflammatory Process in Alzheimer's and Parkinson's Diseases: Central Role of Cytokines. *Curr. Pharm. Des.* **2016**, *22*, 541–548. [CrossRef]
89. Chiang, T.; Messing, R.O.; Chou, W.-H. Mouse Model of Middle Cerebral Artery Occlusion. *J. Vis. Exp.* **2011**, *48*, e2761. [CrossRef]
90. Xu, B.; He, X.; Sui, Y.; Wang, X.; Wang, X.; Ren, L.; Zhai, Y.-X. Ginkgetin Aglycone Attenuates Neuroinflammation and Neuronal Injury in the Rats with Ischemic Stroke by Modulating STAT3/JAK2/SIRT1. *Folia Neuropathol.* **2019**, *57*, 16–23. [CrossRef]
91. Tian, Z.; Tang, C.; Wang, Z. Neuroprotective Effect of Ginkgetin in Experimental Cerebral Ischemia/Reperfusion via Apoptosis Inhibition and PI3K/Akt/MTOR Signaling Pathway Activation. *J. Cell. Biochem.* **2019**, *120*, 18487–18495. [CrossRef]
92. Pan, J.; Li, X.; Guo, F.; Yang, Z.; Zhang, L.; Yang, C. Ginkgetin Attenuates Cerebral Ischemia–Reperfusion Induced Autophagy and Cell Death via Modulation of the NF-KB/P53 Signaling Pathway. *Biosci. Rep.* **2019**, *39*, BSR20191452. [CrossRef]
93. Tang, T.; Wang, X.; Qi, E.; Li, S.; Sun, H. Ginkgetin Promotes M2 Polarization of Microglia and Exert Neuroprotection in Ischemic Stroke via Modulation of PPARγ Pathway. *Neurochem. Res.* **2022**, *47*, 2963–2974. [CrossRef]

94. Deshpande, P.; Gogia, N.; Singh, A. Exploring the Efficacy of Natural Products in Alleviating Alzheimer's Disease. *Neural Regen. Res.* **2019**, *14*, 1321. [CrossRef]
95. Iqbal, U.H.; Zeng, E.; Pasinetti, G.M. The Use of Antimicrobial and Antiviral Drugs in Alzheimer's Disease. *Int. J. Mol. Sci.* **2020**, *21*, 4920. [CrossRef]
96. Citron, M. β-Secretase Inhibition for the Treatment of Alzheimer's Disease—Promise and Challenge. *Trends Pharmacol. Sci.* **2004**, *25*, 92–97. [CrossRef]
97. Sasaki, H.; Miki, K.; Kinoshita, K.; Koyama, K.; Juliawaty, L.D.; Achmad, S.A.; Hakim, E.H.; Kaneda, M.; Takahashi, K. β-Secretase (BACE-1) Inhibitory Effect of Biflavonoids. *Bioorg. Med. Chem. Lett.* **2010**, *20*, 4558–4560. [CrossRef]
98. Ullah, M.A.; Johora, F.T.; Sarkar, B.; Araf, Y.; Ahmed, N.; Nahar, A.N.; Akter, T. Computer-Assisted Evaluation of Plant-Derived β-Secretase Inhibitors in Alzheimer's Disease. *Egypt. J. Med. Hum. Genet.* **2021**, *22*, 26. [CrossRef]
99. Grewal, A.S.; Sharma, N.; Singh, S.; Kanojia, N.; Thapa, K.; Swami, R.; Grover, R. Molecular Docking Guided Screening of Phenolic Compounds from Ginkgo Biloba as Multi-Potent Anti Alzheimer's Agents. *Plant Arch.* **2020**, *20*, 3297–3304.
100. Choi, E.Y.; Kang, S.S.; Lee, S.K.; Han, B.H. Polyphenolic Biflavonoids Inhibit Amyloid-Beta Fibrillation and Disaggregate Preformed Amyloid-Beta Fibrils. *Biomol. Ther.* **2020**, *28*, 145–151. [CrossRef]
101. Zeng, Y.-Q.; Wang, Y.-J.; Zhou, X.-F. Ginkgetin Ameliorates Neuropathological Changes in App/Ps1 Transgenical Mice Model. *J. Prev. Alzheimer's Dis.* **2015**, *3*, 24–29. [CrossRef]
102. Xiong, Y.; Zhu, G.-H.; Wang, H.-N.; Hu, Q.; Chen, L.-L.; Guan, X.-Q.; Li, H.-L.; Chen, H.-Z.; Tang, H.; Ge, G.-B. Discovery of Naturally Occurring Inhibitors against SARS-CoV-2 3CLpro from Ginkgo Biloba Leaves via Large-Scale Screening. *Fitoterapia* **2021**, *152*, 104909. [CrossRef]
103. Li, G.; Wang, G.; Wang, S.; Deng, Y. Ginkgetin in Vitro and in Vivo Reduces Streptococcus Suis Virulence by Inhibiting Suilysin Activity. *J. Appl. Microbiol.* **2019**, *127*, 1556–1563. [CrossRef]
104. Salari, S.; Bagheri, M. In Vivo, in Vitro and Pharmacologic Models of Parkinson's Disease. *Physiol. Res.* **2019**, *68*, 17–24. [CrossRef]
105. Wang, Y.-Q.; Wang, M.-Y.; Fu, X.-R.; Peng-Yu; Gao, G.-F.; Fan, Y.-M.; Duan, X.-L.; Zhao, B.-L.; Chang, Y.-Z.; Shi, Z.-H. Neuroprotective Effects of Ginkgetin against Neuroinjury in Parkinson's Disease Model Induced by MPTP via Chelating Iron. *Free Radic. Res.* **2015**, *49*, 1069–1080. [CrossRef]
106. Wang, Y.; Cheng, R.; Wu, X.; Miao, M. Neuroprotective and Neurotrophic Effects of Ginkgetin and Bilobalide on MPTP-Induced Mice with Parkinson' Disease. *Pharmazie* **2021**, *76*, 27–33. [CrossRef]

Disclaimer/Publisher's Note: The statements, opinions and data contained in all publications are solely those of the individual author(s) and contributor(s) and not of MDPI and/or the editor(s). MDPI and/or the editor(s) disclaim responsibility for any injury to people or property resulting from any ideas, methods, instructions or products referred to in the content.

 *life*

Article

Wound Healing and Anti-Inflammatory Effects of a Newly Developed Ointment Containing Jujube Leaves Extract

Marilena-Viorica Hovaneţ [1,†], Emma Adriana Ozon [2,*], Elena Moroşan [3,*], Oana Cristina Şeremet [4,†], Eliza Oprea [5], Elisabeta-Irina Geană [6], Adriana Iuliana Anghel [1], Carmellina Bădiceanu [7], Ligia Elena Duţu [8], Cristina Silvia Stoicescu [9], Eugenia Nagoda [10] and Robert Ancuceanu [1]

1 Department of Pharmaceutical Botany and Cell Biology, Faculty of Pharmacy, Carol Davila University of Medicine and Pharmacy, 6 Traian Vuia Street, 020945 Bucharest, Romania
2 Department of Pharmaceutical Technology and Biopharmacy, Faculty of Pharmacy, Carol Davila University of Medicine and Pharmacy, 6 Traian Vuia Street, 020945 Bucharest, Romania
3 Department of Clinical Laboratory and Food Safety, Faculty of Pharmacy, Carol Davila University of Medicine and Pharmacy, 6 Traian Vuia Street, 020945 Bucharest, Romania
4 Department of Pharmacology and Clinical Pharmacy, Faculty of Pharmacy, Carol Davila University of Medicine and Pharmacy, 6 Traian Vuia Street, 020945 Bucharest, Romania
5 Department of Botany and Microbiology, Faculty of Biology, University of Bucharest, 1-3 Aleea Portocalelor, 060101 Bucharest, Romania
6 National R&D Institute for Cryogenics and Isotopic Technologies—ICIT, 4th Uzinei Street, 240050 Râmnicu Vâlcea, Romania
7 Department of Pharmaceutical Chemistry, Faculty of Pharmacy, Carol Davila University of Medicine and Pharmacy, 6 Traian Vuia Street, 020945 Bucharest, Romania
8 Department of Pharmacognosy, Phytochemistry and Phytotherapy, Faculty of Pharmacy, Carol Davila University of Medicine and Pharmacy, 6 Traian Vuia Street, 020945 Bucharest, Romania
9 Ilie Murgulescu Institute of Physical Chemistry, Spl. Independentei 202, P.O. Box 12-194, 060021 Bucharest, Romania
10 Botanical Garden "D. Brandza", University of Bucharest, Şos. Cotroceni 32, 060114 Bucharest, Romania
* Correspondence: emma.budura@umfcd.ro (E.A.O.); elena.morosan@umfcd.ro (E.M.)
† These authors contributed equally to this work.

Citation: Hovaneţ, M.-V.; Ozon, E.A.; Moroşan, E.; Şeremet, O.C.; Oprea, E.; Geană, E.-I.; Anghel, A.I.; Bădiceanu, C.; Duţu, L.E.; Stoicescu, C.S.; et al. Wound Healing and Anti-Inflammatory Effects of a Newly Developed Ointment Containing Jujube Leaves Extract. *Life* **2022**, *12*, 1947. https://doi.org/10.3390/life12121947

Academic Editors: Azahara Rodríguez-Luna and Salvador González

Received: 2 November 2022
Accepted: 19 November 2022
Published: 22 November 2022

Publisher's Note: MDPI stays neutral with regard to jurisdictional claims in published maps and institutional affiliations.

Copyright: © 2022 by the authors. Licensee MDPI, Basel, Switzerland. This article is an open access article distributed under the terms and conditions of the Creative Commons Attribution (CC BY) license (https://creativecommons.org/licenses/by/4.0/).

Abstract: *Ziziphus jujuba* Mill. (jujube) is a well-known medicinal plant with pronounced wound healing properties. The present study aimed to establish the chemical composition of the lyophilized ethanolic extract from Romanian *Ziziphus jujuba* leaves and to evaluate the healing and anti-inflammatory properties of a newly developed lipophilic ointment containing 10% dried jujube leaves extract. The ultra-High-Performance Liquid Chromatography Electrospray Ionization Tandem Mass Spectrometry method was used, and 47 compounds were detected, among them the novel epicatechin and caffeic acid. The extract contains significant amounts of rutin (29.836 mg/g), quercetin (15.180 mg/g) and chlorogenic acid (350.96 μg/g). The lipophilic ointment has a slightly tolerable pH, between 5.41–5.42, and proved to be non-toxic in acute dermal irritation tests on New Zealand albino rabbits and after repeated administration on Wistar rats. The ointment also has a healing activity comparable to Cicatrizin (a pharmaceutical marketed product) on Wistar rats and a moderate anti-inflammatory action compared to the control group, but statistically insignificant compared to indomethacin in the rat-induced inflammation test by intraplantar administration of kaolin. The healing and anti-inflammatory properties of the tested ointment are due to phenolic acids and flavonoids content, less because of minor components as apocynin, scopoletin, and isofraxidin.

Keywords: *Ziziphus jujuba* leaves; rutin; quercetin; chlorogenic acid; lipophilic ointment; healing activity; anti-inflammatory properties

1. Introduction

Research concerning healing wound therapy focuses on finding new herbal remedies that are assumed to have fewer side effects, lower cost and similar efficacy compared to conventional synthetic drugs [1,2]. Wound healing involves mainly an inflammation process with vasoconstriction and mediators release, the proliferation of fibroblasts and keratinocytes, formation of granulation tissue, and maturation with collagen fibers remodeling [3–7]. There are different classes of compounds which promote wound healing, such as phenolic derivatives, flavonoids, non-flavonoid polyphenols (phenolic acids as caffeic acid, chlorogenic acid), tannins, lignans, and essential oils such as lavender, chamomile, tea tree, thyme, ocimum oil, isoquinoline alkaloids from the *Papaveraceae* and *Berberidaceae* families, terpenes, saponins, and phloroglucinol derivatives (arzanol) [8–12]. These bioactive substances manifest antioxidant, anti-inflammatory, antimicrobial or antifungal properties, positively influencing wound healing by preventing the development of pathogens, enhancing cell proliferation, increasing collagen production, improving wound contraction, and promoting epithelialization, vascularization and a normal regeneration avoiding fibrosis [13–17].

Among the notable species used throughout the world for their wound healing action are *Plantago major* L., *Plantago lanceolata* L. [18], *Calendula officinalis* L. [19], *Aloe vera* (L.) Burm.f. [20], *Hypericum perforatum* L. [21,22], *Achillea millefolium* L. [23], *Matricaria chamomilla* L. [24,25], *Centella asiatica* (L.) Urb. [26], *Symphytum officinale* L. [27], and *Helichrysum italicum* (Roth) G. Doṅ [28,29]. Another plant with considerable wound healing and anti-inflammatory potential is *Ziziphus jujuba* Mill. (jujube), from the *Rhamnaceae* family. It is a native species of China, found today in temperate and subtropical climates. It is known for its nutritional (from its fruits) and medicinal value, several of its organs (leaves, fruits, seeds, and bark) being used in various ailments [30,31]. Its leaves are traditionally used to treat bleeding, boils, and diarrhea [32], for weight loss purposes [33], and to heal wounds and aphthous ulcers [34]. It contains phenolic derivates, especially phenolic acids, flavonoids, tannins, damarane-type saponins, triterpene acids, and cyclopeptide alkaloids [35,36].

The healing effects and the anti-inflammatory properties of the species' leaves harvested from Romania have been attributed to phenolic acids and flavonoids, which are predominant, and the most active principles have been previously evaluated in two preclinical studies [37,38]. The principles examined include its anti-inflammatory, antioxidant, and anti-allergic property and its pain reducing and boosting of collagen synthesis property [39,40].

This study aimed to characterize chromatographically the phenolic and flavonoid content and other minor compounds of ethanolic dried extract obtained from *Ziziphus jujuba* Mill. leaves harvested from Romania. The study also aimed to evaluate in vivo the healing and anti-inflammatory properties of the dried ethanolic extract leaves after inclusion into a hydrophobic ointment base. For the selection of the ointment base, the aims were to choose safe, biocompatible, emollient, and inexpensive ingredients. Cholesterol is an important component of the extracellular lipophilic matrix of stratum corneum, one that ensures the skin barrier function [41]. It has been proven that it has a beneficial effect on damaged skin [42]. It has a higher melting point than the other components but will dissolve in their mixture. Cetyl alcohol has good moisturizing qualities, and skin-protective characteristics useful for skin irritations caused by stings, bites, and rashes [43]. Vaseline has, also, an emollient role, being recommended by the American Academy of Dermatology for the moisturization of skin injuries [44]. Different compendial tests were used to analyze the semi-solid pharmaceutical dosage form to establish its pharmaco-technical properties. Dermatological irritation testing was also performed. All investigations were carried out in comparison to the base alone, and for in vivo activity assessment, two pharmaceutical products were used as references.

2. Materials and Methods

2.1. Materials

The dried ethanolic extract of *Ziziphus jujuba* Mill. leaves of indigenous plant harvested from Research Institute for Fruit Growing Pitesti, Romania was obtained according to the method indicated in a previously published paper [37]. Briefly, the dry leaves were powdered, then refluxed three times with 70% ethyl alcohol (m/V, the ratio between the herbal product and solvent being 1:10). The solutions thus obtained were mixed and concentrated at 60 °C in an Ingos RVO 004 rotary evaporator. The concentrated solution was subjected to a lyophilization process using a Scanvac CoolSafe Freeze Dryer.

Cetyl alcohol, cholesterol, refined coconut oil and petrolatum were provided by Fagron, Greece. Butylhydroxyanisole was purchased from Merck KGaA, Germany. Kaolin was purchased from Health Chemicals Co., Ltd., Zhangjiagang City, China and urethane from Sigma-Aldrich, Hamburg, Germany. Cicatrizin, produced by Pharmaceutical TIS, Bucharest, Romania, is an ointment that contains extracts of *Hypericum perforatum* (St. John's wort), *Calendula officinalis* (marigold), *Symphytum officinale* (sorrel), *Plantago lanceolata* (plantain) and *Chamomilla recutita* (chamomile) containing as active principles phenolic acids, flavonoids and essential oil.

Indomethacin 40 mg/g ointment, used as a reference product in the anti-inflammatory assessment, is produced by Hyperion, Iași, Romania.

2.2. Methods

2.2.1. Ointment Production and Pharmacotechnical Assessment

Formulation

For the formulation of the lipophilic semi-solid pharmaceutical dosage form containing 10% (*w/w*), *Ziziphus jujuba* Mill. leaves extract, cetyl alcohol, cholesterol and Vaseline were selected to form a single-phase basis suitable for active ingredient suspension. The coconut oil was chosen to adjust the ointment consistency and also for its antioxidant and natural fragrance properties [45–47]. To ensure the stability of the product, butylhydroxyanisole, as an antioxidant and preservative agent, was added [48,49]. The formulation is presented in Table 1.

Table 1. Formulation of the semi-solid pharmaceutical form for cutaneous application.

Ingredients	Quantity (g)
Ziziphus jujuba dried ethanolic leaves extract	10.00
Cetyl alcohol	2.00
Cholesterol	2.00
Petrolatum	80.70
Coconut oil	5.00
Butylhydroxyanisole	0.30
TOTAL	100.00

Production

All ingredients were weighed according to the amounts mentioned in the formulation, using a Mettler Toledo AT261 (0.01 mg sensitivity) balance. The hydrophobic components (cetyl alcohol, cholesterol and Vaseline) were melted together on a water bath heated at about 50 °C, and then the basis was cooled to 35 °C when butylhydroxyanisole was added and dissolved. *Ziziphus jujuba* Mill. leaves extract was first mixed with the coconut oil, and then the lipophilic base was added, continuing the stirring at 700 rpm, at room temperature.

The ointment base to be used alone as a control for the assessment tests, was prepared similarly but without extract inclusion.

Quality Control

Organoleptic Properties and Homogeneity

The organoleptic characterization included appearance, consistency, and homogeneity, together with absence of phase separation, and instabilities of color [50,51]. Grit and consistency were assessed by touch. The homogeneity was determined according to Romanian Pharmacopoeia requirements by spreading 0.5 g of the ointment in a thin layer on a glass slide and examining it with a hand magnifier (4.5×) [52]. The appearance, absence of phase separation and instabilities of color were assessed by visual observation.

pH

The pH measurements were determined as the European Pharmacopoeia recommendations by the potentiometric method [53]. An inoLab level 1 pH meter, produced by WTW GmbH & Co. KG, Weilheim, Germany, was used. It was previously calibrated with a 7.00 pH buffer solution. 0.5 g of ointment was mixed with 10 mL of water by cold stirring, then filtered and the pH of the filtrate was recorded at 19.2 °C. The pH was measured six times for each sample (ointment and base) and the average value and standard deviation were reported.

Ultra-High-Performance Liquid Chromatography Electrospray Ionization Tandem Mass Spectrometry

Target phenolic acids and flavonoids analysis was performed with an UltiMate 3000 UHPLC System, coupled with a Q Exactive Focus Hybrid Quadrupole-Orbitrap mass spectrometer equipped with Heated Electrospray Ionisation (HESI) probe, all from Thermo Fisher Scientific, Bremen, Germany. Separations were performed on Kinetex (C18, 100 × 2.1 mm, 1.7 µm, Phenomenex, Torrance, CA, USA) column (reverse-phase UHPLC column) and a gradient elution of a binary solvent system consisting of solvent A (water with 0.1% formic acid) and solvent B (methanol with 0.1% formic acid). Mass spectra were recorded in the negative ionization mode in the 100–1200 *m/z* range, at 70,000 resolution. Nitrogen was used as collision, sheath, and auxiliary gas at 11–48 arbitrary unit flow rates. The spray voltage was 2.5 kV, and the capillary temperature 320 °C. The energy of the collision-induced dissociation cell was varied in the 30–60 eV range. Calibrations were carried out in the 50–2000 µg/L concentration range, by serial dilution of the 10 mg/L methanolic standard mix. The lyophilized jujube leaves extract was dissolved in methanolic solution and filtered through a 0.45 µm polytetrafluoroethylene membrane before injection into the UHPLC-MS system. Quantitative data were evaluated by the Quan/Qual Browser Xcalibur 2.3 (Thermo Fisher Scientific). The mass tolerance window was set to 5 ppm for the two analysis modes. Individual phenolic acids and flavonoid contents were reported as µg/g lyophilized jujube leaves extract. Also, data processing, analysis, and interpretation using Compound Discoverer v. 2.1 (Thermo Scientific, Waltham, MA, USA) software was performed using an untargeted metabolomics working template.

Spreadability

For materials in the semi-solid form, spreadability is an essential characteristic that reunites the rheological and structural properties. It is an important test in the assessment of topical semi-solid products, as it can accurately predict the behavior during dose disposal and application. The ointments' spreadability mostly depends on the consistency and flowability of the base, but in the case of incorporating high amounts of active ingredients, they can influence the final performance.

The spreadability was determined by using the extensiometric method, analyzing the deformation ability of the product when different weights were applied. The procedure was performed on both the base alone and on the pharmaceutical ointment, in triplicate. The device consists of two square plates of glass with 11 cm sides. The bottom plate is positioned over a millimetric graph paper on which five concentric circles are drawn. On the lower plate in the center of the first circle, 1 g of the sample was brought, then the second glass plate was placed. The diameter of the circle occupied by the ointment, after pressing with the glass plate weighing 145 g, was registered. At intervals of one minute, on the top plate of the extensiometer, weights of 50, 100, 200, and 500 g were gradually applied. The diameters of the circles formed by sample spreading were recorded each time [54–56].

The spreadability is calculated by the equation:

$$S = \pi r^2 \tag{1}$$

where S is the spreading area in mm^2 and r is the radius in mm.

2.2.2. In Vivo Evaluation

The experiments were performed on animals purchased from the Cantacuzino Institute Biobase (Bucharest). The animals were acclimatized to laboratory conditions for five days before the start of the experiments. The room temperature during the treatment was 23 ± 1 °C, and the relative humidity was 50 ± 2%. The lighting was artificial, with a succession of 12 h of light and 12 h of darkness. The animals had unlimited access to conventional laboratory water and food (grains for mice and rats, Cantacuzino Institute, Bucharest).

All the experiments complied with Directive 2010/63/EU of the European Parliament and of the Council of 22 September 2010 on the protection of animals used for scientific purposes and the implementing Law no. 43/2014 on the protection of animals used for scientific purposes and were approved by the Bioethics Commission of the Faculty of Pharmacy, University of Medicine and Pharmacy Carol Davila, Bucharest (997/7 October 2016).

Local Tolerability

Two tests were performed to determine the local tolerability of the newly formulated ointment, in accordance with the relevant OECD guidelines: acute irritation/corrosion and dermal irritation after repeated administration for 21 days. The OECD Guidelines for the Testing of Chemicals is a set of testing methods developed by experts from the OECD and used by various governments, companies, and independent laboratories to identify and characterize potential hazards of chemicals [57] and are often used in the assessment of herbal extracts [58].

Determination of Acute Dermal Irritation/Corrosion (OECD 404)

To determine acute dermal irritation [59], the sample is applied in a single dose to the skin of the experimental animal. Untreated areas of the animal's skin serve as a control. The degree of irritation/corrosion is determined at specified intervals, and is described in detail to provide a full assessment of the effects. The study's duration should be sufficient to assess the reversibility and irreversibility of the irritant/corrosive action.

A male New Zealand albino rabbit (4.55 ± 0.07 kg) was used. About 24 h before the test, the fur on the back was removed very carefully, so as not to damage the skin, using scissors and an electric razor.

The studied ointment (ZIZ-L) (0.5 g) and lipophilic base (B-L) (0.5 g) were applied on a self-adhesive patch on a soft non-woven pad Cosmopore Advance 7.2 × 5 cm (Hartmann, Germany) and then affixed to the back of the rabbit, at a 5 cm distance from each other.

After 4 h of application, the patch was detached, and the ointment was removed by gently wiping it with a cloth soaked in water.

Following patch removal, the rabbit was evaluated for signs of erythema and/or edema immediately, and at 1, 24, 48 and 72 h. The observed dermal reactions were classified according to Table 2.

Table 2. Dermal reactions classification.

Dermal Reactions to the ZIZ-L and B-L Application	
Erythema or ulceration	Score
No erythema	0
Mild erythema (difficult to perceive)	1
Well defined erythema	2
Medium to severe erythema	3
Severe erythema until ulceration is formed	4
Edema occurs	
No edema	0
Very mild edema (difficult to perceive)	1
Mild edema	2
Medium edema (height approx. 1 mm)	3
Severe edema (greater than 1 mm in height and extending beyond the exposed surface)	4
A histopathological examination should be performed to clarify an equivocal response	

Determination of Dermal Irritation after Repeated Administration (OECD 410)

A determination of subchronic dermal toxicity [60] can be performed after obtaining initial information by testing for acute dermal toxicity. The determination of subchronic dermal toxicity provides information on the potential health risks that may result from repeated dermal exposure over a limited period of 21 or 28 days. For the present study, 21 days were used.

If a dose of at least 1000 mg/kg bw of a sample does not produce detectable toxic effects, it is not necessary to use three levels of concentration. Previous research has shown that the dried ethanolic extract obtained from *Ziziphus jujuba* Mill. leaves is virtually nontoxic after single dose administration [38], and therefore, a full study is not required.

The sample is applied daily to the skin of the experimental animals (rats), in graduated doses, using several groups of laboratory animals, one dose for each group, for 21 days. During the application period, the animals are observed daily for signs of toxicity. Rats dying during the test are necropsied, and at the end of the test, surviving and uncropped animals are sacrificed.

A community of 16 rats of 9 week old, Wistar strain, 8 females (214 $\pm$ 10 g) and 8 males (259 $\pm$ 13 g), were subjected to the tests. The animals were distributed into 4 groups, as follows:

- Group 1F: ZIZ-L-F: consisting of 4 females, who received the lipophilic ointment ZIZ-L;
- Group 1M: ZIZ-L-M: consisting of 4 males, who received the lipophilic ointment ZIZ-L;
- Group 2F: B-L-F: consisting of 4 females, who received the lipophilic base B-L;
- Group 2M: B-L-M: consisting of 4 males, who received the lipophilic base B-L.

The fur was removed from the dorsal area of the torso 24 h before the test. This operation was repeated at intervals of about a week.

Ointment and ointment base (an amount of ointment corresponding to the dose of 1000 mg/kg bw plant extract for the test batches and the appropriate amount of ointment base for the control batches, respectively) were applied on a self-adhesive patch on a soft non-woven pad Cosmopore Advance 7.2 $\times$ 5 cm (Hartmann, Germany) and then affixed to the animal's back to prevent it from gaining access to them. The animals were followed, the patch remaining set for at least 6 h after application. As OECD Guide 410 allows, ointments were applied 5 days/week for 21 days.

Animals were monitored, in particular, for changes in the skin, fur, and mucous membranes, as well as in somatic-motor activity and for behavioral changes. The animals were weighed weekly.

Wound Healing Activity

Male Wistar rats weighing 200 ± 10 g were used for the study. The rats were depilated in the dorsal area. After ethyl ether anesthesia, the animals suffered burn wounds using a metal device consisting of a disc with a 1 cm diameter which was heated in water with 5% NaCl at 105 °C. The heated disc was applied to the depilated dorsal area and held for 10 s [16,61–63].

The rats were distributed by the randomization method in groups of 10 animals and were treated as follows:

Group 1—control group, untreated;

Group 2—group treated with lipophilic ointment ZIZ-L;

Group 3—group treated with lipophilic base L-B;

Group 4—group treated with Cicatrizin ointment, taken as a reference product (it contains extracts of St. John's wort, papaya, chamomile, and marigold, herbs recognized for their beneficial effect in wound healing).

The treatment was given daily in a single application for 12 days. The evolution of the wounds was followed every two days by measuring the areas in the treated animals (in mm^2) and comparing them with those of the untreated controls, respectively, with those of the treatment with Cicatrizin, the reference product.

The clinical condition of the rats was also monitored during the study.

Anti-Inflammatory Activity

The anti-inflammatory action of a substance can be quantified by studying the effect of reducing rat paw edema induced by intraplantar administration of kaolin [64].

The animals (30 male rats, Wistar strain, 270 ± 32 g) were divided into three groups of 10 animals each, which were named according to the treatment received, as follows:

- Group ZIZ-L: ZIZ-L: lipophilic ointment;
- Group B-L: B-L lipophilic base;
- Group IND: Indomethacin HYPERION, ointment, 40 mg/g.

The rats were anesthetized with a 13% urethane solution, administered intraperitoneally, in a dose of 130 mg/kg body weight. After the installation of general anesthesia, the initial volume of the right paw was determined.

A quantity of 0.2 g of each ointment was applied to the surface of the right paw and massaged 50 times. Inflammation was generated by intraplantar administration of 0.2 mL kaolin 10% suspension, and the evolution of the induced edema was followed at 1, 2, 3 and 4 h.

The evolution of paw edema was calculated using the following formula (Vx is the paw volume measured x hours after the induction of inflammation, and V0 is the initial paw volume):

$$\% = (V\_x - V\_0)/V\_0 \times 100$$

Statistical Analysis

The statistical analysis was carried out using GraphPad Prism v. 5.0. (GraphPad Software, San Diego, CA, USA) and the computing and programming environment, R v. 4.2.0 (R Foundation for Statistical Computing, Vienna, Austria).

Results were expressed as mean ± standard deviation.

Distribution normality was estimated using the D'Agostino & Pearson global test [65]. The *t* Student test was applied to compare two groups. One-way ANOVA and Tukey's HSD were used to compare multiple groups. The statistical significance threshold was set at 0.05.

3. Results and Discussion

3.1. Ointment Quality Control

3.1.1. Organoleptic Properties and Homogeneity

A greasy, unctuous, dark green ointment with a characteristic coconut and plant odor was obtained. It contains *Ziziphus jujuba* Mill. leaves extract homogeneously suspended in the base, presented as fine particles, without the tendency to agglomerate or phase-separation. All these characteristics remained unchanged during the six months of preservation at room temperature.

3.1.2. pH

The pH of the lipophilic ointment was in the 5.41–5.42 range, while for the base the registered values were between 5.67 and 5.70. According to European Pharmacopoeia specifications, both semi-solid products have an easily tolerated pH, not being irritating when applied to the skin. After maintaining the products for six months at room temperature, a slight decrease in the pH values was remarked for both samples (5.36 for ointment and 5.62 for base).

3.1.3. Ultra-High-Performance Liquid Chromatography Electrospray Ionization Tandem Mass Spectrometry

The analytical approach based on a target UHPLC-ESI/MS analysis allows the quantification of some bioactive compounds responsible for the bioactive potential of jujube leaves extract. The recorded chromatogram is presented in Figure 1.

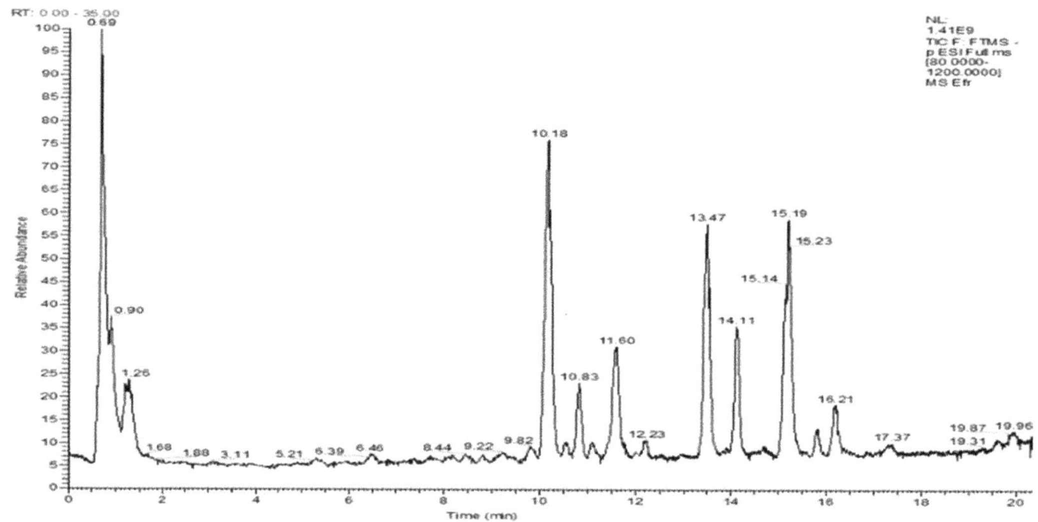

Figure 1. TIC chromatogram of *Ziziphus jujube* leaves extract in negative ionization mode.

Quantitative analysis indicates that chlorogenic, 3,4-dihydroxybenzoic, and syringic acids were the primarily phenolic acids identified, while quercetin and rutin were the main flavonoids in the *Ziziphus jujube* Mill. leaves, similar to results being obtained by Xue (2021) [66]. However, epicatechin and caffeic acid were not identified in the previous study.

The lyophilized jujube leaf extract contains relevant amounts of rutin (29.836 mg/g) and quercetin (15.180 mg/g), and also chlorogenic acid (350.96 µg/g). The compounds which could be responsible for the anti-inflammatory activity of the ointment obtained with *Ziziphus jujuba* Mill. leaves extract are quercetin [67,68], rutin [69], chlorogenic acid [70], catechin [71], pinostrobin [72], and ferulic acid [73].

The analytical approach based on non-target UHPLC-Q-Orbitrap HRMS analysis allows the identification of other bioactive compounds and specialized metabolites that occur in jujube leaf extract, which are also responsible for the anti-inflammatory activity. Phytochemical compounds such as flavonoids, organic acids, fatty acids, and other specific compounds such as apocynin, scopoletin, and isofraxidin show excellent anti-inflammatory activity [74–77].

The compound's name, retention time, exact mass, and accurate mass of m/z adduct ions in negative ESI mode for the identified compounds are shown in Table 3.

Table 3. Identification and quantification of phenolic bioactive compounds in lyophilized *Ziziphus jujuba* leaf extract by UHPLC-Q-Exactive high-accuracy analysis of deprotonated precursors and fragment ions of specific components.

Compound	Retention Time [min]	Exact Mass	Accurate Mass [M-H]$^-$	Experimental Adduct Ion (m/z)	Concentration (μg/g)
		Phenolic acids			
Gallic acid	1.65	170.0215	169.0142	169.0130	11.72
Syringic acid	3.61	198.0528	197.0455	197.0444	80.42
3,4-dihydroxybenzoic acid	3.92	154.0266	153.0193	153.0189	81.03
4-hydroxy benzoic acid	6.72	138.0316	137.0243	137.0230	69.51
p-coumaric acid	8.16	164.0473	163.0400	163.0387	31.39
Ferulic acid	8.84	194.0579	193.0506	193.0495	29.31
Caffeic acid	9.40	180.0422	179.0349	179.0337	24.19
Chlorogenic acid	9.749	354.0950	353.0877	353.0871	350.96
Cinnamic acid	10.18	148.0524	147.0451	147.0440	20.61
		Flavonoids			
Catechin	8.78	290.0790	289.0717	289.0712	96.45
Epicatechin	10.93				20.73
Pinocembrin	19.69	256.0735	255.0662	255.0656	1.29
Pinostrobin	15.39	270.0892	269.0819	269.0821	87.64
Chrysin	20.66	254.0579	253.0506	253.0498	7.66
Apigenin	18.89	270.0528	269.0455	269.0450	1.10
Quercetin	17.29	302.2357	301.0354	301.0347	15180.65
Isorhamnetin	18.01	316.0582	315.0509	315.0500	3.48
Kaemferol	18.57	286.0477	285.0404	285.0399	16.05
Galangin	20.93	270.0528	269.0455	269.0450	2.63
Rutin	15.18	610.1533	609.1460	609.1447	29,836.97
Naringin	16.86	580.1791	579.1718	579.1703	9.79
Quercetin-3-glucoside	15.06	464.0954	463.0881	463.0873	-
Kaempferol-3-glucoside (astragalin)	11.60	448.10056	447.0932	447.0955	-
Kaempferol-7-O-glucoside	16.11	448.10056	447.0932	447.0923	-
Quercetin 3,4′-diglucoside	14.04	626.1483	625.1410	625.1401	-
Procyanidin C	8.51	866.2058	865.1985	865.1974	-
Isorhamnetin-3-rutinoside	16.41	624.1690	623.1617	623.1609	-
Quercetin-3-(6-O-acetyl-beta-glucoside)	15.53	506.1060	505.0987	505.0978	-

Table 3. *Cont.*

Compound	Retention Time [min]	Exact Mass	Accurate Mass [M-H]$^-$	Experimental Adduct Ion (m/z)	Concentration (µg/g)
Quercetin-3-D-xyloside	15.61	434.0849	433.0776	433.0769	-
Kaempferol-3-O-arabinoside	16.36	418.0899	417.0827	417.0822	-
Kaempferol-O-rhamnoside	17.17	432.1056	431.0983	431.0974	-
		Fatty acids			
Trihydroxy octadecadienoic acid	19.69	328.2249	327.2177	327.2170	-
Trihydroxy octadecenoic acid	20.33	330.2406	329.2333	329.2327	-
Hydroxy octadecadienoic acid	24.10	296.2351	295.2278	295.2271	-
Linolenic acid	25.89	278.2245	277.2173	277.2166	-
		Organic acids			
Aconitic acid	0.90	174.0164	173.0091	173.0077	-
Itaconic acid	1.26	130.0266	129.0193	129.0177	-
Uric acid	0.86	168.0283	167.0210	167.0197	-
Quinic acid	0.69	192.0633	191.0561	191.0549	-
Malic acid	0.74	134.0215	133.0142	133.0127	-
Gluconic acid	0.69	196.0583	195.0510	195.0498	-
		Other compounds			
Apocynin	10.18	166.0629	165.0557	165.0542	-
Scopoletin	12.13	192.0422	191.0349	191.0338	-
Isofraxidin	8.44	222.0528	221.0455	221.0445	-
Azelaic acid	15.20	188.1048	187.0975	87.0963	-
3-p-Coumaroylquinic acid	12.23	338.1001	337.0929	337.0924	-

3.1.4. Spreadability

Figure 2 displays the variation of the surface occupied by 1 g of each sample depending on the applied weight, and the registered standard deviations.

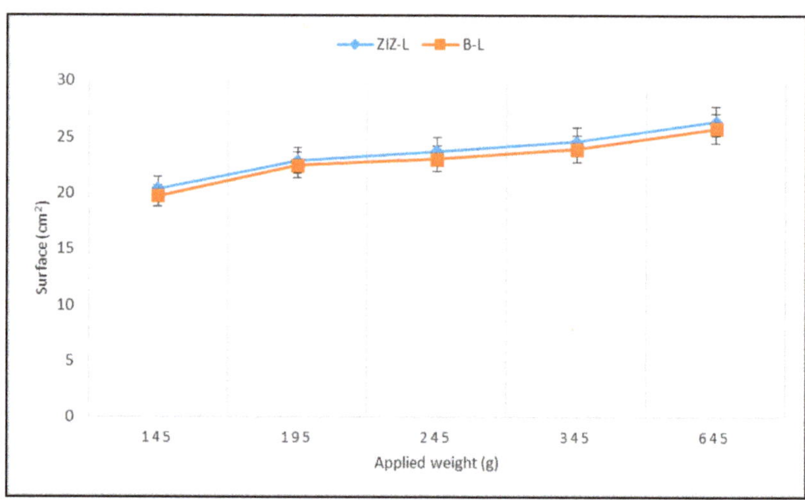

Figure 2. The Spreadability of Ointment and Base.

It is noted that the prepared ointment has proper plasticity, allowing it to easily spread on the skin [78]. When repeating the test after six months of storage, the registered values were similar to the initial ones, confirming the pharmaco-technical stability of the semi-solid product.

3.2. Local Tolerability

3.2.1. Determination of Acute Dermal Irritation/Corrosion (OECD 404)

Areas exposed to the newly formulated ointment, as well as those exposed to the ointment base, were examined 4 h after application on rabbit, immediately after removal of the self-adhesive patches.

In none of the cases was erythema or edema observed. The examination was repeated at 1, 24, 48, and 72 h after patch removal, and no signs of erythema or edema were detected (Table 4).

Table 4. Irritation/corrosion response scores for tested products.

Sample	Erythema/Edema Score				
	Immediately *	**1 h ***	**24 h ***	**48 h ***	**72 h ***
ZIZ-L	0	0	0	0	0
B-L	0	0	0	0	0

* after patches removal.

After testing for acute dermal irritability according to OECD Guideline 404, it can be stated that the ointment obtained from *Ziziphus jujuba* Mill. leaves is not irritating or corrosive following a single cutaneous application.

3.2.2. Determination of Dermal Irritation after Repeated Administration (OECD 410)

For dermal irritation after repeated administration tests, the experimental results regarding the evolution of body weight during the 21 days can be found in Table 5 and Figures 3 and 4.

Table 5. Differences in rats' body weight and statistical interpretation of the differences.

Group		Initially	Day 7	Day 14	Day 21
Group 1	M ± SD	231.4 ± 23.22	254.6 ± 16.72	272.5 ± 21.37	294.4 ± 16.47
	Δ% vs. initial	-	10.03	17.76	27.23
	t Student test (*p*)	-	*** 0.0006	*** <0.0001	*** <0.0001
Group 2	M ± SD	236.9 ± 28.52	253.9 ± 24.66	273.0 ± 20.82	287.8 ± 21.06
	Δ% vs. initial	-	7.18	15.24	21.49
	t Student test (*p*)	-	*** 0.0005	*** <0.0001	*** <0.0001

M = mean; SD = standard deviation; Δ = difference; *** $p < 0.001$.

The experimental results indicated no changes in the external appearance (fur, skin, mucus) or the motor performances of the rats in the two groups tested. Also, there were no alterations in somatic-motor activity or behavior.

For all tested batches, the body weight increased (statistically significant) throughout the treatment, which indicates the lack of toxicity of both the base and the ointment containing the *Ziziphus jujuba* Mill. dried leaves extract.

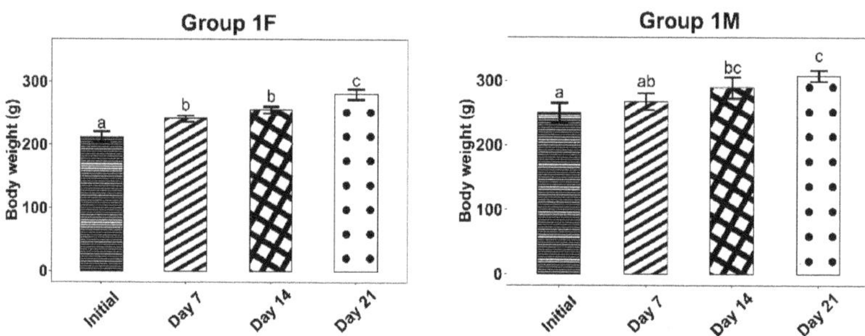

Figure 3. Average body weight evolution of the rats in Group 1, with standard deviations. Letters a–c show statistically significant differences between groups.

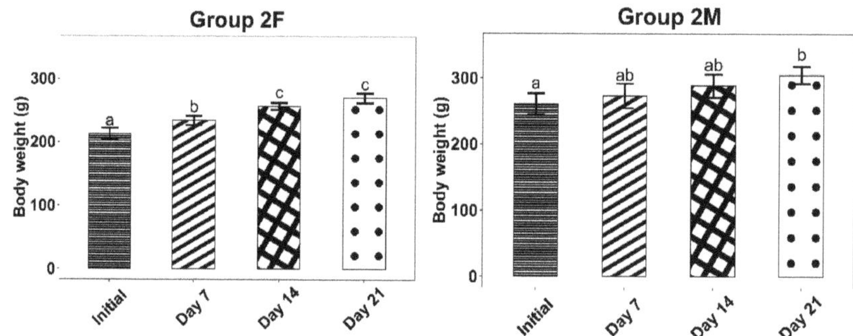

Figure 4. Average body weight evolution of the rats in Group 2, with standard deviations. Letters a–c show statistically significant differences between groups.

3.3. Wound Healing Activity

The results registered for the wound healing effect are shown in Table 6.

Table 6. The evolution of the wound healing effect.

Sample		Wound Surface (mm^2) $\overline{X} \pm SD$						
		Day 1	Day 2	Day 4	Day 6	Day 8	Day 10	Day 12
Group 1—control		99 ± 1.41	82.4 ± 4.33	73.6 ± 4.92	62.8 ± 4.20	49.2 ± 3.27	33.8 ± 5.44	23 ± 7.81
	E%	-	16.76	25.65	37.37	50.30	66.85	76.76
Group 2—ZIZ-L		95.6 ± 5.17	75 ± 4.15	66.6 ± 4.21 *	53.6 ± 2.96 *	43.8 ± 2.04 *	34.4 ± 3.20	20 ± 3.24
	E%	-	21.54	30.33	43.93	54.18	64.01	79.07
Group 3—L-B		96 ± 5.47	79.6 ± 4.92 *	70.8 ± 7.59 *	56.6 ± 3.50 *	46.6 ± 1.812	40 ± 5.14 *	28.2 ± 2.94 *
	E%		17.08	26.25	41.04	51.45	58.33	70.62
Group 4—Cicatrizin		92.6 ± 7.92	77 ± 7.81 *	61.4 ± 7.30 *	47 ± 4.94 *	37 ± 4.11 *	20.2 ± 4.38 **	13.8 ± 5.67 **
	E%	-	16.84	33.69	49.24	60.04	78.18	85.09

$\overline{X}$ ± SD = average ± standard deviation. Group 1—control group, untreated; Group 2—group treated with lipophilic ointment ZIZ-L; Group 3—group treated with lipophilic base L-B; Group 4—group treated with Cicatrizin ointment. Data were analyzed by Student's test. Statistical significance: * $p < 0.05$, ** $p < 0.01$ compared to initial.

The experimental results on the wound healing are summarized in Figures 5 and 6 and Table 6.

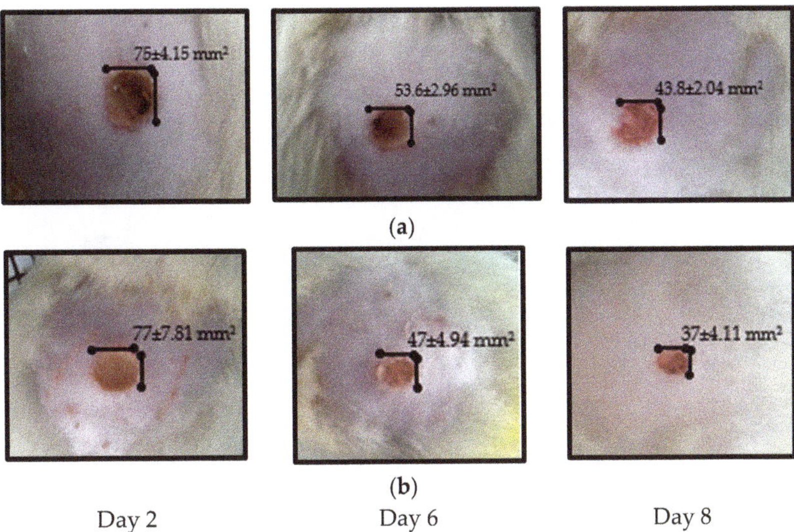

Figure 5. The scarring evolution of the animals treated with (**a**) the lipophilic ointment and (**b**) Cicatrizin.

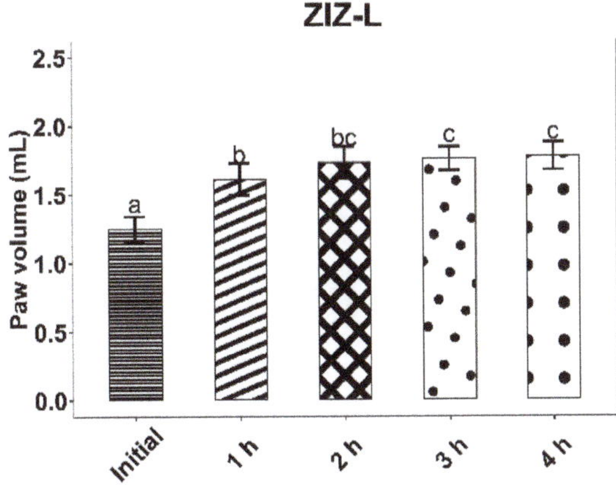

Figure 6. Kaolin-induced edema values, with standard deviations, for the ZIZ-L group. Letters a–c show statistically significant differences between groups.

The control rats showed an initial burn area of 99 mm^2; after 12 days, it was sized 23 mm^2, and showed a cure of 76.76% compared to the primary stage. Total healing occurred after 26 days.

The ointment used in the study as a reference product (Cicatrizin) resulted in 85.09% healing after 12 days of treatment and a burn area of 13.8 mm^2. Complete recovery occurred after 18 days.

The lipophilic *Ziziphus jujuba* Mill. ointment generated wound healing of 79.07% healing compared to the first day of treatment and a burn area of 20 mm^2. Complete recovery occurred after 18 days.

In the case of the animals treated with lipophilic base, 70.62% of healing was registered after 12 days and the burn area reached 28.2 mm^2. Complete recovery occurred after 20 days.

3.4. *Anti-Inflammatory Activity*

The experimental results on the evolution of edema induced by intraplantar injection of 0.2 mL 10% kaolin suspension are found in Figures 6–8 and Table 7.

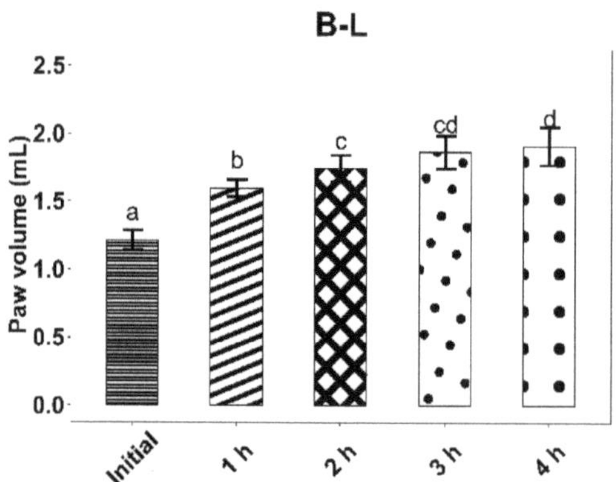

Figure 7. Kaolin-induced edema values, with standard deviations, for the B-L group. Letters a–d show statistically significant differences between groups.

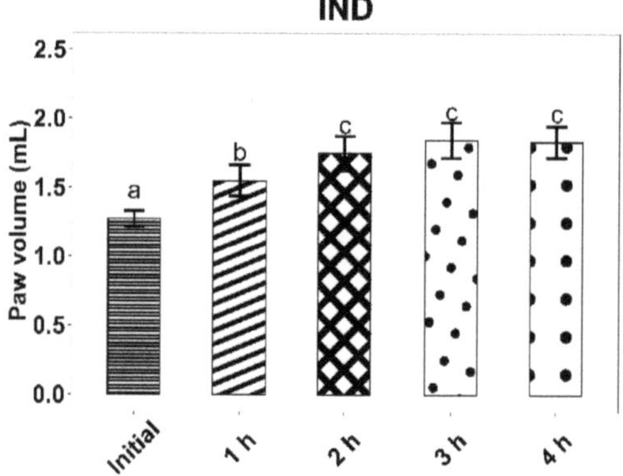

Figure 8. Kaolin-induced edema values, with standard deviations, for the IND group. Letters a–c show statistically significant differences between groups.

The administration of the inflammatory agent produced an increase in the volume of the rat paw for all three tested groups. As expected, the most pronounced increase in volume was observed in the case of rats from the control group, treated with B-L, 4 h after kaolin administration (58.2%, $p < 0.001$). At 4 h after administration of the inflammatory agent, the average increases in rat paws for the ZIZ-L and IND groups were similar (42.24% and 44.33%, respectively, $p < 0.001$).

When compared to B-L activity, ZIZ-L ointment displays an anti-inflammatory effect, with a difference between average paw volume increases of 13.81% ($p < 0.05$) at three hours and 15.78% ($p < 0.05$) at four hours, respectively.

The anti-inflammatory effect of the ZIZ-L ointment occurs about 3 h after the application, whereas the reference product leads to a faster response.

Table 7. The effect of ointments application on the inflammatory process induced by intra-plantar administration of kaolin.

Group	Moment of determination				
	Initial	**1 h**	**2 h**	**3 h**	**4 h**
ZIZ-L (Mean ± SD)	1.25 ± 0.09	1.61 ± 0.12	1.73 ± 0.12	1.76 ± 0.09	1.78 ± 0.10
Paw volume increase (%) #		28.8 ***	38.56 ***	40.56 ***	42.24 ***
B-L (Mean ± SD)	1.21 ± 0.07	1.6 ± 0.06	1.75 ± 0.1	1.87 ± 0.12	1.91 ± 0.14
Paw volume increase (%) #		32.13 ***	44.24 ***	54.37 ***	58.02 ***
IND (Mean ± SD)	1.27 ± 0.05	1.55 ± 0.11	1.75 ± 0.12	1.85 ± 0.13	1.83 ± 0.11
Paw volume increase (%) #		22.05 ***	37.64 ***	45.33 ***	44.33 ***
ZIZ-L vs. B-L (%) a#		3.33	5.68	13.81 *	15.78 *
IND vs. B-L (%) b#		10.08	6.60	9.04	13.69

t Student's test; *** $p < 0.001$; * $p < 0.05$; a difference between paw volume increase (%) seen for ZIZ-L and B-L; b difference between paw volume increase (%) seen for IND and B-L.

The results of our research indicate that the proposed topical dosage form is of appropriate quality and has demonstrated its efficacy and safety in animal models. From an organoleptic point of view, a clear difference was observed in the aspect and the texture between the base alone and the *Ziziphus jujuba* Mill. hydrophobic ointment. Still, both displayed adequate characteristics, typical for the corresponding pharmaceutical semi-solid preparations. Even if a high amount of extract was incorporated into the base, a homogenous and stable ointment was obtained. Hydrophobic ointment bases have been reported in the literature also for other healing extracts of herbal origin, such as *Urtica simensis* Hochst. ex A.Rich. [79], a mixture of herbal extracts from *Salvadora persica* L., *Azadirachta indica* A. Juss, and *Calendula officinalis* L. [80], or an extract from the *Acanthus polystachyus* Delile leaves [81].

Regarding pH, a slight decrease in the value was observed after including the extract in the base, but of little practical significance, as it is within the limits imposed by the compendial standards [52]. Taking into account the results recorded for cutaneous tolerance, the ointment pH seems proper for the recommended use. Additionally, the pH proved to be stable during the evaluated period.

In terms of spreadability (an important parameter for ointments), a minor difference in the behavior was noticed after adding the extract to the base [82]. The final spreading ability of the ointment was mainly influenced by the base and less by the extract included. The spreading properties displayed by the studied ointment reveal suitable structural and viscoelastic attributes and appropriate viscosity. The rheological behavior remains constant over time, confirming the product's stability.

Concerning the tolerance of the developed ZIZ-L ointment, the dermal irritation tests proved that it was well accepted, not inducing any reversible or non-reversible skin damage, and not affecting the somatic-motor functions. As animal skin is considered to be very sensitive to most pharmaceutical ingredients [83], the results obtained provided strong evidence for the lack of risk when applying the proposed formulation even for a prolonged time. This is in part due to the selected lipophilic base, as shown by the results, but also to the plant extract. The dried extract of *Ziziphus jujuba* leaves was included in a high amount and was incorporated by suspension, leading to no deleterious dermal responses, including erythema and edema.

This in vivo experiment confirmed that ZIZ-L ointment treatment significantly accelerated wound healing in Group 2, compared to control Groups 1 (untreated) and 3 (B-L treated). Wound healing resulted in a considerable decrease in the damaged epidermis length compared to controls. Visual examination showed that the epidermis recovery was faster in the wound treated with the ointment containing *Ziziphus jujuba* Mill. extract than those treated with the lipophilic base, suggesting that the extract owns a potent therapeutic effect in the wound healing process. The wound closure percentage displayed by the ZIZ-L ointment application was similar to the one achieved after the reference product administration, confirming that the proposed formulation has the potential for treating skin wounds. This supports the traditional use of this species leaves for wound healing, as reported in India. Whereas previous research reported wound healing activity for extracts prepared from this species' fruits, bark or roots, our research was focused on a leaf extract prepared with 70% ethyl alcohol [84–86].

Ziziphus jujuba Mill leaves extract enhances skin wound healing through multiple mechanisms. The major active ingredients that were identified in the extract were chlorogenic acid, quercetin and rutin. The mechanisms of action are complex. Chlorogenic acid promoted fibroblastic and remodeling phases of wound healing. It accelerated the wound closure in the presence of keratinocyte [87]. Quercetin improves wound healing by inhibition of matrix metalloproteinases, which are normally inhibited by plasminogen activator inhibitor 1 (PAI-1) [88]. Also, it promotes a normal regeneration, not a fibrosis because it influences positive cell migration and proliferation, increases surface αV integrin and decreases β1 integrin in wounds, and increases the production of collagen fibers which are well oriented in sub-epidermal tissue [89,90]. Rutin promotes wound healing by several mechanisms: it enhances the production of antioxidant enzymes in the presence of erythroid 2-related factor 2 (NRF2), inhibits the expression of matrix metalloproteinases (MMPs) and decreases the expression of vascular endothelial growth factor (VEGF). It also induces the expression of the neurogenic-related protein (UCH-L1) [91].

Regarding the in vivo anti-inflammatory effect of *Ziziphus* ointment, the registered results show its potential properties. Even though the activity proved to be slower than the one displayed by the reference product, it was significantly higher than the lipophilic base action. The results support using the ointment as an herbal remedy for curative purposes in various topical inflammatory processes. Considering the chemical composition of *Ziziphus jujuba* Mill. leaves extract, its anti-inflammatory effect is due to its major constituents. Chlorogenic acid has antioxidant properties and because of that, it reduces the expression of inflammatory molecules. It inhibits phospholipase A2, cyclooxygenases and lipoxygenases, and reduces the concentrations of prostanoids and leukotrienes, especially PGE2 (Prostaglandin E2), IL-1β (Interleukin 1 beta), interferon-γ, monocyte chemotactic protein-1, and macrophage inflammatory protein-1α [92,93]. Quercetin inhibits cyclooxygenase (COX) and lipoxygenase (LOX) which catalyzes the production of inflammation molecules especially LTB-4 [94,95]. Also, it inhibits lipopolysaccharide (LPS)-induced tumor necrosis factor α (TNF-α) production in macrophages and LPS-induced IL-8 production [96,97]. It can also inhibit the production of tryptase and histamine and the downregulation of vascular cell adhesion molecule 1 (VCAM-1) and CD80 expression [98]. Rutin inhibits cytokines (e.g., TNF-a, IL-6) that are highly expressed and secreted by macrophages in inflammations [99]. It also activates nuclear factor-κB and extracellular regulated kinases 1/2 by HMGB1 (High mobility group box 1) [100].

We have several reasons to consider that the wound healing process from this study consists of regeneration and not fibrosis. The injuries that we treated with the *Ziziphus* ointment are mild and the epithelial tissue that are involved has an important regenerative potential [101]. The time of complete healing was short, less than a month. Furthermore, the stage of inflammation within the healing process is shortened (several days) by phenolic compounds and flavonoids from the *Ziziphus* extract that have anti-inflammatory properties. This fact may avoid a determination of fibrosis [102,103].

Ziziphus jujuba Mill. leaves' ointment exhibited pronounced wound-healing effects and moderate anti-inflammatory characteristics, thereby supporting its usefulness as a medicinal therapy.

4. Conclusions

The present evaluated an ointment containing 10% dried ethanolic extract from *Ziziphus* jujuba Mill., formulated in a hydrophobic base consisting of petrolatum, but also cetyl alcohol, cholesterol, coconut oil and butylhydroxyanisole. The leaf extract used to prepare the ointment contains chiefly rutin (29.836 mg/g), quercetin (15.180 mg/g), and chlorogenic acid (350.96 μg/g); it also contains various amounts of phenolic acids, other flavonoids, fatty acids, organic acids and other compounds.

The formulated ointment has stable organoleptic properties related to those of the extract, is homogeneous, has a slightly acid pH (5.41–5.42), and appropriate rheological properties. It has demonstrated good tolerability following single and repeated administration in rat experiments. In rat wound models, the ointment resembled Cicatrizin in terms of its healing ability; both products accelerated healing, the effect being comparable.

The tested ointment showed an anti-inflammatory effect compared to the control group. In comparison to indomethacin, although the effect was slightly more pronounced for the herbal ointment, the difference was not statistically significant.

Due to its accessibility, good tolerance, and efficacy demonstrated in experimental models, the developed ointment is a promising therapy for wound healing and would be worth further exploring its benefits in a clinical setting.

Author Contributions: Conceptualization, M.-V.H., R.A., E.A.O., E.M. and A.I.A.; methodology, E.A.O., E.M., O.C.Ș. and E.-I.G.; software, R.A.; validation, all authors.; formal analysis, R.A., O.C.Ș., E.M. and E.A.O.; investigation, E.A.O., E.M., O.C.Ș., E.O., A.I.A., C.B., E.-I.G., L.E.D., C.S.S. and E.N.; resources, M.-V.H., R.A.; data curation, R.A.; writing—original draft preparation, M.-V.H., R.A., E.O., E.A.O., E.M., O.C.Ș., C.B., C.S.S. and E.N.; writing—review and editing, all authors.; visualization, M.-V.H.; supervision, R.A., E.A.O., E.M. and L.E.D.; project administration, M.-V.H., R.A.; funding acquisition, R.A. All authors have read and agreed to the published version of the manuscript.

Funding: This work was supported by UMF Carol Davila—project number 33894/11.11.2014.

Institutional Review Board Statement: The animal study protocol was approved by the Bioethics Commission of the Faculty of Pharmacy, University of Medicine and Pharmacy Carol Davila, Bucharest (997/7 October 2016).

Informed Consent Statement: Not applicable.

Data Availability Statement: The data underlying this article will be shared upon reasonable request to the corresponding author.

Conflicts of Interest: R.A. received consultancy or speakers' fees from UCB, Sandoz, Abbvie, Zentiva, Teva, Laropharm, CEGEDIM, Angelini, Biessen Pharma, Hofigal, AstraZeneca, and Stada. All other authors report no conflict of interest. The funders had no role in the design of the study; in the collection, analyses, or interpretation of data; in the writing of the manuscript; or in the decision to publish the results.

References

1. Shedoeva, A.; Leavesley, D.; Upton, Z.; Fan, C. Wound Healing and the Use of Medicinal Plants. *Evid.-Based Complementary Altern. Med.* **2019**, *2019*, 2684108. [CrossRef] [PubMed]
2. Dorai, A.A. Wound Care with Traditional, Complementary and Alternative Medicine. *Indian J. Plast. Surg.* **2012**, *45*, 418–424. [CrossRef] [PubMed]
3. Jarić, S.; Kostić, O.; Mataruga, Z.; Pavlović, D.; Pavlović, M.; Mitrović, M.; Pavlović, P. Traditional Wound-Healing Plants Used in the Balkan Region (Southeast Europe). *J. Ethnopharmacol.* **2018**, *211*, 311–328. [CrossRef] [PubMed]
4. Kumar, B.; Vijayakumar, M.; Govindarajan, R.; Pushpangadan, P. Ethnopharmacological Approaches to Wound Healing—Exploring Medicinal Plants of India. *J. Ethnopharmacol.* **2007**, *114*, 103–113. [CrossRef]
5. Rhoads, D.D.; Cox, S.B.; Rees, E.J.; Sun, Y.; Wolcott, R.D. Clinical Identification of Bacteria in Human Chronic Wound Infections: Culturing vs. 16S Ribosomal DNA Sequencing. *BMC Infect. Dis.* **2012**, *12*, 321. [CrossRef] [PubMed]
6. Maxson, S.; Lopez, E.A.; Yoo, D.; Danilkovitch-Miagkova, A.; LeRoux, M.A. Concise Review: Role of Mesenchymal Stem Cells in Wound Repair. *Stem Cells Transl. Med.* **2012**, *1*, 142–149. [CrossRef]
7. Reinke, J.M.; Sorg, H. Wound Repair and Regeneration. *Eur. Surg. Res.* **2012**, *49*, 35–43. [CrossRef]
8. Edeoga, H.O.; Okwu, D.E.; Mbaebie, B.O. Phytochemical Constituents of Some Nigerian Medicinal Plants. *Afr. J. Biotechnol.* **2005**, *4*, 685–688. [CrossRef]
9. Chen, W.-C.; Liou, S.-S.; Tzeng, T.-F.; Lee, S.-L.; Liu, I.-M. Effect of Topical Application of Chlorogenic Acid on Excision Wound Healing in Rats. *Planta Med.* **2013**, *79*, 616–621. [CrossRef]

10. Maver, T.; Maver, U.; Stana Kleinschek, K.; Smrke, D.M.; Kreft, S. A Review of Herbal Medicines in Wound Healing. *Int. J. Dermatol.* **2015**, *54*, 740–751. [CrossRef]
11. Sharma, A.; Khanna, S.; Kaur, G.; Singh, I. Medicinal Plants and Their Components for Wound Healing Applications. *Future J. Pharm. Sci.* **2021**, *7*, 53. [CrossRef]
12. Appendino, G.; Ottino, M.; Marquez, N.; Bianchi, F.; Giana, A.; Ballero, M.; Sterner, O.; Fiebich, B.L.; Munoz, E. Arzanol, an Anti-Inflammatory and Anti-HIV-1 Phloroglucinol Alpha-Pyrone from Helichrysum Italicum Ssp. Microphyllum. *J. Nat. Prod.* **2007**, *70*, 608–612. [CrossRef]
13. Wu, F.; Bian, D.; Xia, Y.; Gong, Z.; Tan, Q.; Chen, J.; Dai, Y. Identification of Major Active Ingredients Responsible for Burn Wound Healing of *Centella Asiatica* Herbs. *Evid.-Based Complementary Altern. Med.* **2012**, *2012*, 848093. [CrossRef]
14. Süntar, I.; Akkol, E.K.; Nahar, L.; Sarker, S.D. Wound Healing and Antioxidant Properties: Do They Coexist in Plants? *Free. Radic. Antioxid.* **2012**, *2*, 1–7. [CrossRef]
15. Sen, C.K.; Roy, S. Redox Signals in Wound Healing. *Biochim. Biophys. Acta* **2008**, *1780*, 1348–1361. [CrossRef]
16. Thakur, R.; Jain, N.; Pathak, R.; Sandhu, S.S. Practices in Wound Healing Studies of Plants. *Evid.-Based Complementary Altern. Med.* **2011**, *2011*, 438056. [CrossRef]
17. Manjunatha, B.; Vidya, S.; Rashmi, K.; Mankani, K.; Shilpa, H.; Singh, S.J. Evaluation of Wound-Healing Potency of Vernonia Arborea Hk. *Indian J. Pharmacol.* **2005**, *37*, 223. [CrossRef]
18. Shikov, A.N.; Pozharitskaya, O.N.; Makarov, V.G.; Wagner, H.; Verpoorte, R.; Heinrich, M. Medicinal Plants of the Russian Pharmacopoeia; Their History and Applications. *J. Ethnopharmacol.* **2014**, *154*, 481–536. [CrossRef]
19. Dinda, M.; Mazumdar, S.; Das, S.; Ganguly, D.; Dasgupta, U.B.; Dutta, A.; Jana, K.; Karmakar, P. The Water Fraction of *Calendula officinalis* Hydroethanol Extract Stimulates In Vitro and In Vivo Proliferation of Dermal Fibroblasts in Wound Healing: In Vitro and In Vivo Wound Healing Activity of *Calendula officinalis. Phytother. Res.* **2016**, *30*, 1696–1707. [CrossRef]
20. Givol, O.; Kornhaber, R.; Visentin, D.; Cleary, M.; Haik, J.; Harats, M. A Systematic Review of *Calendula officinalis* Extract for Wound Healing. *Wound Repair Regen.* **2019**, *27*, 548–561. [CrossRef]
21. Quave, C.L. Wound Healing with Botanicals: A Review and Future Perspectives. *Curr. Dermatol. Rep.* **2018**, *7*, 287–295. [CrossRef]
22. Wölfle, U.; Seelinger, G.; Schempp, C. Topical Application of St. John's Wort (*Hypericum perforatum*). *Planta Med.* **2013**, *80*, 109–120. [CrossRef]
23. Idolo, M.; Motti, R.; Mazzoleni, S. Ethnobotanical and Phytomedicinal Knowledge in a Long-History Protected Area, the Abruzzo, Lazio and Molise National Park (Italian Apennines). *J. Ethnopharmacol.* **2010**, *127*, 379–395. [CrossRef]
24. Nayak, B.S.; Raju, S.S.; Rao, A.V.C. Wound Healing Activity of *Matricaria recutita* L. Extract. *J. Wound Care* **2007**, *16*, 298–302. [CrossRef]
25. Niknam, S.; Tofighi, Z.; Faramarzi, M.A.; Abdollahifar, M.A.; Sajadi, E.; Dinarvand, R.; Toliyat, T. Polyherbal Combination for Wound Healing: *Matricaria chamomilla* L. and *Punica granatum* L. *DARU J. Pharm. Sci.* **2021**, *29*, 133–145. [CrossRef]
26. Shetty, B.S.; Udupa, S.L.; Udupa, A.L.; Somayaji, S.N. Effect of *Centella asiatica* L (Umbelliferae) on Normal and Dexamethasone-Suppressed Wound Healing in Wistar Albino Rats. *Int. J. Low. Extrem. Wounds* **2006**, *5*, 137–143. [CrossRef]
27. Araújo, L.U.; Reis, P.G.; Barbosa, L.C.O.; Saúde-Guimarães, D.A.; Grabe-Guimarães, A.; Mosqueira, V.C.F.; Carneiro, C.M.; Silva-Barcellos, N.M. In Vivo Wound Healing Effects of *Symphytum officinale* L. Leaves Extract in Different Topical Formulations. *Pharmazie* **2012**, *67*, 355–360.
28. Antunes Viegas, D.; Palmeira-de-Oliveira, A.; Salgueiro, L.; Martinez-de-Oliveira, J.; Palmeira-de-Oliveira, R. Helichrysum Italicum: From Traditional Use to Scientific Data. *J. Ethnopharmacol.* **2014**, *151*, 54–65. [CrossRef]
29. Sala, A.; Recio, M.; Giner, R.M.; Máñez, S.; Tournier, H.; Schinella, G.; Ríos, J.-L. Anti-Inflammatory and Antioxidant Properties of Helichrysum Italicum. *J. Pharm. Pharmacol.* **2002**, *54*, 365–371. [CrossRef]
30. Hamedi, S.; Shams-Ardakani, M.R.; Sadeghpour, O.; Amin, G.; Hajighasemali, D.; Orafai, H. Designing Mucoadhesive Discs Containing Stem Bark Extract of Ziziphus Jujuba Based on Iranian Traditional Documents. *Iran. J. Basic Med. Sci.* **2016**, *19*, 330. [CrossRef]
31. Guo, S.; Duan, J.; Qian, D.; Tang, Y.; Wu, D.; Su, S.; Wang, H.; Zhao, Y. Content Variations of Triterpenic Acid, Nucleoside, Nucleobase, and Sugar in Jujube (Ziziphus Jujuba) Fruit during Ripening. *Food Chem.* **2015**, *167*, 468–474. [CrossRef] [PubMed]
32. Preeti; Tripathi, S. Ziziphus Jujuba: A Phytopharmacological Review. *IJRDPL* **2014**, *3*, 959–966.
33. Hossain, M.A. A Phytopharmacological Review on the Omani Medicinal Plant: Ziziphus Jujube. *J. King Saud Univ. Sci.* **2019**, *31*, 1352–1357. [CrossRef]
34. Hamedi, S.; Sadeghpour, O.; Shamsardekani, M.R.; Amin, G.; Hajighasemali, D.; Feyzabadi, Z. The Most Common Herbs to Cure the Most Common Oral Disease: Stomatitis Recurrent Aphthous Ulcer (RAU). *Iran. Red Crescent Med. J.* **2016**, *18*, e21694. [CrossRef] [PubMed]
35. Masullo, M.; Cerulli, A.; Montoro, P.; Pizza, C.; Piacente, S. In Depth LC-ESIMSn-Guided Phytochemical Analysis of Ziziphus Jujuba Mill. Leaves. *Phytochemistry* **2019**, *159*, 148–158. [CrossRef]
36. Damiano, S.; Forino, M.; De, A.; Vitali, L.A.; Lupidi, G.; Taglialatela-Scafati, O. Antioxidant and Antibiofilm Activities of Secondary Metabolites from Ziziphus Jujuba Leaves Used for Infusion Preparation. *Food Chem.* **2017**, *230*, 24–29. [CrossRef] [PubMed]
37. Hovaneţ, M.-V.; Oprea, E.; Ancuceanu, R.V.; Duţu, L.E.; Budura, E.A.; Şeremet, O.; Ancu, I.; Moroşan, E. Wound Healing Properties of Ziziphus Jujuba Mill. Leaves. *Rom. Biotechnol. Lett.* **2016**, *21*, 11842–11849.

38. Hovaneţ, M.-V.; Ancuceanu, R.V.; Dinu, M.; Oprea, E.; Budura, E.A.; Negreş, S.; Velescu, B.; Duţu, L.; Anghel, I.A.; Ancu, I.; et al. Toxicity and Anti-Inflammatory Activity of Ziziphus Jujuba Mill. Leaves. *Farmacia* **2016**, *64*, 802–808.
39. Andjić, M.; Božin, B.; Draginić, N.; Kočović, A.; Jeremić, J.N.; Tomović, M.; Milojević Šamanović, A.; Kladar, N.; Čapo, I.; Jakovljević, V.; et al. Formulation and Evaluation of Helichrysum Italicum Essential Oil-Based Topical Formulations for Wound Healing in Diabetic Rats. *Pharmaceuticals* **2021**, *14*, 813. [CrossRef]
40. Han, X.; Beaumont, C.; Stevens, N. Chemical Composition Analysis and in vitro Biological Activities of Ten Essential Oils in Human Skin Cells. *Biochim. Open* **2017**, *5*, 1–7. [CrossRef]
41. Atzmony, L.; Lim, Y.H.; Hamilton, C.; Leventhal, J.S.; Wagner, A.; Paller, A.S.; Choate, K.A. Topical cholesterol/lovastatin for the treatment of porokeratosis: A pathogenesis-directed therapy. *J. Am. Acad. Dermatol.* **2020**, *82*, 123–131. [CrossRef] [PubMed]
42. Murota, H.; Itoi, S.; Terao, M.; Matsui, S.; Kawai, H.; Satou, Y.; Suda, K.; Katayama, I. Topical cholesterol treatment ameliorates hapten-evoked cutaneous hypersensitivity by sustaining expression of 11β-HSD1 in epidermis. *Exp. Dermatol.* **2014**, *23*, 68–70. [CrossRef] [PubMed]
43. Elder, R.L. Final Report on the Safety Assessment of Cetearyl Alcohol, Cetyl Alcohol, Isostearyl Alcohol, Myristyl Alcohol, and Behenyl Alcohol. *J. Am. Coll. Toxicol.* **1988**, *7*, 359–413. [CrossRef]
44. American Academy of Dermatology. Proper Wound Care. 2017. Available online: https://www.aad.org/public/everyday-care/injured-skin/burns/wound-care-minimize-scars (accessed on 10 November 2022).
45. Yeap, S.K.; Beh, B.K.; Ali, N.M.; Yusof, H.M.; Ho, W.Y.; Koh, S.P.; Alitheen, N.B.; Long, K. Antistress and Antioxidant Effects of Virgin Coconut Oil in Vivo. *Exp. Ther. Med.* **2015**, *9*, 39–42. [CrossRef] [PubMed]
46. Ghani, N.A.A.; Channip, A.-A.; Chok Hwee Hwa, P.; Ja'afar, F.; Yasin, H.M.; Usman, A. Physicochemical Properties, Antioxidant Capacities, and Metal Contents of Virgin Coconut Oil Produced by Wet and Dry Processes. *Food Sci. Nutr.* **2018**, *6*, 1298–1306. [CrossRef] [PubMed]
47. Burnett, C.L.; Bergfeld, W.F.; Belsito, D.V.; Klaassen, C.D.; Marks, J.G.; Shank, R.C.; Slaga, T.J.; Snyder, P.W.; Andersen, F.A. Final Report on the Safety Assessment of Cocos Nucifera (Coconut) Oil and Related Ingredients. *Int. J. Toxicol.* **2011**, *30*, 5S–16S. [CrossRef]
48. Gavriloaia, M.-R.; Budura, E.-A.; Toma, C.C.; Mitu, M.A.; Karampelas, O.; Arama, C.; Lupuleasa, D. In Vitro Evaluation of Diffusion and Rheological Profiles for Dexamethasone Inclusion Complexes with β-Cyclodextrin or Hydroxypropyl β-Cyclodextrin. *Farmacia* **2012**, *60*, 895–904.
49. Balaci, T.D.; Ozon, E.A.; Baconi, D.L.; Niţulescu, G.; Velescu, B.; Bălălău, C.; Păunică, I.; Fiţa, C.A. Study on the Formulation and Characterization of a Photoprotective Cream Containing a New Synthetized Compound. *J. Mind Med. Sci.* **2020**, *7*, 193–200.
50. Aiyalu, R.; Govindarjan, A.; Ramasamy, A. Formulation and Evaluation of Topical Herbal Gel for the Treatment of Arthritis in Animal Model. *Braz. J. Pharm. Sci.* **2016**, *52*, 493–507. [CrossRef]
51. Akhtar, N.; Khan, B.; Khan, M.; Mahmood, T.; Khan, H.; Iqbal, M.; Bashir, S. Formulation Development and Moiturising Effects of a Topical Cream of Aloe Vera Extract. *World Acad. Eng. Technol.* **2011**, *51*, 172–179.
52. Comisia Farmacopeei Române. *Farmacopeea Română (Romanian Pharmacopoeia)*, 10th ed.; Editura Medicală: Bucharest, Romania, 1993.
53. European Directorate for the Quality of Medicine & Health Care of the Council of Europe (EDQM). *European Pharmacopoeia*, 10th ed.; EDQM: Strasbourg, France, 2019.
54. Mănescu, O.; Lupuleasa, D.; Miron, D.; Budura, E.; Rădulescu, F. In Vitro Drug Release from Topical Antifungal Pharmaceutical Formulations. *Farmacia* **2011**, *59*, 15–23.
55. Kulawik-Pióro, A.; Drabczyk, A.K.; Kruk, J.; Wróblewska, M.; Winnicka, K.; Tchórzewska, J. Thiolated Silicone Oils as New Components of Protective Creams in the Prevention of Skin Diseases. *Materials* **2021**, *14*, 4723. [CrossRef] [PubMed]
56. Gore, E.; Picard, C.; Savary, G. Spreading Behavior of Cosmetic Emulsions: Impact of the Oil Phase. *Biotribology* **2018**, *16*, 17–24. [CrossRef]
57. Cimino, M.C. New OECD Genetic Toxicology Guidelines and Interpretation of Results. In *Genetic Toxicology and Cancer Risk Assessment*; CRC Press: Boca Raton, FL, USA, 2001; pp. 237–262.
58. Dinu, M.; Anghel, A.-I.; Olaru, O.-T.; Şeremet, O.C.; Calalb, T.; Cojocaru-Toma, M.; Negreş, S.; Hovaneţ, V.; Zbarcea, C.E.; Ancuceanu, R. Toxicity Investigation of an Extract of *Amaranthus retroflexus* L.(Amaranthaceae) Leaves. *Farmacia* **2017**, *65*, 289–294.
59. OECD. *Test No. 404: Acute Dermal Irritation/Corrosion*; OECD Guidelines for the Testing of Chemicals, Section 4; OECD: Paris, France, 2015; ISBN 978-92-64-24267-8.
60. OECD. *Test No. 410: Repeated Dose Dermal Toxicity: 21/28-Day Study*; OECD Guidelines for the Testing of Chemicals; OECD: Paris, France, 1981.
61. Masson-Meyers, D.S.; Andrade, T.A.M.; Caetano, G.F.; Guimaraes, F.R.; Leite, M.N.; Leite, S.N.; Frade, M.A.C. Experimental Models and Methods for Cutaneous Wound Healing Assessment. *Int. J. Exp. Pathol.* **2020**, *101*, 21–37. [CrossRef] [PubMed]
62. Abdullahi, A.; Amini-Nik, S.; Jeschke, M.G. Animal Models in Burn Research. *Cell. Mol. Life Sci.* **2014**, *71*, 3241–3255. [CrossRef]
63. Jassim, R.A.; Mihele, D.; Dogaru, E. Study Regarding the Influence of Vitis Vinifera Fruit (Muscat of Hamburg Species) on Some Biochemical Parameters. *Cancer* **2010**, *1*, 6–7.
64. Mihai, D.P.; Seremet, O.C.; Nitulescu, G.; Ivopol, M.; Sevastre, A.-S.; Negres, S.; Ivopol, G.; Nitulescu, G.M.; Olaru, O.T. Evaluation of Natural Extracts in Animal Models of Pain and Inflammation for a Potential Therapy of Hemorrhoidal Disease. *Sci. Pharm.* **2019**, *87*, 14. [CrossRef]

65. Negres, S.; Dinu, M.; Ancuceanu, R.; Olaru, T.O.; Ghica, M.V.; Seremet, O.C.; Zbarcea, C.E.; Velescu, B.S.; Stefanescu, E.; Chirita, C. Correlations in Silico/in Vitro/in Vivo Regarding Determinating Acute Toxicity in Non-Clinical Experimental Trial, According to Bioethic Regulations Inforced by the European Union. *Farmacia* **2015**, *63*, 877–885.
66. Xue, X.; Zhao, A.; Wang, Y.; Ren, H.; Du, J.; Li, D.; Li, Y. Composition and Content of Phenolic Acids and Flavonoids among the Different Varieties, Development Stages, and Tissues of Chinese Jujube (*Ziziphus jujuba* Mill.). *PLoS ONE* **2021**, *16*, e0254058. [CrossRef]
67. Saeedi-Boroujeni, A.; Mahmoudian-Sani, M.-R. Anti-Inflammatory Potential of Quercetin in COVID-19 Treatment. *J. Inflamm.* **2021**, *18*, 3. [CrossRef]
68. Lin, C.-F.; Leu, Y.-L.; Al-Suwayeh, S.A.; Ku, M.-C.; Hwang, T.-L.; Fang, J.-Y. Anti-Inflammatory Activity and Percutaneous Absorption of Quercetin and Its Polymethoxylated Compound and Glycosides: The Relationships to Chemical Structures. *Eur. J. Pharm. Sci.* **2012**, *47*, 857–864. [CrossRef]
69. Choi, J.K.; Kim, S.-H. Rutin Suppresses Atopic Dermatitis and Allergic Contact Dermatitis. *Exp. Biol. Med.* **2013**, *238*, 410–417. [CrossRef]
70. Girsang, E.; Ginting, C.N.; Lister, I.N.E.; Gunawan, K.Y.; Widowati, W. Anti-Inflammatory and Antiaging Properties of Chlorogenic Acid on UV-Induced Fibroblast Cell. *PeerJ* **2021**, *9*, e11419. [CrossRef]
71. Bae, J.; Kim, N.; Shin, Y.; Kim, S.-Y.; Kim, Y.-J. Activity of Catechins and Their Applications. *Biomed. Dermatol.* **2020**, *4*, 8. [CrossRef]
72. Patel, N.K.; Jaiswal, G.; Bhutani, K.K. A Review on Biological Sources, Chemistry and Pharmacological Activities of Pinostrobin. *Nat. Prod. Res.* **2016**, *30*, 2017–2027. [CrossRef] [PubMed]
73. Cavalcanti, G.R.; Duarte, F.I.C.; Converti, A.; de Lima, Á.A.N. Ferulic Acid Activity in Topical Formulations: Technological and Scientific Prospecting. *Curr. Pharm. Des.* **2021**, *27*, 2289–2298. [CrossRef] [PubMed]
74. Majnooni, M.B.; Fakhri, S.; Shokoohinia, Y.; Mojarrab, M.; Kazemi-Afrakoti, S.; Farzaei, M.H. Isofraxidin: Synthesis, Biosynthesis, Isolation, Pharmacokinetic and Pharmacological Properties. *Molecules* **2020**, *25*, 2040. [CrossRef] [PubMed]
75. Petrônio, M.; Zeraik, M.; Fonseca, L.; Ximenes, V. Apocynin: Chemical and Biophysical Properties of a NADPH Oxidase Inhibitor. *Molecules* **2013**, *18*, 2821–2839. [CrossRef]
76. Jamuna, S.; Karthika, K.; Paulsamy, S.; Thenmozhi, K.; Kathiravan, S.; Venkatesh, R. Confertin and Scopoletin from Leaf and Root Extracts of Hypochaeris Radicata Have Anti-Inflammatory and Antioxidant Activities. *Ind. Crops Prod.* **2015**, *70*, 221–230. [CrossRef]
77. Hovanet, M.-V.; Dociu, N.; Dinu, M.; Ancuceanu, R.; Morosan, E.; Oprea, E. A Comparative Physico-Chemical Analysis of Acer Platanoides and Acer Pseudoplatanus Seed Oils. *Rev. Chim. (Buchar.)* **2015**, *66*, 987–991.
78. Shelke, U.; Mahajan, A. Review on: An Ointment. *Int. J. Pharm. Pharm. Sci.* **2015**, *4*, 170–192.
79. Abeje, B.A.; Bekele, T.; Getahun, K.A.; Asrie, A.B. Evaluation of Wound Healing Activity of 80% Hydromethanolic Crude Extract and Solvent Fractions of the Leaves of Urtica simensis in Mice. *J. Exp. Pharmacol.* **2022**, *7*, 221–241. [CrossRef]
80. Imran, H.; Sohail, T.; Shaukat, S.; Khokar, A. Wound Healing Potential/Activity of Polyherbal Ointment Containing Salvadora persica, Azadirachta indica and *Calendula officinalis* Extracts: An Experimental Study: Wound Healing Potential: An Experimental Study. *Biol. Sci.-PJSIR* **2022**, *65*, 55–61. [CrossRef]
81. Demilew, W.; Adinew, G.M.; Asrade, S. Evaluation of the Wound Healing Activity of the Crude Extract of Leaves of Acanthus polystachyus Delile (Acanthaceae). *Evid.-Based Complementary Altern. Med.* **2018**, *11*, 2047896. [CrossRef]
82. Ghosh, S.; Samanta, A.; Mandal, N.B.; Bannerjee, S.; Chattopadhyay, D. Evaluation of the wound healing activity of methanol extract of *Pedilanthus tithymaloides* (L.) Poit leaf and its isolated active constituents in topical formulation. *J. Ethnopharmacol.* **2012**, *142*, 714–722. [CrossRef]
83. Wang, J.; Li, Z.; Sun, F.; Tang, S.; Zhang, S.; Lv, P.; Li, J.; Cao, X. Evaluation of Dermal Irritation and Skin Sensitization Due to Vitacoxib. *Toxicol. Rep.* **2017**, *4*, 287–290. [CrossRef]
84. Aafi, E.; Reza, M.; Mirabzadeh, M. Jujube (Ziziphus jujuba Mill. (Rhamnaceae)): A review on its pharmacological properties and phytochemistry. *Tradit. Med. Res.* **2022**, *7*, 38. [CrossRef]
85. Soni, H.; Malik, J.K. Phyto-pharmacological potential of Zizyphus jujube: A review. *Sch. Int. J. Biochem.* **2021**, *4*, 1–5. [CrossRef]
86. Chopda, M.Z.; Nemade, N.V.; Mahajan, R.T. Wound healing activity of root of Ziziphus jujuba mill in rat model. *World J. Pharm. Pharm. Sci. (WJPPS)* **2014**, *3*, 830–836.
87. Moghadam, S.E.; Ebrahimi, S.N.; Salehi, P.; Moridi Farimani, M.; Hamburger, M.; Jabbarzadeh, E. Wound Healing Potential of Chlorogenic Acid and Myricetin-3-O-β-Rhamnoside Isolated from Parrotia persica. *Molecules* **2017**, *22*, 1501. [CrossRef] [PubMed]
88. Lim, H.; Kim, H.P. Inhibition of mammalian collagenase, matrix metalloproteinase-1, by naturally-occurring flavonoids. *Planta Med.* **2007**, *73*, 1267–1274. [CrossRef] [PubMed]
89. Doersch, K.M.; Newell-Rogers, M.K. The impact of quercetin on wound healing relates to changes in αV and β1 integrin expression. *Exp. Biol. Med.* **2017**, *42*, 1424–1431. [CrossRef] [PubMed]
90. Osama, M.A.; Tarek, M.; Hala, M.; Hany, H.; Rasha, R.A.; Ebtsam, A. Quercetin and low level laser therapy promote wound healing process in diabetic rats via structural reorganization and modulatory effects on inflammation and oxidative stress. *Biomed. Pharmacother.* **2018**, *101*, 58–73. [CrossRef]
91. Chen, L.Y.; Huang, C.N.; Liao, C.K.; Chang, H.M.; Kuan, Y.H.; Tseng, T.J.; Yen, K.J.; Yang, K.L.; Lin, H.C. Effects of Rutin on Wound Healing in Hyperglycemic Rats. *Antioxidants* **2020**, *9*, 1122. [CrossRef]

92. Moreira de Castro, M.E.; Pereira, R.G.F.A.; Danielle Ferreira Dias Gontijo, V.S.; Vilela, F.C.; Isac de Moraes, G.; Giusti-Paiva, A.; Henrique dos Santos, M. Anti-inflammatory effect of aqueous extracts of roasted and green *Coffea arabica* L. *J. Funct. Foods* **2013**, *5*, 466–474. [CrossRef]
93. Bagdas, D.; Gul, Z.; Meade, J.A.; Cam, B.; Cinkilic, N.; Gurun, M.S. Pharmacologic Overview of Chlorogenic Acid and its Metabolites in Chronic Pain and Inflammation. *Curr. Neuropharmacol.* **2020**, *18*, 216–228. [CrossRef]
94. Kim, H.P.; Mani, I.; Iversen, L.; Ziboh, V.A. Effects of naturally-occurring flavonoids and biflavonoids on epidermal cyclooxygenase and lipoxygenase from guinea-pigs. *Prostaglandins Leukot. Essent. Fat. Acids* **1998**, *58*, 17–24. [CrossRef]
95. Lee, K.M.; Hwang, M.K.; Lee, D.E.; Lee, K.W.; Lee, H.J. Protective effect of quercetin against arsenite-induced COX-2 expression by targeting PI3K in rat liver epithelial cells. *J. Agric. Food Chem.* **2010**, *58*, 5815–5820. [CrossRef]
96. Manjeet, K.R.; Ghosh, B. Quercetin inhibits LPS-induced nitric oxide and tumor necrosis factor-alpha production in murine macrophages. *Int. J. Immunopharmacol.* **1999**, *21*, 435–443. [CrossRef]
97. Geraets, L.; Moonen, H.J.; Brauers, K.; Wouters, E.F.; Bast, A.; Hageman, G.J. Dietary flavones and flavonoles are inhibitors of poly(ADP-ribose)polymerase-1 in pulmonary epithelial cells. *J. Nutr.* **2007**, *137*, 2190–2195. [CrossRef]
98. Kempuraj, D.; Madhappan, B.; Christodoulou, S.; Boucher, W.; Cao, J.; Papadopoulou, N.; Cetrulo, C.L.; Theoharides, T.C. Flavonols inhibit proinflammatory mediator release, intracellular calcium ion levels and protein kinase C theta phosphorylation in human mast cells. *Br. J. Pharmacol.* **2005**, *145*, 934–944. [CrossRef]
99. Kim, J.H.; Park, S.H.; Beak, E.J.; Han, C.H.; Kang, N.J. Anti-oxidant and Anti-inflammatory Effects of Rutin and Its Metabolites. *Curr. Res. Agric. Life Sci.* **2013**, *31*, 165–169.
100. Yoo, H.; Ku, S.K.; Baek, Y.D.; Bae, J.S. Anti-inflammatory effects of rutin on HMGB1-induced inflammatory responses in vitro and in vivo. *Inflamm. Res.* **2014**, *63*, 197–206. [CrossRef]
101. Adler, M.; Mayo, A.; Zhou, X.; Franklin, R.A.; Meizlish, M.L.; Medzhitov, R.; Kallenberger, S.M.; Alon, U. Principles of Cell Circuits for Tissue Repair and Fibrosis. *iScience* **2020**, *23*, 100841. [CrossRef]
102. Wynn, T.A. Cellular and molecular mechanisms of fibrosis. *J. Pathol.* **2008**, *214*, 199–210. [CrossRef]
103. Wynn, T.A.; Vannella, K.M. Macrophages in Tissue Repair, Regeneration, and Fibrosis. *Immunity* **2016**, *44*, 450–462. [CrossRef]

 life

Article

Chemical Composition, Antioxidant, Anti-Diabetic, Anti-Acetylcholinesterase, Anti-Inflammatory, and Antimicrobial Properties of *Arbutus unedo* L. and *Laurus nobilis* L. Essential Oils

Samiah Hamad Al-Mijalli [1], Hanae Naceiri Mrabti [2], Hayat Ouassou [3], Rachid Flouchi [4], Emad M. Abdallah [5], Ryan A. Sheikh [6], Mohammed Merae Alshahrani [7], Ahmed Abdullah Al Awadh [7], Hicham Harhar [8,*], Nasreddine El Omari [9], Ahmed Qasem [10], Hamza Assaggaf [10], Naif Hesham Moursi [11], Abdelhakim Bouyahya [12], Monica Gallo [13,*] and Moulay El Abbes Faouzi [2]

1 Department of Biology, College of Sciences, Princess Nourah bint Abdulrahman University, P.O. Box 84428, Riyadh 11671, Saudi Arabia
2 Laboratory of Pharmacology and Toxicology, Bio Pharmaceutical and Toxicological Analysis Research Team, Faculty of Medicine and Pharmacy, University Mohammed V in Rabat, Rabat BP 6203, Morocco
3 Faculty of Sciences, University Mohammed First, Boulevard Mohamed VI BP 717, Oujda 60000, Morocco
4 Laboratory of Microbial Biotechnology and Bioactive Molecules, Science and Technologies Faculty, Sidi Mohamed Ben Abdellah University, Fez BP 2202, Morocco
5 Department of Science Laboratories, College of Science and Arts, Qassim University, Ar Rass 51921, Saudi Arabia
6 Biochemistry Department, Faculty of Science, King Abdulaziz University, Jeddah 21589, Saudi Arabia
7 Department of Clinical Laboratory Sciences, Faculty of Applied Medical Sciences, Najran University, Najran 61441, Saudi Arabia
8 Laboratory of Materials, Nanotechnology and Environment LMNE, Faculty of Sciences, Mohammed V University in Rabat, Rabat BP 1014, Morocco
9 Laboratory of Histology, Embryology, and Cytogenetic, Faculty of Medicine and Pharmacy, Mohammed V University in Rabat, Rabat 10000, Morocco
10 Department of Laboratory Medicine, Faculty of Applied Medical Sciences, Umm Al-Qura University, Makkah 21955, Saudi Arabia
11 King Fahad Armed Forced Hospital, Jeddah 23311, Saudi Arabia
12 Laboratory of Human Pathologies Biology, Department of Biology, Faculty of Sciences, Mohammed V University in Rabat, Rabat BP 6203, Morocco
13 Department of Molecular Medicine and Medical Biotechnology, University of Naples Federico II, Via Pansini 5, 80131 Naples, Italy
* Correspondence: hichamoo79@yahoo.fr (H.H.); mongallo@unina.it (M.G.)

Citation: Al-Mijalli, S.H.; Mrabti, H.N.; Ouassou, H.; Flouchi, R.; Abdallah, E.M.; Sheikh, R.A.; Alshahrani, M.M.; Awadh, A.A.A.; Harhar, H.; Omari, N.E.; et al. Chemical Composition, Antioxidant, Anti-Diabetic, Anti-Acetylcholinesterase, Anti-Inflammatory, and Antimicrobial Properties of *Arbutus unedo* L. and *Laurus nobilis* L. Essential Oils. *Life* **2022**, *12*, 1876. https://doi.org/10.3390/life12111876

Academic Editors: Azahara Rodríguez-Luna and Salvador González

Received: 18 October 2022
Accepted: 9 November 2022
Published: 14 November 2022

Publisher's Note: MDPI stays neutral with regard to jurisdictional claims in published maps and institutional affiliations.

Copyright: © 2022 by the authors. Licensee MDPI, Basel, Switzerland. This article is an open access article distributed under the terms and conditions of the Creative Commons Attribution (CC BY) license (https://creativecommons.org/licenses/by/4.0/).

Abstract: The objectives of this work were to determine the phytochemical composition and antioxidant, anti-diabetic, antibacterial, anti-inflammatory, and anti-acetylcholinesterase properties of *Arbutus unedo* L. and *Laurus nobilis* L. EOs. The antioxidant effects were estimated using four complementary methods. In addition, the anti-diabetic activity was assessed by targeting three carbohydrate-hydrolyzing enzymes, namely α-amylase, α-glucosidase, and lipase. The anti-inflammatory and anti-acetylcholinesterase effects were evaluated by testing the inhibitory potential of both plants on lipo-oxygenase and acetylcholinesterase (AChE), respectively. The antimicrobial activity of these oils was evaluated using disc-diffusion, minimum inhibitory concentration (MIC), and minimum lethal concentration (MLC) tests. The chemical composition of *L. nobilis* essential oil (EO) was dominated by eucalyptol (36.40%), followed by α-terpineole (13.05%), α-terpinyl acetate (10.61%), linalool (10.34%), and northujane (5.74%). The main volatile compounds of *A. unedo* EOs were decenal (13.47%), α-terpineol (7.8%), and palmitic acid (6.00%). *L. nobilis* and *A. unedo* EOs inhibited α-amylase with IC_{50} values of 42.51 ± 0.012 and 102 ± 0.06 μg/mL, respectively. Moreover, both oils inhibited the activity of α-glucosidase (IC_{50} = 1.347 ± 0.021 μg/mL and IC_{50} = 76 ± 0.021 μg/mL) and lipase (IC_{50} = 21.23 ± 0.021 μg/mL and IC_{50} = 97.018 ± 0.012 μg/mL, respectively). In addition, *L. nobilis* EO showed an anti-AChE activity (IC_{50} = 89.44 ± 0.07 μg/mL) higher than that of *A. unedo* EO (IC_{50} = 378.57 ± 0.05 μg/mL). Regarding anti-inflammatory activity, in vitro assays showed that

L. nobilis significantly inhibits (IC_{50} = 48.31 ± 0.07 μg/mL) 5-lipoxygenase compared to A. unedo (IC_{50} = 86.14 ± 0.05 μg/mL). This was confirmed in vivo via a notable inhibition of inflammation recorded after 6 h of treatment in both plants at a dose of 50 mg/kg. The microbiological results revealed that EOs from both plants inhibited the growth of all tested organisms except *P. aeruginosa*, with the highest antimicrobial effect for *L. nobilis*. The results of these tests showed that these two plants possess remarkable biological and pharmacological properties, explaining their medicinal effects and suggesting them as promising sources of natural drugs.

Keywords: *Laurus nobilis; Arbutus undeo*; volatile compounds; anti-diabetic; anti-inflammatory; antimicrobial

1. Introduction

Despite the current emphasis on synthetic pharmaceuticals, medicinal plants have always been and will continue to be the primary source of drugs [1]. Even today, medicinal plants are believed to be the main source of health care for up to 80% of the world's population, most of whom live in developing countries [2]. Plant extracts, especially essential oils (EOs), include several phytochemicals with diverse physiological effects on the body [3–13]. EOs are mixtures of molecules extracted from plants primarily through steam distillation, which collects the major aromatic compounds such as terpenoids and phenolic compounds [13,14].

Indeed, different methods of extraction have been developed recently for the isolation of secondary metabolites [15–18]. The secondary metabolites have many medical uses, including antioxidants, antibacterial, antifungal, antiviral, anticancer, anti-inflammatory, and antiprotozoal [1,19–22].

Antioxidants have recently gained scientific attention following recent studies demonstrating various health benefits, including anti-inflammatory and anti-aging properties [23]. Moreover, free radicals are considered pathophysiological agents. Hence, antioxidant intake protects against oxidative stress by preventing the formation of reactive species [24]. Experimental evidence shows that high levels of antioxidants may be beneficial in inhibiting several types of free radical damage related to the development of diabetes mellitus (DM) [25–27]. Medicinal plants are widely popular in many countries and are used in alternative medicine and as supplementary foods [28,29].

With the recent spread of microbial infections and the growing concern that antibiotics may not be able to inhibit the growth of antibiotic resistant pathogens in the future, there is an urgent need for new sources of drugs such as natural products rich in antimicrobial compounds [30,31]. Many studies have been conducted on the antimicrobial properties of extracts and EOs from medicinal plants or their isolated compounds such as phenolics, flavonoids, lactones, terpenes, naphtoquinones, or alkaloids [32–36]. Some of these phytochemicals were identified after discovering antibacterial activity in the plant, and this process is called bioguided isolation and determination of phytochemicals [13]. In addition, recent studies have been able to decipher the molecular mechanisms through which plants or their active ingredients act. Indeed, natural antimicrobial molecules can act at several levels; sub-cellular (cell wall and membranes), cellular (intracellular signaling of microbes), and molecular (DNA replication and transcription, as well as protein synthesis) [37].

On the other hand, scientific research on the management of DM has attracted worldwide attention because it is a growing global health problem and often significantly increases the risk of many cardiovascular issues, such as coronary artery disease with chest discomfort, heart attacks, strokes, arterial constriction, and nerve damage [38]. Sulfonylureas, α-glucosidase inhibitors, biguanides, and thiazolidinediones are some of the currently available anti-diabetic drugs that are often used to manage hyperglycemia. However, the progression of diabetes complications is not dramatically changed by these medications. Due to unfavorable clinical circumstances and high risks of subsequent failure, they are only sometimes used. It is, therefore, crucial to seek more effective antidiabetic treatments with fewer adverse effects [39]. Today, there are more than 410 medicinal plants with antidiabetic activities whose effectiveness in the treatment of hyperglycemia, a metabolic disorder, has been scientifically proven in various ways (in vitro, in vivo, and in clinical studies) [40].

Arbutus unedo L. (Ericaceae), is widely distributed in Mediterranean countries, such as Morocco, Algeria, Tunisia, Spain, Portugal, France, Syria, Turkey, Greece, and Croatia [41]. In Morocco, it is commonly called "sasnou" or "Bakhano". It has been used in folk medicine as an antiseptic, anti-diabetic, anti-hypertensive, diuretic, and laxative [42–45]. *A. unedo* is already known as a good source of organic acids and antioxidants, including phenolic compounds, vitamins (C and E), and carotenoids [44,46]. *Laurus nobilis* L. (Lauraceae), is an evergreen tree cultivated in many warm regions of the world, especially in Mediterranean countries such as Morocco, Algeria, Spain, Portugal, Turkey, and Greece [47]. It has long been used to treat diabetes, rheumatism, dermatitis, stomach problems, snakebites, and migraines [48].

Therefore, the aims of this study were to determine the chemical profile and investigate the antioxidant, anti-diabetic, anti-inflammatory, anti-acetylcholinesterase, and antimicrobial properties of *Arbutus unedo* EO (AUEO) leaves and *Laurus nobilis* EO (LNEO) leaves. Although the biological activities of these two plants have been previously studied, certain biological tests such as the anti-inflammatory effects remain to be developed. Moreover, the two studied plants were collected from a region where they have yet been studied.

2. Material and Methods

2.1. Chemicals and Reagents

2,2-diphenyl-1-picrylhydrazyl (DPPH), 2,2-azino-bis-3-ethylbenzothiazoline-6-sulfonic acid (ABTS), 6-hydroxy-2,5,7,8-tetramethylchroman-2-carboxylic acid (Trolox), and ascorbic acid were purchased from Sigma-Aldrich (Saint-Quentin-Fallavier, France). Lipoxygenase (5-LOX) and linolenic acid were purchased from Sigma-Aldrich (St. Louis, MO, USA). Mueller–Hinton Agar (MHA), Sabouraud dextrose agar (SA), Tryptone Soy agar, dimethyl sulfoxide (DMSO), and chloramphenicol were purchased from Biokar (Beauvais, France). All the other reagents were of analytical grade.

2.2. Collection of Medicinal Plants

Leaves of both (2 kg) used medicinal plants *(A. unedo* and *L. nobilis)* were collected from the region of Taza, Morocco. The botanical identification of both plants was carried out at the botany department of the Scientific Institute (Mohammed V University in Rabat). RAB10549 and RAB10143 are the attributed voucher specimens for *A. unedo* and *L. nobilis*, respectively.

2.3. Isolation of Essential Oils

Essential oils of *A. unedo* and *L. nobilis* leaves were isolated by a hydrodistillation procedure using Clevenger apparatus. The dry leaves are deposited in the Clevenger apparatus with water for 3 h. The volatile compounds (essential oils) are vaporized with the water and subsequently separated after cooling by density difference.

2.4. Identification of Chemical Compounds

The analytical technique of gas chromatography-mass spectrometry (GC-MS) was used to characterize and identify the chemical compounds of AUEO and LNEO as published by Al-Mijalli et al. [49].

A 5% phenylmethyl silicone HP-5MS capillary column (30 m $\times$ 0.25 mm $\times$ film thickness of 0.25 μm) was used in GC. The temperature of the column was increased from 50 °C for 5 min to 200 °C with a 4 °C/min rate. The used carrier gas was helium with a 1.5 mL/min flow rate and a split mode (flow: 112 mL/min, ratio: 1/74.7). The hold time was 48 min, and the injector and detector were both at 250 °C. MS operating conditions functionned at 70 eV ionization voltage, 230 °C ion source temperature, and a 35–450 (m/z) scanning range. The identification of chemical composition was done by comparing the MS spectra with the library and matching the Kovats index (Library of NIST/EPA/NIH MASS SPECTRAL LIBRARY Version 2.0, 1 July 2002). Moreover, an internal normalization of the total area of peaks detected in each chromatogram was done for the quantification of each compound.

2.5. Determination of Antioxidant Activity

2.5.1. Hydroxyl Radical Scavenging Assay

Hydroxyl radical scavenging activity was carried out according to a slightly modified method by Basak et al. [48]. The hydroxyl radicals produced by the Fe^{3+}/ascorbate/EDTA/H_2O_2 system were measured to assess the hydroxyl radical scavenging activity. Deoxyribose is attacked by the hydroxyl radical, which causes the production of compounds that react with thiobarbituric acid (TBARS). A volume of 100 μL of each sample (1000 μg/mL) was added to a reaction mixture containing 100 μL of 3.0 mM deoxyribose, 100 μL of 0.1 mM $FeCl_3$, 100 μL of 0.1 mM EDTA, 100 μL of 0.1 mM ascorbic acid, 100 μL of 1 mM H_2O_2, and 20 mM phosphate buffer (pH = 7.4) in a final volume of 1.0 mL. Then, the absorbance was measured at 532 nm against a blank containing deoxyribose and buffer. The percentage inhibition (I) of deoxyribose degradation was calculated as follows:

$$\%\,I = (A_0 - A_1/A_0) \times 100$$

where A_0 is the absorbance of the control reaction and A_1 is the absorbance of the test compound.

2.5.2. Inhibition of Superoxide Radical Assay

Superoxide radical generation by the xanthine/xanthine oxidase system was determined according to Basak et al. [48]. Briefly, a 100 μL of each sample (1000 μg/mL) was added to a reaction mixture containing 100 μL of 2 nM xanthine, 100 μL of 12 nM NBT, 100 μL of 1.0 U/mL xanthine oxidase, and 0.1 M phosphate buffer (pH = 7.4), making a final volume of 2.0 mL. After incubating the mixture at 25 °C for 10 min, the percent inhibition of superoxide anion was calculated using the following equation:

$$\%\ \text{Inhibition} = (A_0 - A_1/A_0) \times 100$$

where A_0: the absorbance of the control and A_1: the absorbance of the samples.

2.5.3. DPPH Assay

The method (DPPH), described by Basak et al. [48], follows the bleaching of a purple methanol solution of DPPH. Briefly, 50 μL of the essential oil (1000 μg /mL) was added to 5 mL of a 0.004% solution of DPPH in methanol. The absorbance was measured at 517 nm after 30 min of incubation at dark conditions. The DPPH radical scavenging activity was calculated using the following formula:

$$\text{DPPH inhibition percentage} = (A_0 - A_1/A_0) \times 100$$

where A_0 is the absorbance of the control and A_1 is the absorbance of the presence of samples.

2.5.4. Lipid Peroxidation Inhibition Assay

Non-enzymatic lipid peroxidation assay was realized according to the procedure described by Basak et al. [48], with minor changes. The reaction mixture contained 100 μL of each sample (1000 μg/mL), 100 μL of supernatant, 20 μL of 1 mM $FeCl_3$, and 20 μL of 1 mM ascorbic acid to induce hydroxyl radical generation. After a 1 h incubation period at 37 °C, the extent of lipid peroxidation was measured by the TBA reaction. After cooling,

2.5 mL of *n*-butanol was added and the samples were centrifuged at 3500 rpm for 5 min. The absorbance was read at 532 nm. The percentage inhibition of activity was calculated using the formula:

$$\% \text{ inhibition} = (A_0 - A_1/A_0) \times 100$$

In this equation, A_0 is the absorbance of the control (without sample) and A_1 is the absorbance of the samples.

2.6. *In Vitro Anti-Diabetic Assay*

The potential of oils of *A. unedo* and *L. nobilis* to inhibit the enzymatic activity of α-amylase and α-glucosidase was assessed according to previously published research [50,51], and the assay of the lipase inhibitory activity was conducted according to the method described by Hu et al. [52].

2.7. *Anti-Acetylcholinesterase Activity*

Inhibition of acetylcholinesterase (AChE) activity was measured using an adaptation of the method described by Ingkaninan et al. [53], with some modifications. Briefly, 10 µL solution of EOs in Tween 80 (0.5% *v*/*v*) at different concentrations and 25 µL of AChE enzyme were mixed with 0.1 M of TrisHCl buffer at 0.1 M concentration and pH 8. The solutions were incubated for 15 min at room temperature. After incubation, 10 µL of a solution of acetylthiocholine iodide (ASCh) of 0.5 mM concentration and 10 µL of 5,5′-dithio-bis-2-nitrobenzoic acid (DTNB) of 3 mM concentration were added. The absorbance of the mixture was measured at 412 nm in a UV-visible spectrophotometer.

2.8. *In Vitro Anti-Inflammatory Assays*

The in vitro anti-inflammatory effect of AUEO and LNEO were assessed by the 5-Lipoxygenase (5-LOX) inhibitory activity, according to the previously published method [48]. Briefly, 20 µL of EOs and 20 µL of 5-LOX from Glycine max (100 U/mL) were pre-incubated with 200 µL of phosphate buffer (0.1 M, pH 9) at room temperature for 5 min. Then, 20 µL of linolenic acid (4.18 mM in ethanol) was added in order to start the reaction, which was followed for 3 min at 234 nm.

2.9. *In Vivo Anti-Inflammatory Assay*

The anti-inflammatory effects were investigated using a rat model of carrageenan-induced paw edema (Rege et al. [54]). Briefly, Wistar rats (150 to 180 g) were fasted for 18 h and then randomly divided into eight groups (n = 6 per group). The groups of rats received different concentrations of the studied drugs (AUEO and LNEO, (1:1)) (50 and 100 mg/kg). The control group received distilled water while the other groups received indomethacin (10 mg/kg) as the reference drug. After 60 min, all rats were injected subcutaneously with carrageenan solution (0.05 mL of 1% carrageenan suspended in 0.9% NaCl) into the subplantar region of the left hind paw. The paw volumes of the tested rats were recorded using a LE 7500 digital plethysmometer controlled by SeDaCOM software before the injection of carrageenan (V_0), and after the carrageenan injection at three different times 1 h, 3 h, and 6 h (V_t). The anti-inflammatory effect is calculated using the following equation:

$$\% \; inhibition = \frac{(V_t - V_0) \; control - (V_t - V_0) \; treated \; group}{(V_t - V_0) \; control} \times 100$$

2.10. *Antimicrobial Activity*

2.10.1. Tested Microorganisms

Microorganisms were used to test for antimicrobial activity, including four Gram-negative bacterial strains: *Escherichia coli* (ATCC 25934), *Proteus mirabilis* (ATCC 25933), *Salmonella typhimurium* (ATCC 700408), and *Pseudomonas aeruginosa* (ATCC 27853); three Gram-positive bacterial strains: *Bacillus subtilis* (ATCC 6633), *Listeria monocytogenes* (ATCC 13932); one yeast strain: *Candida albicans* (clinical isolate); and two fungal strains: *Trichophy-*

ton rubrum (clinical isolate) and *Aspergillus niger* (food-spoilage isolate). Microorganisms were generously provided by the Laboratory of Microbial Biotechnology and Bioactive Molecules, Science and Technologies Faculty, Sidi Mohamed Ben Abdellah University, Morocco. All isolates were preserved at −20 °C until used. The bacterial strains were sub-cultured on MHA medium (brought from the Laboratory of Microbial Biotechnology and Bioactive Molecules) at 37 °C for 18 h. Yeast and fungal isolates were sub-cultured on SA medium (brought from the Laboratory of Microbial Biotechnology and Bioactive Molecules) at 25 °C for 48 h and five days, respectively.

2.10.2. Inoculum Preparation

Fresh microbial cultures were adjusted to 0.5 McFarland Standard's turbidity. To make the McFarland standard, 99.5 mL of 1% (vol/vol) sulfuric acid was combined with 0.5 mL of a solution of barium chloride dihydrate ($BaCl_2 \bullet 2H_2O$), which has a weight-to-volume ratio of 1.175%. The adjusted sample of the suspension contains about 10^8 colony forming units (CFU/mL) for bacteria, approximately 10^6 CFU/mL for yeast, and around 10^4 spore forming/mL for fungi [13,55].

The adjustment of the microbial suspensions for reactivation was carried out by inoculating MHA medium with a loopful of the frozen (−20 °C) stock and incubating it at 37 °C for 24 h for bacteria and for 48 to five days for yeast and fungi. The antimicrobial testing was conducted using modified microbial innocula.

2.10.3. Disc-Diffusion Assay

The antimicrobial screening of AUEO and LNEO was performed using the disc diffusion technique according to previously reported methods [56,57]. The culture suspension was inoculated on MHA medium for bacteria and SA medium for yeast and fungal strains. Sterile paper discs, 6 mm in diameter sterile paper discs soaked in 10 µL of each EO (combined with 5% DMSO) were then placed on each plate. The positive control for bacteria was chloramphenicol (30 µg), the positive control for yeasts and fungi was nystatin (100 I.U.), and the negative control was DMSO (10 µL; 5%). The bacterial plates were incubated at 37 °C for 24 h, while the fungal and yeast plates were incubated at 25 °C for 48 and 72 h, respectively. The widths of the inhibitory zones were measured in millimeters after incubation, and the findings were expressed as the mean and standard deviation of three replicates.

2.10.4. Determination of Minimum Inhibitory Concentration

The minimum inhibitory concentration (MIC) was determined using the microtube dilution method as described in [58]. Briefly: in sterile microtubes consisting of 100 µL of MHA broth (for bacteria) or SA broth (for yeast and fungi) and containing 5% Tween 80 (emulsifier), a decreasing concentration of EOs from 4% to 0.0625% (v/v) were prepared. Then, 5 µL of the adjusted microbial suspension (adjusted to 0.5 McFarland as previously described) were added to each tube. Another set of tubes containing liquid media without EOs and inoculated with the adjusted microbial suspensions, and un-inoculated tubes containing liquid media plus EOs served as negative controls. A microtube set containing serial dilutions of chloramphenicol and nystatin instead of EOs served as positive controls. After incubation, microtubes with low concentrations and without visible microbial growth were considered the MIC [57].

2.10.5. Determination of Minimum Lethal Concentration

The minimum lethal concentration (MLC) was carried out following the MIC test. In aseptic conditions, 5 µL of each tube that had no apparent growth on the MIC test was poured onto plates containing MHA (for bacteria) or SA agar (for yeast and fungi) and incubated at 35–37 °C for 24 h for bacteria, or at 25 °C for 48 and 72 h for yeast and fungi, respectively. Then, incubated plates were inspected. The lowest concentration of EO at which a microorganism can be killed (no visible growth) was deemed the MLC [59,60].

2.11. Statistical Analysis

The findings of each experiment were run in triplicate, and they are presented as mean ± standard deviations (SD). The data were analyzed using IBM SPSS Statistics for Windows, Version 21.0 Armonk, NY, USA, and one-way ANOVA in addition to the Tukey test were used to compare means. When $p < 0.05$, the differences between the means were deemed significant.

3. Results

3.1. Chemical Composition

Chemical analyses of AUEO and LNEO constituents, the percentage content of each compound, elution order, structural subclass, and retention index are presented in Table 1. Chemical analysis of EOs from *A. undeo* and *L. nobilis* found 15 and 18 chemical compounds, which make up 53.33% and 89% of the total composition of these oils, respectively.

Table 1. Chemical composition of AUEO and LNEO.

Number	AUEO			LNEO		
	R_T	Compounds	%	R_T	Compounds	%
1	0.479	Caryophyllene	0.26	2.036	α-Thujene	0.43
2	0.805	Myrtenol	0.78	2.115	α-Pinene	3.34
3	1.481	Geraniol	0.57	2.262	Camphene	0.47
4	1.741	β-Eudesmol	1.28	2.791	β-Thujene	5.74
5	2.439	γ-Eudesmol	1.43	4.132	Eucalyptol	36.40
6	3.228	Nonanoicacid	4.38	5.947	Linalool	10.34
7	4.445	Decenal	13.47	7.479	α-terpineol	13.05
8	5.099	Palmitic acid	6.00	10.048	4-Thujen-2-α-yl acetate	0.91
9	5.211	(E,Z)-2,6-Nonadienal	0.62	10.150	Bornyl acetate	0.87
10	5.420	(E)-2-Undecenal	0.7	12.426	α-Terpinyl acetate	10.61
11	8.186	(E)-Geranylacetone	3.36	12.595	Eugenol	1.58
12	10.057	α-Terpineol	7.8	14.071	Methyleugenol	3.74
13	10.125	Linalool	1.82	15.018	Naphthalene	0.64
14	10.226	Nonanal	3.8	17.441	β-Neoclovene	0.50
15	10.384	Dodecanoicacid	1.2	17.666	Isoelemicin	0.38
16	11.409	β-Ionone	1.26	-	-	-
17	11.815	Octanol	0.64	-	-	-
18	17.167	Myristic acid	3.96	-	-	-
	Total identified compounds %		53.33%	**Total identified compounds %**		89%
	Monoterpene hydrocarbons %		-	**Monoterpene hydrocarbons %**		4.88%
	Oxygenated monoterpenes %		24.44%	**Oxygenated monoterpenes %**		72.18%
	Sesquiterpene hydrocarbons %		1.52%	**Sesquiterpene hydrocarbons %**		6.24%
	Oxygenated sesquiterpenes %		2.71%	**Oxygenated sesquiterpenes %**		0.38%

R_T: retention time.

The AUEO was characterized by a high level of decenal (13.47%) accompanied by other constituents with variable content, such as α-terpineol (7.8%) and palmitic acid (6.00%). For LNEO, monoterpenes constituted the largest class of terpenes (72.18% oxygenated monoterpenes and 4.88% monoterpene hydrocarbons). Moreover, the EO of this plant was dominated by eucalyptol (36.40%), followed by α-terpineole (13.05%), α-terpinyl acetate (10.61%), linalool (10.34%), and β-thujene (5.74%).

3.2. Antioxidant Activity

In this study, four different in vitro assays, namely hydroxyl radical, superoxide radical, lipid peroxidation, and DPPH radical scavenging activity, were adopted to determine the antioxidant properties of AUEO and LNEO. The IC_{50} values in Table 2 showed significant differences ($p < 0.05$) between the antioxidant activities of AUEO, LNEO, and butylated hydroxytoluene (BHT), used as a positive control. Based on the antioxidant potency achieved using the hydroxyl radical (OH) test, it was found that the IC_{50} value of AUEO (0.527 ± 0.01 μL/mL) and LNEO (0.354 ± 0.02 μL/mL) exhibited the highest inhibitory activity against hydroxyl radicals compared to the antioxidant BHT (12.027 ± 0.01 μL/mL).

Table 2. Antioxidant activity of AUEO and LNEO.

EOs	Hydroxyl IC_{50} (μL/mL)	Superoxide IC_{50} (μL/mL)	Lipid Peroxidation IC_{50} (μL/mL)	DPPH IC_{50} (μL/mL)
Laurus nobilis	0.354 ± 0.02 [a]	0.133 ± 0.01 [a]	0.101 ± 0.05 [a]	0.489 ± 0.07 [a]
Arbutus unedo	0.527 ± 0.01 [b]	0.275 ± 0.07 [b]	0.207 ± 0.03 [b]	0.711 ± 0.04 [b]
BHT	12.027 ± 0.01 [c]	43.307 ± 0.001 [c]	2.022 ± 0.031 [c]	21.057 ± 0.051 [c]

Different letters in the same column represent significant differences at $p < 0.05$.

In this sense, LNEO has been reported to be able to scavenge hydroxyl radicals generated by an in vitro Fe^{3+}/ascorbate/EDTA/H_2O_2 system [48]. Moreover, AUEO and LNEO were found to scavenge superoxide with an IC_{50} of 0.275 ± 0.07 μL/mL and 0.133 ± 0.01 μL/mL, respectively. In comparison, the standard BHT had a superoxide radical scavenging activity of 43.307 ± 0.001 μL/mL. For the lipid peroxidation method, the IC_{50} values were found to compare favorably (0.207 ± 0.03 μL/mL and 0.101 ± 0.05 μL/mL for AUEO and LNEO, respectively) with that obtained by BHT (IC_{50} = 2.022 ± 0.031 μL/mL). In addition, strong DPPH free radical scavenging activity was recorded for AUEO and LNEO with IC_{50} of 0.711 ± 0.04 μL/mL and 0.489 ± 0.07 μL/mL, respectively. These results may be compared to BHT showing an IC_{50} equal to 21.057 ± 0.051 μL/mL.

3.3. Anti-Diabetic Activity

One of the most widely used approaches to reducing postprandial hyperglycemia and managing diabetes is to inhibit digestive enzymes such as α-amylase and α-glucosidase [61]. Table 3 shows the results of the α-amylase, α-glucosidase, and pancreatic lipase inhibition activities of AUEO and LNEO. Indeed, AUEO and LNEO were discovered to have a significant inhibitory effect on the α-amylase enzyme, with IC_{50} values of 102 ± 0.06 μg/mL and 42.51 ± 0.012, respectively. The results were compared to those of acarbose, a reference antidiabetic, which showed an IC_{50} of 32.14 ± 0.016 μg/mL. Interestingly, LNEO showed a much higher α-glucosidase inhibitory activity (IC_{50} = 1.347 ± 0.021 μg/mL) than that of the control (IC_{50} = 22.0 ± 0.005 μg/mL), while AUEO showed an activity of 76 ± 0.021 μg/mL against the same enzyme. With respect to lipase inhibition, AUEO and LNEO inhibited pancreatic lipase activity with IC_{50} values of 97.018 ± 0.012 μg/mL and 21.23 ± 0.021 μg/mL, respectively, when compared to the reference compound, orlistat, which had lipase inhibitory activity with an IC_{50} value of 14.12 ± 0.023 μg/mL (Table 3).

Table 3. The inhibition of the digestive enzymes α-glucosidase, α-amylase, and lipase by AUEO and LNEO.

IC_{50} (μg/mL)	α-Amylase	α-Glucosidase	Lipase
Arbutus unedo EO	102 ± 0.06 [c]	76 ± 0.021 [c]	97.018 ± 0.012 [c]
Laurus nobilis EO	42.51 ± 0.012 [b]	1.347 ± 0.021 [a]	21.23 ± 0.021 [b]
Acarbose	32.14 ± 0.016 [a]	22 ± 0.005 [b]	_
Orlistat	_	_	14.12 ± 0.023 [a]

Different letters in the same column represent significant differences at $p < 0.05$. AUEO: *Arbutus unedo* EO. LNEO: *Laurus nobilis* EO.

3.4. *Anti-Acetylcholinesterase Activity*

Acetylcholinesterase (AChE) is an enzyme implicated significantly in neurodegenerative diseases, particularly Alzheimer's disease (AD). Moreover, its inhibition constitutes a major therapeutic and preventive pathway. In this work, the inhibitory effects of the two EOs and the control (rivastigmine) were tested against AChE. The results are expressed in IC_{50} and presented in Table 4. As indicated in the table, LNEO showed a greater inhibitory effect (IC_{50} = 89.44 ± 0.07 μg/mL) than AUEO (IC_{50} = 378.57 ± 0.05 μg/mL); rivastigmine showed high activity (IC_{50} = 2.24 ± 0.03 μg/mL).

Table 4. AChE inhibitory activities of EOs compared to the standard drug rivastigmine.

IC_{50} (μg/mL ± SEM)	AUEO	LNEO
AChE	378.57 ± 0.05 [b]	89.44 ± 0.07 [a]
Rivastigmine	-	2.24 ± 0.03 [c]

Different letters in the same column represent significant differences at $p < 0.05$. AUEO: *Arbutus unedo* EO. LNEO: *Laurus nobilis* EO.

3.5. *Anti-Inflammatory Activity*

A large number of aromatic medicinal plant species contain various bioactive compounds with beneficial health properties, especially anti-inflammatory effects. Therefore, using EOs as natural additives is a good approach to prevent abnormal inflammation [62,63].

In this context, the determination of anti-inflammatory effect of the AUEO and LNEO in our work was investigated both in vitro, using 5-LOX enzyme, and in vivo using carrageenan-induced mouse paw edema.

Both EOs tested showed more or less promising inhibitory activity of 5-LOX with IC_{50} values of 86.14 ± 0.05 μg/mL for AUEO and 48.31 ± 0.07 μg/mL for LNEO in comparison with the control (quercetin), which showed an IC_{50} value of 17.59 ± 0.01 μg/mL (Table 5). Moreover, the results of an in vivo experiment showed a significant inhibition of carrageenan-induced hind paw edema volume after 6 h of treatment with these oils at a dose of 50 mg/kg, with a percentage inhibition of 58.82% for AUEO and 70.59% for LNEO compared to the anti-inflammatory drug, indomethacin (72.55%), used as a positive control (Table 6).

Table 5. In vitro anti-inflammatory activity of AUEO and LNEO.

Assays.	(IC_{50} μg/mL)		Control
	AUEO	LNEO	Quercetin
5-Lipoxygenase	86.14 ± 0.05 [c]	48.31 ± 0.07 [b]	17.59 ± 0.01 [a]

Different letters in the same column represent significant differences at $p < 0.05$. AUEO: *Arbutus unedo* EO. LNEO: *Laurus nobilis* EO.

Table 6. Inhibition percentage of the left hind paw volume in rats treated with AUEO and LNEO.

Compounds	Dose (mg/kg)	Carrageenan-Induced Hind Paw Edema Volume (mL; Mean) and % of Inhibition						
		T_0	1 h	% inh.	3 h	% inh.	6 h	% inh.
Control	-	0.84	1.43	-	1.63	-	1.86	-
L. nobilis	50	0.89	1.31	28.81	1.27	51.90	1.19	70.59
	100	0.89	1.23	42.37	1.15	67.09	1.01	88.26
A. unedo	50	0.76	1.24	18.64	1.25	37.97	1.18	58.82
	100	0.82	1.26	25.42	1.23	48.10	1.19	63.72
Indomethacin	10	0.85	1.14	50.85	1.16	60.76	1.13	72.55

3.6. Antimicrobial Activity

The disc diffusion method was carried out to evaluate the antimicrobial activity of EOs from AUEO and LNEO. The results are shown in Table 7. The findings revealed that EOs from both plants had significant antimicrobial activity against all tested microorganisms compared to the conventional antibiotics; chloramphenicol for bacteria and nystatin for yeasts and fungi (ANOVA, $p < 0.05$). It should be noted that, although superior to chloramphenicol, *P. aeruginosa* showed the weakest activity (in vitro). In general, the Gram-positive bacteria recorded the highest mean zone of inhibition (varying between 15.4 ± 0.2 and 19.3 ± 0.2 mm) compared to Gram-negatives (varying between 8.0 ± 0.0 and 16.2 ± 0.1 mm), and the mean inhibition zones of yeasts (ranging from 16.2 ± 0.2 to 19.8 ± 0.3 mm) were higher than the fungal strains (ranging from 13.0 ± 0.3 to 16.2 ± 0.2 mm), although antibiotics remained the most effective (Table 7).

Table 7. The antimicrobial activities of *A. unedo* and *L. nobilis* EOs assessed using disc diffusion test (diameter equals 6.0 mm means no inhibition).

Microorganism	*Arbutus unedo* EO (100%)	*Laurus nobilis* EO (100%)	Chloramphenicol (30 µg/mL)	Nystatin (100 I.U.)
Escherichia coli ATCC 25922	14.6 ± 0.2	16.2 ± 0.1	22.9 ± 0.1	0.0
Proteus mirabilis ATCC 25933	14.2 ± 0.1	15.6 ± 0.2	22.6 ± 0.2	0.0
Salmonella typhimurium ATCC 700408	11.0 ± 0.1	12.7 ± 0.8	13.6 ± 0.0	0.0
Pseudomonas aeruginosa ATCC 27853	8.0 ± 0.0	8.0 ± 0.0	6.0 ± 0.0	0.0
Bacillus subtilis ATCC 6633	16.2 ± 0.3	18.0 ± 0.2	16.3 ± 0.1	0.0
Staphylococcus aureus ATCC 29213	15.4 ± 0.2	18.3 ± 0.2	25.6 ± 0.1	0.0
Listeria monocytogenes ATCC 13932	16.9 ± 0.2	19.3 ± 0.2	28.6 ± 0.2	0.0
Candida albicans	16.2 ± 0.2	19.8 ± 0.3	0.0	28.8 ± 0.3
Trichophyton rubrum	13.0 ± 0.3	15.6 ± 0.3	0.0	25.0 ± 0.02
Aspergillus niger	13.4 ± 0.2	16.2 ± 0.2	0.0	25.8 ± 0.1

Diameter equals 6.0 mm means no inhibition, mean ± standard deviation.

The results of the MIC and MLC tests are shown in Table 8. According to the MIC values, the lowest concentration of AUEO and LNEO inhibiting a microbe's visible growth ranged between 2 and 1% for the tested Gram-negative bacteria (except for *P. aeruginosa* which was >4%), 1 and 0.5% for all Gram-positive bacteria tested, 0.5 and 0.25% for yeasts (*Candida albicans*), and 2 and 1% for the tested fungal strains, respectively. On the other hand, the MLC values are represented, in this study, by the lowest concentration of the EOs killing the test organism under in vitro conditions.

Table 8. MIC and MLC values of AUEO and LNEO.

Microorganisms	EOs % (*v*/*v*)				Controls (µg/mL)	
	AUEO		LNEO		Chloramphenicol	Nystatin
	MIC	MLC	MIC	MLC	MIC	MIC
Escherichia coli ATCC 25922	1.0	2.0	1.0	1.0	4.0	NT
Proteus mirabilis ATCC 25933	2.0	2.0	1.0	1.0	4.0	NT
Salmonella typhimurium ATCC 700408	2.0	4.0	1.0	2.0	64.0	NT
Pseudomonas aeruginosa ATCC 27853	>4.0	>4.0	>4.0	>4.0	>64.0	NT
Bacillus subtilis ATCC 6633	1.0	2.0	0.5	1.0	32.0	NT
Staphylococcus aureus ATCC 29213	1.0	2.0	0.5	1.0	4.0	NT
Listeria monocytogenes ATCC 13932	1.0	1.0	0.5	1.0	2.0	NT
Candida albicans	0.5	NT	0.25	NT	NT	4.0
Trichophyton rubrum	2.0	NT	1.0	NT	NT	16.0
Aspergillus niger	2.0	NT	1.0	NT	NT	16.0

NT: not tested.

In the current study, the MLC values of AUEO and LNEO ranged between 2 and 1% (except with *P. aeruginosa* which was >4%); the Gram-positive bacteria ranged between 2–1 and 1%, respectively. Interestingly, *P. aeruginosa* was not susceptible to AUEO and LNEO.

4. Discussion

4.1. Chemical Composition

The analysis of phytochemical composition is an indispensable tool in the development of active ingredients derived from plant tissues. In our study, several natural substances have been identified from the studied plants that could be responsible for their various biological activities.

Indeed, the AUEO was characterized by a high rate of decenal (13.47%), accompanied by other constituents with variable content, such as α-terpineol (7.8%) and palmitic acid (6.00%). It should also be noted that the EO of the same species in Algeria showed a significant amount of some major chemical constituents such as palmitic acid (35.2%), linoleic acid (18.8%), and 2,6-di-tert-butyl-p-cresol (6.2%) [64]. Likewise, AUEO from Turkey showed a significant amount of various chemical constituents. Indeed, thirty-seven constituents were characterized as present, including (E)-2-decenal (12.0%), α-terpineol (8.8%), hexadecanoic acid (5.1%), and (E)-2-undecenal (4.8%) as the major constituents [65]. This variation in chemical profiles of essential oils could be attributed to several factors, such as harvest season, plant age, soil composition, and geographical variation [50,51,66,67].

In our work, from a practical point of view, our results are comparable to other studies on EOs extracted from Moroccan *L. nobilis* (northern Morocco) [47], identifying about 26 compounds, with the predominance of eucalyptol (52.43%), followed by α-terpinyl acetate (8.96%), sabinene (6.13%), limonene (5.25%), β-pinene (3.72%), linalool (3.14%), terpinene-4-ol (2.56%), α-terpinene (2.12%), β-terpineol (1.56%), bornyl acetate (1.89%), α-phellandrene (1.28%), myrcene (1.13%), camphene (1.05%), p-cymene (0.94%), α-terpinene (0.98%), and eugenol (0.56%).

Numerous studies have evaluated the chemical composition of LNEO in different countries. In Argentina, Lira et al. [68] showed a predominance of eucalyptol (45.1%), then linalool (11.9%), sabinene (9.3%), α-terpinyl acetate (8.0%), and methyl eugenol (2.8%).

In Bulgaria, Fidan et al. [69] revealed the presence of eucalyptol (41.0%), followed by α-terpinyl acetate (14.4%), sabinene (8.8%), methyl eugenole (6.0%), β-linalool (4.9%), and α-terpineol (3.1%), and in Iran, Mohammadreza [70] identified a preponderance of eucalyptol (55.80%), α-terpinyl acetate (15.14%), terpinene-4-ol (5.27%), α–pinene (5.26%), p-cymene (2.70%), linalool (1.40%), and terpinene-4-yla cetate (1.13%). In Turkey, Dadalioğlu and Evrendilek [71] found eucalyptol (60.72%), α-terpinene (12.53%), sabinene (12.12%), and α-pinene (6.11%) as major constituents.

In Algeria, Nabila et al. [72] revealed chemical variations of LNEO with a predominance of oxygenated monoterpenes (59%), eucalyptol (30.1%), and α-terpynil acetate (21.6%) as major compounds. The second-most-abundant class was phenylpropanoids (18.7%), which were made up of methyl eugenol (16.9%), elemicin (0.9%), and apiole (0.9%).

4.2. Antioxidant Activity

The evaluation of the antioxidant activity was carried out using several in vitro tests. This allows one to have complementary of results. Indeed, the antioxidant reaction mechanisms in the different tests are variable, and ideally an antioxidant should respond positively to different mechanisms. In this context, the antioxidant power of *L. nobilis* and *A. unedo* was evaluated by four tests, namely hydroxyl radical, superoxide radical, lipid peroxidation, and DPPH radical scavenging activity. In fact, the ability to prevent lipid peroxidation could only be determined in lipophilic samples, including essential oils [29].

From our results, it was clear that AUEO and LNEO possess important antioxidant characteristics and, therefore, could be considered as a promising source of natural antioxidants. This capacity of *A. unedo* leaf extracts has already been reported by other authors [29,73].

It is well known that the antioxidant effect of an extract or compound is generally associated with their redox properties, allowing them to act as reducing agents [74]. This significant antioxidant activity of both plants could be attributed to the presence of antioxidant compounds. Some monoterpenes and oxygenated sesquiterpenes have been reported to have an inhibitory oxidation power [75]. Thus, the antioxidant activity of AUEO and LNEO might be attributed to the presence of high concentrations of, mainly, eucalyptol (1,8-cineole) (36.40%), linalool (10.34%), α-terpinen-4-ol (13.05%), α-terpinyl acetate (10.61%) for LNEO, and (+)-isomenthone (4.23%), α-terpineol (7.8%), and myristic acid (3.96%) for AUEO. Our findings are similar to those reported by previous studies on the antioxidant activity of AUEOs and LNEOs [29,48,76,77]. Hence, these results confirm the role of EOs as natural antioxidants as well as their natural protective role for human health to preserve many physiological functions.

4.3. Anti-Diabetic Activity

The evaluation of the anti-diabetic activity of both selected plants was carried out by testing their capacity to inhibit the activity of two carbohydrate-hydrolyzing enzymes (α-amylase and α-glucosidase) and also that of lipase, allowing them to slow down the absorption of fatty acids and, subsequently, induce the degradation of internal glucose.

From our results, essential oils obtained from *A. unedo* and *L. nobilis* showed remarkable anti-diabetic activity against the three selected enzymes. Furthermore, it has been reported that *A. unedo* aqueous extract exhibits a potent inhibitory effect against α-amylase and α-glucosidase activity with IC_{50} values of 730.15 ± 0.25 and 94.81 ± 5.99 μg/mL, respectively [78].

On the other hand, previous studies have reported that LNEO possesses a potent inhibitory effect of more than 90% on α-glucosidase inhibitory activity, with an IC_{50} value of 1.748 ± 0.021 μL/mL [48]. Thus, this supports our suggestion that AUEO and LNEO have significant anti-hyperglycemic activities.

Based on the results of the lipase test and the fact that anti-diabetic activity was found, it seems likely that AUEO and LNEO could be natural sources of agents that stop fat from being absorbed and could be used to treat obesity. Despite numerous studies on the chemical composition and antimicrobial activity of AUEOs and LNEOs, there is little information on their anti-diabetic activity. In this study, our results revealed that both EOs exhibit high inhibitory activities towards α-amylase, α-glucosidase, and pancreatic lipase. This interesting anti-hyperglycemic effect may be due to several bioactive compounds, in particular those with high concentrations. Previous work showed that 1,8-cineole, α-pinene, and limonene inhibit α-glucosidase and α-amylase activities [48,79,80]. The combination of limonene and linalool has been reported to provide potent anti-hyperglycemic activity [81]. Other studies have revealed that terpinen-4-ol has remarkable anti-diabetic activity [75]. On the other hand, and as mentioned above, the most abundant volatile components in *A. unedo* were (+)-isomenthone, α-terpineol, and myristic acid and therefore the inhibition of the enzymes could be related to the presence of these components in the EOs. Moreover, AUEOs showed a highly significant correlations between their bioactive compounds and the percentage of α-glucosidase inhibition [82]. Similarly, the in vitro inhibitory activity of LNEOs against pancreatic lipase might be due to their volatile compounds. Our results are promising and suggest that EOs may be able to treat diabetes, but more research is needed to prove this assumption.

4.4. Anti-Acetylcholinesterase Activity

AChE is an important pathogenic factor in AD and its main role is to stimulate the hydrolysis of acetylcholine (ACh) to choline. Indeed, ACh deficiency is responsible for AD pathogenesis [83]. Several pharmacological investigations have focused on AChE inhibitors to reduce cholinergic deficits and improve neurotransmission [84]. However, these inhibitors have certain limitations, such as their short half-lives and the associated hepatotoxicity, which represents the main side effect [85]. This prompted several researchers to develop new anti-AChE agents [86].

In our study, EOs from *A. unedo* and *L. nobilis* showed significant anti-AChE activity that could be associated with their main compounds. Indeed, other studies have already shown that terpenes have anti-cholinesterase properties [87–90]. The mechanisms of action often involve competitive inhibition of the enzyme following the binding of bioactive molecules to the active site of the enzyme [91,92].

4.5. Anti-Inflammatory Activity

Inflammation is a natural response of the immune system, characterized by a mechanism that represents a chain of organized and dynamic responses comprising both cellular and vascular events with specific humoral secretion. These pathways involve physical changes in the localization of white blood cells (monocytes, basophils, eosinophils, and neutrophils), plasma and fluids at the inflamed site [93]. It is characterized by leukocyte activation, increased vascular permeability, edema, and pain [94].

The results of our tests (both in vivo and in vitro) show that the EOs we obtained from the leaves of *A. unedo* and *L. nobilis* have strong anti-inflammatory properties. These effects could be attributed to the presence of various bioactive molecules such as eucalyptol (1,8-cineole), α-terpinen-4-ol, camphor, α-terpinyl acetate, linalool, limonene, α-pinene, and camphene. Indeed, AUEO showed less 5-LOX inhibitory activity compared to LNEO. Moreover, AUEO exhibited remarkable activity (IC_{50} = 86.14 ± 0.05 μg/mL), which may be due to the presence of certain chemical compounds.

Furthermore, AUEO contains other molecules such as geraniol, which have already demonstrated a significant inhibitory effect, inhibiting pro-inflammatory cytokines as well as NF-κB signaling pathways [95]. In another study, Su and his colleagues [96] found that geraniol has a promising effect on nitric oxide and prostaglandin E2 (PGE_2), which are pro-inflammatory molecules. On the other hand, previous studies have reported the anti-inflammatory effects of *A. undo* extracts [97,98]. However, the mechanisms underlying this effect have yet to be clarified.

Eucalyptol was reported as a major compound of LNEO (36.40%), and may provide gastroprotection via anti-inflammatory mechanisms; it could possess an inhibitory activity against 5-LOX, one of the key mediators involved in inflammatory responses [99]. Moreover, eucalyptol in interaction with limonene has been shown to cause partial potentiation of anti-inflammatory action [100]. In fact, limonene has been shown to inhibit lipopolysaccharide-induced inflammation and inflammatory cell migration [99].

Furthermore, monoterpenoid compounds have demonstrated potential anti-inflammatory effects. Indeed, several studies have shown that linalool is an anti-inflammatory agent, controlling the release of anti-inflammatory cytokines and regulating the activation of transcription factor-κB (NF-κB) and its translocation in the nucleus [101]. Similarly, camphene appears to be responsible for inhibiting the incorporation of arachidonic acid into the active site of the corresponding enzymes, thus inhibiting the production of inflammatory prostaglandins and leukotrienes [32].

Therefore, LNEO contains interesting bioactive compounds responsible for anti-inflammatory activities. Indeed, several studies have examined the effectiveness of LNEO as an effective anti-inflammatory agent [102–104]. Our results showed that LNEO compounds can be effective at inhibiting 5-LOX.

5. Antimicrobial Activity

In recent years, antibiotic resistance has become one of the most serious threats to global health, food security, and development. It is a natural phenomenon, but the misuse of antibiotics in humans and animals accelerates the process. The search for anti-microbial agents of natural origin will therefore provide a promising alternative. As has already been indicated, the tested EOs exerted significant anti-microbial activities against the different microorganisms.

Indeed, there have been many scientific publications on medicinal plant research indicating the remarkable sensitivity of Gram-positive bacteria to plant extracts or EOs compared to Gram-negative bacteria, which can initially be attributed to the cell wall structure [13,105,106]. Some previous studies have reported that AUEO has remarkable antimicrobial activity. It has been published that the methanol, ethanol, and ethyl acetate extracts of *A. unedo* leaves exhibit different levels of antibacterial activity against *E. coli*, *S. aureus*, *L. monocytogenes*, and *P. aeruginosa* strains that cause food-borne diseases; however, only the methanol extract showed antibacterial activity against *P. aeruginosa* with no effect from the ethanol and ethyl acetate extracts [107]. This is in harmony with the current results, which showed that *P. aeruginosa* was the least susceptible bacterium to the tested EOs. On the other hand, a previous study showed that the fruits, twigs, and leaves of *L. nobilis* recorded good antimicrobial activity against almost all microorganisms tested, including Gram-positive and Gram-negative bacteria, yeasts, and fungi; however, *P. aeruginosa* showed the lowest response [69]. Indeed, *P. aeruginosa* is a difficult-to-treat microorganism. It is highly resistant to numerous antibiotics, including ciprofloxacin, gentamicin, tobramycin, imipenem, and ceftazidime [108]. The general resistance of *P. aeruginosa* is well documented, and its genes can express a wide range of resistance mechanisms; mutations in chromosomal genes that control resistance genes can occur, and it can acquire resistance genes from other organisms through plasmids, transposons, and bacteriophages [109]. Therefore, *P. aeruginosa* was classified by the WHO as a resistant pathogen on a list of antibiotic-resistant bacteria that are in urgent need of new alternative treatments [110].

Our findings regarding MIC and MLC are consistent with previously published reports, suggesting that EOs from these plants can be used as nutraceuticals and functional foods as well as antimicrobial agents, The MIC and MBC of the leaves of *Laurus nobilis* was reported to be as low as 2.5 g/L for *Escherichia coli* and 1.25 g/L (MIC) and 2.5 g/L (MBC) for *Yersinia enterocolitica*, where the *Arbutus Unedo* root extracts showed relatively high MIC compared with *Laurus nobilis* leaves against *S. aureus* (MIC = 12.5 mg/mL) [111–114]. Essential oils mostly have higher inhibitory activity against microorganisms than crude extracts as they have more aromatic compounds with antimicrobial effects [115].

Overall, the findings of our current investigation are consistent with those of previous publications using the diffusion method that showed potent broad-spectrum antibacterial ability [116,117]. It is claimed that the mode of action of the EOs against target microorganisms is different from the regular antibiotics; when exposed to EOs, bacterial cells might be destroyed by the irregular disruption of the intracellular structure and the bursting of cell walls and membranes [105]. This aspect needs much future study, since so far medicinal plants have not been used as a drug alternative to antibiotics.

6. Conclusions

Here, we currently report the identification of phytochemicals as well as some biological properties of AUEO and LNEO. Phytochemical analysis revealed a diversity of volatile compounds in both EOs. The results showed significant antioxidant, anti-diabetic, and antimicrobial effects with very low EO concentrations. Other results are needed to highlight the mechanisms by which these EOs act. The major compounds of these oils must also be investigated for their biological effects. In addition, toxicity studies should be performed to verify the safety of EOs and bioactive compounds.

Author Contributions: Conceptualization, M.E.A.F., A.B. and H.N.M.; methodology, H.N.M., A.B. and N.H.M. software, H.O.; validation, M.E.A.F., A.B., H.H. and S.H.A.-M.; formal analysis, R.F. and E.M.A.; investigation, E.M.A., R.A.S., M.M.A., N.E.O., M.G. and A.A.A.A.; data curation, H.A., A.B. and A.Q.; writing—original draft preparation, S.H.A.-M., H.O., R.F., E.M.A. and A.B.; writing—review and editing, H.N.M., R.A.S., M.M.A., N.E.O., M.G. and A.A.A.A.; visualization, H.N.M., A.B., H.H., M.E.A.F. and N.H.M. All authors have read and agreed to the published version of the manuscript.

Funding: This research received no external funding.

Institutional Review Board Statement: Not applicable.

Informed Consent Statement: Not applicable.

Data Availability Statement: Not applicable.

Acknowledgments: This study was supported by the Princess Nourah Bint Abdulrahman University Researchers Supporting Project number (PNURSP2022R158) Princess Nourah Bint Abdulrahman University, Riyadh, Saudi Arabia.

Conflicts of Interest: The authors declare no conflict of interest.

Abbreviations

ABTS	2,2-azino-bis-3-ethylbenzothiazoline-6-sulfonic acid
Ach	Acetylcholine
AChE	Acetylcholinesterase
AD	Alzheimer's Disease
ASCh	Acetylthiocholine Iodide
BHT	Butylated Hydroxytoluene
AUEO	*Arbutus unedo* Essential Oil.
DM	Diabetes Mellitus
DMSO	Dimethyl Sulfoxide
DNA	Deoxyribonucleic Acid
DPPH	2,2-diphenyl-1-picrylhydrazyl
DTNB	5,5′-dithio-bis-2-nitrobenzoic acid
EO	Essential Oil
EDTA	Ethylenediaminetetraacetic acid
$FeCl_3$	Chlorure ferrique
GC-MS	Gas Chromatography-Mass Spectrometry
LNEO	*Laurus nobilis* Essential Oil.
MHA	Mueller-Hinton Agar
MFC	Minimum Fungicidal Concentration
MIC	Minimum Inhibitory Concentration
MLC	Minimum Lethal Concentration
NBT	Nitro Blue Tetrazolium

NF-κB	Transcription Factor-κB
OH	Hydroxyl Radical
PGE_2	Prostaglandin E_2
R_T	Retention Time
SA	Sabouraud dextrose agar
TBA	Thiobarbituric Acid
TBARS	Thiobarbituric Acid Reactive Substances
TCA	Trichloroacetic Acid
Trolox	6-hydroxy-2,5,7,8-tetramethylchroman-2-carboxylic acid
WHO	World Health Organization
5-LOX	Lipoxygenase
TBA	Thiobarbituric Acid

References

1. Abdallah, E.M. Plants: An Alternative Source for Antimicrobials. *J. Appl. Pharm. Sci.* **2011**, *1*, 16–20.
2. Ekor, M. The Growing Use of Herbal Medicines: Issues Relating to Adverse Reactions and Challenges in Monitoring Safety. *Front. Pharmacol.* **2014**, *4*, 177. [CrossRef]
3. Abdelaali, B.; El Menyiy, N.; El Omari, N.; Benali, T.; Guaouguaou, F.-E.; Salhi, N.; Naceiri Mrabti, H.; Bouyahya, A. Phytochemistry, Toxicology, and Pharmacological Properties of Origanum Elongatum. *Evid. Based Complement. Altern. Med.* **2021**, *2021*, 6658593. [CrossRef] [PubMed]
4. Al-Mijalli, S.H.; Assaggaf, H.; Qasem, A.; El-Shemi, A.G.; Abdallah, E.M.; Mrabti, H.N.; Bouyahya, A. Antioxidant, Antidiabetic, and Antibacterial Potentials and Chemical Composition of Salvia Officinalis and Mentha Suaveolens Grown Wild in Morocco. *Adv. Pharmacol. Pharm. Sci.* **2022**, *2022*, 2844880. [CrossRef] [PubMed]
5. Benali, T.; Bouyahya, A.; Habbadi, K.; Zengin, G.; Khabbach, A.; Hammani, K. Chemical Composition and Antibacterial Activity of the Essential Oil and Extracts of Cistus Ladaniferus Subsp. Ladanifer and Mentha Suaveolens against Phytopathogenic Bacteria and Their Ecofriendly Management of Phytopathogenic Bacteria. *Biocatal. Agric. Biotechnol.* **2020**, *28*, 101696. [CrossRef]
6. Bouyahya, A.; Chamkhi, I.; Benali, T.; Guaouguaou, F.-E.; Balahbib, A.; El Omari, N.; Taha, D.; Belmehdi, O.; Ghokhan, Z.; El Menyiy, N. Traditional Use, Phytochemistry, Toxicology, and Pharmacology of *Origanum majorana* L. *J. Ethnopharmacol.* **2021**, *265*, 113318. [CrossRef]
7. Bouyahya, A.; Belmehdi, O.; El Jemli, M.; Marmouzi, I.; Bourais, I.; Abrini, J.; Faouzi, M.E.A.; Dakka, N.; Bakri, Y. Chemical Variability of Centaurium Erythraea Essential Oils at Three Developmental Stages and Investigation of Their in Vitro Antioxidant, Antidiabetic, Dermatoprotective and Antibacterial Activities. *Ind. Crops Prod.* **2019**, *132*, 111–117. [CrossRef]
8. El Omari, N.; Guaouguaou, F.E.; El Menyiy, N.; Benali, T.; Aanniz, T.; Chamkhi, I.; Balahbib, A.; Taha, D.; Shariati, M.A.; Zengin, G. Phytochemical and Biological Activities of Pinus Halepensis Mill., and Their Ethnomedicinal Use. *J. Ethnopharmacol.* **2021**, *268*, 113661. [CrossRef]
9. Khouchlaa, A.; Talbaoui, A.; El Idrissi, A.E.Y.; Bouyahya, A.; Ait Lahsen, S.; Kahouadji, A.; Tijane, M. Determination of Phenol Content and Evaluation of in Vitro Litholytic Effects on Urolithiasis of Moroccan Zizyphus Lotus L. Extract. *Phytothérapie* **2017**, *16*, 14–19. [CrossRef]
10. Marmouzi, I.; Bouyahya, A.; Ezzat, S.M.; El Jemli, M.; Kharbach, M. The Food Plant *Silybum Marianum* (L.) Gaertn.: Phytochemistry, Ethnopharmacology and Clinical Evidence. *J. Ethnopharmacol.* **2021**, *265*, 113303. [CrossRef]
11. Sharifi-Rad, J.; Dey, A.; Koirala, N.; Shaheen, S.; El Omari, N.; Salehi, B.; Goloshvili, T.; Cirone Silva, N.C.; Bouyahya, A.; Vitalini, S. Cinnamomum Species: Bridging Phytochemistry Knowledge, Pharmacological Properties and Toxicological Safety for Health Benefits. *Front. Pharmacol.* **2021**, *12*, 600139. [CrossRef] [PubMed]
12. Boukhatem, M.N.; Sudha, T.; Darwish, N.H.; Nada, H.G.; Mousa, S.A. Rose-Scented Geranium Essential Oil from Algeria (Pelargonium Graveolens L'Hérit.): Assessment of Antioxidant, Anti-Inflammatory and Anticancer Properties against Different Metastatic Cancer Cell Lines. *Ann. Pharm. Fr.* **2021**, *80*, 383–396. [CrossRef] [PubMed]
13. Nada, H.G.; Mohsen, R.; Zaki, M.E.; Aly, A.A. Evaluation of Chemical Composition, Antioxidant, Antibiofilm and Antibacterial Potency of Essential Oil Extracted from Gamma Irradiated Clove (Eugenia Caryophyllata) Buds. *J. Food Meas. Charact.* **2022**, *16*, 673–686. [CrossRef]
14. Raut, J.S.; Karuppayil, S.M. A Status Review on the Medicinal Properties of Essential Oils. *Ind. Crops Prod.* **2014**, *62*, 250–264. [CrossRef]
15. Baldino, L.; Scognamiglio, M.; Reverchon, E. Supercritical Fluid Technologies Applied to the Extraction of Compounds of Industrial Interest from *Cannabis Sativa* L. and to Their Pharmaceutical Formulations: A Review. *J. Supercrit. Fluids* **2020**, *165*, 104960. [CrossRef]
16. Sahraoui, N.; Boutekedjiret, C. Innovative Process of Essential Oil Extraction: Steam Distillation Assisted by Microwave. In *Progress in Clean Energy*; Springer: Berlin/Heidelberg, Germany, 2015; Volume 1, pp. 831–841.

17. Gonzalez-Coloma, A.; Martín, L.; Mainar, A.M.; Urieta, J.S.; Fraga, B.M.; Rodríguez-Vallejo, V.; Díaz, C.E. Supercritical Extraction and Supercritical Antisolvent Fractionation of Natural Products from Plant Material: Comparative Results on Persea Indica. *Phytochem. Rev.* **2012**, *11*, 433–446. [CrossRef]
18. Macıas-Sánchez, M.D.; Mantell, C.; Rodrıguez, M.; de La Ossa, E.M.; Lubián, L.M.; Montero, O. Supercritical Fluid Extraction of Carotenoids and Chlorophyll a from Nannochloropsis Gaditana. *J. Food Eng.* **2005**, *66*, 245–251. [CrossRef]
19. Janssen, A.M.; Scheffer, J.J.C.; Svendsen, A.B. Antimicrobial Activity of Essential Oils: A 1976-1986 Literature Review. Aspects of the Test Methods. *Planta Med.* **1987**, *53*, 395–398. [CrossRef]
20. Ma, L.; Yao, L. Antiviral Effects of Plant-Derived Essential Oils and Their Components: An Updated Review. *Molecules* **2020**, *25*, 2627. [CrossRef]
21. Bhalla, Y.; Gupta, V.K.; Jaitak, V. Anticancer Activity of Essential Oils: A Review. *J. Sci. Food Agric.* **2013**, *93*, 3643–3653. [CrossRef]
22. Ibrahim, F.A.; Usman, L.A.; Akolade, J.O.; Idowu, O.A.; Abdulazeez, A.T.; Amuzat, A.O. Antidiabetic Potentials of Citrus Aurantifolia Leaf Essential Oil. *Drug Res.* **2019**, *69*, 201–206. [CrossRef] [PubMed]
23. Zehiroglu, C.; Ozturk Sarikaya, S.B. The Importance of Antioxidants and Place in Today's Scientific and Technological Studies. *J. Food Sci. Technol.* **2019**, *56*, 4757–4774. [CrossRef] [PubMed]
24. Matough, F.A.; Budin, S.B.; Hamid, Z.A.; Alwahaibi, N.; Mohamed, J. The Role of Oxidative Stress and Antioxidants in Diabetic Complications. *Sultan Qaboos Univ. Med. J.* **2012**, *12*, 5. [CrossRef] [PubMed]
25. Mohora, M.; Greabu, M.; Muscurel, C.; Duta, C.; Totan, A. The Sources and the Targets of Oxidative Stress in the Etiology of Diabetic Complications. *Rom. J. Biophys.* **2007**, *17*, 63–84.
26. Hong, J.-H.; Kim, M.-J.; Park, M.-R.; Kwag, O.-G.; Lee, I.-S.; Byun, B.H.; Lee, S.-C.; Lee, K.-B.; Rhee, S.-J. Effects of Vitamin E on Oxidative Stress and Membrane Fluidity in Brain of Streptozotocin-Induced Diabetic Rats. *Clin. Chim. Acta* **2004**, *340*, 107–115. [CrossRef]
27. Coleman, H.N.; Greenfield, W.W.; Stratton, S.L.; Vaughn, R.; Kieber, A.; Moerman-Herzog, A.M.; Spencer, H.J.; Hitt, W.C.; Quick, C.M.; Hutchins, L.F. Human Papillomavirus Type 16 Viral Load Is Decreased Following a Therapeutic Vaccination. *Cancer Immunol. Immunother.* **2016**, *65*, 563–573. [CrossRef]
28. Jamila, F.; Mostafa, E. Ethnobotanical Survey of Medicinal Plants Used by People in Oriental Morocco to Manage Various Ailments. *J. Ethnopharmacol.* **2014**, *154*, 76–87. [CrossRef]
29. Boulanouar, B.; Abdelaziz, G.; Aazza, S.; Gago, C.; Miguel, M.G. Antioxidant Activities of Eight Algerian Plant Extracts and Two Essential Oils. *Ind. Crops Prod.* **2013**, *46*, 85–96. [CrossRef]
30. Coates, A.; Hu, Y.; Bax, R.; Page, C. The Future Challenges Facing the Development of New Antimicrobial Drugs. *Nat. Rev. Drug Discov.* **2002**, *1*, 895–910. [CrossRef]
31. Sadeek, A.M.; Abdallah, E.M. Phytochemical Compounds as Antibacterial Agents a Mini Review. *Saudi Arab. Glob. J. Pharm. Sci.* **2019**, *53*, 555720.
32. Hachlafi, N.E.; Aanniz, T.; Menyiy, N.E.; Baaboua, A.E.; Omari, N.E.; Balahbib, A.; Shariati, M.A.; Zengin, G.; Fikri-Benbrahim, K.; Bouyahya, A. In Vitro and in Vivo Biological Investigations of Camphene and Its Mechanism Insights: A Review. *Food Rev. Int.* **2021**, 1–28. [CrossRef]
33. El Menyiy, N.; Guaouguaou, F.-E.; El Baaboua, A.; El Omari, N.; Taha, D.; Salhi, N.; Shariati, M.A.; Aanniz, T.; Benali, T.; Zengin, G.; et al. Phytochemical Properties, Biological Activities and Medicinal Use of Centaurium Erythraea Rafn. *J. Ethnopharmacol.* **2021**, *276*, 114171. [CrossRef] [PubMed]
34. El Omari, N.; Jaouadi, I.; Lahyaoui, M.; Benali, T.; Taha, D.; Bakrim, S.; El Menyiy, N.; El Kamari, F.; Zengin, G.; Bangar, S.P.; et al. Natural Sources, Pharmacological Properties, and Health Benefits of Daucosterol: Versatility of Actions. *Appl. Sci.* **2022**, *12*, 5779. [CrossRef]
35. El Omari, N.; Akkaoui, S.; El Blidi, O.; Ghchime, R.; Bouyahya, A.; Kharbach, M.; Yagoubi, M.; Balahbib, A.; Chokairi, O.; Barkiyou, M. HPLC-DAD/TOF-MS Chemical Compounds Analysis and Evaluation of Antibacterial Activity of Aristolochia Longa Root Extracts. *Nat. Prod. Commun.* **2020**, *15*, 1934578X20932753. [CrossRef]
36. Bouyahya, A.; Mechchate, H.; Benali, T.; Ghchime, R.; Charfi, S.; Balahbib, A.; Burkov, P.; Shariati, M.A.; Lorenzo, J.M.; Omari, N.E. Health Benefits and Pharmacological Properties of Carvone. *Biomolecules* **2021**, *11*, 1803. [CrossRef]
37. Rios, J.-L.; Recio, M.C. Medicinal Plants and Antimicrobial Activity. *J. Ethnopharmacol.* **2005**, *100*, 80–84. [CrossRef]
38. Kim, D.J. The Epidemiology of Diabetes in Korea. *Diabetes Metab. J.* **2011**, *35*, 303–308. [CrossRef]
39. Surya, S.; Salam, A.D.; Tomy, D.V.; Carla, B.; Kumar, R.A.; Sunil, C. Diabetes Mellitus and Medicinal Plants-a Review. *Asian Pac. J. Trop. Dis.* **2014**, *4*, 337–347. [CrossRef]
40. Prabhakar, P.K.; Doble, M. A Target Based Therapeutic Approach towards Diabetes Mellitus Using Medicinal Plants. *Curr. Diabetes Rev.* **2008**, *4*, 291–308. [CrossRef]
41. Serçe, S.; Özgen, M.; Torun, A.A.; Ercişli, S. Chemical Composition, Antioxidant Activities and Total Phenolic Content of Arbutus Andrachne L. (Fam. Ericaceae) (the Greek Strawberry Tree) Fruits from Turkey. *J. Food Compos. Anal.* **2010**, *23*, 619–623. [CrossRef]
42. Bnouham, M.; Merhfour, F.Z.; Ziyyat, A.; Aziz, M.; Legssyer, A.; Mekhfi, H. Antidiabetic Effect of Some Medicinal Plants of Oriental Morocco in Neonatal Non-Insulin-Dependent Diabetes Mellitus Rats. *Hum. Exp. Toxicol.* **2010**, *29*, 865–871. [CrossRef] [PubMed]

43. Legssyer, A.; Ziyyat, A.; Mekhfi, H.; Bnouham, M.; Herrenknecht, C.; Roumy, V.; Fourneau, C.; Laurens, A.; Hoerter, J.; Fischmeister, R. Tannins and Catechin Gallate Mediate the Vasorelaxant Effect of Arbutus Unedo on the Rat Isolated Aorta. *Phytother. Res. Int. J. Devoted Pharmacol. Toxicol. Eval. Nat. Prod. Deriv.* **2004**, *18*, 889–894.
44. Ruiz-Rodríguez, B.-M.; Morales, P.; Fernández-Ruiz, V.; Sánchez-Mata, M.-C.; Camara, M.; Díez-Marqués, C.; Pardo-de-Santayana, M.; Molina, M.; Tardío, J. Valorization of Wild Strawberry-Tree Fruits (*Arbutus Unedo* L.) through Nutritional Assessment and Natural Production Data. *Food Res. Int.* **2011**, *44*, 1244–1253. [CrossRef]
45. Ziyyat, A.; Mekhfi, H.; Bnouham, M.; Tahri, A.; Legssyer, A.; Hoerter, J.; Fischmeister, R. Arbutus Unedo Induces Endothelium-Dependent Relaxation of the Isolated Rat Aorta. *Phytother. Res. Int. J. Devoted Pharmacol. Toxicol. Eval. Nat. Prod. Deriv.* **2002**, *16*, 572–575. [CrossRef]
46. Alarcão-e-Silva, M.; Leitão, A.E.B.; Azinheira, H.G.; Leitão, M.C.A. The Arbutus Berry: Studies on Its Color and Chemical Characteristics at Two Mature Stages. *J. Food Compos. Anal.* **2001**, *14*, 27–35. [CrossRef]
47. Derwich, E.; Benziane, Z.; Boukir, A.; Mohamed, S.; Abdellah, B. Chemical Composition and Antibacterial Activity of Leaves Essential Oil of Laurus Nobilis from Morocco. *Aust. J. Basic Appl. Sci.* **2009**, *3*, 3818–3824.
48. Basak, S.S.; Candan, F. Effect of Laurus Nobilis L. Essential Oil and Its Main Components on α-Glucosidase and Reactive Oxygen Species Scavenging Activity. *Iran. J. Pharm. Res.* **2013**, *12*, 367.
49. Hamad Al-Mijalli, S.; ELsharkawy, E.R.; Abdallah, E.M.; Hamed, M.; El Omari, N.; Mahmud, S.; Alshahrani, M.M.; Mrabti, H.N.; Bouyahya, A. Determination of Volatile Compounds of *Mentha Piperita* and *Lavandula Multifida* and Investigation of Their Antibacterial, Antioxidant, and Antidiabetic Properties. *Evid. Based Complement. Altern. Med.* **2022**, *2022*, e9306251. [CrossRef]
50. Assaggaf, H.M.; Naceiri Mrabti, H.; Rajab, B.S.; Attar, A.A.; Alyamani, R.A.; Hamed, M.; El Omari, N.; El Menyiy, N.; Hazzoumi, Z.; Benali, T.; et al. Chemical Analysis and Investigation of Biological Effects of Salvia Officinalis Essential Oils at Three Phenological Stages. *Molecules* **2022**, *27*, 5157. [CrossRef]
51. Al-Mijalli, S.H.; Mrabti, H.N.; Assaggaf, H.; Attar, A.A.; Hamed, M.; Baaboua, A.E.; Omari, N.E.; Menyiy, N.E.; Hazzoumi, Z.; Sheikh, R.A. Chemical Profiling and Biological Activities of Pelargonium Graveolens Essential Oils at Three Different Phenological Stages. *Plants* **2022**, *11*, 2226. [CrossRef]
52. Hu, B.; Cui, F.; Yin, F.; Zeng, X.; Sun, Y.; Li, Y. Caffeoylquinic Acids Competitively Inhibit Pancreatic Lipase through Binding to the Catalytic Triad. *Int. J. Biol. Macromol.* **2015**, *80*, 529–535. [CrossRef] [PubMed]
53. Ingkaninan, K.; Temkitthawon, P.; Chuenchom, K.; Yuyaem, T.; Thongnoi, W. Screening for Acetylcholinesterase Inhibitory Activity in Plants Used in Thai Traditional Rejuvenating and Neurotonic Remedies. *J. Ethnopharmacol.* **2003**, *89*, 261–264. [CrossRef] [PubMed]
54. Rege, M.G.; Ayanwuyi, L.O.; Zezi, A.U.; Odoma, S. Anti-Nociceptive, Anti-Inflammatory and Possible Mechanism of Anti-Nociceptive Action of Methanol Leaf Extract of Nymphaea Lotus Linn (Nymphaeceae). *J. Tradit. Complement. Med.* **2021**, *11*, 123–129. [CrossRef] [PubMed]
55. Guinea, J.; Recio, S.; Escribano, P.; Torres-Narbona, M.; Peláez, T.; Sánchez-Carrillo, C.; Rodríguez-Créixems, M.; Bouza, E. Rapid Antifungal Susceptibility Determination for Yeast Isolates by Use of Etest Performed Directly on Blood Samples from Patients with Fungemia. *J. Clin. Microbiol.* **2010**, *48*, 2205–2212. [CrossRef]
56. Doudach, L.; Al-Mijalli, S.H.; Abdallah, E.M.; Mrabti, H.N.; Chibani, F.; Faouzi, M.E.A. Antibacterial Evaluation of The Roots of Moroccan Aristolochia Longa Against Referenced Gram-Positive and Gram-Negative Bacteria. *Adv. Life Sci.* **2022**, *9*, 116–121.
57. Hu, F.; Tu, X.-F.; Thakur, K.; Hu, F.; Li, X.-L.; Zhang, Y.-S.; Zhang, J.-G.; Wei, Z.-J. Comparison of Antifungal Activity of Essential Oils from Different Plants against Three Fungi. *Food Chem. Toxicol.* **2019**, *134*, 110821. [CrossRef]
58. Ed-Dra, A.; Filali, F.R.; Presti, V.L.; Zekkori, B.; Nalbone, L.; Bouymajane, A.; Trabelsi, N.; Lamberta, F.; Bentayeb, A.; Giuffrida, A. Chemical Composition, Antioxidant Capacity and Antibacterial Action of Five Moroccan Essential Oils against Listeria Monocytogenes and Different Serotypes of Salmonella Enterica. *Microb. Pathog.* **2020**, *149*, 104510. [CrossRef]
59. Abdallah, E.M. Antibacterial Activity of *Hibiscus Sabdariffa* L. Calyces against Hospital Isolates of Multidrug Resistant Acinetobacter Baumannii. *J. Acute Dis.* **2016**, *5*, 512–516. [CrossRef]
60. Safian, S.; Majid, H.; Swift, S.; Silva, F.V. Antimicrobial Properties against Human Pathogens of Medicinal Plants from New Zealand. *Appl. Microbiol.* **2022**, *2*, 357–366. [CrossRef]
61. Chipiti, T.; Ibrahim, M.A.; Singh, M.; Islam, M.S. In Vitro α-Amylase and α-Glucosidase Inhibitory Effects and Cytotoxic Activity of Albizia Antunesiana Extracts. *Pharmacogn. Mag.* **2015**, *11*, S231.
62. Al-Reza, S.M.; Yoon, J.I.; Kim, H.J.; Kim, J.-S.; Kang, S.C. Anti-Inflammatory Activity of Seed Essential Oil from Zizyphus Jujuba. *Food Chem. Toxicol.* **2010**, *48*, 639–643. [CrossRef] [PubMed]
63. Chou, S.-T.; Lai, C.-P.; Lin, C.-C.; Shih, Y. Study of the Chemical Composition, Antioxidant Activity and Anti-Inflammatory Activity of Essential Oil from Vetiveria Zizanioides. *Food Chem.* **2012**, *134*, 262–268. [CrossRef]
64. Bessah, R.; Benyoussef, E.-H. Essential Oil Composition of Arbutus Unedo L. Leaves from Algeria. *J. Essent. Oil Bear. Plants* **2012**, *15*, 678–681. [CrossRef]
65. Kivcak, B.; Mer, T.; Demirci, B.; Baser, K.H.C. Composition of the Essential Oil of Arbutus Unedo. *Chem. Nat. Compd.* **2001**, *37*, 445–446. [CrossRef]

66. Bouyahya, A.; El Omari, N.; Elmenyiy, N.; Guaouguaou, F.-E.; Balahbib, A.; Belmehdi, O.; Salhi, N.; Imtara, H.; Mrabti, H.N.; El-Shazly, M. Moroccan Antidiabetic Medicinal Plants: Ethnobotanical Studies, Phytochemical Bioactive Compounds, Preclinical Investigations, Toxicological Validations and Clinical Evidences; Challenges, Guidance and Perspectives for Future Management of Diabetes Worldwide. *Trends Food Sci. Technol.* **2021**, *115*, 147–254.
67. Bouyahya, A.; Guaouguaou, F.-E.; El Omari, N.; El Menyiy, N.; Balahbib, A.; El-Shazly, M.; Bakri, Y. Anti-Inflammatory and Analgesic Properties of Moroccan Medicinal Plants: Phytochemistry, in Vitro and in Vivo Investigations, Mechanism Insights, Clinical Evidences and Perspectives. *J. Pharm. Anal.* **2021**, *12*, 35–57. [CrossRef]
68. Di Leo Lira, P.; Retta, D.; Tkacik, E.; Ringuelet, J.; Coussio, J.D.; van Baren, C.; Bandoni, A.L. Essential Oil and By-Products of Distillation of Bay Leaves (*Laurus Nobilis* L.) from Argentina. *Ind. Crops Prod.* **2009**, *30*, 259–264. [CrossRef]
69. Fidan, H.; Stefanova, G.; Kostova, I.; Stankov, S.; Damyanova, S.; Stoyanova, A.; Zheljazkov, V.D. Chemical Composition and Antimicrobial Activity of *Laurus Nobilis* L. Essential Oils from Bulgaria. *Molecules* **2019**, *24*, 804. [CrossRef]
70. Mohammadreza, V. Chemical Composition and Larvicidal Activity of the Essential Oil of Iranian *Laurus nobilis* L. *J. Appl. Hortic.* **2010**, *12*, 155–157. [CrossRef]
71. Dadalioğlu, I.; Evrendilek, G.A. Chemical Compositions and Antibacterial Effects of Essential Oils of Turkish Oregano (*Origanum Minutiflorum*), Bay Laurel (*Laurus Nobilis*), Spanish Lavender (*Lavandula Stoechas* L.), and Fennel (*Foeniculum Vulgare*) on Common Foodborne Pathogens. *J. Agric. Food Chem.* **2004**, *52*, 8255–8260. [CrossRef]
72. Nabila, B.; Piras, A.; Fouzia, B.; Falconieri, D.; Kheira, G.; Fedoul, F.-F.; Majda, S.-R. Chemical Composition and Antibacterial Activity of the Essential Oil of Laurus Nobilis Leaves. *Nat. Prod. Res.* **2022**, *36*, 989–993. [CrossRef] [PubMed]
73. Oliveira, I.; Baptista, P.; Malheiro, R.; Casal, S.; Bento, A.; Pereira, J.A. Influence of Strawberry Tree (*Arbutus Unedo* L.) Fruit Ripening Stage on Chemical Composition and Antioxidant Activity. *Food Res. Int.* **2011**, *44*, 1401–1407. [CrossRef]
74. Liang, T.; Yue, W.; Li, Q. Comparison of the Phenolic Content and Antioxidant Activities of *Apocynum Venetum* L. (Luo-Bu-Ma) and Two of Its Alternative Species. *Int. J. Mol. Sci.* **2010**, *11*, 4452–4464. [CrossRef] [PubMed]
75. Bouyahya, A.; Lagrouh, F.; El Omari, N.; Bourais, I.; El Jemli, M.; Marmouzi, I.; Salhi, N.; Faouzi, M.E.A.; Belmehdi, O.; Dakka, N. Essential Oils of Mentha Viridis Rich Phenolic Compounds Show Important Antioxidant, Antidiabetic, Dermatoprotective, Antidermatophyte and Antibacterial Properties. *Biocatal. Agric. Biotechnol.* **2020**, *23*, 101471. [CrossRef]
76. Mssillou, I.; Agour, A.; El Ghouizi, A.; Hamamouch, N.; Lyoussi, B.; Derwich, E. Chemical Composition, Antioxidant Activity, and Antifungal Effects of Essential Oil from Laurus Nobilis L. Flowers Growing in Morocco. *J. Food Qual.* **2020**, *2020*, 8819311. [CrossRef]
77. Ramos, C.; Teixeira, B.; Batista, I.; Matos, O.; Serrano, C.; Neng, N.R.; Nogueira, J.M.F.; Nunes, M.L.; Marques, A. Antioxidant and Antibacterial Activity of Essential Oil and Extracts of Bay Laurel Laurus Nobilis Linnaeus (Lauraceae) from Portugal. *Nat. Prod. Res.* **2012**, *26*, 518–529. [CrossRef]
78. Mrabti, H.N.; Sayah, K.; Jaradat, N.; Kichou, F.; Ed-Dra, A.; Belarj, B.; Cherrah, Y.; Faouzi, M.E.A. Antidiabetic and Protective Effects of the Aqueous Extract of Arbutus Unedo L. in Streptozotocin-Nicotinamide-Induced Diabetic Mice. *J. Complement. Integr. Med.* **2018**, *15*. [CrossRef]
79. Ahamad, J.; Uthirapathy, S.; Mohammed Ameen, M.S.; Anwer, E.T. Essential Oil Composition and Antidiabetic, Anticancer Activity of Rosmarinus Officinalis L. Leaves from Erbil (Iraq). *J. Essent. Oil Bear. Plants* **2019**, *22*, 1544–1553. [CrossRef]
80. Paul, K.; Bhattacharjee, P. Process Optimization of Supercritical Carbon Dioxide Extraction of 1,8-Cineole from Small Cardamom Seeds by Response Surface Methodology: In Vitro Antioxidant, Antidiabetic and Hypocholesterolemic Activities of Extracts. *J. Essent. Oil Bear. Plants* **2018**, *21*, 317–329. [CrossRef]
81. More, T.A.; Kulkarni, B.R.; Nalawade, M.L.; Arvindekar, A.U. Antidiabetic Activity of Linalool and Limonene in Streptozotocin-Induced Diabetic Rat: A Combinatorial Therapy Approach. *Int. J. Pharm. Pharm. Sci.* **2014**, *6*, 159–163.
82. Macchioni, V.; Santarelli, V.; Carbone, K. Phytochemical Profile, Antiradical Capacity and α-Glucosidase Inhibitory Potential of Wild Arbutus Unedo L. Fruits from Central Italy: A Chemometric Approach. *Plants* **2020**, *9*, 1785. [CrossRef] [PubMed]
83. Qadir, S.A.; Kwon, M.C.; Han, J.G.; Ha, J.H.; Chung, H.S.; Ahn, J.; Lee, H.Y. Effect of Different Extraction Protocols on Anticancer and Antioxidant Activities of Berberis Koreana Bark Extracts. *J. Biosci. Bioeng.* **2009**, *107*, 331–338. [CrossRef] [PubMed]
84. Munoz-Torrero, D. Acetylcholinesterase Inhibitors as Disease-Modifying Therapies for Alzheimer's Disease. *Curr. Med. Chem.* **2008**, *15*, 2433–2455. [CrossRef] [PubMed]
85. Bores, G.M.; Huger, F.P.; Petko, W.; Mutlib, A.E.; Camacho, F.; Rush, D.K.; Selk, D.E.; Wolf, V.; Kosley, R.W.; Davis, L. Pharmacological Evaluation of Novel Alzheimer's Disease Therapeutics: Acetylcholinesterase Inhibitors Related to Galanthamine. *J. Pharmacol. Exp. Ther.* **1996**, *277*, 728–738.
86. Ren, Y.; Houghton, P.J.; Hider, R.C.; Howes, M.-J.R. Novel Diterpenoid Acetylcholinesterase Inhibitors from Salvia Miltiorhiza. *Planta Med.* **2004**, *70*, 201–204.
87. Kennedy, D.O.; Scholey, A.B. The Psychopharmacology of European Herbs with Cognition-Enhancing Properties. *Curr. Pharm. Des.* **2006**, *12*, 4613–4623. [CrossRef]
88. Owokotomo, I.A.; Ekundayo, O.; Abayomi, T.G.; Chukwuka, A.V. In-Vitro Anti-Cholinesterase Activity of Essential Oil from Four Tropical Medicinal Plants. *Toxicol. Rep.* **2015**, *2*, 850–857. [CrossRef]

89. Smeriglio, A.; Alloisio, S.; Raimondo, F.M.; Denaro, M.; Xiao, J.; Cornara, L.; Trombetta, D. Essential Oil of Citrus Lumia Risso: Phytochemical Profile, Antioxidant Properties and Activity on the Central Nervous System. *Food Chem. Toxicol.* **2018**, *119*, 407–416. [CrossRef]
90. Bonesi, M.; Okusa, P.N.; Tundis, R.; Loizzo, M.R.; Menichini, F.; Stévigny, C.; Duez, P.; Menichini, F. Chemical Composition, Antioxidant Properties and Anti-Cholinesterase Activity of Cordia Gilletii (Boraginaceae) Leaves Essential Oil. *Nat. Prod. Commun.* **2011**, *6*, 1934578X1100600225. [CrossRef]
91. Grobler, J.A.; Stillmock, K.; Hu, B.; Witmer, M.; Felock, P.; Espeseth, A.S.; Wolfe, A.; Egbertson, M.; Bourgeois, M.; Melamed, J. Diketo Acid Inhibitor Mechanism and HIV-1 Integrase: Implications for Metal Binding in the Active Site of Phosphotransferase Enzymes. *Proc. Natl. Acad. Sci. USA* **2002**, *99*, 6661–6666. [CrossRef]
92. López, M.D.; Pascual-Villalobos, M.J. Mode of Inhibition of Acetylcholinesterase by Monoterpenoids and Implications for Pest Control. *Ind. Crops Prod.* **2010**, *31*, 284–288. [CrossRef]
93. Abdulkhaleq, L.A.; Assi, M.A.; Abdullah, R.; Zamri-Saad, M.; Taufiq-Yap, Y.H.; Hezmee, M.N.M. The Crucial Roles of Inflammatory Mediators in Inflammation: A Review. *Vet. World* **2018**, *11*, 627. [CrossRef] [PubMed]
94. Ferrero-Miliani, L.; Nielsen, O.H.; Andersen, P.S.; Girardin, S. Chronic Inflammation: Importance of NOD2 and NALP3 in Interleukin-1β Generation. *Clin. Exp. Immunol.* **2007**, *147*, 227–235. [CrossRef] [PubMed]
95. Medicherla, K.; Sahu, B.D.; Kuncha, M.; Kumar, J.M.; Sudhakar, G.; Sistla, R. Oral Administration of Geraniol Ameliorates Acute Experimental Murine Colitis by Inhibiting Pro-Inflammatory Cytokines and NF-KB Signaling. *Food Funct.* **2015**, *6*, 2984–2995. [CrossRef] [PubMed]
96. Su, Y.-W.; Chao, S.-H.; Lee, M.-H.; Ou, T.-Y.; Tsai, Y.-C. Inhibitory Effects of Citronellol and Geraniol on Nitric Oxide and Prostaglandin E2 Production in Macrophages. *Planta Med.* **2010**, *76*, 1666–1671. [CrossRef] [PubMed]
97. Miguel, M.G.; Faleiro, M.L.; Guerreiro, A.C.; Antunes, M.D. *Arbutus Unedo* L.: Chemical and Biological Properties. *Molecules* **2014**, *19*, 15799–15823. [CrossRef] [PubMed]
98. Moualek, I.; Aiche, G.I.; Guechaoui, N.M.; Lahcene, S.; Houali, K. Antioxidant and Anti-Inflammatory Activities of Arbutus Unedo Aqueous Extract. *Asian Pac. J. Trop. Biomed.* **2016**, *6*, 937–944. [CrossRef]
99. Frum, Y.; Viljoen, A.M. In Vitro 5-Lipoxygenase Activity of Three Indigenous South African Aromatic Plants Used in Traditional Healing and the Stereospecific Activity of Limonene in the 5-Lipoxygenase Assay. *J. Essent. Oil Res.* **2006**, *18*, 85–88. [CrossRef]
100. de Cássia da Silveira e Sá, R.; Andrade, L.N.; de Sousa, D.P. A Review on Anti-Inflammatory Activity of Monoterpenes. *Molecules* **2013**, *18*, 1227–1254. [CrossRef]
101. Mohamed, N.; Lg, M.; Sa, R. Linalool Glycosides from Flowers of Lantana Montevidensis with Promising AntiInflammatory Potentials. *Nat. Prod. Chem. Res.* **2020**, *8*, 1–8.
102. Maleš, I.; Dragović-Uzelac, V.; Jerković, I.; Zorić, Z.; Pedisić, S.; Repajić, M.; Garofulić, I.E.; Dobrinčić, A. Non-Volatile and Volatile Bioactives of *Salvia officinalis* L., *Thymus serpyllum* L. and *Laurus nobilis* L. Extracts with Potential Use in the Development of Functional Beverages. *Antioxidants* **2022**, *11*, 1140. [CrossRef] [PubMed]
103. Guedouari, R.; Nabiev, M. Anti-Inflammatory Activity of Different Extracts from Laurus Nobilis Growing in Algeria. *Algerian J. Environ. Sci. Technol.* **2021**, *7*, 2115–2120.
104. Valdés, C.; Laurido, C.; Morales, B.; Jaimes, L.; Vinet, R.; Martínez, J.L. Complete Essential Oils of Laurus Nobilis Inducing Antinociceptive Action by Opioid Mechanism in C-Reflex and Spinal Wind-Up Model in Rat. *Bol. Latinoam. Caribe Plantas Med. Aromáticas* **2020**, *19*, 420–427. [CrossRef]
105. Biswas, B.; Rogers, K.; McLaughlin, F.; Daniels, D.; Yadav, A. Antimicrobial Activities of Leaf Extracts of Guava (*Psidium Guajava* L.) on Two Gram-Negative and Gram-Positive Bacteria. *Int. J. Microbiol.* **2013**, *2013*, 746165. [CrossRef] [PubMed]
106. Soniya, M.; Kuberan, T.; Anitha, S.; Sankareswari, P. In Vitro Antibacterial Activity of Plant Extracts against Gram Positive and Gram Negative Pathogenic Bacteria. *Int. J. Microbiol. Immunol. Res.* **2013**, *2*, 1–5.
107. Bouyahya, A.; Abrini, J.; El-Baabou, A.; Bakri, Y.; Dakka, N. Determination of Phenol Content and Antibacterial Activity of Five Medicinal Plants Ethanolic Extracts from North-West of Morocco. *J. Plant Pathol. Microbiol.* **2016**, *7*, 2. [CrossRef]
108. Cunha, B.A. Pseudomonas Aeruginosa: Resistance and Therapy. *Semin. Respir. Infect.* **2002**, *17*, 231–239. [CrossRef]
109. Lambert, P. Mechanisms of Antibiotic Resistance in Pseudomonas Aeruginosa. *J. R. Soc. Med.* **2002**, *95*, 22.
110. Shrivastava, S.R.; Shrivastava, P.S.; Ramasamy, J. World Health Organization Releases Global Priority List of Antibiotic-Resistant Bacteria to Guide Research, Discovery, and Development of New Antibiotics. *J. Med. Soc.* **2018**, *32*, 76. [CrossRef]
111. Brahmi, F.; Achat, S.; Guendouze-Bouchefa, N.; Benazzouz-Smail, L.; Elsebai, M.F.; Madani, K. Recent Advances in the Identification and the Study of Composition and Activities of Medicinal Plants. *J. Coast. Life Med.* **2016**, *4*, 983–999.
112. da Silveira, S.M.; Luciano, F.B.; Fronza, N.; Cunha Jr, A.; Scheuermann, G.N.; Vieira, C.R.W. Chemical Composition and Antibacterial Activity of Laurus Nobilis Essential Oil towards Foodborne Pathogens and Its Application in Fresh Tuscan Sausage Stored at 7 C. *LWT-Food Sci. Technol.* **2014**, *59*, 86–93. [CrossRef]
113. Mrabti, H.N.; Bouyahya, A.; Ed-Dra, A.; Kachmar, M.R.; Mrabti, N.N.; Benali, T.; Shariati, M.A.; Ouahbi, A.; Doudach, L.; Faouzi, M.E.A. Polyphenolic Profile and Biological Properties of Arbutus Unedo Root Extracts. *Eur. J. Integr. Med.* **2021**, *42*, 101266. [CrossRef]
114. Nastasi, J.R.; Kontogiorgos, V.; Daygon, V.D.; Fitzgerald, M.A. Pectin-Based Films and Coatings with Plant Extracts as Natural Preservatives: A Systematic Review. *Trends Food Sci. Technol.* **2022**, *120*, 193–211. [CrossRef]

115. Herman, A.; Herman, A.P.; Domagalska, B.W.; Młynarczyk, A. Essential Oils and Herbal Extracts as Antimicrobial Agents in Cosmetic Emulsion. *Indian J. Microbiol.* **2013**, *53*, 232–237. [CrossRef] [PubMed]
116. Morgado, S.; Morgado, M.; Plácido, A.I.; Roque, F.; Duarte, A.P. Arbutus Unedo L.: From Traditional Medicine to Potential Uses in Modern Pharmacotherapy. *J. Ethnopharmacol.* **2018**, *225*, 90–102. [CrossRef]
117. Sırıken, B.; Yavuz, C.; Güler, A. Antibacterial Activity of Laurus Nobilis: A Review of Literature. *Med. Sci. Discov.* **2018**, *5*, 374–379. [CrossRef]

MDPI
St. Alban-Anlage 66
4052 Basel
Switzerland
www.mdpi.com

Life Editorial Office
E-mail: life@mdpi.com
www.mdpi.com/journal/life

Disclaimer/Publisher's Note: The statements, opinions and data contained in all publications are solely those of the individual author(s) and contributor(s) and not of MDPI and/or the editor(s). MDPI and/or the editor(s) disclaim responsibility for any injury to people or property resulting from any ideas, methods, instructions or products referred to in the content.

www.ingramcontent.com/pod-product-compliance
Lightning Source LLC
LaVergne TN
LVHW071543170726
843515LV00008B/2203

* 9 7 8 3 0 3 6 5 9 4 7 7 4 *